BIOLOGICAL
AND MEDICAL PHYSICS SERIES

Springer
Berlin
Heidelberg
New York
Hong Kong
London
Milan
Paris
Tokyo

BIOLOGICAL AND MEDICAL PHYSICS SERIES

The field of biological and medical physics is a broad, multidisciplinary, and dynamic one, touching on many areas of research in physics, biology, chemistry, and medicine. The Biological and Medical Physics Series is intended to be comprehensive, covering a broad range of topics important to the study of biological and medical physics. Its goal is to provide scientists, medical doctors and engineers with text books, monographs and reference books to address the growing need for information.

John Milton Peter Jung (Eds.)

Epilepsy as a Dynamic Disease

With 170 Figures, 24 Color Plates, and 11 Tables

Springer

Dr. John Milton
University of Chicago, Department of Neurology
Chicago, IL 60637, USA
e-mail: sp1ace@ace.bsd.uchicago.edu

Dr. Peter Jung
Ohio University, Department of Physics and Astronomy
and Institute for Quantitative Biology
Athens, OH 45701, USA
e-mail: jungp@ohio.edu

Library of Congress Cataloging in Publication Data: Epilepsy as a dynamic disease/ John Milton, Peter Jung (eds.) p. cm. – (Biological and medical physics series, ISSN 1618-7210) (Physics and astronomy online library) Includes bibliographical references and index. ISBN 3540427627 (alk. paper) 1. Epilepsy. 2. Brain stimulation. 3. Convulsions. 4. Electroencephalography. 5. Electric countershock. I. Milton, John, 1950– II. Jung, Peter, 1957– III. Series. IV. Series: Physics and astronomy online library. RC372.E6632 2002 616.8'53–dc21 2002030470

ISSN 1618-7210

ISBN 3-540-42762-7 Springer-Verlag Berlin Heidelberg New York

Springer-Verlag Berlin Heidelberg New York
a member of BertelsmannSpringer Science+Business Media GmbH

http://www.springer.de

Cover concept by eStudio Calamar Steinen using a background picture from The Protein Databank (1 Kzu). Courtesy of Dr. Antoine M. van Oijen, Department of Molecular Physics, Huygens Laboratory, Leiden University, The Netherlands. Reprinted with permission from Science 285 (1999) 400–402 ("Unraveling the Electronic Structure of Individual Photosynthetic Pigment-Protein Complexes", by A. M. van Oijen et al.) Copyright 1999, American Association for the Advancement of Science.

Typesetting by the editors
Data conversion: Frank Herweg, Leutershausen
Cover production: *design & production* GmbH, Heidelberg

Printed on acid-free paper SPIN 10855588 57/3141/ba - 5 4 3 2 1 0

Preface

Since the time of Hippocrates, epilepsy has been felt to be related to an imbalance in the regulatory activity of the brain; caused by either a relative paucity of inhibition or an excess of excitation. Present-day readers will find it hard to believe that only 125 years ago, the management of persons with intractable seizures included confinement to life on a farm [1]. The peace and quiet of country life was felt to be necessary to prevent the extremes of emotion and intellectual activity that predisposed to the convulsive state. Following this rationale physicians would advise patients to avoid intellectual thought, extremes in hot and cold, wine, constipation, sex, and to lie in bed with the head raised to assist the circulation of the brain!

Two developments that have had a major impact on the lives of people with epilepsy are the discovery of anti-convulsant medications and the invention of the automobile. Anticonvulsant medications save lives and have enabled people with epilepsy to re-enter the main stream of society. However, the drug regulatory bodies have a very different definition for an effective treatment from that of a typical person with epilepsy. Regulatory approval requires that the treatment result in a statistically significant reduction in seizure frequency. To the sufferer the only acceptable definition of a good treatment is no seizures, i.e. the seizure frequency must become zero. Typically the reduction in seizure frequency is not sufficient for patients with epilepsy to be able to legally operate an automobile. Thus many occupations remain unattainable; the social stigma remains high. This limitation together with the non-trivial cognitive side effects of anti-convulsant medications results in a quality of life of people with epilepsy which is as dismal as those suffering terminal cancer or the disabilities that accompany chronic medical conditions [2,3].

There is an inherent illogicality to the use of medications to treat a paroxysmal disorder such as epilepsy. Why should a person take medications everyday to prevent an event which might occur once every six months? The present day answer is that medications must be taken everyday since we can not predict on which day the seizure will occur. However, recent research has suggested that it may be possible to predict a seizure up to 30 min. before it occurs (Chaps. 12 and 13). This observation suggests the feasibility of an alternative therapeutic strategy. In this text we consider the possibility of constructing an implantable device that would be capable of detecting when a seizure occurs and then, at the appropriate time, delivering a carefully honed stimulus designed to abort the seizure. Thus, very much in the spirit of the implantable cardiac defibrillators, could a "brain defibrillator" be developed to treat epilepsy? The advantage of this type of strategy is that the treatment would be called upon only when needed.

The philosophy of this text is in the spirit of Herbert H. Jasper and his colleagues innovative text, *Basic Mechanisms of the Epilepsies* which was first published in 1969 [4]. Experts in epilepsy, bio-engineering, and dynamical systems theory have been brought together to discuss the treatment of epilepsy. Each of these experts has been encouraged by us to do their best to apply their knowledge to a topic that in many cases is unfamiliar to them and for which there may be, in fact, little known. In the next section we give an overview of the text so that the reader will better appreciate the flow of the discussion. We anticipate that this text will be used by individuals with as varied a background as the contributors. Thus we have also included some chapters that provide essential background to make the text self-contained.

The production of a working prototype of a "brain defibrillator" based on principles of the nonlinear dynamics of neurons and neural populations may seem far-fetched. However, we hope that in reading this text readers will come to our realization that there is no obvious reason why working prototypes of brain defibrillators do not already exist. On the other hand, perhaps this book will make at least one reader angry enough at our navieté to go to the laboratory, do the necessary experiments, and demonstrate to us how it really should be done!

Chicago, Athens,
June 2002

John Milton
Peter Jung

Contents

List of Contributors

Fawaz Ahmad
Department of Neurology
The University of Chicago
Chicago, IL 60637
USA

Paul Bressloff
Department of Mathematics
University of Utah
Salt Lake City, UT 84112
USA
bressloff@math.utah.edu

Sozari A. Chkhenkeli
Department of Neurology
The University of Chicago
Chicago, IL 60637
USA
sozari@medicine.bsd.uchicago.edu

S. Coombes
Nonlinear and Complex Systems Group
Department of Mathematical Sciences
Loughborough University
Loughborough LE11 3TU, Leicestershire
UK
s.coombes@lboro.ac.uk

A.H. Cornell-Bell
Viatech Imaging/Cognetix
58 Main Street
Ivoryton, CT 06442
USA
ACornell-Bell@cognetix.com

Ton deGrauw
Division of Neurology
Children's Medical Center
333 Burnet Avenue
Cincinnati, OH 45229-3039
USA
ton.degrauw@chmcc.org

William L. Ditto
Georgia Tech/Emory Biomedical Engineering Dept.
Georgia Institute of Technology
315 Ferst Drive
Atlanta, Georgia 30332-0535
USA
william.ditto@bme.gatech.edu

Kevin Dolan
Institut für Medizin
Forschungszentrum Jülich, GmbH
52425 Jülich
Germany
k.dolan@fz-juelich.de

John S. Ebersole
Department of Neurology
The University of Chicago
Chicago, IL 60637
USA
jebersol@medicine.bsd.uchicago.edu

Jennifer Foss
Department of Neurology
The University of Chicago
Chicago, IL 60637
USA
foss@midway.uchicago.edu

Joseph T. Francis
Krasnow Institute
for Advanced Study
George Mason University
Fairfax, VA 22030
USA
jfranci@gmu.edu

Bruce J. Gluckman
Department of Physics
and Astronomy
and
The Krasnow Institute
for Advanced Study
George Mason University
Fairfax, VA 22030
USA
bgluckma@gmu.edu

Kurt Hecox
Department of Pediatrics
The University of Chicago
Chicago, IL 60637
USA
khecox@peds.bsd.uchicago.edu

John R. Hetling
Department of Bioengineering
University of Illinois at Chicago
Chicago, IL 60607-7052
USA
jhetli1@uic.edu

John D. Hunter
Department of Neurology
The University of Chicago
Chicago, IL 60637
USA
jdhunter@ace.bsd.uchicago.edu

Kristin Jerger
The Krasnow Institute
for Advanced Study
George Mason University
Fairfax VA 22030
USA
kjerger@gmu.edu

Peter Jung
Department of Physics
and Astronomy
Ohio University
Athens, OH 45701
USA
jungp@ohio.edu

Michael Kohrman
Department of Pediatrics
The University of Chicago
Chicago, IL 60637
USA
mkohrma@peds.bsd.uchicago.edu

Herbert Levine
Department of Physics
University of California, San Diego
La Jolla, CA 92093
USA
hlevine@physics.ucsd.edu

Alexander S. Mikhailov
Department of Physical Chemistry
Fritz-Haber-Institut
der Max-Planck-Gesellschaft
14195 Berlin
Germany
mikhailov@fhi-berlin.mpg.de

John Milton
Department of Neurology
The University of Chicago
Chicago, IL 60637
USA
splace@ace.bsd.uchicago.edu

Frank Moss
Center for Neurodynamics
University of Missouri
at St. Louis
8001 Natural Bridge Rd.
St. Louis, Missouri 63121
USA
mossf@umsl.edu

Theoden J. Netoff
The Krasnow Institute
for Advanced Study
and
George Mason University
Fairfax, VA 22030
USA
tnetoff@gwis2.circ.gwu.edu

L. Pecora
Naval Research Laboratory
Washington, D.C., 20375
USA
Louis.Pecora@nrl.navy.mil

T. Sauer
Department of Mathematics
George Mason University
Fairfax, VA 22030
USA
tsauer@gmu.edu

Robert Savit
Department of Physics
University of Michigan
Ann Arbor, MI 48109
USA
savit@umich.edu

Steven J. Schiff
The Krasnow Institute
for Advanced Study
and
Department of Psychology
George Mason University
Fairfax, VA 22030
USA
sschiff@gmu.edu

Mark L. Spano
NSWC, Carderock Laboratory
9500 MacArthur Blv
W. Bethesda, Maryland 20817
USA
spanoml@nswccd.navy.mil

R. Strawsburg
Division of Neurology
Children's Medical Center
333 Burnet Avenue
Cincinnati, OH 45229-3039
USA
richard.strawsburg@chmcc.org

Vernon L. Towle
Department of Neurology
The University of Chicago
Chicago, IL 60637
USA
vtowle@neurology.bsd.uchicago.edu

Steven L. Weinstein
Children's National Medical Center
Department of Neurology
111 Michigan Avenue, NW
Washington, DC 20010
USA

Arthur T. Winfree
Department of Ecology and
Evolutionary Biology
University of Arizona
Tucson, AZ 85721
USA
winfree@email.arizona.edu

Roadmap

Temporal lobe epilepsy is not a static entity, but continuously evolves, and changes its character over the lifetime of the sufferer. The dynamic nature of epilepsy implies the need for dynamic therapies. The overall theme of this collection of articles is to discuss the various aspects involved in the design of an implantable brain defibrillator. There are three basic components: 1) *Detection*: what is the best way to detect seizures ?; 2) *Treatment*: what is the best strategy to abort seizures?; and 3) *Treatment delivery*: at what sites(s) should the seizure-aborting stimuli be delivered?

The text is divided into five sections. It is anticipated that readers of this book will have varied backgrounds. Therefore in order to increase the accessibility to this text we have prepared a synopsis for Chaps. 1–18. The purpose of these synopses is to help the reader appreciate all aspects of the arguments. For example, the non-mathematically oriented reader will be able to understand the take home message of a more mathematically oriented chapter by referring to the appropriate synopsis. Mathematically oriented readers may appreciate recourse to the synopsis of chapters that emphasize clinical and neuro-anatomical aspects of epilepsy. In Chap. 19 we critically assess the progress that has been made towards the construction of a brain defibrillator, identify unsolved problems, and give suggestions for future work.

Special attention is drawn to the chapters written by Professor Sozari A. Chkhenkeli (i.e. Chaps. 1, 3, 14). It is becoming increasingly recognized that the work of his group in Tbilisi of the Georgian Republic, and his colleague Professor M. Sramka of Slovakia, represent the major pioneering work in the field of the treatment of epilepsy using deep-brain stimulation and feedback control of seizures. However, this work has been largely inaccessible to an English speaking audience since it has been written in Russian and has, up to now, only appeared in presentations at international symposia.

1 Clinical Aspects

The first section (Chaps. 1–4) examines the clinical and surgical aspects of seizures that arise from the temporal lobe complex. These chapters form the foundation for the discussion in this text. We have chosen to focus on this type of epilepsy because it is the epilepsy most commonly treated with surgical procedures and hence the epilepsy that has been most intensively studied.

Chapter 1: Temporal lobe epilepsy cannot be simply described as a static disease characterized by stereotyped seizures which recurrently and unpredictably arise from the temporal lobe complex (i.e. temporal lobe, amygdala, hippocampus). Throughout their lifetime, the sufferers develop worsening behavioral and memory-related problems leading to a decline in socio-economic status which is often out of proportion to the seizure frequency. This disease evolution is reflected by the response of seizure frequency to anticonvulsant medication: seizures are controlled for varying periods of time following introduction of a new anticonvulsant, only to eventually break through: a "boom-bust" cycle in which the seizure-free intervals become shorter and shorter. This cycle eventually culminates in medically intractable epilepsy (MIE), a stage in which the seizure frequency is no longer sensitive to changes in anticonvulsant medication(s). This evolving clinical picture suggests that temporal lobe epilepsy should be regarded as a dynamically changing disease and implies that therapeutic strategies need to take into account that patients may be in different stages of the evolution of this illness. Temporal lobe epilepsy involves many of the same brain regions that are responsible for learning. Is MIE the result of learning processes gone awry? This possibility is supported by observations in animal models of epilepsy which demonstrate that regions of the brain spatially distinct from the initial epileptic focus can be trained to become themselves self-sufficient epileptic foci. This process of "secondary epileptogenesis" is accompanied by a reorganization in the synaptic circuitry of the neural population which possibly even involves the glia support cells (Chap. 4). Computer simulations of populations of model neurons support the conjecture that changes in neuronal connectivity play an important role in shaping spatiotemporal dynamics. Although it remains controversial whether secondary epileptogenesis can lead to the formation of new epileptic foci in the human brain, it is quite possible that the same mechanisms could produce MIE. The important point is that the mechanisms which likely play a major role in the establishment of MIE reside in the deeply located nuclei of the subcortex and brainstem. These observations provide the first clues to suggest that deep brain stimulation will likely play an important role in the development of a brain defibrillator. This discussion is continued in Chaps. 2, 3 and 14.

Chapter 2: It is possible that a seizure arising in the temporal lobe complex can secondarily generalize to involve the whole cortex. Anatomical considerations suggest that there are three pathways available for seizure generalization: 1) intracortical spread localized to the gray matter; 2) white matter mediated spread over pathways that interconnect adjacent as well as distant areas of cortex; and 3) spread via reciprocal connections between neurons in the gray matter and the subcortical and brainstem nuclei. In this chapter the relative importance of these pathways for seizure generalization is assessed by comparing their propagation velocities to estimates of seizure propagation. Studies with brain slices suggest that intracortical neuron-to-neuron propagation occurs via unmyelinated axons at $\leq$ 6–18 cm/s. Involvement of the larger-diameter myelinated axons of the white matter produces seizure propagation rates of $\geq$ 200–300 cm/s. However, 40–60% of the white matter is composed of very-small-diameter myelinated and unmyelinated axons which introduce significant conduction delays, e.g. 100–200 ms between the two hemispheres. Although the faster conducting fibers likely dominate activation of distant cortical regions, the subsequent spread of delayed neural activity may have important modulatory effects. It is difficult to measure seizure propagation in the human brain because its surface is highly convoluted and the EEG is relatively insensitive to seizure activity which arises from the depths of the sulci (Chap. 5). Typical estimates are $\sim$ 10–40 cm/s, i.e. a propagation velocity intermediate to that of gray matter and white matter mediated seizure propagation. However, seizure propagation is not constant: often slow propagation, initially confined to the temporal lobe complex, is followed by a very rapid generalization. Observations of this type suggest that the third pathway may be the major route for seizure generalization: cortical neurons connect to subcortical nuclei, such as the thalamus, by large-diameter myelinated axons, and then the subcortical nuclei project diffusely to the cortex. Thus the fastest pathway for seizure generalization is not the most direct one. Although direct electrical stimulation of the subcortical structures does not elicit a seizure, it is becoming increasingly recognized that many of these deeply located structures play important roles for cortical seizure propagation as permissive gating mechanisms and in the generalization and cessation of seizures. Yet another clue which directs attention for developers of a brain defibrillator to the potential advantages of stimulation of the deep brain structures. This theme is continued in the next chapter.

Chapter 3: An epileptic system is a spatially distributed system of groups of neurons located in the deep regions of the brain which cooperate to control the onset, propagation, and arrest of epileptic seizures. This chapter reviews the results of depth electrode recordings in patients with seizures which arise from the temporal lobe complex. The limited success and even failure of epilepsy surgery in some patients, in particular those with longstanding MIE, suggest that the nature of seizure generation and propagation is likely more complex than implied by the concept of an epileptic focus. To underscore this point an example is presented in which three consecutive seizures in the same patient differ in both their mode of onset and in their subsequent propagation. The amygdala and hippocampus are the best characterized epileptic foci. However, other regions in the temporal lobe complex are more excitable, e.g. the entorhinal cortex and area tempestas. These regions do not act as epileptic foci themselves, but rather appear to act as triggers. The interhemispheric propagation time for hippocampal onset seizure is surprisingly slow ($>$ 50 s in patients who experience the best outcome for temporal lobectomy). The major mechanism for spread is not via a commissural link, but rather via the intrahemispheric and vertical spread of seizure activity by indirect pathways involving many deep brain structures including the thalamus, cingulum and mesencephalon. In addition, areas of inhibitory influences tessellate throughout the epileptic system, including the caudate nucleus, cerebellar cortex and deep cerebellar nuclei, and multiple brainstem structures (Chap. 14). Often overlooked is the fact that symmetrical deep temporal lobe structures can also react to each other in a suppressive, or inhibitory, manner. For example, it is possible that epileptic activity can actively suppress that in the contralateral temporal lobe (active inhibition). Moreover, the activation of a dormant epileptic focus in the contralateral temporal can be seen following epilepsy surgery or transiently as a post-ictal phenomenon or following suppression of a temporal lobe using intra-carotid injection of amytal (WADA test). A number of clinical observations support the concept that the epileptic system is a dynamic ever-changing entity. These include the increased risk of seizures at times of stress and anxiety, and the effects of the menstrual cycle and the sleep-wake cycle on seizure timing. The deeply located groups of neurons which comprise the epileptic system are obvious candidates for the site of the effector arm of a brain stimulator.

Chapter 4: Traditionally epilepsy has been assumed to be solely a disease of neurons and of their interactions with other neurons. This chapter reviews the evidence that another cell type, a neuroglia cell referred to as the star-shaped astrocyte, likely plays an important role in the maintenance and perhaps development of the epileptic focus. Astrocytes have an intimate partnership with neurons in taking care of the neuronal microenvironment. Their functions include making adjustments to extracellular K^+ concentrations, preserving neuronal synapses, and responding to changing levels of excitatory neurotransmitters. The contributions of neuroglia to the EEG are discussed in Chap. 5. The observation that the epileptic focus is typically depleted of neurons suggests that astrocytes must be involved in the maintenance of the epileptic focus. A number of differences have been observed in astrocytes that are located in epileptogenic brains, including an increased density of voltage-gated Na^+ channels, increased gap junction coupling, altered morphology, and a hyperexcitable response to neurotransmitters. Particular attention is given to Ca^{2+} signalling between astrocytes which takes the form of propagating waves in cultures of astrocytes taken from rat hippocampus and human epileptogenic brain tissue. Video imaging techniques using a calcium-sensitive fluoroprobe together with space-time cluster decomposition methods can be used to monitor the propagation of these Ca^{2+} waves. The wave pattern is a target wave with each wave having a characteristic finite lifetime and spreading distance: the waves arise spontaneously and randomly at different sites and times. Statistical methods, such as the cluster entropy, are required to characterize the observed spatiotemporal dynamics. In epileptogenic brains both the speed of propagation of the calcium wave fronts and the number of cells involved in the wave is markedly are increased over those measured in rat hippocampal cultures. The cluster entropy is decreased, suggesting that, at least temporarily, there is strong coupling between the astrocytes so that synchrony is increased. All said and done, in this text we will continue to focus on collections of neurons as the target for a brain defibrillator. However, this chapter should serve to implant a nagging doubt at the back of the reader's mind that the true culprit may be the astrocyte.

2 Neural Synchronization

At its most fundamental level, a seizure represents synchronization of a massively large population of neurons. This section describes how this synchronization is reflected by changes in the electro-encephalogram (EEG), how neurons become synchronized, and how this synchronization can be altered with electrical stimuli.

Chapter 5: The main tool available to the epileptologist to evaluate patients with epilepsy and to monitor seizure propagation is the EEG. This chapter reviews the anatomical and electrophysiological principles that underlie the generation of the EEG. The scalp EEG is not a phenomenon of a single neuron, but is a measure of the activity of a very large population of pyramidal cortical neurons. Two anatomical properties of these neurons are critically important: 1) their vertical orientation with respect to the cortical surface; and 2) the spatial segregation of inhibitory and excitatory inputs on their surface. The generators of the EEG are not neural action potentials, but are the excitatory and inhibitory postsynaptic potentials which summate on the neural surface. Considerations of the extracellular current flows lead to the concept that a single neuron can be represented as a fluctuating dipole; the vertical alignment of the dipoles permits large numbers of dipoles to summate in order to generate a potential large enough (μVs) so that it can be measured at the scalp. An important clinical application is that the potential field generated by this dipole layer can be used to determine where the EEG generators are located and thus to localize the epileptic focus. The main limitations of the EEG stem from the fact that the potential generated by a dipole falls off rapidly as the distance to the dipole increases. Thus the scalp EEG is relatively insensitive to the dynamic interplay between the cortex and sub-cortex, alluded to in Chaps. 2 and 3. However, the most important fact about the EEG is that, at its most fundamental level, it is a measure of neural synchrony in selected neural populations. Thus the characteristic waveforms of the interictal EEG and the changing rhythms and amplitudes of the ictal EEG reflect changes in neural synchrony. Three basic patterns of ictal discharge have been observed for seizures which originate in the temporal lobe complex. Although these patterns can be correlated with the location of the epileptic focus, they are not a reflection of cortical activity at ictal onset, but rather arise from the subsequent spatiotemporal dynamics and development of synchrony as the seizure evolves. From the point of view of a brain defibrillator, these observations suggest that strategies which alter neural synchrony will likely be the most effective for rapidly aborting a seizure.

Chapter 6: The propagation of epileptic activity from an epileptic focus involves the synchronization of massive numbers of neurons. In Chap. 5 it was argued that the changes recorded by the EEG during a seizure primarily reflect changes in neural synchrony. Thus, measurements of EEG coherence between different cortical sites should be useful for monitoring seizure evolution. In this chapter the authors present data from two patients with medically intractable epilepsy which illustrate the variability in seizure development. The EEG activity is monitored from grids of subdural electrodes implanted on the cortical surface and the coherence is measured between pairs of electrodes. In the first case seizure propagation is heralded by a sudden change in EEG activity which is recorded almost simultaneously in a number of electrodes (Fig. 6.2). Analysis of EEG coherence shows that the seizure activity remains highly coherent across all of the involved electrodes (Figs. 6.3–6.4). In the second case, measurements of EEG give the impression that seizure activity migrates across the subdural grid of electrodes (Fig. 6.5). However, the coherence analysis shows that this impression is misleading. In fact the seizure involves two spatially distinct regions which are not coherent with respect to each other. During the initial portion of the seizure one region predominates, and as the seizure evolves this region decreases in importance and the second region dominates (Fig. 6.6). It is argued that for some patients the interictal pattern of EEG coherence appears to herald the ictal patterns. These observations are important from a theoretical point of view since they emphasize that the propagation of seizures as monitored by EEG may be fundamentally different from wave propagation in excitable media. From a clinical point of view, measurements of EEG coherence may provide a convenient way to identify regions which are candidates for surgical seizure ablation. Finally, from the point of view of the brain defibrillator, measurements of EEG coherence may provide a method to target and delineate the epileptic network.

Chapter 7: This chapter reviews the synchronization of populations of regularly spiking model neurons as a function of the synaptic coupling strength. In these mathematical models, the coupling strength, ϵ, appears as a parameter which scales the effect that an incoming spike input has on the total synaptic conductance of the neuron. An extensive theory of synchronization has been developed for the case when ϵ is very small (i.e. weak coupling). An important concept is that of the phase resetting curve (PRC). The PRC can be experimentally measured for any spiking, or bursting, neuron and summarizes the effect that an inhibitory or excitatory spike has on the timing, or more generally the phase, of the next spike produced by the neuron (Chap. 16). In the case of weak coupling the interaction between neurons can be regarded as a convolution of the PRC and the post-synaptic current over one period of oscillation. Using averaging theory it is possible to reduce network dynamics to a system of equations in which phase differences between oscillators become the relevant dynamical variable. However, the number of excitatory spike inputs necessary to cause a hippocampal neuron to fire is small and, moreover, these neurons are coupled electrically by gap junctions. Thus it is quite likely that ϵ may be too large in populations of real neurons to justify the mathematical assumption of weak coupling. In the case of strong coupling the interaction between neurons involves the direct product of the PRC with the post-synaptic current, such that in the resulting mathematical models phases no longer appear as differences. However, under these circumstances the velocity of a wave of neural activity is essentially independent of the ionic details of the membrane and depends mainly on the integration rise time from the resting potential to the firing threshold. Consequently, in the particular case of networks of integrate-and-fire (IF) neurons, it is possible to study synchronization outside the weak coupling regime in terms of the underlying firing time dynamics. The properties of wave propagation in these networks depend intimately on the magnitude of the delay, τ, introduced by finite axonal conduction times. For example, in 1-D networks of IF neurons with excitatory and inhibitory connections, once τ exceeds a critical value smooth wave propagation is replaced by a "lurching" wave phenomenon reminiscent of that reported in the context of the generation of thalamo-cortical spindle waves. It is anticipated that in 2- and 3-D, conduction delays could lead to the generation of more complex wave phenomena than the planar waves observed in standard excitable media, e.g. target waves and spirals (Chaps. 9–11). Particularly relevant to the developers of a brain defibrillator is the observation that under many circumstances, desynchronization of the activity of IF neuron networks can occur. These observations form part of the theoretical justification for the potential use of injected periodic stimuli to abort epileptic seizures (Chap. 8).

Chapter 8: This chapter discusses the effects of periodic stimulation on populations of neurons. In principle a periodic stimulus having frequency f could act as a rate controller to desynchronize a synchronized population or it could enhance intrinsic brain inhibitory mechanisms. The concept of the rate control of synchrony is based on the fact that the entrainment between two oscillators depends on their relative frequencies and the strength of the coupling between them (Chap. 7). For some modulation amplitudes and frequencies, it is possible that for each n cycles of the stimulus there are m cycles of the spontaneous rhythm ($n{:}m$ phase locking). The results are typically summarized in the form of an Arnold tongue diagram. In general, the $1{:}m$ phase-locking regions generate the highest degree of synchronization in a population of uncoupled neurons receiving a common input. Decreasing the frequency, f_o, of the neuron either by injecting hyperpolarizing currents or IPSCs from an inhibitory interneuron increases the relative frequency, f/f_o, and hence produces a rightward shift in the Arnold tongue diagram (similarly, excitatory stimuli produce leftward shifts). Experiments involving invertebrate neurons are used to illustrate that rate control can increase or decrease the degree of synchrony in a population of uncoupled neurons. In populations of uncoupled neurons, there is a close association between those periodic inputs which increase synchronization and those which result in precise and reliable spike timing. These observations suggest that aperiodic inputs may work as rate controllers of neural synchrony in an analogous manner to periodic inputs. The concept of enhancement of local inhibitory mechanisms is based on the observation that laterally inhibited networks of neurons with conduction delay, τ, preferentially amplify periodic inputs having frequency $1/2\tau$. This effect, together with the phenomenon of rate control of synchrony, may explain the observation that certain frequencies of Mossy fiber stimulation have been shown to reduce the number of seizures generated by epileptic hippocampal slices. Despite the fact that the theory for these effects is well established, current applications of deep brain stimuli to treat neurological diseases, such as epilepsy, typically do not take into account the potential advantages for more careful design of the stimulating parameters, such as frequency. The observations in this chapter support the notion that it may be possible to construct a brain defibrillator that uses periodic stimulation of a deep brain site to abort seizures.

3 Propagation of Epileptic Seizures

This section examines whether seizure propagation takes the form of waves of neural activity.

Chapter 9: The propagation of activity in a spatially extended population of neurons takes the form of waves of neural activity. The next three chapters review wave propagation in excitable media. A population of locally connected neurons is an example of an excitable medium. Each neuron exists in one of three states: quiescent, spiking, and refractory. When a quiescent neuron receives sufficient excitation it generates a brief spike and then enters a relatively longer refractory state. The neuron is not able to spike again until the refractory state has expired. Under appropriate circumstances, it is anticipated that wave propagation in populations of neurons will share many of the properties seen in other excitable media.

This chapter discusses the formation and properties of traveling spiral waves by reviewing the impact that the pioneering work of N. Wiener and A. Rosenblueth has had on this field (a critique is given in Chap. 10). Traveling spiral waves are of particular interest because they represent the simplest form of wave propagation in excitable media which is self-maintained, i.e. once initiated, spiral waves can persist indefinitely in the absence of external inputs (see Chap. 1). The formation of spiral waves requires the presence of inhomogeneities in the medium which can take the form of either obstacles to wave propagation or spatial inhomogeneities in neuronal connectivity or refractoriness (Chap. 1). The properties and the response of spiral waves to external stimuli are essentially determined by what happens in a narrow region close to the spiral tip. Inclusion of a relative refractory period in the form of a propagation velocity which is a function of the time lapsed since the last excitation leads to a meandering of the spiral tip. Introduction of propagation of anisotropies into the medium produces new wave phenomena including the formation of edges and of traveling wave fragments. Periodic stimulation of the spiral tip at the same frequency as the rotation period causes the spiral to drift. One mechanism to disrupt spiral waves involves the introduction of spatially distributed noise-like stimuli in order to introduce appropriate spatial inhomogeneities. Other possibilities could be based on an exploitation of the relevant Hopf bifurcation (Chaps. 1 and 7). An epileptic focus is often characterized by reduced numbers of neurons and decreased interneuronal interconnectivity. Thus it is tempting to speculate that an epileptic focus could provide the necessary "obstacle" to trigger spiral wave formation. The message to future developers of a brain defibrillator is that it may be necessary to consider the use of strategies designed to alter the propagation of neural activity waves.

Chapter 10: The most important clinical application of the theory of wave propagation in excitable media has been the insights it has provided into the nature of another medical emergency, cardiac fibrillation. This chapter discusses the extent to which analogies can be drawn between wave propagation in cardiac and neural tissues. Any suggestion that the heart and brain are simply two sides of the same coin is likely to prompt a heated debate between, at the very least, cardiologists and neurologists. Nonetheless, all differences in opinion aside, it must be admitted that similarities do exist. Seizures and fibrillation both involve electrically coupled cells connected to neighbors, are electrical modes which appear as alternatives to the normal modes which exist in the same tissue with the same parameters, can be initiated with periodic electrical stimuli, and stopped by the application of a single brief stimulus. The electrical and chemical diffusion which occur via gap junctions between adjacent myocytes has its analogy with Pathway 1 propagation in Chap. 1, long-range fast conduction via the Bundle of His with Pathway 2, and the autonomous nervous system together with the brainstem nuclei that regulate the cardiac rhythm and blood pressure with Pathway 3. Studies with optical imaging techniques support the wavelike properties of epileptic propagation in localized regions of cortex. The often emphasized difference that the brain contains both excitatory and inhibitory connections pales in view of the fact that inhibition is mathematically equivalent to a kind of relative refractory period. The important differences between cardiac and neural tissue turn out to be more subtle. For example, an important question is whether the physical extent of the cortical gray matter ($\sim$ 2–6 mm) necessitates an approximation as a 2-D or a 3-D excitable media. In normal 3-D heart walls, fibrillation arises promptly after a rotor (the 3-D equivalent of a spiral wave) is started;

however, this transition does not occur if the heart wall is artificially thinned. The diameter of a rotor can be estimated from either the distance that the activation front travels or the distance the membrane potential diffuses in one period of the rotor. Estimates place the rotor size at $\sim$ 3–10 mm, i.e. slightly larger than the width of the cortex. Resolving this issue will require the development of techniques capable of monitoring spatiotemporal dynamics with a spatial resolution better than 1 mm, and a temporal resolution of the order of 1 ms. On the other hand, the presence of significant time delays in neural tissue is expected to increase the complexity of the wave phenomena that occur (Chaps. 2 and 7). The second part of this chapter discusses how existing theories will need to be modified in order to begin to address the issues raised by the propagation of waves of neural activity. The final section addresses the mystery of why a single DC shock can be effective at causing cardiac defibrillation. This chapter makes it very clear why work towards the development of 'smart' cardiac defibrillators is presently so advanced: it remains to be seen whether practical analogies can be drawn to jump start the development of a brain defibrillator.

Chapter 11: The nervous system is not a static excitable medium. The development of the epileptic syndrome takes place on time-scales ranging from seconds to years and results in recognizable changes in the morphology of neurons, in the distribution of receptors on their surface, and changes in neural-glial cell interactions (see Chaps. 1 and 4). An important problem is determining which aspects of epilepsy arise from instabilities in an excitable medium and which arise, for example, from molecular changes as a consequence of altered gene expression. In other words, which aspects of a seizure have a dynamic origin amenable to treatment with a brain defibrillator and which have a molecular biology origin that might be better managed using, for example, gene therapy.

This chapter reviews work on wave propagation and pattern formation in populations of the soil-living amoebae, *Dictyostelium discoideum.* Populations of this organism undergo a dramatic and much studied change in behavior as these amoebae cooperate in the face of starvation. The analogy to the nervous system arises because with respect to pattern formation these populations of amoebae behave in all respects similar than excitable media. Investigations of the spatiotemporal dynamics of *Dictyostelium* cultures provide one of the clearest expositions delineating the relative role of dynamic and molecular biology factors in shaping spatiotemporal dynamics. Three stages of pattern formation have been analyzed in detail: 1) wavefield evolution; 2) cell streaming and mound formation; and 3) cell sorting and tip formation. At each stage it is tempting to invoke the expression of new genes as the mechanism causing the new behavior; however, in each case it is shown that the behaviors likely have a simpler and dynamical explanation. The investigations do not imply that techniques of modern molecular biology have no role to play in the analysis of biological excitable media. Indeed studies of certain mutants serve as important tests of proposed theories, to point in the direction of the essential dynamic mechanisms, and to identify new puzzles. Although it will likely be many decades before it will be possible to study the development of epileptogenic brain in as much detail as *Dictyostelium*, this chapter should alert investigators to the care that must be taken in identifying the mechanism which lies behind the appearance of new spatiotemporal dynamics in a biological excitable medium.

4 Seizure Prediction

An essential task to be performed by a brain defibrillator is the detection of the occurrence of a seizure. Obviously, if a seizure cannot be detected, then it will not be possible to trigger the application of an aborting stimulus. However, in any practical application there will always exist a significant time interval between seizure detection and the application of the aborting stimulus. For example, sufficient time is required to charge up a capacitor in order to be able to deliver a DC current pulse or to activate an anticonvulsant drug delivery mechanism. If, in order to abort a seizure, these stimuli must be given at, or before, seizure onset to be effective, then seizure detection alone is not what is required. What is required is the ability to predict seizure occurrence sufficiently far in advance so that the technology inherent to the brain defibrillator has time to work. The next two chapters review the progress in seizure prediction from time series analysis of the EEG. There is reason to believe that it may be possible to predict the occurrence of a seizure 2–30 min in advance.

Chapter 12: It is now well established that exceedingly complex, noise-like dynamics (often referred to as "chaotic" by aficionados) can be generated by deterministic nonlinear mathematical models. This observation has prompted a surge in research directed at the development of time series methods for seizure prediction based on the dynamical properties of nonlinear dynamical systems. This chapter discusses the underlying issues, successes and limitations of these tools for seizure prediction. A commonly employed strategy is to search for changes in values of certain quantities that are loosely based on the correlation integral and that are topologically invariant. Examples include the Lyapunov exponent and the Hausdorff, or other, dimensions. Experience using these measures indicates that human temporal lobe epilepsy may be so heterogeneous that a single quantity may not be robust enough to predict seizure occurrence in all instances. One possibility is that this heterogeneity reflects differences in the dynamical route to a seizure. However, more likely, this heterogeneity arises because the EEG is not stationary. Quantities based on the correlation integral lose their theoretical significance if the EEG is non-autonomous, i.e. non-stationary. Of course it may be useful to compute these quantities as empirical measures for seizure prediction. However, a sounder approach would be to develop techniques to directly characterize the EEG's non-autonomy. One possibility is to use cross-correlation approaches to determine the extent to which two temporal windows of a time series are dynamically similar. In this way it is shown that there is a persistent type of nonautonomy in EEG recordings of patients with epilepsy that is not seen in the EEGs of healthy subjects. The nature of this non-autonomy itself changes with time in a patient with epilepsy and, in particular, undergoes changes minutes to tens of minutes prior to a seizure. Time series methods which are based on the detection of non-autonomy appear to be more robust for seizure prediction than those based on the correlation integral. However, despite their promise, a number of methodological problems plague the field of seizure prediction, including the design of appropriate surrogates (see Chap. 17) and of appropriate null hypotheses. Moreover, only a limited number of EEGs which have been analyzed and few studies have appeared in which which different techniques of seizure prediction are compared by different laboratories on the same data sets. The ability to reliably predict seizures may provide insights into the nature of the responsible neural mechanisms and hence serve as a guide towards the design of effective treatment strategies.

Chapter 13: It is presently commonplace in a modern epilepsy monitoring unit to make use of computerized time series analysis of the EEG in order to automatically detect seizure occurrence. This chapter examines the ability of these methods to predict seizure occurrence relative to that of a trained EEG reader and to the more recently developed methods of nonlinear time series analysis. Collectively the traditional methods of seizure detection are "linear methods". This terminology is used to emphasize that with these approaches it is possible to represent an arbitrary time series as a sum of terms, each of which is formed by the product of a term related to amplitude and another related to frequency. This philosophy is embodied, for example, in the use of the theorems of Laplace and Fourier. Five methods of linear time series analysis used in seizure detection, i.e. power spectral analysis, cross-correlation, principal components analysis, wavelets, and phase correlation, are shown to be equally effective at predicting seizure occurrence by up to ~ 2.5 min. Methods of nonlinear time series analysis are designed to detect special properties of nonlinear dynamical systems. It is important to realize that it is always possible to describe the dynamics of a nonlinear dynamical system using a linear method such as power spectral analysis. The key point is whether a nonlinear time series method is capable of extracting new information, i.e. information not accessible to a linear approach to analysis. Two methods of nonlinear time series analysis are examined: correlation dimension and mutual prediction. It was found that the threshold for seizure detection for the linear methods of analysis often occurred earlier than or at the same time as the nonlinear methods. In fact it is suggested that since methods for nonlinear time series analysis have typically been developed for systems with few degrees of freedom, they may be at a disadvantage when it comes to the analysis of a signal, such as the EEG, which likely contains many degrees of freedom. The implementation of a linear method for seizure prediction is expected to place less demand on the limited computational and power resources of a brain defibrillator (Chap. 18). A lead time of 2–3 min before seizure occurrence is likely to be adequate for the activation of a brain defibrillator, particularly if the seizure aborting strategy is based on the delivery of carefully honed and timed electrical stimuli (Chaps. 16 and 17). Thus the observations in this chapter support the use of a linear method for seizure prediction in the development of a brain defibrillator.

5 Feedback Control of Epileptic Seizures

The concept of the feedback control of seizures is not new. For centuries it is quite likely that mothers of children with primary generalized epilepsy realized that shaking their child at seizure onset was often sufficient to abort the seizure. At one time the medical profession applied ligatures to the affected limb as a treatment for partial simple motor seizures. This section discusses the present state of the art for the feedback control of seizures.

Chapter 14: One of the earliest clinical attempts at the feedback control of seizures in patients with MIE using deep brain stimulation took place nearly 25 years ago in Tbilisi. In this chapter the principal investigator of these trials discusses the rationale and outcomes for these studies. A total of 54 patients had deep brain stimulators implanted for periods ranging from 1 to 25 months. The patient was able to assess the level of activity in their epileptic focus by monitoring the output of chronically implanted electrodes using a specially designed integrator of electrical activity. This method of feedback control is technically open-loop, since once the parameters of the deep brain stimulator are set they are not further modified by ongoing neural activity. The sites chosen for the placement of the stimulating electrode were the head of the caudate nucleus and the dentate nucleus. These choices were based on studies of patients with MIE which demonstrated that stimulation of these sites was effective in aborting seizures. Stimulation of the caudate or the dentate nucleus was more effective for seizure control than stimulation of the centromedian nucleus of the thalamus: stimulation of the caudate nucleus was better than dentate nucleus. These studies demonstrate that chronic deep brain stimulation can be a safe and effective therapeutic option for seizure control. An implication of these studies for the development of a brain defibrillator is that the selection of the site of stimulation may be more important than the details of the stimulus protocol.

Chapter 15: A brain defibrillator is a closed-loop feedback control device: the delivery of the seizure stimulus is controlled automatically by a programmed feedback controller that monitors neural activity. This chapter describes the development of the first device capable of feedback control of seizures. The device uses extracellular measurements to monitor neural activity and reduces neural activity by the application of an electric field. The basis for this approach is the observation that an electrical field oriented parallel to the somatic-dendritic axis of a neuron whose spike initiation zone is asymmetrically placed with respect to its centroid will modulate the neuron's firing. The magnitude of the electric field (typically 1–5 mV/mm) required to modify the activity of a neuron is much smaller than the magnitude required to provoke neuronal firing from rest. Preliminary results suggest that electric fields as small as 100 μV/mm may be effective: that is, electric fields considerably smaller than those emitted by a typical cellphone! An experimental paradigm employing a rat hippocampal slice is used to illustrate the feasibility of this approach. A simple feedback law is used: the applied field is proportional to the positive difference between the root-mean-square of the recorded neural activity and a threshold value. Clear suppression of epileptic activity was seen in 20 out of 30 seizing slices. A limitation of this technique arises because neural tissue is conductive and thus application of an electric field is accompanied by electric currents which can potentially create biological harmful products with prolonged and chronic use. Although these effects can be greatly minimized by the use of non-polarizing electrodes and non-rectified control (Chap. 18), it is clear that the development of brain defibrillators based on electric field suppression will have to allow for these limitations within their control laws. The beauty of this approach is its simplicity: synchronization of large populations of neurons, and hence a seizure, obviously cannot occur without the participation of spiking neurons. Of course it could be argued that without neural spiking it is unlikely that any other function of the nervous system is possible. However, since this approach critically depends on the orientation of the electric field with respect to the axis of the neuron it should be possible to focus it to interact with only spatially restricted regions of the brain, such as the site of the presumed epileptic focus.

Chapter 16: A brief sensory or electrical stimulus given at seizure onset can sometimes abort the event. This chapter discusses the possibility that epilepsy arises in a multistable dynamical system. It has been well established that multistability can arise in isolated neurons as well as neural populations: EEG analyses using implanted subdural electrodes indicates that there may be a switching between attractors at the onset of seizures arising from the temporal lobe complex. The nervous system, even at its most macroscopic level, has a recurrently looped structure in which time delays are an integral feature. These time delays arise because of, for example, finite axonal conduction velocities and integration times (Chap. 2). An experimental paradigm of a delayed recurrent loop is constructed from a computer and a single periodically spiking invertebrate motoneuron. The membrane potential of the motoneuron is monitored through a recording electrode and an analog signal proportional to the membrane potential is passed to a computer. Each time that the computer detects a spike, after a time delay, τ, it delivers a single square hyperpolarizing pulse to the motoneuron. The dynamics of this delayed recurrent loop can be completely understood in terms of two experimentally measurable quantities: τ and a function called the phase resetting curve (PRC) which summarizes the phase resetting properties of the motoneuron to hyperpolarizing pulses. The condition for multistability is that τ is longer than the motoneuron spiking period. The fact that the dynamics of the delayed recurrent loop are similar for a variety of different PRCs suggests that it might be possible to develop generic methods for the control of multistable delayed recurrent loops and new insights into EEG dynamics. In the presence of noise, dynamical systems with retarded variables can exhibit a phenomenon known as statistical periodicity: the probability density itself cycles periodically. The identification of statistical periodicity in the EEG will require careful analyses of its complex spatiotemporal patterns. These observations, taken together with those in Chap. 17, emphasize that an important consideration for the design of a brain defibrillator will be the precise time that the stimulus is applied relative to the ongoing seizure dynamics.

Chapter 17: Even dynamical systems which exhibit exceedingly complex dynamics, such as chaos (Chap. 12), can in principle be controlled by the application of carefully honed and timed perturbations. These so-called "control of chaos" techniques have already been successfully applied to the control of epileptic seizures in rat hippocampal seizures. Theoretically a chaotic dynamical system can be considered to be composed of a very large number of unstable periodic orbits (UPOs). Provided that a UPO can be identified it is possible to stabilize it. This chapter reviews the current state of the art for the detection of UPOs in short and noisy data sets. In low-dimensional dynamical systems it may be possible to identify a UPO by first converting the dynamics into an appropriately constructed recurrence map, referred to as the Poincaré section, and then searching the time series for a characteristic "marching in, marching out" phenomenon: when a trajectory is near a UPO it always approaches the UPO from a stable direction, and departs via an unstable direction. However, in general, it can be difficult to construct the required Poincaré section and thus recourse to statistical methods for detecting UPOs, such as the topological recurrence and periodic orbit transform methods, is necessary. The use of these statistical methods, in turn, necessitates the need to test against appropriate surrogates. The strengths and weaknesses of various surrogates for detecting UPOs is discussed. The observations illustrate the point that analyses of neural dynamics as reflected, for example, in the EEG may provide important clues to guide the development of a brain defibrillator.

Chapter 18: The successful development of a brain defibrillator requires a strong interplay between technology and theory. The properties, strengths, and limitations of the various components of such a device serve to select those theoretical strategies for seizure prediction and control which can be most feasibly incorporated into a clinically useful therapeutic device. This chapter reviews the technical challenges faced in the development of a brain defibrillator. The state of the art for fabrication of micro-devices is rapidly advancing. However, since the details are generally protected by patent laws, it becomes necessary to know what can be done, rather than how it can be done. A variety of devices for therapeutic stimulation already exist: cardiac pacemakers, vagal nerve stimulators, deep brain stimulators, cochlear implants, and implantable drug pumps. Thus, the essential component parts of an implantable cortical stimulator for epilepsy, namely a set of electrodes connected to control circuitry housed in a suitable casing, have been previously developed and FDA approved for other clinical applications. A novel possibility which avoids the use of electrodes is to use the intrinsic optical signals of neural tissue to monitor brain activity. Indeed an entire optical bench can be fabricated on a single silicon device measured in millimeters. All said and done, the two most important problems to be addressed for the construction of a brain defibrillator are: 1) its computational requirements; and 2) its biocompatibility. As a yardstick, modern pacemakers incorporating complementary metal oxide semiconductor (CMOS) circuit technology requiring ≤ 500 bytes of memory can be housed with battery and telemetry circuitry in a case approximately 60 mm in diameter and 10 mm thick. On the other hand, all of the computations could certainly be carried out on a desktop PC. It is possible that RF telemetry techniques could be used to decrease the amount of computational power that must be placed within the body much in the same way as currently proposed retinal neuroprostheses. The emerging field of interfacial molecular bioengineering is aimed at developing techniques to minimize problems of biocompatibility by forming artificial substrates or matrices in which the surface chemistry is modified by coating with selected proteins, and then seeded with cells to form a seamless interface between the artificial and the biological. The benefit of this approach is that implants are not just tolerated, but actually have improved functionality, becoming integrated into the natural physiology of the brain. Once the theoretical issues discussed in the remainder of this text are resolved, the development of a working prototype of a brain defibrillator is likely to be not far down the road.

1 Medically Intractable Epilepsy*

J. Milton

A seizure is a paroxysmal and uncontrolled discharge of a large population of neurons [5]. The term "epilepsy" is used when seizures have the tendency to recur. This book examines the sub-group of patients who have medically intractable epilepsy (MIE). The term "medically intractable" means that seizures recur despite the use of anti-convulsant medications. Important caveats are that the patient be compliant with medication(s), that the medications most appropriate for their type of epilepsy have been tried, and that the patient does not have psychogenic seizures.

Many different epilepsies can become medically intractable. However, it is patients who have seizures that arise from the temporal lobe complex (i.e. temporal lobe-amygdala-hippocampus) who most often benefit from surgical treatment. Consequently this is the epileptic syndrome for which most is known. In this textbook, unless otherwise stated, we use the term "medically intractable epilepsy (MIE)" to refer to patients with seizures that involve the temporal lobe complex.

1.1 Temporal Lobe Epilepsy

There are extensive classifications of both epileptic seizures [6] and of the epilepsies [7]. To the non-clinician the terminology used by neurologists to describe epilepsy can be quite confusing. Here we give a simplified overview.

The classification of seizures is based on how the seizure begins in the first split second. If the seizure begins in a focal region of the brain it is called *partial*, if it begins generally everywhere at the same time it is called *generalized*. Partial seizures are further sub-divided into those in which the level of consciousness is not impaired (partial simple seizures) and those in which it is (partial complex). The term "aura" is used to refer to that portion of a partial seizure that occurs before consciousness is lost and for which memory is retained afterwards. In clinical practice it can be very difficult to apply these classifications [8]. Partial onset seizures may generalize so rapidly that their focality is obscured. How is a very mild impairment of consciousness to be detected? Many patients may be unable to describe their auras (i.e. those who are poorly educated or mentally retarded) or unwilling (e.g. auras

* This research was supported by grants from the National Institutes of Mental Health and the Brain Research Foundation.

associated with sexual feelings). Thus if the letter of the classification were to be precisely adhered to then it would be impossible to distinguish a partial simple from a partial complex seizure or a primary generalized seizure from a secondarily generalized seizure.

On the other hand, the classification of the epilepsies is based on the clinical description of the seizure and the clinical context in which the seizures recur. The term "psychomotor seizure" describes seizures characterized by automatisms or automatic behavior, somatomotor activity, and various degrees of impairment in the level of consciousness. A typical psychomotor seizure might proceed as follows: sudden onset of unresponsiveness, staring, lip smacking, picking movement of both hands lasting 90–120 s followed by 10–15 min of confusion. Often psychomotor seizures are preceded by an aura. The aura most commonly reported by patients in my clinic is an epigastric rising sensation; however, other commonly emphasized types of auras include olfactory sensations and feelings of "deja vu". For a given patient there is typically only one type of aura: the stereotypy of their seizures provides an important clue that points to the existence of an epileptic focus, i.e. an epileptogenic region of brain from which seizures emanate.

A difficulty arises because many clinicians use the terms "partial complex" and "psychomotor" seizures to be synonymous and to mean seizures which arise from the temporal lobe complex. Although this is likely true for the majority of patients with psychomotor seizures seen in a clinical practice, it is also known that for some, the seizures actually initiate at an extratemporal site. If the seizure spreads with sufficient rapidity to the temporal lobe complex then its origin may be obscured by the clinical manifestations of a seizure generated by the temporal lobe complex. Well-known masqueraders of temporal lobe seizures include those with onset in the orbito-frontal cortex [9] and in the occipital lobe [10]. Indeed one of the major goals of the presurgical evaluation of a patient with epilepsy is directed at pinpointing the site of electrographic seizure onset.

We will use the term "temporal lobe epilepsy" to refer to those seizures for which there is clear clinical and/or electrographic evidence for involvement of the temporal lobe complex in the initial stages of the seizure. It is important to keep in mind that our present-day concept of an epileptic focus may be too simplistic; in other words perhaps there can be an epileptic system that involves structures in the temporal lobe complex as well as extratemporal lobe structures. Thus an advantage of using the term "temporal lobe seizures" as described above is that it leaves this possiblity open.

1.2 Prevalence of MIE

How many seizures does it takes before one determines that they have MIE? The definition of MIE differs depending on whether you ask the sufferer or the care giver. Physicians who treat patients with epilepsy as either a part of their practice and/or in the context of a clinical trial have been accustomed to accepting a low frequency seizure rate, e.g. less than 2–4 seizures per month, as a definition of a well-controlled seizure disorder. In patient drug trials the generally accepted benchmark for a new anti-convulsant medication is a 50 % reduction in seizure frequency. In contrast, sufferers have a very different opinion. To them the only acceptable therapeutic goal is a seizure frequency which is precisely zero.

Table 1.1 gives the seizure frequency of 1851 consecutive patients with temporal lobe epilepsy seen in my epilepsy clinics at The University of Chicago. For each patient the clinical history and description of the seizures by a witness was compatible with a diagnosis of temporal lobe epilepsy. All patients had an interictal scalp EEG recording demonstrating focal slowing and/or epileptic activity (spike-slow wave discharges) in one or both temporal lobes. Seizure frequencies were estimated from seizure diaries collected over a period of six months after the patient had been in the clinic for a year during which time anti-convulsant medication regimens were optimized.

A seizure frequency of <4 per month is often regarded as good seizure control for a patient with long-standing temporal lobe seizures. Thus from the caretaker point of view this epilepsy clinic has been quite successful since 91% of the patients have good control using this definition. On the other hand, the picture is clearly not so rosy from the point of view of the patient. Only 9% have been seizure free for six months and hence would qualify, for example, for a driver's license. From the sufferer's point of view their epilepsy is not well controlled. Clearly there is a huge difference in opinion!

The limitations of medical therapy to obtain complete remission of patients with MIE has only been recently emphasized. Indeed the traditional viewpoint has been to emphasize that one-third to one-half of patients are able to obtain long-term remission (for a review see [11]). However, recent studies have tended to support the dismal picture portrayed in Table 1.1, i.e. the remission rate is much, much lower [12–14]. Moreover, the quality of life of patients with even "well-controlled" epilepsy may be no better than that of patients who suffer chronic medical conditions, such as cancer and severe congestive heart failure [2]. These observations emphasize the need for alternative therapeutic modalities for managing patients with epilepsy.

Table 1.1. Prevalence of MIE in 1851 patients with partial complex epilepsy

Seizure frequency	Number of Patients	%
$f_s > 4$	167	9.0
$0 < f_s \leq 4$	1555	84.0
0	129	7.0

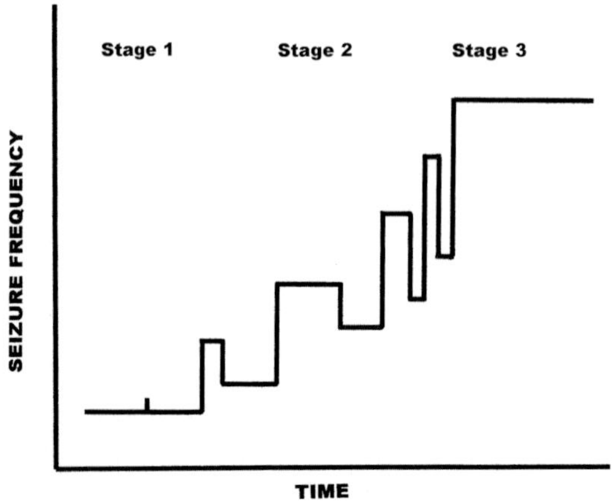

Fig. 1.1. Schematic representation of the development of medically intractable epilepsy

1.3 The Evolving Epileptic Syndrome

A retrospective review of the clinical histories of patients with temporal lobe epilepsy (Table 1.1) suggests that many of them evolve to the MIE state through a number of fairly recognizable stages (Fig. 1.1). Initially (Stage 1) the seizure frequency is low and the response to anti-convulsant medication is good. A typical story is that an adolescent presents with a single generalized tonic-clonic seizure (or even an episode of status epilepticus) followed by a period of time (even years) during which seizures either did not occur at all or which responded to a single anti-convulsant medication. The next stage (Stage 2) is marked by an increase in seizure frequency and a decreased efficacy of anti-convulsant medications. A not uncommon scenario is that the patient seems to respond to a particular anti-convulsant for a period of time, then their seizures "break through". At this point replacing the anti-convulsant medication with another often results in better seizure control; however, eventually the seizures break through. This cycle of "seizure control-seizure break-through" continues, the seizure-free periods becoming shorter and shorter. Finally the patient enters the stage of medically intractable seizures (Stage 3).

Figure 1.2a shows seizure diaries for a number of patients who have developed medically intractable epilepsy [15]. Two generalizations can be made: 1) the rate of seizure recurrence is constant for patients with MIE; and 2) the seizure recurrence rate differs for each patient. Figure 1.2b shows the effect of changes in anti-convulsant medication of seizure recurrence of a patient with MIE. It is not the absolute seizure frequency that defines a patient as having MIE, but rather the observation that the seizure frequency does not change following changes in anti-convulsant medications.

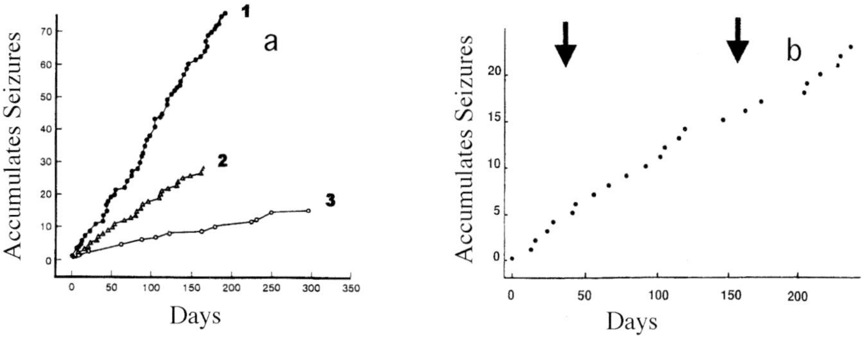

Fig. 1.2. Accumulated number of seizures for five patients with MIE. (**a**) shows that the seizure frequency (i.e. the slope) differs for different patients (data from [15]). (**b**) shows that making changes in medications (indicated by ↓) does not alter the seizure frequency

Admittedly the delineation between these stages is not precise. Moreover there is considerable variability in the rates of progression of different patients between the different stages. For example, some patients never seem to evolve beyond Stage 1; others are in Stage 3 almost from the time of their first seizure. Do patients in Stage 1 have the same disease as those in Stage 3? Obviously not. The failure to recognize that different patients can be at different stages of evolution towards the MIE state lies at the basis of much of the confusion that exists in both the basic science and clinical literature regarding epilepsy.

1.4 Lesional Epilepsy

The cornerstone in the management of patients with MIE has been to try to identify the epileptic brain region (i.e. epileptic focus), and if successful, to remove it surgically in the hopes of rendering the patient seizure free. Typically epileptogenic brain regions lie close to brain regions that have structural abnormalities that can be identified at the gross and light microscope level, hence the term "lesional epilepsy". In some cases the lesions are related to trauma or brain tumors. However, by far the most common pathological lesion associated with temporal lobe epilepsy is mesial temporal sclerosis. Approximately two-thirds of these patients who undergo surgical resection have pathological evidence of mesial temporal sclerosis [16].

Mesial temporal sclerosis is the term used by pathologists to describe the anatomical changes that are most prevalent in patients with temporal-lobe onset partial complex seizures [17,18]. There is a loss of pyramidal neurons in the CA1 and CA3 hippocampal regions with a relative sparing of neurons in the CA2 region accompanied by neuronal loss in the dentate gyrus (in particular the hilus) and the subiculum. Although mesial temporal sclerosis has long been felt to be the sequelae of anoxia/ischemic events that occur early in life (for a review see [19]), it is also frequently seen in patients in whom no obvious antecedent event can be identified. With the development of better radiological imaging techniques, cortical migrational defects have become increasingly recognized in patients with mesial temporal sclerosis [20]. Thus, mesial temporal sclerosis can be a non-specific marker that points to pre-existing pathologies.

The existence of pathological abnormalities, such as mesial temporal sclerosis, have supported the concept of an epileptogenic focus, i.e. a region of the brain from which focal onset seizures begin. However, pathological lesions, such as gliosis, obviously cannot themselves cause seizures. Viable neurons are required to produce seizures! Moreover, static pathological lesions do not explain the slow temporal evolution of MIE nor do they readily account for the paroxysmal nature of seizure recurrence. Presumably these features of MIE must be related to the changes that occur in the surviving neurons, the supporting cell populations and the ionic micro-environment.

1.5 Do Seizures Begat Seizures?

What is the force that drives the evolution towards the MIE state? Is it simply the occurrence of seizures, or is something else involved? Why for so many patients does the MIE syndrome seem to develop despite apparently good seizure control?

The strongest evidence in favor of the hypothesis that seizures begat seizures has been obtained in animal models. Neuronal damage can result from changes in intracellular Ca^{++}. Under normal circumstances brief changes in intracellular Ca^{++} are necessary for synaptic transmission and act as a second messenger for a number of metabolic functions. However, during severe seizures, Ca^{++} overload can occur and this impairs neuronal viability leading to oxidative damage, cytoskeletal degeneration, microtubule dysfunction and protein aggregation [21]. In addition, clustering of procaspases occurs which in turn promotes the activation of caspases, thus initiating the execution phase of apotosis [22]. Early clinical studies seemed to support this concept by suggesting that children with histories of febrile seizures or status epilepticus were at increased risk for the development of MIE.

Further support for the hypothesis that seizures begat seizures comes from experiments with animal models of epilepsy that demonstrate that the synaptic "bombardment" of brain structures functionally connected to the primary epileptic focus can change the action of these structures so that they themselves become involved in the epileptic process [23–30]. This process is referred to as "secondary epileptogenesis" [25,31] and "kindling phenomena" [32]. The facilitation of the synaptic transmission associated with this long-lasting bombardment can serve as a possible mechanism for the evolving epileptic process. This view is supported by the data concerning the activation of synaptic terminals in the electrocorticographically and polarographically confirmed cortical epileptogenic focus in the patients with brain tumors [33,34] and in the hippocampus of the kindled experimental animals [31].

In contrast, data collected from patients with epilepsy have, in general, not supported the hypothesis that seizures begat seizures [19]. First, the seizure-recurrence rate of patients with chronic epilepsy is not an ever-increasing one as might be expected if seizures recruited more neurons into an expanding epileptogenic region (see Fig. 1.2). Second, the hippocampus in the immature brain is not particularly sensitive to status epilepticus [35] and aggressive treatment of febrile seizures does not seem to reduce the risk of subsequent development of MIE [36–38]. Third, the issue as to whether secondary epileptogenesis occurs in humans remains controversial [23].

1.6 Neuronal Re-organization

It is currently felt that the individual neurons in epileptogenic brain are intrinsically normal [39]. Thus recent emphasis has been placed on the re-organization of the synaptic circuitry that occurs in epileptic brain [40]. Morphological changes occur both in the axons and the dendritic trees of neurons. First there can be a sprouting of axon branches and terminals in response to neuronal loss. An example is the re-organization of CA3 mossy fibers into the inner molecular layer of the dentate gyrus [41–44]. Mossy fiber re-organization has been described in patients with temporal lobe epilepsy [43,45] and can occur even in the absence of hippocampal pathology [46]. Axonal sprouting, synaptic re-organization, and formation of new recurrent excitatory circuits can also form as a result of continuous neuronal bombardment [24,47,48]. Second, the arborization of dendritic spines and branches of neurons in or nearby epileptic brain can be greatly stunted. Indeed the loss of spines and fine dendritic branches of neurons can be so severe as to leave neurons with only their proximal branches intact [49,50]. Similar observations have been made on neurons that lie adjacent to cortical tubers in the brains of patients with tuberous sclerosis [51]. Modeling and experimental studies have shown that even modest morphological alterations to the synaptic circuitry can greatly alter the excitability of a cortical region [52].

One experimental approach that has demonstrated the extent of neuronal (synaptic) plasticity that occurs in progressive epileptogenesis involves delivering a single electrical pulse to one of the structures and registering its response in the other [53,54]. Stimulation of amygdala in non-epileptic patients elicits a response in the ipsilateral hippocampus in all cases. However, 13/16 non-epileptic patients did not show an amygdalar response when the hippocampus was stimulated [53]. One reason for this asymmetry in spread of activity between amygdala and hippocampus is anatomical: there are rich anatomical connections of amygdala to hippocampus via the entorhinal cortex [55–57] and septum [58]; however, pathways from hippocampus to amygdala are scarce, although a few pathways pass through the septal area [59,60] and temporal neocortex [61]. In contrast, for 34/44 patients with epilepsy, hippocampal stimulation evoked an amygdalar response.

The spatio-temporal dynamics of networks of neurons depend on the interplay between 1) the magnitude of the neural inputs, 2) the time course of the relative refractory state of the neuron, and 3) the tightness of the interneuronal connectivity [62–71]. To illustrate these concepts let us consider a 2-D network of integrate-and-fire neurons[1] with excitatory interconnections. In order to mimic an epileptic focus we assume that the most centrally placed neuron fires periodically. Finally the probability, $P_{\hat{r}}$, of neural interconnectivity decreases exponentially with distance, $\hat{r}$, i.e.

$$P_{\hat{r}} \sim \beta \exp(-\alpha \hat{r}) , \tag{1.1}$$

where α, β are positive constants. This type of connectivity was initially measured in visual cortex [73], but has also been reported in other cortical regions [51].

The right-hand side of Fig. 1.3 shows examples of the spatio-temporal patterns of activity in this type of neural network as a function of the strength, ϵ, of the excitatory inputs. Changes in ϵ can arise because of changes in the neuron's morphology. Neurons are electrically compact [74–76]. This means that if the dendrites of a pyramidal neuron were shortened in such a way that its complement of synaptic contacts were maintained, then the effects of a given synaptic input would be enhanced.

When ϵ is sufficiently small (Fig. 1.3a), circular waves regularly form and travel away from the centrally placed periodically firing neuron. These waves are referred to as target waves because of their appearance. Target waves represent transient wave phenomena. This is demonstrated by the fact that these waves cease to appear once the centrally placed neuron stops firing periodically (see $\downarrow$ on right-hand side of Fig. 1.3). Once ϵ becomes sufficiently large two new types of spatial patterns of neural activity appear: travelling

[1] Integrate-and-fire neurons add inputs and when the membrane potential exceeds a certain firing threshold they generate a spike. The integrate-and-fire neurons used in the computer simulations shown in Figs. 1.3 and 1.4 incorporate two additional features [65,68,72]: 1) They are "leaky" integrators, i.e. when the membrane potential is augmented to a sub-threshold value it subsequently decays exponentially towards the resting membrane potential; and 2) They possess both an absolute and a relative refractory state. Once the neuron produces a spike the threshold elevates immediately to a very large number and for a time t_R the neuron is absolutely refractory, i.e. it cannot be made to fire irrespective of the input. The absolute refractory period is typically ~1–3 ms. Subsequently, the threshold decays towards its resting value. During this period the neuron is relatively refractory. In other words it can be made to generate a spike; however, a larger input that would be used for a resting neuron is necessary. Typical values for a relative refractory state are 50–150 ms.

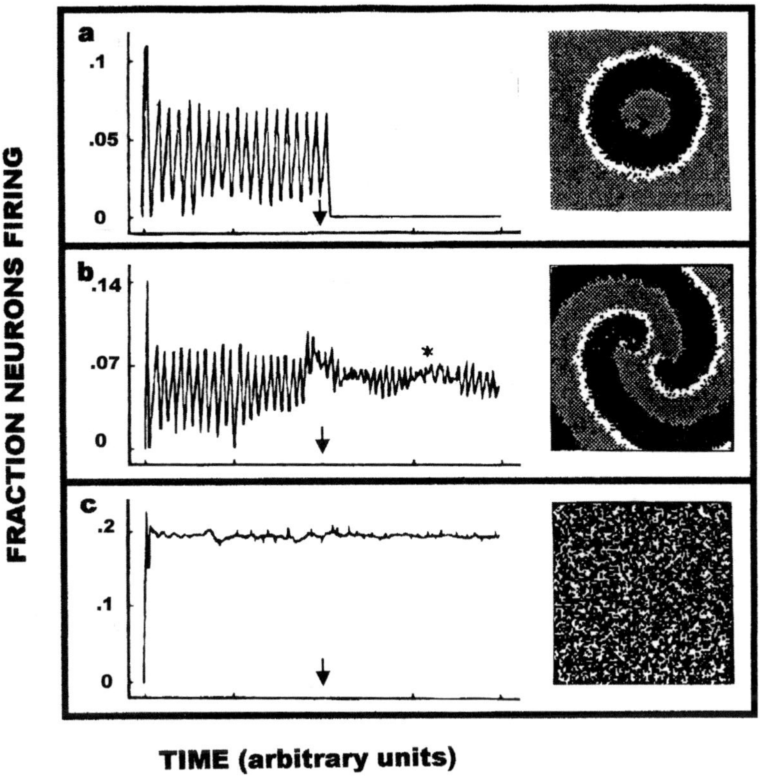

Fig. 1.3. Dynamics of a 2-D network of integrate-and-fire neurons as a function of the magnitude of the excitatory input, ϵ (see text for more details): (**a**) $\epsilon = 0.45$; (**b**) $\epsilon = 0.58$, and (**c**) $\epsilon = 0.72$. The left-hand side shows a plot of the fraction of neurons firing per unit time for the same values of ϵ. The right-hand side shows the spatio-temporal patterns of neural activity. At $t = 0$ all neurons except the central source are quiescent. At $t = 500$ (indicated by $\downarrow$) the source is shut off. Data is taken from [65,72]

spiral waves (Fig. 1.3b) and more complex patterns (Fig. 1.3c). These wave patterns are self-maintaining, i.e. they continue even though the centrally placed neuron ceases to fire periodically (see left-hand side Fig. 1.3).

When the network is more "loosely connected" and the recovery dynamics are adjusted so that refiring in the refractory trough becomes possible, then it is possible for the whole network to undergo synchronous bulk oscillations [66]. In some cases these bulk oscillations continue; in others, they cease spontaneously. Numerical [65,72] and theoretical [70,71] studies suggest that a sub-critical Hopf bifurcation can occur. In addition to these large-amplitude bulk oscillations, migrating localized oscillatory bursts of activity can occur. As the network becomes tighter and the relative refractory state more prolonged, the network produces both traveling circular and spiral waves [64–66,72]. The importance of the relative refractory period for these wave phenomena in networks with excitatory connections is emphasized by the fact that spiral waves do not arise in such models when this period is absent.

Computer simulations of networks with excitatory connections indicate that the formation of traveling spiral waves is favored in more tightly packed neuronal networks [65–67]. Figure 1.4 shows the effect of the "tightness" of inter-neuronal connectivity. As can be seen, P_{W} increases with the tightness of the connectivity.

A third possibility is that changes in the excitability of a neural network reflect alternations in the relative balance between excitatory and inhibitory connections [63,64,69–71]. The age-old belief that the epileptic focus is a brain region that is relatively excitatory has been favored by a number of animal studies (for a review see [47]). Increases in excitation can result from an enhanced function of NMDA receptors [77,78] and the formation of new recurrent excitatory loops [41,44]. Decreased inhibition is thought to result from either loss of inhibitory neurons [79] or as a result of the loss of excitatory input to inhibitory neurons ("dormant basket cell hypothesis") [80,81]. However, there has always been the nagging suspicion that the real picture is not quite this simple. For example, some forms of epilepsy develop in the presence of enhanced inhibitory mechanisms [82].

The role of inhibition in epileptogenic brain has been an enigma. On the one hand, it is often very difficult to identify inhibitory interneurons, particularly GABAergic neurons [83]. However, with the development of better techniques it has become clear that GABAergic neurons are actually relatively spared in the epileptogenic hippocampus [84]. Indeed histological studies [85]

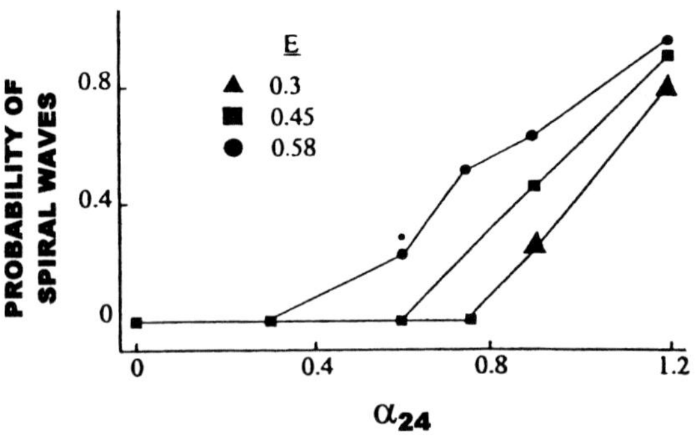

Fig. 1.4. Probability of the formation of a travelling spiral wave, P_{w}, as a function of the tightness of the interneuronal connectivity. The average total number of connections per neuron, C_{T}, for neurons arranged on a square lattice is $C_{\mathrm{T}} \sim d\beta/\alpha^2$, where α, β are constants (1.1) and $d = 4\sqrt{2}$. The role of the tightness of the interneuronal connectivity on spiral wave formation can be investigated by varying α, β while keeping C_{T} fixed [65]

as well as depth electrode studies in humans with MIE [86] have suggested that synaptic inhibition may be actually increased in epileptogenic cortex. On the other hand, experimental observations [87] and computer simulations studies [88,89] have shown that synaptic inhibition is very important for synchronization. Since synchronization of the activity of a large population of neurons is the essence of epilepsy (see Chaps. 5 and 6) it is quite possible that the role of inhibitory mechanisms in epilepsy has been grossly underevaluated.

Inhibition can be thought of as a kind of refractoriness [69]. indexinhibition!versus refractoriness Thus networks of excitatory and inhibitory neurons without refractoriness resemble very much those described in Figs. 1.3 and 1.4. A key parameter is the relative spatial extent of the excitatory and inhibitory connections. In general standing wave patterns predominate when the spatial range of inhibition exceeds that of excitation, whereas travelling wave phenomena arise in the converse case [62,71].

Finally, it must be emphasized that it is inherently difficult to infer the dynamical properties of a neural network solely from morphological measurements. For example, the phase resetting curves for neurons, including cortical neurons, contain regions of phase advance and phase delay [90–93]. Thus, depending on the timing (phase) of inputs, inhibition can be excitatory and vice versa (see Chap. 16). This observation stresses the need for the development of realistic models before the mechanisms that generated experimentally observed dynamics can be correctly inferred [94]. For an example of the successful intermixing of experiment and theory the reader is referred to the work of Traub, Miles and their collaborators [75].

1.7 Epilepsy as a Learning Disorder

There is a tendency in the epilepsy literature to think of an epileptic network as a localized phenomenon, e.g. a large cluster of neurons restricted to sharply delineated region such as the hippocampus. However, the picture is clearly not this simple. For example, in contrast to seizures that occur in isolated hippocampal slices, in humans depth electrode recordings during the occurence of a temporal lobe seizure indicate that less than one-third of pyramidal neurons are firing [95]. One possible explanation for this observation is that the epileptic system may be much more spatially distributed perhaps in the form of a reverbatory circuit [96,97] with activity propagating through it in the form of traveling waves [98].

No brain region alone can be singled out as the area solely responsible for chronic epileptogenesis [96,99]. The amygdala, hippocampus and temporal cortex are themselves part of a much larger cortical network, the limbic system. Evidence to support the concept of a dynamic, distributed epileptic network is discussed in Chap. 3. Here we draw attention to an important consequence of an epileptic system that is distributed throughout the same neural substrates that are important for memory and learning.

Since the electrical discharges generated by neurons are remarkably similar in amplitude and morphology, the source of a neuronal spike cannot be identified by a neuron that receives the spike. In other words a neuron in the limbic system cannot determine whether the spike it receives is part of a normal learning process or was generated by an epileptic network. If neurons cannot distinguish between spikes that should or should not be learned, then we are forced to consider the possibility that the brain can learn to have seizures [48,26]. There are a number of obvious similarities between the neural basis of learning and partial complex seizures. For example, in both cases the temporal lobe and hippocampus play important roles and physical changes in the brain occur, such as changes in the arborization of dendritic trees.

Learning proceeds through stages and is a continually changing process. The development of MIE also develops through stages. A number of inter-ictal abnormalities can be detected in the EEG of patients with partial complex seizures, e.g. focal spike-slow wave discharges, slow-sharp waves, intermittent focal slowing. Let us suppose that the brain learns from these phenomena in order to become a seizure-producing machine.

How long would it take the brain to learn this task ? Obviously, it is difficult to answer this question precisely. The best estimates are obtained from the time it takes to acquire highly skilled motor coordination. For example, consider learning to be proficient at hand knitting a sweater. Suppose one must knit 20 sweaters before one would be considered skilled. There are approximately 75,000 stitches per sweater and hence 1.5×10^6 repetitions are acquired. In a similar manner investigations have obtained estimates of the same order of magnitude for tasks that include rug making, violin playing, basketball playing, baseball throwing, and cigar making (Karl-H. Mauritz, unpublished observations). Thus it takes about $1 - 3 \times 10^6$ repetitions to develop a skilled motor pattern.

Using this same approach we can estimate the number of epileptic spikes that it would take before the brain would become skilled in this activity. Suppose an epileptic spike occurs once every 6 minutes. This means that there will be 10 spikes per hour, 240 spikes per day, and 87,600 spikes in a year. At this rate it would 11.4 years to have 10^6 spikes and hence for the brain to "learn" this skilled task.

Of course, different spiking rates mean different learning times. The main point is that the time scales are on the order of years. Similar time scales are involved in in the development of MIE for patients with partial complex seizures. For example, even in patients in whom an initial insult to the temporal lobe complex was suspected to have occurred early in life, for many the complex seizures do not begin until after puberty. Another example that emphasizes that the development of MIE is often preceded by a prolonged period of quiescence occurs in the setting of epilepsy surgery. Even after apparently successful epilepsy surgery, up to 35% of patients experience seizure recurrence within 5 years of the surgery; many of these were seizure free for a few years [100].

What is being learned? The major tenet of the secondary epileptogenesis hypothesis of Morrell and co-workers is that a previously normal region of the brain can learn to have seizures [25,27,34]. However, in humans, as we have just seen, the idea that seizures kindle other seizures is not particularly well supported by clinical observations [19]. In contrast, my impression is that the most typical picture of a patient who develops MIE is not that mirror foci develop but rather that the seizures remain very stereotyped. The change that heralds the development of MIE is the resistance of the seizures to medical and sometimes even surgical therapy (Fig. 1.1). This observation suggests that the epileptic system must be learning to become a better (i.e. more resistant) seizure generator.

If we accept the hypothesis that the brain can learn to be a more effective seizure producing machine, then we must also entertain the possibility that this task can be unlearned. It has been known since antiquity that behavioral strategies offer seizure control for certain patients (for a review see [101]). It is not uncommon to hear of patients who are "able to fight off" their seizures by thinking certain thoughts. Patients have also obtained relief from seizures by binding their affected limbs [102], the application of ligatures [101,102] ,and by the use of a variety of sensory stimuli, such as odors [103] and auditory stimuli [104]. The use of biofeedback techniques to treat patients also supports the notion that the brain can unlearn to have seizures [101,104]. The efficacy of these approaches has been difficult to understand from the perspective of the molecular biology and neuro-physiology of the nervous system. Hence the tendency by most neurologists has been to discount these reports. However, the very fact that such cases exist raises troublesome issues. For example, if neurons in an epileptogenic brain region could be retrained for other purposes, then it becomes increasingly difficult to justify their surgical removal.

1.8 Dynamical Aspects of Epilepsy

As we have just discussed, the current "state of the art" in the conceptualization of the development of the medically intractably epileptic state is that changes in network circuitry are important. However, the time scale on which these network changes occur and are reinforced appears to be on the order of months to years. The major issue left unsettled is the explanation for the paroxysmal nature of seizure recurrence. In other words why does a seizure occur when it does? This subject is in fact one of the subjects of this book. The reason why this topic is so important is that if it is known why the control of neural activity can suddenly go awry, then perhaps it might be possible to develop a smart therapeutic strategy that would re-establish this control.

At its most fundamental level of description the occurrence of an epileptic seizure represents a qualitative change in the dynamics of neuronal firing [105]. A sudden change in qualitative dynamics, presumably in response to a change in a critically important control parameter, is the hallmark of a dynamic disease. Dynamic diseases can arise from alterations in underlying physiological control mechanisms [94,106–111]. By analogy with mathematical models, changes of certain physiologically important parameters into critical ranges result in the sudden appearance of qualitatively different dynamical behaviors. In the mathematical models these qualitative changes in dynamics correspond to bifurcations. As we have seen the critical control parameter could be a change in the tightness of the connectivity of the network or the level of excitation. Other possibilities include transient changes in ionic fluxes across neuronal membranes or in the ionic environment of the neurons [112].

The significance of identifying epilepsy as a dynamic disease is that it raises the possibility of devising strategies to treat seizures that are based on manipulations of the underlying dynamics. The precise nature of the control strategies that are possible depends on the type of the bifurcation that has occurred and the nature of the abnormal dynamics that arose from the bifurcation. The advantage of this approach is that it is known that there are only a small number of types of bifurcations and dynamical behaviors that are possible.

The simplest therapeutic strategy is to manipulate the altered control parameters back into a range associated with healthy dynamics. For example, it might be possible to use molecular genetic techniques to change the critical parameters (see, for example, Chap. 11). However, we do not know the nature of the critical control parameters in an epileptic brain region. Nonetheless, it may still be possible to devise control strategies that exploit the properties of the abnormal dynamics that arise, for example, multistable dynamics can be controlled using brief perturbations ([113], see Chap. 16), and chaotic dynamics can be controlled using control of chaos techniques (see Chap. 17).

1.9 Conclusions

The picture that evolves from the observations in this chapter is that there is a dynamic nature to the generation and propagation of seizure as well as to the evolution of the epileptic syndrome. The importance of this perspective is that it implies that the most effective therapeutic strategies will be those that are also dynamic in nature. In other words a good way to control seizures is to control the mechanisms which in turn control the seizures. Moreover, therapeutic strategies drawn from the dynamic aspects of learning may be the only avenue available to allow the brain to unlearn the task of generating epileptic seizures.

2 Insights into Seizure Propagation from Axonal Conduction Times*

J. Milton

The term "secondarily generalized seizure" refers to a generalized seizure that was initiated in a spatially localized epileptic focus and then spread to eventually involve the whole cortex. Clearly the occurrence of such a seizure implies that seizure activity must be capable of propagating throughout the brain. Along which pathways does this propagation occur?

Figure 2.1A shows a schematic representation of the human brain. Roughly speaking the brain is organized anatomically as a double sphere (Fig. 2.1B): 1) the neurons of the cortical gray are located on the surface of the outer sphere; 2) the sub-cortical nuclei and brainstem structures are enclosed by the inner sphere; and 3) the region between the two spheres is occupied by the axons of the cortical white matter that interconnect distant neurons located either in the cortical gray matter and sub-cortical nuclei and also convey the sensory inputs and motor outputs of the central nervous system. It follows from Fig. 2.1 that there are three possible routes for seizure propagation.

1. Intracortical spread (Path 1 in Fig. 2.1): epileptic activity may spread from the epileptic focus like a "diffusing oil spot" progressively engulfing increasingly large numbers of neurons [62,114];
2. White matter mediated spread (Path 2 in Fig. 2.1): seizure activity propagates over the white matter pathways that interconnect adjacent as well as distant areas of cortex in both the ipsilateral and contralateral hemispheres;
3. Sub-cortical nuclei mediated spread (Path 3 in Fig. 2.1): seizure activity may spread via reciprocal connections (mainly myelinated) between the cortical gray matter and the underlying sub-cortical nuclei.

In this chapter we examine the propagation velocity of epileptic activity via neuronal-neuronal pathways. Since neural glia, in particular astrocytes, play a major role in the regulation of extracellular ion concentrations in the brain, it is possible that these cells may also participate in seizure propagation [112]. Waves of spreading depression have been observed in retina [115], cortex [116,117], and cultured hippocampal slices [118] and are thought to reflect diffusion of extracellular potassium ions [119,120]. Although spreading depression may play a role in migraine and certain phenomena in the stroke penumbra, its propagation velocity, i.e. ~ 0.002–$0.05\,\mathrm{cm/s}$, it is currently

* This research was supported by grants from the National Institutes of Mental Health and the Brain Research Foundation.

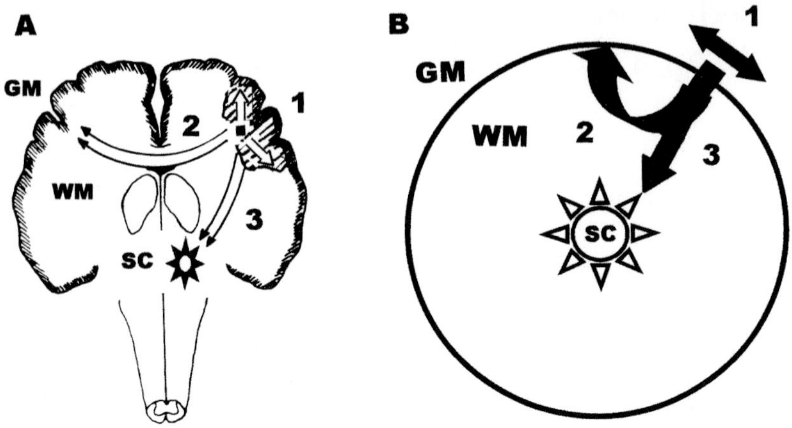

Fig. 2.1. Schematic representation of three seizure propagation pathways (**A**). A simplification of these pathways is shown in (**B**). The starburst symbol for the sub-cortical nuclei (SC) indicates that they have a diffuse projection. WM: white matter, GM: gray matter

thought to be too slow to be a primary pathway for seizure propagation. Calcium waves occur in both neurons [121] and astrocytes [122,123] and have been shown to travel faster and preceed waves of spreading depression [118]. The possible role of calcium waves in astrocytes to epilepsy is discussed in Chap. 4.

2.1 Axonal Conduction Velocities

The intracortical, or transverse, axons of the cortex are largely unmyelinated, whereas those axons that exit the cortical gray to form the commissural and association fibers are largely myelinated.

2.1.1 Unmyelinated Axons (Path 1)

Conduction velocities of the intracortical pathways for seizure propagation have been directly measured in hippocampal slices [124,125] and in isolated neocortical slices [126–128]. In hippocampal slices, propagation of spikes occurs at 30–40 cm/s [125], which is in agreement with the conduction velocities of the unmyelinated Schaeffer collateral axons (20–50 cm/s [125,129–132]). On the other hand seizure propagation occurs at the slower rate of 12–18 cm/s [124]. It has been suggested that this 2–3 fold reduction in the rate of seizure propagation reflects the sequential recruitment of neurons [124]. The propagation rate of epileptic activity in cortical slices obtained from somatosensory cortex ranges from 6–9 cm/s [126] for the propagation of seizure activity via the recruitment of neurons within a layer of a cortical column to 10–15 cm/s for recruitment of connecting cortical columns [127,128]. Conduction velocities along unmyelinated intracortical axons is 10–30 cm/s [133,134]. The reduction in the seizure propagation rate compared to the axonal conduction rate is thus of about the same order as observed in hippocampal slices.

Seizure propagation in neocortical slices has several characteristics that were not appreciated in the studies of hippocampal slices [126,135,136]: 1) the conduction velocity in several cortical regions is not uniform but instead shows a spatially periodic pattern, with dominant frequencies of the order of 1 mm^{-1}; 2) conduction velocity is direction dependent; 3) different architectonic areas of cortex exhibit different propagation patterns; and 4) patterns of propagation in the opposite direction across the same region of cortex are often phase-inverted. "Herky-jerky", or lurching patterns of seizure propagation have also been observed in computer models of neuronal populations in which inhibitory mechanisms play an important role [89].

It is important to remember that since these tissues have been surgically isolated they are devoid of all controlling influences from other cortical and sub-cortical regions. Moreover, in many of these preparations local inhibitory influences have been removed by pharmacological blockade. Thus the only avenue available for the propagation of epileptic activity in these preparations is in the form of the excitatory axonal connections between the neurons. Re-institution of inhibition significantly slows seizure propagation [89,124,137]. Since inhibition is not absent (and in fact may be normal (see Chap. 1)) in epileptogenic brain we must assume that seizure propagation via unmyelinated axons is very probably $\leq$ 6–18 cm/s.

2.1.2 Myelinated Axons (Path 2)

In all of the above studies the conduction velocities of the unmyelinated axons has been directly measured using microelectrode techniques. In contrast, neuro-anatomists typically estimate the conduction velocity, v, of myelinated axons in the central nervous system indirectly from measurements of their diameter, d. For a myelinated axon

$$v \sim dg\sqrt{-\ln g}\,, \tag{2.1}$$

where g is equal to the ratio of axon diameter to overall fiber diameter, and for an unmyelinated axon

$$v \sim \sqrt{d}\,. \tag{2.2}$$

For experimentally measured values of g, these relations predict that conduction velocity in myelinated axons exceeds that of unmyelinated axons provided that the diameter of the axon is greater than 0.2 µm [138].

In the peripheral nervous system most axons with diameters < 1 µm are unmyelinated; however, in the white matter axons with diameters as small as 0.2 µm are myelinated [138]. A reliance on morphological measurements to estimate a physiological parameter, such as conduction velocity, clearly introduces the potential for error. For example, the above equations lead to the conclusion that for small diameter fibers the conduction velocity is faster for the unmyelinated ones! (this paradox was clarified by the work of [138]). Moreover, aside from the obvious concerns to the possible effects of the fixatives and stains, the measurements tend to be biased in favor of the larger diameter axons. In the few studies that have determined v using microelectrodes [139–142], it is obvious that it is the size of the electrode that determines which axons can be measured. Miller [143] has estimated that the selection bias introduced by these problems is significant. Indeed, he suggests that 40–60% of the fibers in the corpus callosum may introduce a 100–200 ms conduction delay between the two hemispheres! If we take the average length of a cortico-cortical axon to be 5 cm [144], then $v \sim$ 25–50 cm/s. Moreover, despite the terminology, it is important to note that the cortical white matter also contains considerable numbers of small-diameter, unmyelinated axons. For example, in the human corpus callosum it has been shown that 30–45% of the axons are unmyelinated [145] with conduction velocities of the order of 10–20 cm/s [133,134]. Exploring the possible significance of these slow conducting pathways for cortical function is experimentally difficult since it has been estimated that conduction on these time scales would produce contributions to the EEG in the low-frequency range (i.e. $\leq$ 1 Hz) [144]. This is the frequency range in which the most troublesome artifacts arise, i.e. electrocardiogram, movements, respiration. Nonetheless, Wright and Sergejew [146] have shown the presence of traveling waves in the EEG with $v \sim$ 10–30 cm/s.

The above observations suggest that there exists an important mechanism for signal propagation in the central nervous system that involves relatively long time delays. However, it seems reasonable to assume that for epileptic activity to be able to activate other cortical and sub-cortical regions, the faster conducting fibers would be most significant. This conclusion derives from the observation that following excitation a neuron becomes refractory (absolutely for 1–3 ms, relatively for up to 100 ms) [68]. Of course neural activity arriving 100–200 ms after the fastest impulses could have an important modulatory effect [142].

Nunez [144,147] used morphological measurements from several sources to conclude that the average conduction velocity of the larger-diameter myelinated cortico-cortical fibers was $\sim$ 600–900 cm/s (with a range of 250–2800 cm/s). Measurements of spike propagation in human temporal lobes [148–150] and between the two hemispheres [151,152], the spread of electrically induced after-discharges in animal models [114], and of dispersive waves over the cortical surface [144] provide convincing evidence that neural activity can indeed propagate via the relevant large-diameter interconnecting axons. If we use the same reduction for seizure propagation rate estimated for unmyelinated axons, we obtain an estimate for the rate of myelinated axonal pathway-mediated seizure propagation of 200–300 cm/s.

2.2 Human Seizure Propagation Rates

In this section we interpret human seizure-propagation rates in terms of the conduction rates estimated in the preceding section. Determination of seizure-propagation rates is problematic because it is difficult to measure with certainty the distance the seizure travels per unit time. The difficulty arises because the surface of the brain is highly convoluted. Indeed, approximately 2.5–3 times the total cortical surface area is located in the sulci [153]. In order to estimate human seizure-propagation velocities we will imagine that the cortical surface has been spread onto a flat surface into a circular shape. Estimates of the cortical surface area range from $1469\,\mathrm{cm}^2$ to $2275\,\mathrm{cm}^2$ [153] with a typical consensus of $\sim 2200\,\mathrm{cm}^2$, i.e. the circle has a radius of $\sim 26.5\,\mathrm{cm}$. Next we assume that the epileptic focus is located at the center of the circle and that during a seizure activity propagates uniformly and radially from the central point. With these assumptions we have

$$SA = \pi(vt_s)^2 , \tag{2.3}$$

or

$$v = \sqrt{\frac{SA}{\pi t_s^2}} , \tag{2.4}$$

where v is the seizure propagation rate and SA is the total cortical surface area involved by the seizure at time t_s. Using (2.4) we see that for propagation via unmyelinated fibers ($v \sim 6$–$18\,\mathrm{cm/s}$), $t_s \sim 1.6$–$53\,\mathrm{s}$, and for propagation via myelinated fibers ($v \sim 200$–$300\,\mathrm{cm/s}$), $t_s \sim 0.1$–$0.5\,\mathrm{s}$. It is often stated that the seizure propagation rate in intact cortex is ~ 10–$40\,\mathrm{cm/s}$ both in animal models [154,155] and in humans [4,156]. However, the clinical spectrum of seizures and their propagation is much more diverse than this consensus would seem to imply. Here we discuss three examples.

First, seizures can propagate at rates much slower than 10–40 cm/s. Slow seizure propagation typically occurs in the setting of partial simple motor or sensory seizures that arise from primary sensorimotor cortex. Clinically the seizures progress in the form of a "Jacksonian march", for example, clonic movements may begin in the hand and subsequently march up the arm. It has been shown that a small cortical region containing, at most, a few thousand neurons in layer V is sufficient to initiate, generate and propagate epileptiform activity [135,157,158]. The success of the use of multiple sub-pial transection techniques to control seizures arising from sensorimotor cortex emphasizes the dominant role played by the horizontal (mainly unmyelinated) intracortical fibers for the generation and elaboration of these seizures [159,160]; i.e. Path 1 (Fig. 2.1). The clinical progression of motor or sensory symptoms during a Jacksonian march provides an estimate of seizure-propagation velocities. Such estimates are typically of the order of 0.05 mm/s [4,160]. Although slow propagation velocities of this magnitude point to spreading depression, these seizures do not respond to therapeutic modalities that work in other conditions in which spreading depression plays a prominent role, e.g. migraine. In fact these seizures are notoriously refractive to anti-convulsant therapy. Thus it is more likely that the slow propagation velocity is related to polysynaptic pathways acting in concert with local inhibitory mechanisms.

At the other end of the spectrum, seizures can propagate so quickly from an epileptic focus that their focal nature is obscured [161,162]. For example, temporal lobe onset seizures can be misidentified on the EEG as a primary, even 3-Hz spike-and-wave, form of epilepsy [161]. However, by using appropriate techniques, small time differences of 5–50 ms can often be demonstrated between the two hemispheres [151]. This suggests a seizure-propagation rate of the order of 60–600 cm/s. Thus Path 2 (Fig. 2.1) can be the predominant route for seizure propagation in some patients.

Between these extremes are focal seizures that remain confined to the region of the epileptic focus for a period of time and then either dissipate or rapidly generalize. All focal onset seizures have the potential to secondarily generalize, but many do not. The propensity for generalization depends on the location of the epileptic focus. Seizures that arise from primary sensorimotor cortex have a much lower tendency to generalize than those arising from anterior frontal lobe. In patients with partial complex seizures arising from the temporal lobe complex, it is more likely that the seizure remains localized (psychomotor seizure) than it is to give rise to a generalized tonic-clonic seizure. The length of time that the seizure remains localized is also variable. Seizures arising in the frontal lobe anterior to the primary sensorimotor cortex typically generalize with great rapidity. On the other hand, in patients with partial complex seizures arising from the hippocampus, the best surgical outcomes are obtained for patients in whom the inter-hemispheric propagation time is > 50 s [163]. In a rare form of focal epilepsy called epilepsia partialis continuans, seizures can remain confined to a focal region for months to even years [161].

The paths for seizure propagation can also change dynamically. In the evaluation of patients with suspected focal onset seizures an important clinical point is whether or not the clinical symptomatology is stereotyped: stereotypy favors focal onset. However, it is important to realize that stereotypy can breakdown, particularly when seizures occur frequently (see Chap. 3). Consider a patient with a focal onset seizure which arises from the amygdala-hippocampal complex. The seizure may remain confined to this region, or spread to supplementary motor cortex on the route to secondary generalization, or spread to involve homologous regions in the contralateral hemisphere. As seizure frequency increases, focal cortical areas unrelated to the initial seizure focus can take over and temporarily become sites for seizure onset, only to give away to other sites, even the initial one again.

The observations in this section emphasize that unmyelinated (Path 1) and myelinated (Path 2) axons likely play different roles in seizure propagation. The complexity and variability in seizure phenomenology implies that the pathways for seizure propagation must be under the influence of controlling mechanisms. Where are these controlling mechanisms located and what is their nature? The next section makes the argument that these mechanisms are, in part, located in the deep sub-cortical nuclei (Path 3 in Fig. 2.1). It is this realization that emphasizes the potential therapeutic value of deep brain stimulation and which opens the door for the development of dynamic control methods for seizures.

2.3 Sub-cortical Mediated Spread (Path 3)

What is the possible role for the third pathway shown in Fig. 2.1, i.e. propagation via reciprocal cortical-sub-cortical connections (Path 3)? At first sight it does not seem to be a very attractive idea that a long, circuitous route from cortex to sub-cortex back to cortex would be very relevant. However, focal electrical stimulation studies in animals strongly show that cortically generated after-discharges (AD) spread preferentially to sub-cortical structures rather than directly to other cortical regions [164,165]. The lone exception appears to be primary sensorimotor cortex that spreads to homologous regions in the contralateral hemisphere.

In humans the cortex to thalamus conduction time is 14 ms [166]. If we take a cortex to thalamus distance of ~ 5 cm [144], allow 0.5–0.9 ms for the synaptic delay [167] and ~ 10 ms to generate an excitatory synaptic potential in the thalamus [168], then we obtain a conduction velocity of ~ 1.7 m/s. As expected, conduction between cortex and thalamus occurs via myelinated fibers. Many sub-cortical structures, including the thalamus, project diffusely to the cortex. Indeed, reciprocal thalamocortical pathways participate intimately in the generation of generalized seizures [169,170]. Thus we see that the pathway that corresponds to the shortest time for seizure propagation between two points on the cortex is not necessarily the pathway having the shortest length! Intuitively it would seem that the fastest pathway for seizure generalization available to an epileptic focus is the one that involves the sub-cortical structures. Differences in the time for generalization for different epileptic foci and seizures may simply be related to the times taken to overcome local controlling mechanisms, spread to the deep sub-cortex, and then to overcome the sub-cortical regulatory mechanisms. The importance of sub-cortical structures for the development and propagation of seizures is now well recognized [114,171,172]. Indeed, thalamocortical connections are also involved in the synchronization and propagation of seizures from an epileptic focus [173].

Direct stimulation of sub-cortical structures does not elicit seizures [164]. However, electrical stimulation of a variety of sub-cortical structures has effects on both partial and generalized seizures: cerebellum [174], caudate nucleus [175,176], centromedian nucleus of the thalamus [177], and the nucleus of the solitary tract via stimulation of the vagal nerve [178]. The traditional viewpoint is that these observations indicate that these sub-cortical structures exert powerful inhibitory influences on seizure generation and generalization.

A more recent concept is the recognition that certain sub-cortical regions can act as a permissive gating mechanism. The best-characterized gating region is the dentate gyrus of the hippocampal formation [179]. Although the dentate is a relatively unexcitable region of the hippocampus (i.e. as compared to CA1 and CA3), maximal activation of the dentate is required for seizure spread and generalization. Other gating regions are thought to exist: the substantia nigra pars reticulata [180], superior colliculus [181], and the centromedian nucleus of the thalamus [182].

An important question is why do seizures stop once they have begun. In 1965 Kreindler [114] wrote

> 'The epileptic attack progresses by continuously recruiting new reverberating circuits. Its arrest probably occurs when the epileptic discharge tends to migrate from one structure to another, at a given moment penetrating a circuit having desynchronizing properties and hence extinguishing the attack.'

The observation that generalized seizures, even those that were initiated in an epileptic focus, stop suddenly and simultaneously over the cortex strongly implies the intervention of an active abolishing mechanism [114]. The important seizure abolishing sub-cortical mechanisms appear to reside in thalamo-caudate circuits [183–187] and brainstem (Chap. 14).

Thus we see that although sub-cortical structures do not function as epileptic foci themselves, they nevertheless play an important role for the generation, propagation and cessation of seizures.

2.4 Discussion

Our examination of the propagation rate for human seizures has clearly undermined the simple notion of an epileptic focus from which seizures propagate. The occurrence of a seizure is clearly a very complicated phenomenon that involves the dynamic cooperation between cortical and subcortical neurons. Some of these regions are important as seizure triggers (for a review see [39]), others control the activity of the triggers, and still others control the subsequent spread of activity. Even this complexity represents a gross over-simplification of the richness of the phenomena that can be observed during a seizure.

The realization that much of this complexity may be regulated by subcortical structures is at the same time fascinating and frustrating. The fascination arises because of the possibility of developing novel therapeutic strategies for epilepsy based on their stimulation [110,188]. The frustration arises because there is no technique presently available capable of monitoring the reciprocal information flow between cortex and sub-cortex. The electroencephalogram (EEG) is most sensitive to changes arising at the free cortical surface (Chap. 5), whereas optical imaging techniques [189] are necessarily restricted to the free surface. Although functional magnetic resonance imaging (fMRI) may potentially offer the required spatio-temporal resolution [68], it is ultimately limited by the fact that the changes recorded do not directly reflect changes in neural activity by rather arise primarily from local changes in blood flow [190]. The lone exception are studies of seizure propagation measured in humans who have had implanted depth electrodes (Chap. 3), but even in this case the information is limited by the small number of depth electrodes which can be safely implanted.

3 Dynamic Epileptic Systems Versus Static Epileptic Foci?*

S.A. Chkhenkeli and J. Milton

In 1881 William R. Gowers published his seminal book "Epilepsy and Other Chronic Convulsive Disorders" [102]. Since that time a major emphasis in clinical epileptology has been determining the site of origin of the epileptic discharge. Indeed the concept that a very limited area of brain is capable of producing and maintaining the epileptic discharge forms the cornerstone of current neuro-surgery procedures that treat epilepsy by either total resection [191–208] or stereotactic destruction [191,209–221] of the offending cortical region. However, the limited success and even failure of epilepsy surgery in a number of patients has drawn attention to the possibility that the nature of epileptic generators may be much more sophisticated than initially suspected, especially in patients with long-standing medically refractory epilepsy [40,191,210,222–229]. This chapter reviews the reasons for failure of epilepsy surgery and introduces the concept of epileptic systems [191,230], i.e. networks involving many distinct regions of the brain that are active in the brains of patients with medically refractory epilepsy.

3.1 Failures of Epilepsy Surgery

It is easy to understand why epilepsy surgery fails in situations when only a partial resection of the epileptic focus is possible. If the epileptic focus lies very close to regions of eloquent cortex, then the risk of loss of important functions, such as memory and language, outweighs the benefit of seizure cure. However, more puzzling are those cases in which the resected suspicious epileptic region is incongruent with its true localization or when the removal of brain tissue with the highest epileptogenicity leads to the emergence of a new epileptic focus from an entirely different cortical region. A number of explanations have been offered to explain this observation:

1. There can be a discordance between the location of the epileptic focus as suggested by the clinically observed seizure and its actual origin. For example, the typical manifestations of temporal lobe epilepsy (e.g. staring, automatisms) can be due to an extratemporal epileptic foci located, for example, in orbital frontal cortex [10,228,231,232].

* This research was supported by grants from the National Institutes of Mental Health and the Brain Research Foundation.

2. Incongruities can arise between the apparent localization of an epileptic foci as determined by electro-physiological measurements and its true location [233–235]. For example, the location of the foci determined from scalp recordings can have a migratory appearance ("jack-in-the-box-seizures" [236], "events at a distance" [237], "projected" epileptiform activity [238,239]).
3. It is possible that when multiple epileptic foci are present that there is a dominance of one over the others. In other words the subjugated epileptic foci become activated when the dominant focus is removed [240–242].

3.2 Dynamic Epileptic Systems

The final reason is that the concept of an epileptic focus is itself misleading. In cases of intractable epilepsy, the brain of patients with epilepsy appears to be "abnormal in many different areas and in many different ways" [40,226]. The symptoms in patients with temporal lobe epilepsy are not the expression of the epileptic focus, but rather of its circuits [243]. The concept that there is a single epileptic focus from which seizures emanate via a preferred pathway is an "over-schematized simplicity" [39,228].

Figures 3.1–3.3 illustrate this point by showing three consecutive seizures recorded in the same patient using implanted depth electrodes. The first seizure (Fig. 3.1) originates in the hippocampus and projects to the ipsilateral fronto-temporal cortex without involvement of the ipsilateral amygdala. The second seizure (Fig. 3.2) rapidly spreads from the hippocampus into the fornix with the late involvement of the ipsilateral amygdala and without visible reflection in the cortical leads. In the third seizure (Fig. 3.3) there is instantaneous involvement of the ipsilateral amygdala and hippocampus with spread of the discharge into the fornix. These three seizures are distinct both in terms of their mode of onset and their subsequent propagation.

The observations in this chapter support the following concept for the development of medically intractable epilepsy. The epileptic focus is not limited to a group of hyper-synchronized firing neurons, but involves the functional cooperation of such groups located in the different structures of the brain that are joined to form an epileptic system. This cooperation undergoes dynamic changes and re-organization during chronic disease and years of unsuccessful medical treatment. It is possible that the degree of the mutual subjugation of the brain structures joined in the epileptic system change depending on the internal (motivations, endocrine changes, sleep-awake cycle etc.) and external (some physical factors, concomitant diseases, trauma, etc.) factors. The dynamic inter-relationship between the structures of the epileptic system is reflected by variations in progression of the disease, the polymorphism of clinical seizures, and their electrographic patterns. The complexity of this dynamic entity cannot be under-emphasized (Fig. 3.1–3.3). The concept that seizures in patients with intractable epilepsy arise from a large distributed

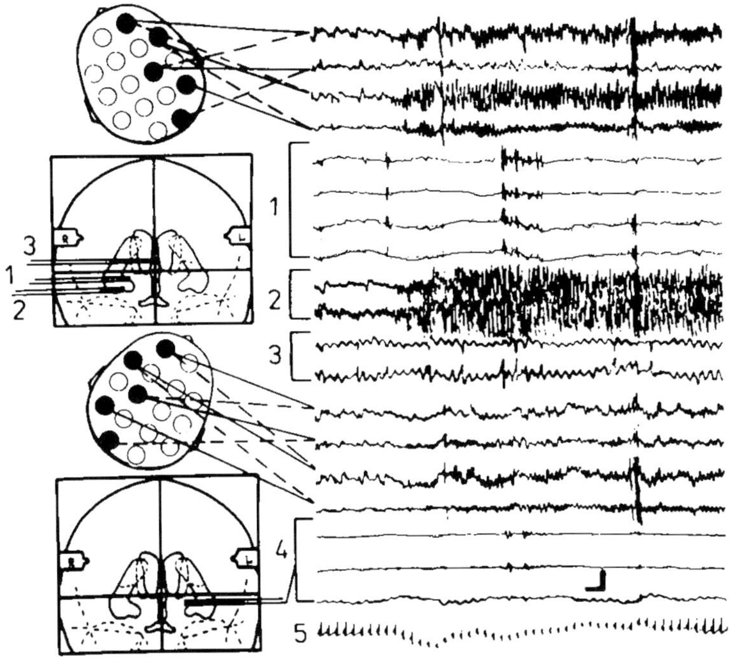

Fig. 3.1. Seizure arising in the hippocampus that projects to the ipsilateral fronto-temporal cortex. There is no involvement of the ipsilateral amygdala. This seizure is recorded in the same patient as the seizures shown in Figs. 3.2 and 3.3 (see text for discussion). 1, 2 and 3: electrical activity recorded, respectively, from the right amygdala, hippocampus and fornix; 4: activity recorded from the left amygdala

network has been borne out both in animal models of epilepsy [222] and in patients with temporal lobe epilepsy [206].

It is important to realize that the essential components of this dynamic epileptic system lie in the deep regions of the brain far from the cortical surface. Consequently their role in epilepsy cannot easily be evaluated using EEG recordings made at the cortical surface (Chap. 5) or other techniques such as those involving the use of optical dyes [189]. Functional magnetic resonance (fMRI) techniques [244] have limited applicability because of the risks associated with leaving a patient undergoing a seizure in the depths of a scanner. The lone exception in humans in which it has been possible to be able to peek into the complexities of the epileptic system has been in patients in whom chronic depth electrodes have been implanted as part of the presurgical evaluation. In the remainder of this chapter we present examples obtained from such studies that most clearly illustrate the complexity of the control and regulation of epileptic seizures.

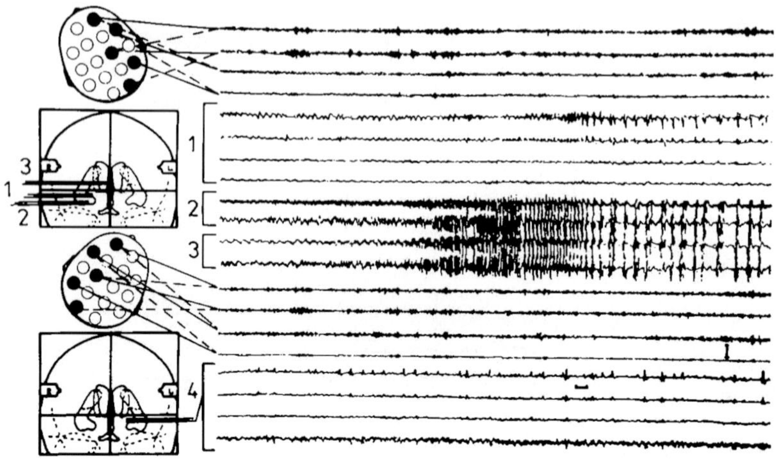

Fig. 3.2. Seizure that arises from the hippicampus and spreads into the fornix with late involvement of the ipsilateral amygdala. There is no clear involvement in the cortical electrodes. This seizure is recorded from the same person as shown in Fig. 3.1 (see caption of Fig. 3.1 for more details)

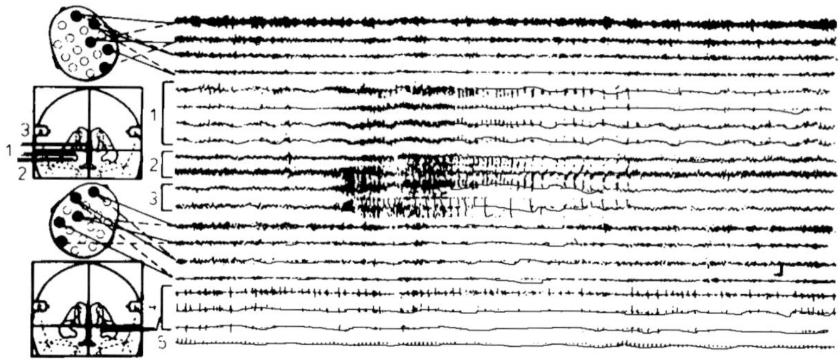

Fig. 3.3. Seizure with instantaneous involvement of the ipsilateral amygdala and hippocampus and subsequent spread into the fornix. This seizure is recorded from the same person as shown in Fig. 3.1 (see caption of Fig. 3.1 for more details)

3.3 Amygdalo-Hippocampal Interactions

The amygdala nuclear complex and hippocampal formation are the structures of the temporal lobe which are most often considered to be the source of epileptic discharges. In clinical investigations it can be difficult to distinguish whether the epileptic pacemaker is the hippocampus or the amygdala. Studies involving stereotactically placed electrodes indicate that seizures more commonly begin in the hippocampus than in the amygdala [245–247]. However, even in these cases, the activity of the amygdala and ipsilateral hippocampus can be so precisely coordinated there can be the impression that these struc-

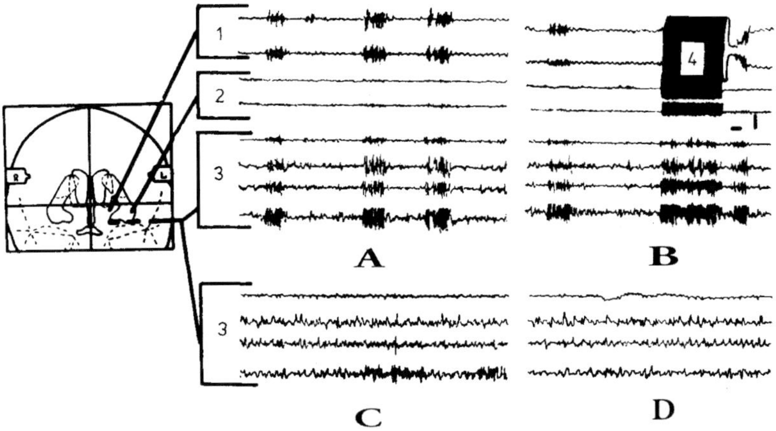

Fig. 3.4. The pacemaker role of amygdala in the activity of the whole amygdalo-hippocampal complex. (**A**) spontaneous visually synchronous discharges in the dorsomedial part of amygdala and hippocampus; (**B**) diagnostic electrostimulation of amygdala causes the development of discharges in hippocampus and shortening of the interdischarge intervals; (**C**) disappearance of the hippocampal discharges during the reversible cooling of the ipsilateral dorsomedial amygdala; (**D**) absence of epileptic activity in the hippocampus after the cryodestruction of amygdala. 1, 2 and 3- electrical activity (SEEG) of dorsomedial, basolateral amygdala and anterior hippocampus respectively; 4- artifact of stimulation

tures are integrated into a single functioning complex [248,249]. For example, even with coherence measurements it can be difficult to distinguish the leader from the follower. One explanation for this difficulty stems from considerations of the cytoarchitecture of amygdala and hippocampus [250]. There are a number of peculiarities that can lead to differences in the electrical field created by the discharging neurons in these structures and hence erroneous conclusions about the emanating point of discharge [250].

The interplay between the amygdala and hippocampus for seizure onset is clearly complex (see Figs. 3.1–3.3). However, in some cases it is absolutely clear that the amygdalar complex can play a leading role. For example, rhythmic stimulation of amygdala leads to the acquisition of this rhythm by the whole amygdalo-hippocampal complex [251] indicating that the amygdala can function as a pacemaker [252,253]. Figure 3.4 shows a recording of the EEG activity in the dorsomedial amygdala (electrode 1), basolateral amygdala (electrode 2) and anterior hippocampus (electrode 3) obtained during intrasurgical evaluation of a patient with epilepsy. Spontaneous epileptic discharges synchronously arise from the anterior hippocampus and dorsomedial amygdala, but not the basolateral amygdala (Fig. 3.4A). Simulation of the dorsomedial amygdala causes the development of discharges in the hippocampus and a shortening of the interdischarge interval (Fig. 3.4B). The hippocampal epileptic discharges disappear following cooling of the dorsomedial amygdala (Fig. 3.4C) and its surgical removal (Fig. 3.4D).

The hippocampus and amygdala are certainly not the most epilepsy-prone areas of the temporal lobe complex. For example, in slice preparations that contain both hippocampus and entorhinal cortex (i.e. the posterior part of the pyriform gyrus), it is the entorhinal cortex which is the primary seizure generator [254,255]. In these situations the hippocampus seems to be the site for the generation of interictal epileptic discharges, but during the ictal event it acts as a follower to the entorhinal cortex. Even more excitable are regions located more anteriorly in the pyriform gyrus, in particular, the area tempestas [256,257]. The confounding issue is that the regions that are especially prone to seizure generation in response to electrical stimulation, such as area tempestas, do not constitute epileptic foci, but rather seem to act as triggers. Thus, even in the case of the best-characterized epileptic focus, its nature is far from simple.

3.4 Propagation Pathways

Which neural pathways are involved in the inter-hemispheric propagation of epileptic discharges that originate in the mesial temporal lobe? This is the question introduced in Chap. 2. The answer to this question is important for two reasons. First, it places bounds on the possible physical extent of an epileptic network, i.e. brain structures located along the path of propagation are those that can potentially be recruited into the epileptic system. Second, the identification of propagation pathways provides a possible avenue for surgical intervention: if the propagation pathway can be interrupted then it might be possible to confine the epileptic activity to a small region of cortex [258].

Temporal-Temporal Propagation. It is likely that the scalp EEG of every patient with medically intractable temporal lobe epilepsy shows epileptic discharges (e.g. spike-slow wave discharges) that arise independently from both temporal lobes. Numerous depth electrode studies have shown that the epileptic dischargees that arise in one temporal lobe do not spread to the other [53,223,224,248,259–266]. In our experience for patients who suffer from medically intractable epilepsy, independent interictal epileptic abnormalities are observed in the deep structures of both temporal lobes in $\geq 86\%$ of cases (385 patients). This preponderance of bilateral, independent inter-ictal epileptic activity is observed even though patients may have a clear unilateral interictal and ictal epileptic focus on scalp EEG. The likelihood of involvement of contralateral symmetrical structures is dependent on the intensity and duration of the epileptic syndrome.

On the other hand it is very clear that during ictal events, seizures that emanate from one temporal lobe spread to the contralateral temporal lobe. However, the inter-hemispheric propagation time for hippocampal onset seizures is surprisingly slow [163]. Propagation times > 50 s are observed in those patients who experience a beneficial result from surgical removal of one temporal lobe. In contrast, an interhemispheric propagation time in the range of 0.5–5.0 s was found to be a negative prognostic sign even in the absence of bilateral EEG abnormalities. The inter-hemispheric propagation time is significantly longer in patients with histological evidence of hippocampal sclerosis [267].

Intuitively one might expect that the anterior commissure would be the likely route for spread of seizure activity from one temporal lobe to the other (see Chap. 2). However, it is now felt that, in contrast to certain animal models of epilepsy [258], this forebrain commissure is unlikely to be of importance for the inter-hemispheric propagation of seizures arising from the deep temporal lobe structures in humans [268–270]. This hypothesis is supported by anatomical investigations. Phylogenetically there is a gradual reduction of the amygdala contribution to the anterior commisure such that it is essentially eliminated in evolved primates [250,271,272]. If the commisural afferents to amygdala exist, they connect not the symmetric structures, but probably to the contralateral inferior temporal cortex [273]. It is possible that the differences in the degree of anatomical and functional communications between the hippocampi in humans or primates as compared to sub-primates may be reflected in the inter-hemispheric differences and lateralization of function [274].

One forebrain commissure that has received some attention as a possible avenue for inter-hemispheric propagation is the hippocampal commissure. Case 2 in [261] describes a patient with epilepsy, in which a focal discharge arising in the right hippocampus subsequently involved the cortical regions of both hemispheres without evidence of crossing through anterior commissure or psalterium. It was suggested that the propagation pattern involved transcallosal cortico-cortical propagation of discharge that initially appeared in the right temporal cortex. Spencer et al. [275] described the different variants of involvement of the ipsi- and contralateral neocortex in epileptic discharge originating from the hippocampus in patients with implanted electrodes. Although the region of hippocampus in which the seizure arose was constant, the patterns of spread varied in the same patient. When contralateral involvement occurred, it occurred after or with the involvement of contralateral hippocampus, but never before. These findings were interpreted as evidence for the existence of an operational hippocampal commissure in humans. The dorsal, not the ventral [275], hippocampal commissure appears to be the main functional pathway in humans [250]. Gloor and colleagues [250] found that at least 26 seizures in 12 patients with implanted electrodes, the dorsal hippocampal commissure was the only pathway that could have conveyed the seizure discharge to the contralateral side. Moreover, it appeared that repetitive stimulation may be necessary to effectively activate this pathway [250].

It must be emphasized that from a clinical point of view the issue of whether or not the forebrain commissural pathways play a substantial role in the inter-hemispheric propagation of seizure activity is of little practical importance. The selective, isolated stereotactic operations on the deep minor forebrain commissures of yesteryear have been largely abandoned because of their low efficacy. The more important point is that some spread of activity along these pathways probably exists and hence these pathways form the anatomo-functional substrate for progressive epileptogenesis and subsequent formation of the epileptic system (see Chap. 1).

3.5 Extratemporal Seizure Propagation

The major mechanism for spread of seizure activity between the temporal lobes is not via a direct commissural link, but rather the intra-hemispheric and vertical spread of seizure activity by indirect pathways involving many deep brain structures including the brainstem [28,267,269,270,276–280]. This observation is not entirely unanticipated. Essentially the same conclusion was reached in Chap. 2 by comparing seizure propagation times with the conduction properties of myelinated and unmyelinated axons in the nervous system. The following anatomical considerations strongly support this concept (Fig. 3.5).

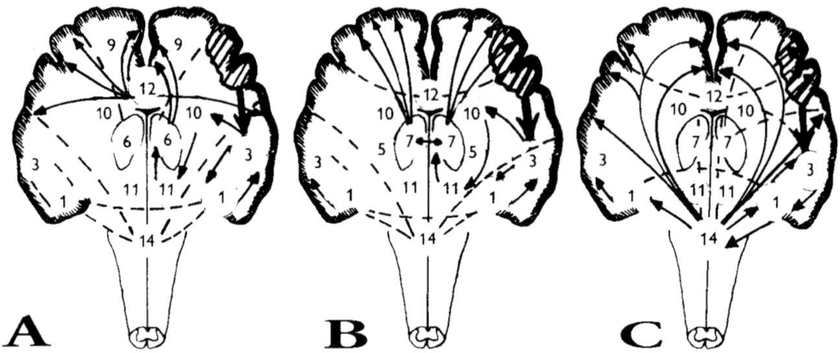

Fig. 3.5. Diagramatic representation of the main pathways of the spread of epileptic discharge from the right amygdalo-hippocampal complex: (**A**) spread involving the anterior thalamic nuclei, gyrus cinguli and corpus callosum; (**B**) spread involving the non-specific thalamic nuclei; (**C**) spread involving the mesencephalic reticular formation. The *solid lines* show the "main" directions of spread and generalization of the epileptic discharge and the *dotted lines* show the auxillary routes. 1- amygdala, 3- hippocampus, 4- gyrus hippocampus, 5- dorsomedial thalamic nuclei, 6- anterior thalamic nuclei, 7- centrum median thalamic nucleus, 9- gyrus cinguli, 10- fornix, 11- hypothalamus, 12- corpus callosum, 14- mesencephalic reticular formation

The major projections of the amygdala and hippocampus are not to the symmetric contralateral structures, but rather to ipsilateral deep-brain structures from which the seizure activity subsequently generalizes. These structures include certain nuclei of the thalamus, certain cortical regions (in particular the cingular gyrus), and the midbrain. The amygdala and hippocampal project to different thalamic relay nuclei, hence have differing projections to neocortex implying that there are different routes of seizure generalization (Fig. 3.5). The hippocampal route is hippocampus → fornix → mamillary body → mammillothalamic tract (Vicq d'Azir's bundle) → anterior thalamic nuclei (anteromedial and anteroventral ipsilaterally and anterodorsal bilaterally) → posterior (areas 23 and 29) cingulate gyrus [281–285]. The amygdalar route is amygdala → magnocellular part of the mediodorsal thalamic nucleus → anterior insular, frontal and anterior cingular (area 24) cortex [61,250,286–288].

The most common mode of spread of mesial temporal lobe seizures is initiating temporal lobe → ipsilateral frontal lobe → contralateral frontal lobe → contralateral temporal lobe [276]. The orbito-frontal cortex is strongly influenced by mesial temporal ictal activity (these regions are connected by the uncinate fasciculus). It is likely that an important way station for seizure generalization via this pathway lies on the medial surface of the hemispheres [289,290], in particular the cingulate gyrus. The typical pattern of spread for seizures with onset in the cingulum is instantaneous spread of focal pericallosal discharge over the convexity of both frontal poles with subsequent, rapid generalization [291]. The rapidity for generalization from this site is probably related to the abundant intercingular transcallosal fibers that interconnect both frontal lobes [292]. It has been demonstrated that fronto-cingular and cingular convulsive discharges have a more prominent tendency for interhemispheric propagation than paleocortical discharges [293].

Intrasurgical investigations have confirmed the close relationship between the anterior cingulum (area 24 of Brodman) with both frontal regions as well as with the mesiobasal temporal lobe structures [191,294–298]. Bilateral paroxysmal bursts of epileptic activity can be seen over anterior head regions not only in cases in which the cingular region is the initiating point, but also when activity spreads from a mesiobasal temporal lobe focus to involve the ipsilateral anterior cingulum.

A variant of seizure generalization from a focal deep mesiobasal focus is a generalized discharge that develops practically instantaneously following a very short period of generalized cortical and subcortical desynchronization of electrical activity [229]. This variant of generalization of focal epileptic discharge that emanates from the mesiobasal temporal lobe structures is most readily explained by the involvement of the mesencephalic reticular formation. The important role of the brainstem in the inter-hemispheric spread and generalization of epileptic activity is supported by experimental observations [28,277–280]. In rats amygdalar discharge propagates by the amygdala-

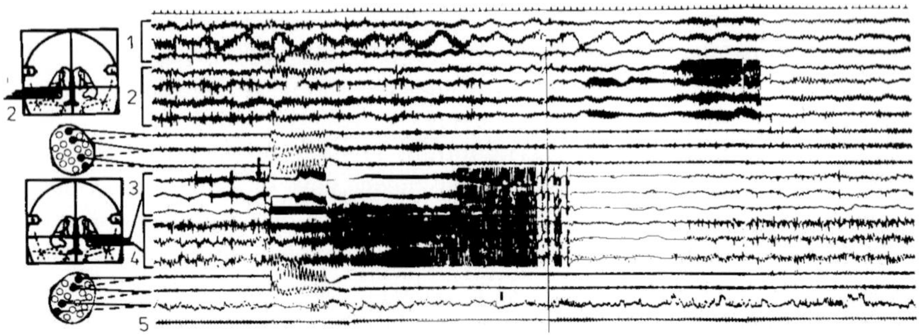

Fig. 3.6. Suppression of the epileptic discharge (1,2) by the development of electrographic seizure (3,4) in the symmetric temporal lobe structures. Electrical stimulation of the left amygdala (electrode 1, *left side* of figure) completely abolishes epileptic activity in the right amgdala (electrodes 3, *right side* of figure). The epileptic activity recurs in the right amygdala following the cessation of epileptic activity on the left. Electrodes 2 and 4 are from, respectively, the left and right hippocampi. Parameters for the electrical stimulation were 5 mA, 0.2 ms, 50 Hz

brainstem route and not via the frontal and motor cortex areas [270]. Recently the role of limbic-brainstem connections for the spread of epileptic activity as an alternate to forebrain commissural pathways has been emphasized [28,299].

The above structures (e.g. thalamus, cingulum, mesencephalon) are essential components of the epileptic system since without them effective generalization of seizure activity would not be possible. Thus from an anatomical point of view the necessity for an epileptic network is both obvious and self-evident.

3.6 Inhibitory Mechanisms

Inhibitory mechanisms are also intrinsic components of the epileptic network. Of course, the existence of intrinsic inhibitory mechanisms underscores the feasibility of developing a "brain defibrillator". These inhibitory mechanisms can be responsible for some of the clinical manifestations of seizures [230,300] and for the failure of epilepsy surgery. In addition to "surround" inhibition [301], areas of inhibitory influences tessellate throughout the human epileptogenic region [87]. These inhibitory influences contribute to the generation of a hyper-synchronous ictal onset of the "disinhibitory" type of seizure [40,227]. The interaction of excitatory and inhibitory processes is seen not only in the restricted epileptic focus, but also between different brain structures. The role of thalamocortical and brainstem pathways for the active arrest of seizures has already been discussed (see Chap. 1). Here we draw attention to the inhibitory mechanisms that exist in the temporal lobe complex.

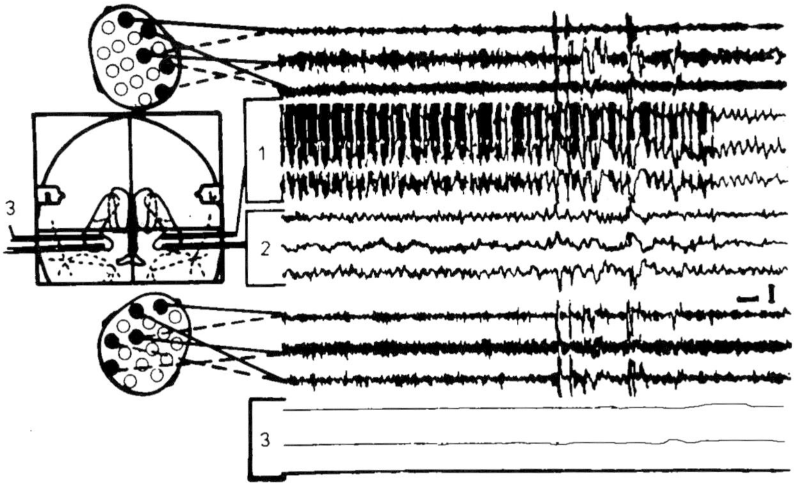

Fig. 3.7. The "passive" inhibitory relationships between symmetric amygdalar complexes. A powerful focal epileptic discharge in the left amygdala (electrode 1) developes immediately after the cryodestruction of the right amygdala (electrode 3). Electrodes 2 show the EEG activity in the left hippocampus

Studies on patients with epilepsy using implanted electrodes indicate that symmetrical deep temporal lobe structures can not only produce the epileptic discharges independently or take part in the spread of seizure activity but also react with each other in the inhibitory (or suppressive) way [300]. These inhibitory influences may be active (Fig. 3.6) when the epileptic discharge developed in one of the temporal lobes, suppresses the background activity or epileptic discharge in the contralateral temporal lobe; or it may be passive (Fig. 3.7) when the functional deactivation of one of the temporal lobes (or its structures) leads to the outburst of epileptic activity in the symmetrical structure [300,302]. The passive activation of a dormant epileptic focus can also be seen transiently as a post-ictal phenomenon or following suppression of a temporal lobe using an intra-carotid injection of amytal (WADA test).

3.7 Conclusion

An epileptic system is a dynamic, ever-changing and evolving, distributed system of neurons that controls the onset, propagation, and arrest of epileptic seizures. Although in many patients there is a dominant epileptic focus, or "pacemaker", which acts as the driving force for the epileptic network [191,297,300], the site of this driving force does not always correlate with the site of the primary or even secondary epileptic focus indicated by radiological or histo-pathological evidence [268]. Perhaps it is more correct to think of a succession of pacemaker regions, each ascending to dominance for a short time, only to be subsequently refined and perhaps replaced by other regions. This undulating landscape of epileptic foci clearly evolves over many time scales: ranging from minute to minute (Figs. 3.1–3.3) to the lifetime of the patient.

Many brain regions play a role in guiding the dynamic character of the epileptic network. For example, certain sites appear to act as gating mechanisms and in this way play important roles in seizure generation and generalization. One such gating region is the dentate gyrus of the hippocampal formation [179]. Although the dentate is a relatively inexcitable region of the hippocampus (i.e. as compared to CA1 and CA3), maximal activation of the dentate is required for seizure spread and generalization. Other gating regions include the substantia nigra pars reticulata [180], superior colliculus [181], and the centromedian nucleus of the thalamus [182].

A number of clinical observations lend support to the concept of a dynamically evolving epileptic network. The fact that in some patients with epilepsy, seizure occurrence is correlated with levels of stress and anxiety [303] suggests the existence of controlling mechanisms associated with functions of the limbic system. In addition, the observation that in some women the risk of seizure occurrence is strongly correlated with hormonal levels (i.e. catamenial epilepsy) [304] implicates the hypothalamic-pituitary axis. Perhaps the strongest clinical evidence for controlling influences is the connection between the timing of seizure occurrence and the state of arousal [305]. Sleep deprivation is a commonly employed method to increase the likelihood that epileptic activity is seen on an EEG and prolonged sleep deprivation can result in a seizure in suspectible patients. Moreover, it has long been recognized that seizures are more likely to occur in certain sleep stages than in others [305]. There is general agreement that the light stages of non-REM nocturnal sleep enhance paroxysmal activity and even contribute to the generalization of focal epileptic discharge [306,307]. More recently it was demonstrated that REM sleep could facilitate the focal temporal lobe discharge [191,298,308,309]. In animals destruction of the thalamic non-specific nuclei that are suspected in taking part in the synchronization of brain rhythms during sleep as well as during epileptic seizures was observed to suppress epileptic activity [310–314]. These observations probably reflect the fact that the important brain structures for both regulating sleep and wakefulness and seizure generalization are located in the thalamus and brainstem.

The recognition that groups of neurons located in the deep regions of the brain play dominant roles in the regulation of the epileptic system is an important cornerstone upon which to develop a "brain defibrillator". Many of these deep structures have diffuse projections to widely distributed regions of the brain. These deep structures are the obvious candidates for the site of the effector arm of a brain stimulator (see Chaps. 14–18).

4 Neuroglia, the Other Brain Cells*

P. Jung, A.H. Cornell-Bell, A. deGrauw, and R. Strawsburg

The name *glia* derives from the Greek work for *glue* and suggests what role these cells have been thought to play, namely holding together the rest of the brain tissue. While it is probably true that glial cells provide structural support for the neurons, which they outnumber in the brain by a factor of 9, research during recent years has revealed many more important functions these cells play in the central and peripheral nervous system. It is now understood that glia directs the neuronal growth in the developing brain. Oligodendrocytes produce coatings of myelin that electrically insulates axons and thus speed up propagation of electric pulses and microglia serve as immune cells to the brain.

An important species of glial cells are the start-shaped astrocytes. They appear as close partners to the neurons as they regulate the neuronal microenvironment, listen to the neuronal chatter, and feedback to the neuronal dynamics. Astrocytes pick up the neurotransmitter *glutamate* from synapses and convert it into *glutamine*. Glutamine is returned to the neurons where it can be converted back into glutamate when needed. It has been demonstrated recently that the synapses of neurons are disrupted minutes after the functions of glial cells in their surroundings had been disrupted with a chemical. Another important function of the astrocytes (and other glial cells) is to absorb K^+ from the extracellular space. This is achieved by a large conductance for K^+ and interconnectivity between astrocytes through *gap junctions* that permit rapid distribution of ionic products throughout the astrocytic network. This is particularly important for the neuronal system since K^+ is released from the neurons after the action potential. The failure of the astrocytes of doing so, leads to neuronal depolarization and misfunction; cortical depression waves and migraine are being discussed as a consequence of astrocytic failure in tissues such as the visual cortex where astrocytes are rarer in favor of more neurons. The influence of glial cells on the EEG is discussed in Chap. 5.

Astrocytes sport a variety of receptors at their surface, receptors for the excitatory neurotransmitter glutamate, the inhibitory neurotransmitter GABA and the rarer kainate. These receptors allow the astrocyte to read neuronal messages. Although astrocytes do not develop long extensions such

* This material is based upon work supported by the National Science Foundation under Grant No. IBN-0078055.

as the neurons, they send shorter extensions to envelope neuronal clefts where they sense and bind the neurotransmitter.

4.0.1 Voltage-gated Na^+ Channels in Astrocytes and Action Potentials

For a long time glial cells have been believed to be electrically passive since full cell recordings have indicated this. Patch clamp recordings from cultured rabbit Schwann cells [315] and rat astrocytes [316,317] have revealed that astrocytes do express voltage-gated Na^+, K^+ and Ca^{2+} channels. For the formation of action potentials in neurons, the voltage-gated Na^+ channels are essential. There is evidence that the expression of Na^+ channels is influenced by the neuronal environment. Removal of neuronal tissue has led to almost complete loss of Na^+ channels in nearby astrocytes [318,319]. These findings suggest that the expression of ion channels in glial cells is *dynamic* and controlled by the micro-environment. In Schwann cells, the glial cell in which neurons of the peripheral nervous system are embedded in, the density of voltage-gated Na^+ channels is particularly large close to the nodes of Ranvier [320], the segments of axons that are not myelinated and active propagation re-shapes the dispersing electric signal. This suggests that Schwann cells serve as a local source of Na^+ channels for the nodal region of the neuronal axon. This theory is consistent with the observation that Schwann cells contain an intracellular pool of the Na^+ channel protein [320].

The astrocytes and the neurons in the central nervous system (CNS) are engaged in a similar symbiosis as are the Schwann cells and the neurons in the PNS [321,322]. Electrophysiological recordings from astrocytes from the optic nerve have revealed voltage activated transient Na^+ currents [323,324]. The presence of voltage-activated Na^+ currents suggest that astrocytes may elicit action potentials. This, however, is normally not the case for the following reasons [325]:

1. The density of Na^+ channels is normally much lower than in neurons.
2. The K^+ conductivity is higher than in neurons. The resting potential is therefore more negative and an even smaller portion of the smaller number of existing Na^+ channels is open.
3. The gap junction coupling between the astrocytes tends to clamp the voltage to a common fixed value and suppresses voltage variations and therefore, for example the formation of an action potential.

In pathological tissues, however, the situation can be different. In patients with temporal lobe epilepsy astrocytes from the seizure focus (where the seizure focus was in the hippocampus) have an approximately two orders of magnitude larger density of voltage-gated Na^+ channels. Careful studies of these astrocytes have revealed the following results [326]:

1. The resting membrane potential ($\approx -55\,\mathrm{mV}$) was significantly more depolarized than in normal cells ($\approx -80\,\mathrm{mV}$).

2. The Na^+ currents are much larger than in normal astrocytes.
3. It is possible to trigger action-potential-like responses when sufficient current was injected.

It is worth mentioning that action potential like responses of astrocytes have also been reported from human oligodendroglial tumor cells [327], neurocytoma cells [328], and human glioblastomas [329].

Often, the area around the seizure focus is depleted of neurons since the intracellular concentration of Ca^{2+} in hyperexcitable neurons is elevated and can be toxic for neurons. Nevertheless, the focus appears to initiate the seizure. In such a situation astrocytes may be involved in maintaining the seizure focus (they take over when the neurons have died), or may aid in propagating excitation through neuron-loss interrupted $CA1$ hippocampal pathways [326].

4.0.2 Calcium Signaling in Glial Cells

Some glial cells, including astrocytes are equipped with voltage-gated Ca^{2+} channels. These are not a constitutive physiological property of the astrocytes because, for example, Bergman glial cells are not equipped with voltage-gated Ca^{2+} channels. The expression of Ca^{2+} channels is controlled by the neuronal environment of the astrocytes since Ca^{2+} currents were only detected when the glial cells had been co-cultured with neurons [330,331]. These voltage gated Ca^{2+} channels can be activated by depolarizing the membrane. Depolarizing of the membrane can be caused by an elevated level of K^+ ions in the aftermath of neuronal activity. Physiologic levels of K^+ are sufficient to depolarize the glial membrane potential to the threshold of activation of the Ca^{2+} channels [332,333]. But also activation of GABA and glutamate receptors depolarize the membrane sufficiently to activate the Ca^{2+} channels. Another pathway for Ca^{2+} uptake is ionotropic receptors, such as for example the highly Ca^{2+} permeable kainate, preferring glutamate receptors, and non-selective channels.

The most prominent signaling pathway for Ca^{2+} *within* glial cells is mediated by the production of inositol receptor, IP_3. The IP_3 receptor has three binding sites, one for IP_3, and two for Ca^{2+}. The IP_3-receptor acts as a Ca^{2+} channel controlling the flow of Ca^{2+} from the endoplasmic reticulum to the intracellular space. It consists of four subunits. The channel is open if three of the four subunits are activated. A subunit is activated when the IP_3 and the calcium activation site is bound. If Ca^{2+} binds to the second binding Ca^{2+} site, the subunit is deactivated. Thus, as IP_3 is bound, and Ca^{2+} opens the channel, more Ca^{2+} enters the intercellular space which, in turn, opens more channels. This autocatalytic increase of Ca^{2+} (Ca^{2+} induced Ca^{2+} release) gives rise to a large amount of Ca^{2+} in the intracellular space. As the Ca^{2+} level increases, the units become deactivated and the rise of Ca^{2+} stops. Elevated intracellular levels of calcium are removed by Ca^{2+} pumps and Na^+/Ca^{2+} exchangers to the extracellular space and re-absorbed into intracellular storage by pumps to the endoplasmic reticulum.

The time course of a Ca^{2+} signal is determined by all of these transporting mechanisms. The activity of the pump to the endoplasmic reticulum is determined by the degree of filling of the calcium pools. The activity of the Na^+/Ca^{2+} depends on the gradients of Ca^{2+} and Na^+ across the cell membrane. This couples the Ca^{2+} signaling to the Na^+ dynamics. As described above, the intracellular Na^+ concentration is controlled, for example, by the transmembrane potential which is close to the K^+-voltage for the high K^+-conductance of the membrane. Elevated concentrations of K^+ due to, for example, neuronal activity can depolarize the membrane potential and open a portion of the Na^+ channels. Subsequently Na^+ enters the cytosol by passive transport. Na^+ can enter the cell also by activation of the glutamate and kainate receptors. High levels of intracellular Na^+ will damp or even invert the Na^+/Ca^{2+} exchanger blocking an important pathway of cellular Ca^{2+} release.

Short extensions of the astrocytes contact other astrocytes by *gap junctions*, an array of proteins that form a tube that connects the astrocytes. Connected with the gap junctions, the astrocytes form an interconnected – low impedance – cellular network. An increased concentration of Ca^{2+} in one cell can cause Ca^{2+} elevation in the neighbor cells and so on. This intercellular Ca^{2+} transfer requires, however, another messaging mechanism. The nature of this messaging mechanism may depend on the nature of the Ca^{2+} wave. For example, mechanically stimulated Ca^{2+} waves do not require extracellular Ca^{2+}; only release of Ca^{2+} from intracellular pools is involved [334]. Ca^{2+} waves stimulated by glutamate in contrast require extracellular Ca^{2+} for propagation [335]. It has been suggested that IP_3 is the messaging molecule for intercellular Ca^{2+} waves [336]. According to this model, the Ca^{2+} wave rides on the back of IP_3 diffusing through the gap junctions. Intercellular Ca^{2+} thus provides the means of long-distance signaling in the brain by forming intercellular calcium waves that involve sometimes thousands of astrocytes.

We have observed calcium wave patterns in cultures of hippocampal astrocytes of rat brain. The culture under investigation includes about 500 astrocytes. First the cells are loaded with calcium-sensitive fluoroprobe, Fluo-3M. Minutes after kainate is perfused homogeneously over the culture, calcium waves all across the specimen can be observed. The fluorescent activity reflects the calcium concentration in the astrocytes. Snapshots of the calcium activity have been taken at intervals of $\Delta t = 1.7\,\mathrm{s}$ (see Fig. 4.1a). The Ca^{2+} concentration is encoded in gray-scale; lighter color indicates higher Ca^{2+} levels. The gradients of Ca^{2+} concentration are small and thus the contrast of the gray-scale is also small. To emphasize the wave fronts and to eliminate nonstationary background levels of Ca^{2+}, Fig. 4.2 shows the positive (discrete) gradients of the gray-scale images, i.e. we subtract the gray-scales

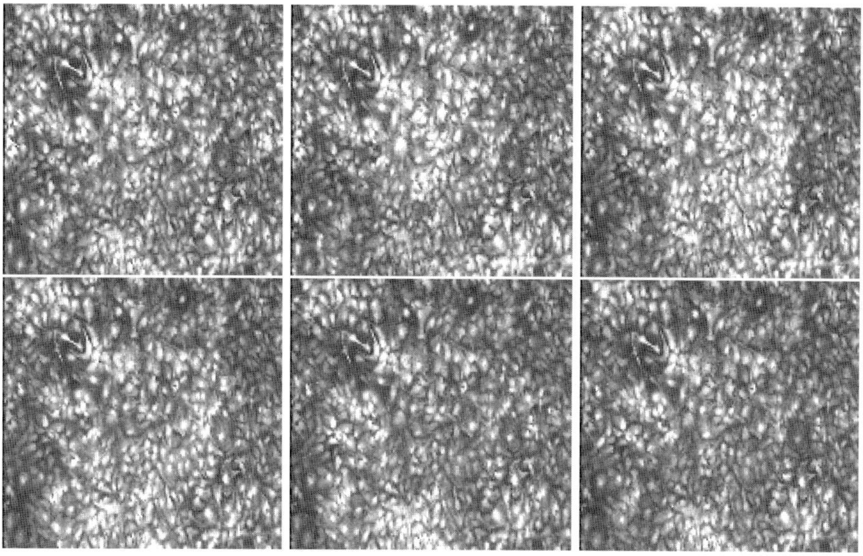

Fig. 4.1. Fluorescent images of a culture of rat hippocampal astrocytes are shown. The snapshots are taken 17 s apart

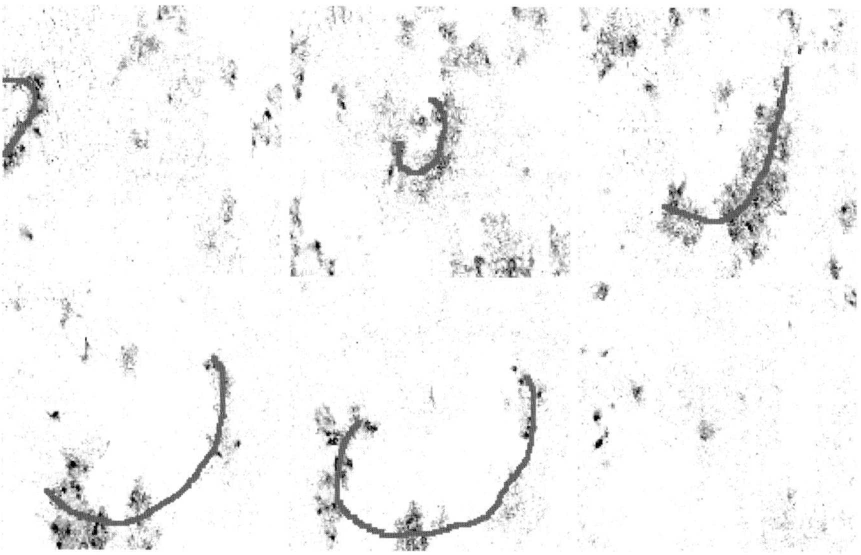

Fig. 4.2. Three positive-gradient images (corresponding to the images in Fig. 4.1 with an image delay of $n = 10$ images (17 s) are shown. The gray-scale has been inverted: *darker pixels* correspond to higher calcium activity. The waves are strongly fractured; the *red lines* indicate how the fragments belong to one large wave that fades out in the last frame (The colour plate of this figure appears on page 353)

at time t with the gray-scales of a previous snapshot at time $t - n\Delta t$. If the Ca^{2+}-level decreased from $t - \Delta t$ to t we plot the corresponding pixel white. The wave fronts (dark) are now clearly visible. The wave patterns observed can be described best as *target waves* with a finite lifetime and spreading distance. The waves seem to be sparked at random sites and randomly in time.

4.1 Cluster Decomposition and Cluster Entropy of Ca^{2+} Waves

Given the video-imaged fluorescence patterns, we attempt to characterize these quantitatively. The goal is to obtain a measures of how the cells communicate across the entire culture. Ca^{2+} waves have a statistical component and therefore need to be analyzed with a statistical method. Comparison between theory and experiment for the very same reason has to be done based on a statistical method. We utilize an event-based method [337,338] that decomposes the spatiotemporal evolution into space-time clusters. The method works as follows:

First, we stack a temporal sequence of N_t snapshots of the array, taken at times $t_n = n\Delta t$ to obtain a large space-time cube, which carries all the spatiotemporal information within the time interval $N_t \Delta t$. In the second step, we draw a small cube around each excited cell with a spatial side-length d_s and a temporal side-length d_t. The temporal and spatial side lengths can be varied to analyze the system on different length and time scales. In this chapter, we use for the temporal side length d_t the elementary time step Δt and for the spatial side length d_{S} the pixel-distance, i.e. only the nearest neighbors produce overlapping small cubes. Overlapping small cubes in *the time-forward direction* form objects that we term *coherent space-time clusters*. Collisions of waves need to be treated with a set of rules:

- When two waves merge, they can propagate after the collision as one wave. In this case, the younger cluster is terminated at the merging point and the outgoing wave is counted to the cluster of the older incoming wave. This rule separates off small wave fragments that merge into a larger wave.
- When two waves merge, they can branch off following the collision and propagate as separate waves. In this case, the younger cluster is terminated at the merging point and both outgoing waves count to the cluster of the older incoming wave. Two branching structures occur after collision of two open waves (wave fragments). Thus, the reason for two instead of one outgoing wave lies in earlier cluster events.
- Local propagation failure can cause the breakup of a wave into two fragments. Since both fragments originate from the same *mother wave*, they count as one cluster. This rule avoids artificial additional clusters when a

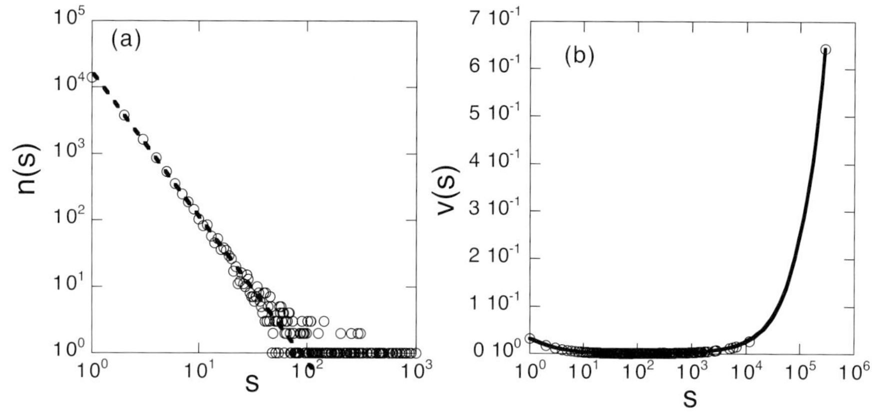

Fig. 4.3. The cluster size distribution $p(s)$ (**a**) and the corresponding coverage distribution $v(s)$ (**b**) are shown for the 360 subsequent video images of calcium activity obtained from rat hippocampal astrocyte cultures

wave collides with a boundary of the system. Although the wave fractures, all the propagating pieces count as one cluster.

The lifetime of a cluster is limited by collision with another cluster or spontaneous disappearance. The size, s, of a coherent cluster is defined as its space-time volume (in units of the small cubes) throughout its entire lifetime. A first characterization of the patterns is the histogram of cluster sizes $n(s)$, i.e. the number of clusters of size s. The cluster size distribution is obtained after normalization as

$$
\begin{aligned}
p(s) &= \frac{1}{Z} n(s) \\
Z &= \sum_{n=1}^{\infty} n(s) \, .
\end{aligned}
\tag{4.1}
$$

The clusters are recorded over time intervals where we belief that the perfused neurotransmitters are stationary. The cluster-size distribution obtained from 360 subsequent video images (six of them are shown in Fig. 4.1 is shown in Fig. 4.3. For small cluster sizes (1–10 pixels), the cluster size distribution follows a power law. This is indicative of randomness in the structure due to noise-triggered cluster-lets or due to spatiotemporal chaos [339].

The cluster-size distribution is oblivious of the relevance of the clusters for the entire spatiotemporal dynamics. For example, a cluster of size 1 counts the same way as one of size 10000. One way to improve this is to weigh the clusters by their relative contribution to the entire pattern. The groups of all clusters of size s cover the volumes $n(s)s$. Normalizing this coverage to the entire volume of the pattern, i.e. $v_{\mathrm{tot}} = n(1) + 2n(2) + 3n(3) + \ldots$ we obtain the relative coverage distribution

$$v(s) = \frac{sn(s)}{\sum_{s'} sn(s')} \, . \tag{4.2}$$

The coverage distribution describes the probability that an excited cell, arbitrarily picked at any time during the time interval where clusters are recorded, belongs to a space-time cluster of size s. A measure for the uncertainty of membership in a cluster of a certain length is thus the cluster entropy [339]

$$S = -\sum_{s} v_s \ln v_s = \ln < s > - \frac{1}{< s >} \sum_{s} sp_s \ln(sp_s) \, . \tag{4.3}$$

The entropy vanishes if the excited sites belong to only one cluster class c_s ($v_s = 1$). Thus, the spatiotemporal pattern consists of either a single cluster or clusters of only one size. A vanishing entropy does not necessarily require a "simple" initial state but can also arise from complex initial conditions, such as many spiral waves with randomly selected cores, if an inherent process for generating new cluster sources is lacking. The collision and merging of uncorrelated clusters then occurs over a transient time to yield one surviving cluster. The artificial fracturing of waves at boundaries does not affect the entropy, since the fractured waves belong to the same coherent parent structure. The entropy reflects the generation of new clusters due to incoherent processes such as fluctuations or spatiotemporal chaos.

4.2 Calcium Signaling in Hyperexcitable Tissues

Physiological properties of glial cells cultured from resected medial temporal lobe epilepsy patients retain a hyperexcitable response to neurotransmitters [340]. Hyperexcitable glutamate-induced Ca^{2+} responses were produced by cells derived from tissue regions that exhibited hyperexcitable EEGs in the operating room. Astrocytes from epileptic foci exhibited altered physiology such as increased gap-junction coupling [341] and altered morphology and changed channel expression [342]. The underlying causes of seizure development remain unclear. Excitatory neurotransmitters such as glutamate, aspartate and glycine and the inhibitory neurotransmitter gamma amino butyric acid (GABA) have been implicated in the pathophysiology of seizures [343]. A number of studies in animal models and in humans have shown that glutamate concentration is increased extracellularly and GABA may be decreased in the epileptic area during and after seizures (see [344] for review). Involvement of glial cells in epilepsy remains an uncharted area. It is possible that the astroglial glutamate transporters are impaired in epileptic foci [345]. In another study these authors showed that astrocytes from epileptic foci have a much higher density of Na^{+} channels and as a consequence, they are capable of generating much larger currents than astrocytes from normal tissue [346].

Additionally there is a dramatic alteration in the K^+ channel properties that may also indicate destruction of the potassium spatial buffering that normally protects tissue from excessive buildup of K^+ in the extracellular space [342]. Genetic studies of familial epilepsies on the other hand have shown mutations in ion channel genes. A newly described epilepsy, *generalized epilepsy with febrile seizures plus* [347], was linked to the β_1 auxiliary subunit of the voltage-gated sodium channel (SCN1B) on chromosome 19 [348]. "Benign neonatal familial convulsions (BNFC)" [349] has been linked to chromosome 8 and 20. These two genes encode highly homologous potassium channel subunits, KCNQ2 and KCNQ3 [350,351]. It is fair to say that epilepsy derives from multiple causes yet shared properties of excitotoxicity and cellular excitability are manifested in all of these conditions. Here we demonstrate that calcium signaling in cultures of epileptic astrocytes differs significantly from cultures of healthy astrocytes. This difference can be quantified using the space-time cluster decomposition method.

4.2.1 Cultures of Human Nervous Tissue

Tissue culture methods for human neuro-surgical specimens have evolved in the Viatech Imaging laboratory [340,341]. Inclusion of protease inhibitors in the transfer media has provided an effective preservation solution for the transport of epileptic specimens to the laboratory for culturing. Specimens are obtained from patients at Children's Medical Center, Department of Pediatric Neurology, Cincinnati, Ohio during the course of neuro-surgical resection for intractable epilepsy. Electroencephalographic (EEG) recordings of the tissue regions are established in the neuro-surgery operating room. Hyperexcitable versus normal EEG activities are recorded for the tissues that will be characterized in the imaging experiments. Viable cultures that exhibit behaviors characteristic of the diseased tissues can be produced even after two days in transit. Cells cultured from regions of hyperexcitable EEG activity show abnormal responses to neurotransmitter. Cultures have been established for several forms of epilepsy including medial temporal lobe disease, cortical tumors, hyperexcitable regions of the cortex and for the purpose of these studies we have also cultured a Tuberous Sclerosis (TS) case. In this latter condition, *giant cell astrocytomas* can develop within the brain. A tumorous lesion within the cortex of a 17-month-old patient was cultured following surgical resection to alleviate epileptiform seizures. *Subependymal giant cells* are apparent in these cultures (see Fig. 4.4).

The giant cells generally reside on top of a normal-appearing astrocyte layer. One such giant cell is localized within an asterisk in the left panel of Fig. 4.4A. Extensive process formation and a large nucleolus, which has stained heavily with the Ca^{2+} dye, characterize this cell. These cells are common in these cultures. A second giant cell (asterisk in Fig. 4.4B) has well-developed projections to neighboring cells that are reminiscent of the end-feet of normal astrocytes. This cell, however, is not integrated into the

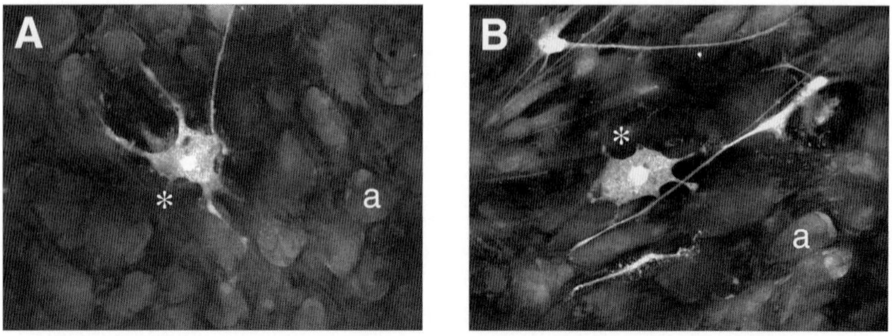

Fig. 4.4. A two-year-old patient underwent neuro-surgery for treatment of intractable seizures associated with tuberous sclerosis. Surgical specimens from the posterior temporal lobe were cultured [341] and imaged using confocal microscopy. (**A**) shows relatively normal astrocytes (a) with a reactive astrocyte (*) on the upper surface of the astrocyte syncytium. In (**B**) several neurons and a second reactive astrocyte (*) are also superficial in the culture. The reactive astrocytes share the morphology of both neurons and astrocytes. Extensive process formation (see (**A**), or elaborate endfeet (see (**B**)) are characteristic of reactive astrocytes. The frequent polymorphic astrocytes in the culture are laden with eosinophilic granules. In addition, neurons can exhibit normal morphology (*upper left*, (**B**)) or varying degrees of abnormal development (**B**)

astrocyte syncytia that it sits on top of. Neurons show extensive changes in morphology (dysmorphism) including extensive small processes as seen in the neuron below the giant cell in Fig. 4.4B. The neuron to the right of the giant cell in this panel exhibits an unusual cell body as well. In contrast, the neuron at the top in this panel has a normal appearance with a rounded cell body and long projection off to the right. Numerous immuno-histochemical studies of the giant cells have shown they contain astrocyte as well as neuronal markers (Hirose et al., 1995). Often giant cells respond to neurotransmitter exposure with an increase in intracellular Ca^{2+}, however, a significant number of these cells do not respond at all to neurotransmitter.

Multiple identical coverslips (32 coverslips) were cultured with the TS case cells described above. All coverslips were treated identically, i.e. they were rinsed with the same saline, stained with the same Ca^{2+} dye (2 μM Fluo3AM), imaged within a 48-h period of each other using an identical imaging paradigm (25 frames in saline, 400 frames in neurotransmitter, followed by 50 frames washout) and exposed to equivalent neurotransmitter solutions. Images were collected to the hard drive of our confocal scanning laser microscope (Nikon PCM2000) every 1.7 s. These images were then exported to a writable CD drive.

In Fig. 4.5 a sequence of snapshots are shown. We have applied subtraction of a sliding reference frame (10 frames gap). For most of the time, calcium activity includes only a few cells. This pattern, however, is interrupted by

Fig. 4.5. A sequence of subtracted snapshots of calcium activity of a culture of astrocytes from TS-tissue is shown from *left* to *right*. The time interval between two snapshots is 1.7 s. Higher calcium activity is coded in larger gray-scale. One can observe local islands of calcium activity but no large-scale organization

drastic events where one or two waves involving virtually all cells on the coverslip are elicited (see Fig. 4.6). We have not observed this behavior in cultures of the healthy rat hippocampal astrocytes. This observation indicates at least temporary strong coupling between the astrocytes in the culture that leads to a fast and strong wave. During these episodes, the cells in the entire culture appear synchronized in their calcium activity. After these events the synchrony is lost and the calcium activity is local. Applying the cluster decomposition, we correspondingly see a difference in the cluster size distribution compared to that of healthy rat brain cells (see Fig. 4.7). There are only very few large clusters, corresponding to the dramatic events when a calcium wave occupies the entire culture.

The speed of the wave well exceeds the speed of the wave fronts observed in the rat-brain tissue in the previous section. This indicates a more rapid response of the diseased astrocytes in comparison with astrocytes from normal tissue. The cluster entropy can be calculated based on the normalized cluster size distribution using (4.3). For an image gap of 1 (subtraction of subsequent images) we obtain during typical time intervals with no dramatic wave events a cluster entropy of $S = 2.9$. In comparison, the cluster entropy of of the healthy rat brain cultures was about $S = 4.3$. Including the rare wave events, the entropy of the epileptic culture increased to about $S = 3.3$, which is still well below the cluster entropy of the rat-brain culture.

4.3 Neuron-Glial Signaling

Neuronal activity influences Ca^{2+} signaling by modulating the gap junction permeability (through, for example, binding of neurotransmitters) and vice versa, Ca^{2+} signaling modulates neuronal activity. Recent experiments in neurone-glia co-cultures suggest that Ca^{2+} signaling in glial cells can generate a Ca^{2+} response in neurons in close proximity [352–355]. The mechanism for this, however, is still under investigation and there might be more than

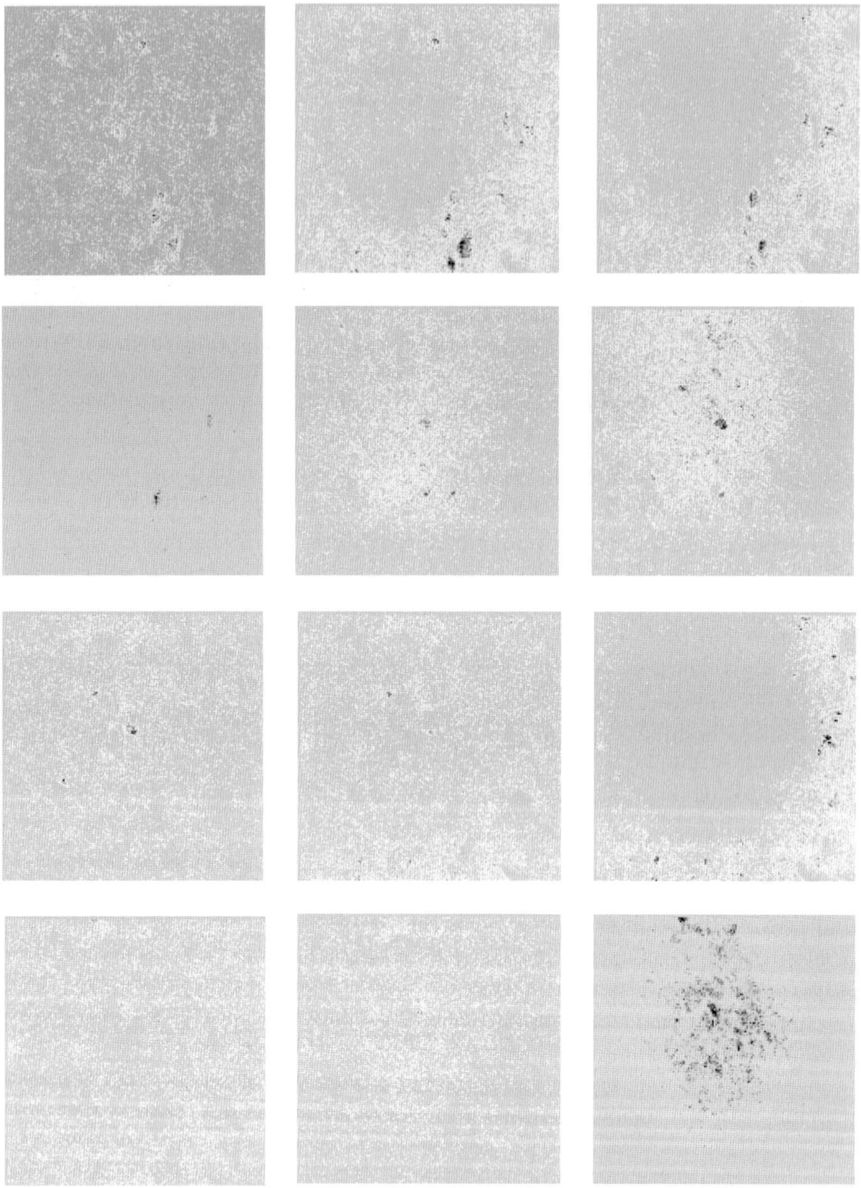

Fig. 4.6. A sequence of subtracted snapshots of calcium activity in a culture of astrocytes from TS-tissue is shown from *left top* to *right bottom*. The time interval between two snapshots is 1.7 s. Higher calcium activity is coded in larger gray-scale. One can observe two dramatic events in the form of a large-scale organization of calcium activity in form of circular waves (The colour plate of this figure appears on page 354)

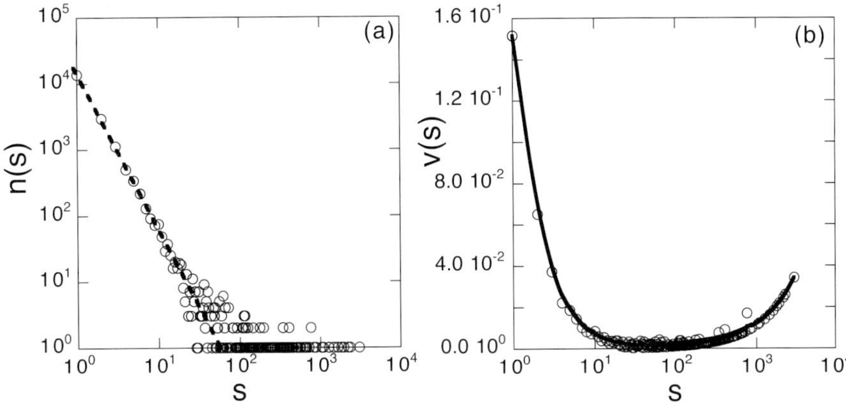

Fig. 4.7. The cluster size distribution (**a**) obtained from the TS-data at 40 µM Kainate is shown. All the clusters larger than $s = 1000$ belong to the rare dramatic wave events. In comparison with Fig. 4.3, many smaller clusters are observed indicating mostly local astrocyte signaling. The coverage distribution $v(s)$ (**b**) differs dramatically from that obtained from the rat-brain tissue. The relative coverage by smaller clusters is almost twice as high. The coverage by large clusters is smaller. If one ignores these rare dramatic wave events, these results imply that the incoherent activity in the epileptic case is higher as in the normal case

one mechanism depending on the type of astrocyte and calcium wave. One hypothesis suggest that gap junction coupling between neurons and astrocytes [352] is responsible while another proposal [353] suggests that glutamate produced in the astrocytes binds to a receptor in the neuron, causing Na^+ influx and thus depolarization. The released calcium in the neuron affects the membrane potential and thus couples to its electrophysiology. Charles et al. [356] have shown that neurons approached by a calcium wave became so depolarized that they fired action potentials.

4.4 Conclusions

Astrocytes are likely to play a role in human epilepsy. This is manifest in altered electrophysiologic behavior and gap junction density in astrocytes from epileptic foci. In recent research the complex neuro-glia interaction has been started to be explored and its alteration in epileptic tissues is very little understood. Contrary to the intuitive expectations, calcium signaling patterns are less coherent in cultures of human epileptic astrocytes than in cultures of normal control tissues obtained from rats brain.

5 The Electroencephalogram (EEG): A Measure of Neural Synchrony*

J.S. Ebersole and J. Milton

Our current understanding of epilepsy is very much linked to our understanding of the electroencephalogram (EEG). The EEG is a measurement of the potential difference between two points on the surface of the brain or scalp (Fig. 5.1). The potential differences measured are very small (tens of microvolts (μV's)) compared to the resting potential across the nerve membrane (10s of millivolts (mV's)). It is important to understand how these potential differences arise and to assess the relevant strengths and limitations of the EEG for providing information concerning the generation and spread of epileptic activity. Here we present an overview of the EEG with emphasis given to those aspects which will be most relevant for this text. Readers interested in a more complete discussion of the EEG are referred to a number of excellent monographs [357–360].

5.1 Cortical Architecture

The EEG is a measure of the activity of neurons which are located in a thin circumferential band measuring 1.5–4.5 mm in thickness under the cortical surface referred to as the gray matter. Approximately 75% of these neurons are excitatory; the remainder are inhibitory. The gray matter is characterized by two major organizational features: horizontal lamination and vertical columnation.

5.1.1 Laminar Organization

From the point of view of the input-output relationships of the gray matter it is convenient to divide in into three regions (Fig. 5.2): respectively from superficial to deep, the supragranular layer, the granular layer and the infragranular layer. Three layer cortex is characteristic of the olfactory cortex, the hippocampal formation, and the dentate gyrus (collectively referred to as the allocortex). The remainder of the cortex (i.e. the neocortex) is further subdivided into six layers: supragranular layers (Layers I, II, III), the granular layer (IV), and the infragranular layers (V, VI). These sub-divisions are made on the basis of the types, density and arrangement of the cells.

* This research was supported by grants from the National Institutes of Mental Health and the Brain Research Foundation.

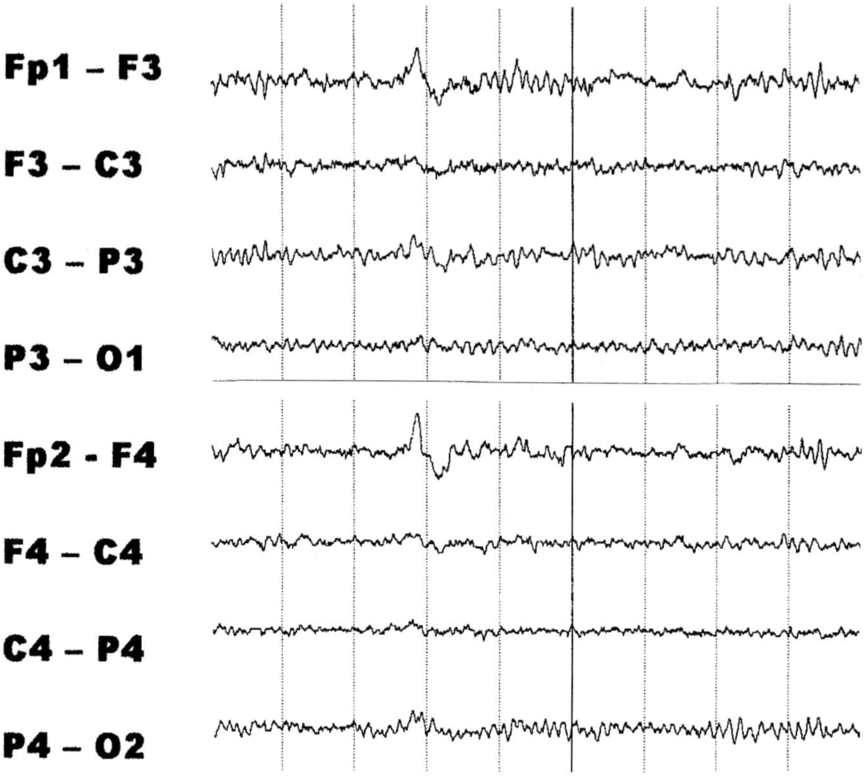

Fig. 5.1. EEG of an awake, healthy adult

The afferent and efferent connections to the gray matter tend to support the concept that the supragranular and granular regions of the cortex (layers II, III and IV) are primarily a receptive and cortical processing area, whereas the infragranular region (layers V and VI) are specialized for output (e.g. [361]). Although pyramidal neurons have been described that have axons that do not reach the white matter, the vast majority of pyramidal axons enter the white matter. The projection targets of the pyramidal neurons are a function of the neuron's laminar position [362]: neurons in layers II and III have axons that project to other cortical regions, whereas the neurons in deeper layers project, in addition, to various sub-cortical regions, such as the thalamus, striatum and brainstem nuclei.

The afferent inputs to the cortex tend to reciprocate the efferent outputs. For example, connections between cortical regions and between cortical regions and specific thalamic nuclei are reciprocal. There are important exceptions to this rule. In primary sensory cortex input from the thalamus (i.e. thalamocortical afferents) preferentially terminate on specialized multipolar neurons referred to as the spiny stellate neurons. In contrast, the axons and

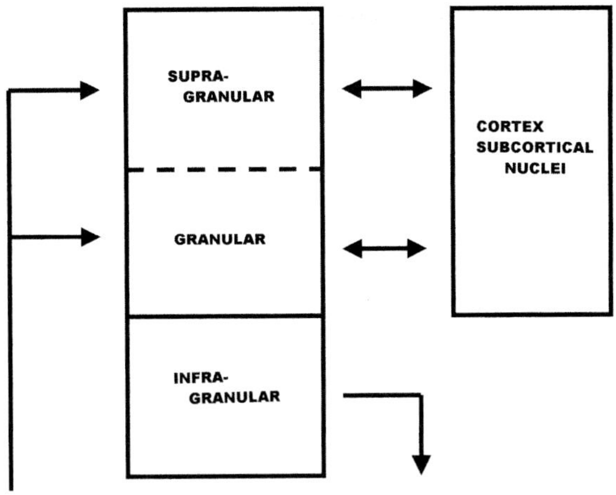

Fig. 5.2. Schematic representation of the horizontal layers of the cortex in relation to the major inputs and outputs

dendrites of the spiny stellate neurons tend to be locally confined to layer IV and the immediately adjacent layers. In addition, there is no direct afferent input from basal ganglia to cortex; the output of the basal ganglia travels along long polysynaptic pathways before finally reaching the neurons of the gray matter.

5.1.2 Columnar Organization

Vertically, the cortex is thought to have a columnar organization: the diameter of each column is 0.25–0.5 mm and contains $\sim 10^5$ neurons [57,362–365]. This columnar arrangement is most apparent in primary sensory cortical areas, but has been shown to exist in other regions as well, including human entorhinal cortex [366]. On physiological grounds it is thought that all neurons comprising a column share a common input and as a group perform a common function [367]. Because of the short length of the dendrites and axons of the non-pyramidal neurons it is thought that these neurons play an important role in the local circuit dynamics of the cortical column. Many of these non-pyramidal neurons are inhibitory (e.g. basket cells and Chandelier cells), but some are excitatory (e.g. certain bipolar neurons).

Many investigators have emphasized models for neuronal networks in which the role of short-range inhibitory and long-range excitatory connections are emphasized [63,66,67,70,89]. However, the situation is far from being this simple.

The major short-range excitatory input pobably arises from the axon collaterals of the pyramidal neuron itself [362]. Axon collaterals arise at any point along the course of the axon and travel horizontally or obliquely through the laminae. Moreover, side branches arise from the collaterals and ramify both upwards and downwards to create a rich arborization of local excitatory input. These collaterals synapse on other pyramidal neurons as well as the non-pyramidal neurons many of which are inhibitory. These synaptic connections provide the anatomical substrate for the feed-forward excitation of pyramidal neurons and recurrent inhibition (both feed-forward and feedback) via connections with inhibitory interneurons. The fact that axon collaterals are unmyelinated, and hence slow conducting, implies that conduction delays may be an important component of the dynamics. An early version of a mathematical model describing the dynamics of a cortical column is the "netlet" concept introduced by Harth et al. (e.g. [368]); a more recently emphasized attempt is the canonical cortical circuit proposed in [369](see also Chap. 7).

The major long-range excitatory input to neurons within a cortical column is from pyramidal neurons located in other cortical regions within the same hemisphere as well as the contralateral hemisphere. An additional cortical input, which is important for discussions of epileptogenesis, comes from the sub-cortical nuclei. For example, the thalamic nuclei are reciprocally connected to all cortical regions and it is now well established that thalamocortical afferents synapse with both pyramidal and non-pyramidal neurons distributed in layers III–VI (for a review see [362]). Moreover, very long polysynaptic pathways exist that involve the cortex, sub-cortical nuclei and the brainstem, for example, cortex → striatum and basal ganglia → brainstem nuclei → thalamus → cortex. Thus the anatomical substrate exists for very long range recurrent excitatory and inhibitory feedback. The very length of these pathways, plus the fact that they can be polysynaptic, provides yet another indication that conduction delays are probably of importance for EEG dynamics [68,370].

5.2 Generation of the EEG

To a first approximation the EEG is generated by the large vertically oriented pyramidal neurons located in layers II, III, V and VI (Fig. 5.3). These neurons have two features that are critically important for the EEG: 1) they are oriented vertical to the cortical surface; and 2) the inhibitory and excitatory inputs are spatially segregated over the surface of these neurons.

Although a variety of morphological subtypes of pyramidal neurons have been described they all share the following features (Fig. 5.3) [371]: 1) a single, dominant apical dendrite which extends vertically from the cell body to the cortical surface; 2) several basal dendrites that radiate more or less horizontally from the base of the cell body; 3) dendritic spines that have the greatest density in the middle regions of the dendrite; 4) a single axon that projects downwards and gives off a number of collaterals.

A distinguishing feature of pyramidal neurons is that only inhibitory synapses are made at the soma. On the other hand dendrites receive both excitatory and inhibitory synapses: most of the synapses onto dendritic shafts

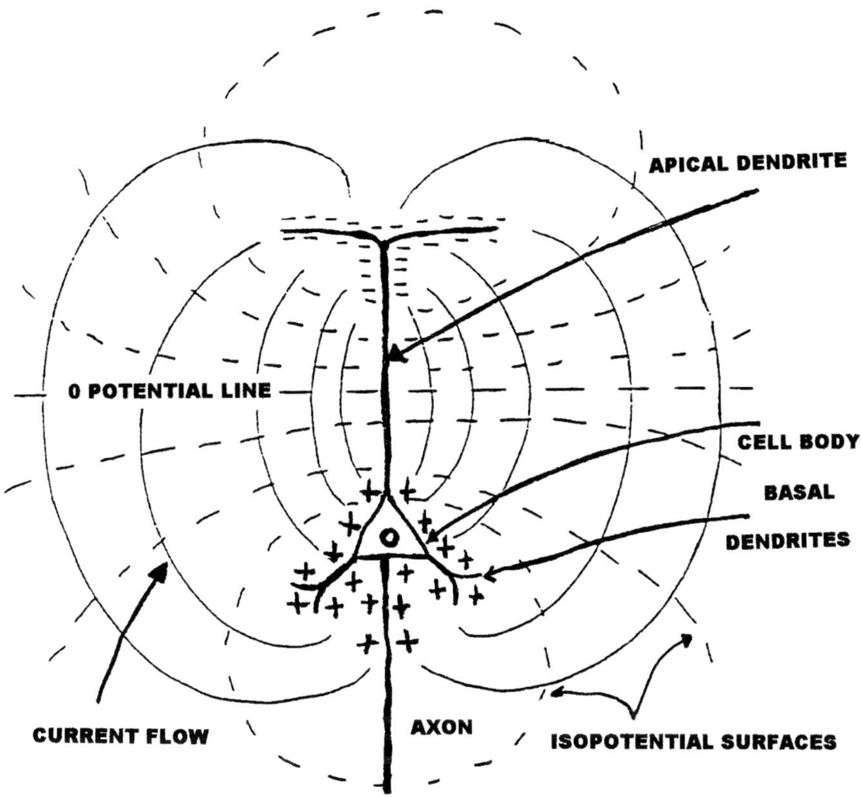

Fig. 5.3. Schematic representation of a pyramidal cortical neuron

(axodendritic) are inhibitory, whereas those onto spines (axospinous) are usually excitatory. Excitatory synaptic inputs to pyramidal neurons outnumber the inhibitory ones by a factor of about 6.5 [372]. This does not imply that inhibition is not important for the functioning of the brain. Indeed, it has been estimated that the average cortical neuron spends $> 98.6\%$ of its time silent [373]. This observation implies that the role of inhibitory neurons must be much greater than suggested by their number.

The generators of the EEG are the excitatory and inhibitory postsynaptic potentials, respectively EPSPs and IPSPs, that summate on the pyramidal neurons (Fig. 5.3). In order to appreciate why the spatial segregation of the excitatory and inhibitory synapses is of importance, consider the effect of increasing excitation on the more distal part of the apical dendritic tree in the supragranular region. By definition, the direction of current flow is taken with respect to the direction that the positive ions flow. Since an EPSP causes an inward flux of positive ions, the extracellular compartment at this point in space becomes more negative. To maintain neutrality this inward flux of positive ions at the superficial layer must be balanced by an outward flux at deeper cortical layers. Hence at this point in space the extracellular compartment becomes relatively more positive. Consequently the pyramidal neuron can be thought of as a vertical dipole oriented so that the sink (negative pole) is near the cortical surface and the source (positive pole) far from the surface. If, on the other hand, excitatory input is increased to synapses located more closely to the soma, the dipole has the reverse orientation.

The same potential changes can be induced by increasing inhibitory inputs to the pyramidal neuron. By repeating the above arguments it is easy to appreciate that the effect of increasing IPSPs at deeper cortical levels is identical to increasing EPSPs at more superficial levels; increasing IPSPs at superficial levels is the same as increasing EPSPs at deeper levels.

It is important to note that the EEG does not detect action potentials of the pyramidal neurons [374]. Simultaneous activation of the whole cell during the generation of an action potential does not set up a significant external potential field in spite of the large change of transmembrane potential since the current flow is the same through all parts of the membrane at once. Large external electric fields are set up only when potential differences occur between two or more regions on the surface of the neuron.

5.3 Neuronal Dipole Layers

The potential produced by a single neuronal dipole is too small to be measured by an EEG electrode. The reason why these very small extracellular currents can be measured is because the pyramidal neurons are aligned in a parallel fashion throughout the gray matter. This anatomical feature means that individual dipoles can summate to produce an equivalent large dipole large enough to be measured. Thus the EEG is generated by a dipole layer of neurons.

There are a number of important implications that follow from the concept of a dipole layer. First, the number of neuronal dipoles whose activity is monitored by a single electrode is large. It has been estimated that the cortical area required to produce a detectable EEG signal is $\sim 6\,\mathrm{cm}^2$ [375].[1] If we take an estimate of $\sim 10^5$ neurons per $0.008\,\mathrm{cm}^2$ (i.e. the horizontal extent of a typical cortical column), then this corresponds to $\sim 10^8$ neurons. Hence the EEG is a measure of the spatially averaged activity of a large population of neurons as a function of time [68,370].

Second, the EEG recorded from a cortical region in which neurons are not parallel will be much smaller than that recorded from cortical regions having vertically oriented neurons since the dipoles of the individual neurons will tend to cancel. An example of a cortical region in which the neurons are not arranged in a parallel fashion is the amygdala. The amygdala is thought to be

[1] In contrast, when the EEG is measured directly from the cortical surface it has been estimated that a cortical surface area of 12–25 mm^2 is sufficient to produce an epileptic spike discharge [376–378].

the site of the epileptic focus in many patients with partial complex seizures. Thus the EEG will not easily detect seizures that arise in the amygdala; the changes recorded by the EEG reflect activity that has spread into nearby areas of cortex (e.g. hippocampal formation) in which neurons are arranged in a parallel manner.

Finally, the EEG is essentially a measure of the degree of synchrony in a large population of oscillating dipoles. It can be shown that the time-averaged potential, $|\overline{\Phi}|$, for m parallel dipoles which oscillate in phase is [147]

$$|\overline{\Phi}| \sim m\,,$$

whereas if they oscillate out of phase we have

$$|\overline{\Phi}| \sim \sqrt{m}.$$

It follows from these observations that the EEG emphasizes the contribution of the synchronously oscillating dipoles. For example, in the case of a scalp electrode there are no local sources and the relative contribution of the m synchronous and n asynchronous oscillating dipoles to $|\overline{\Phi}|$ is $m/\sqrt{n}$. Suppose that the scalp electrode records from 10^8 neurons of which only 1% are synchronously active. Then the contribution of this minority of synchronously oscillating neurons to the EEG is 100 times greater that the number of asynchronously oscillating dipoles!

5.4 Role of Glial Cells

Neurons are not the only cell type which contribute to the EEG. In addition to neurons the cortex contains glial cells (see also Chap. 4). Glial cells do not generate action potentials or postsynaptic potentials [379,380]. However, the glial cell membrane potential behaves very much like a potassium electrode and hence is very sensitive to changes in the extracellular concentration of potassium ions. When neuronal activity is increased there is an outflow of potassium ions into the extracellular space. If the potassium concentration does not affect the entire glial cell membrane and remains increased only locally, then potential gradients build up along the glial cell, giving rise to intra- and extracellular current flows similar to those that arise in neurons. Since glial cells are electrically coupled it is quite likely that they can have an amplifying effect on the genesis of extracellular field potentials by dipole layers.

5.5 Epileptic Spikes

Spikes represent the basic element of paroxysmal activity measured by the EEG in patients with epilepsy (Figs. 5.4 and 5.5). The potential changes reflected by the morphology of the spike are generated by changes in the underlying dipole layer. During the spike the EEG first undergoes a small positive deflection, then a large negative one, and then a longer positive deflection (by convention upward deflections of the EEG pen are negative). In order to localize the sources and sinks underlying the epileptiform spike it is necessary to use electrodes that are oriented perpendicular to the cortical surface [381,382]. These studies have been done for the special case when epileptic spikes are generated following the focal application of penicillin. This antibiotic selectively and irreversibly binds to GABA receptors on neurons thus eliminating GABAergic inhibitory inputs. Measurements of the current source density during the spike indicate that first a brief sink transiently arises in layer V, then a large sink appears in layers II/III, which is replaced by a smaller sink in layers IV/V. Thus, from the point of dipole layer theory this implies that the orientation of the dipoles "flip-flops" during the generation of a spike.

A number of observations indicate that the sink in layers II/III is generated by neurons in layer V ([382]; for a review see [135]). Thus in a neural preparation in which there is no inhibitory input, the sequence of events must be an initial increase of EPSPs in layer V, followed by increase of EPSPs in layer II/III, then finally increased EPSPs to layers IV/V.

It is far from clear whether the same sequence of events underlies spikes recorded from the EEGs of patients with epilepsy. The very same sequence of potential changes could have been generated by inhibitory mechanisms (i.e. increased inhibition in layers II/III, then increased inhibition in layers IV/V, then finally increased inhibition in layers II/III). Although this inhibitory scenario is rarely emphasized, it may actually be the more correct one, given the recently recognized importance of inhibitory mechanisms for synchronizing neural populations [89].

From a practical point of view the importance of the fact that field potentials are generated by an oscillating dipole layer is that they can be measured and used to locate the region of the brain that generates a spike or a seizure. Consider that the individual dipoles in this layer are located perpendicular to the cortical surface (Fig. 5.4). Thus corresponding to a dipole layer with the sinks near the cortical surface there is a mountain of negativity in the field potential. If the dipole layer is oriented in the opposite direction, there is a valley of positivity. Viewed in this way interpreting an EEG is very much like reading a topological map [383]. The difference is that whereas continuous contours on the map indicate similar altitudes, on the EEG the peaks and valleys of potential must be estimated from a few vantage points. These vantage points correspond to the locations at which electrodes are placed on the scalp (typically 8–32 scalp electrodes are used in a routine EEG which are located using the International 10–20 Electrode System [384]).

In principle, there are two ways in which a peak or valley in a potential field can be located (Fig. 5.5). The simplest way is to measure the potential at every point relative to a common reference electrode (Fig. 5.5A): for example,

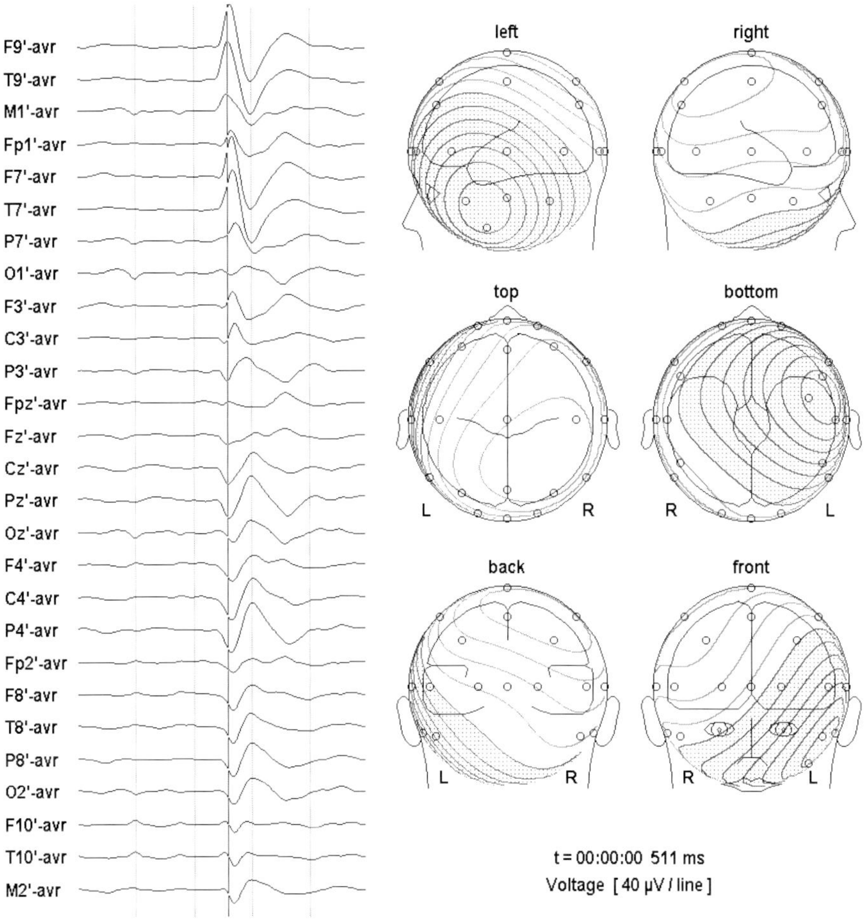

Fig. 5.4. A left temporal spike is displayed in a common average reference montage. Topographic maps illustrate the voltage field of the spike at the cursor. Voltage field contours are marked by isopotential lines for every 40 µV. Negative polarity field is speckled. Note the dipolar configuration of the voltage topography. Negative maximum is anterior inferior temporal and the positive maximum has a contralateral parietal location. A three-dimensional line between the two maxima defines the orientation of the field, which is orthogonal to the orientation of the cortical source. The negative field maximum is of greater amplitude and has a steeper field gradient because the spike source is located nearer to it

the electrode closest to the peak of the negativity has the most positive amplitude. The usefulness of these so-called monopolar electrode montages is limited by difficulties in obtaining a relatively inactive reference electrode in the head and neck. The second approach avoids the problem of obtaining an inactive reference electrode by estimating the slope of the potential field

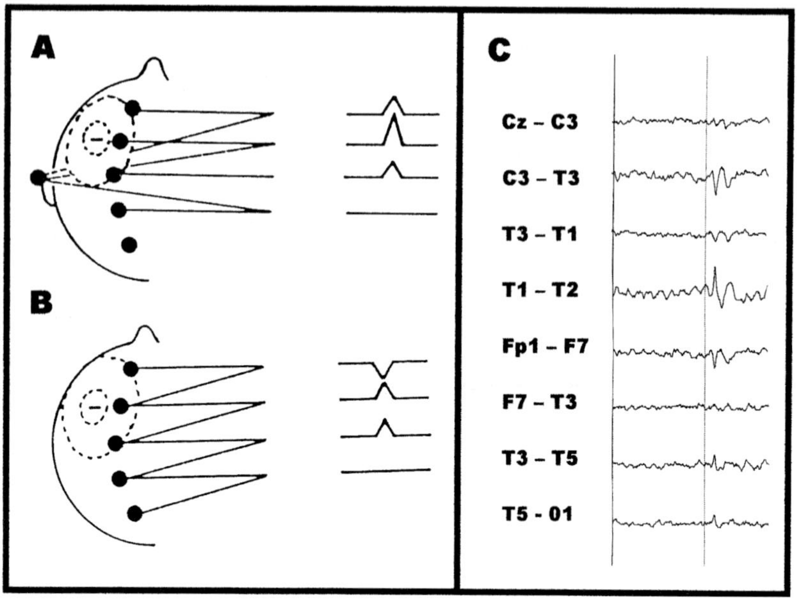

Fig. 5.5. Locating the peak of the potential field using either a (**A**) monopolar electrode montage or (**B**) a bipolar electrode montage. (**C**) An epileptic spike measured using a bipolar montage. The spike is equipotential at T3-T1 and F7-T3. In other words the shape of the electrical field is quite flat between, respectively, T3,T1 and F7,T3

as the difference in potential between two electrodes (bipolar montages). In this case a peak or valley is identified as the point at which the slope of the potential field changes sign, i.e. by a phase reversal (Fig. 5.5B) (The unfortunate choice of this term by EEG workers will hopefully not cause too much confusion for the more mathematically oriented readers of this text!).

5.6 Seizures

Spikes are seen with increased frequency in the EEGs of patients who have focal-onset epilepsy. Indeed localization of these spikes onto the surface of the brain by the EEG worker often provides an important clue pointing to the identification of the epileptic focus, i.e. the region from the brain where the seizures emanate, and hence as a starting point for investigations leading to its surgical removal for cure. However, the relationship between spikes and the mechanisms for the generation of epileptic seizures is not a straight forward one. Indeed it has even been suggested that spikes may actually be a post-ictal phenomenon [385]. The reason for this uncertainty is because typically the onset of focal seizures is not heralded by increasing frequency of spiking as might be expected by the reader, but rather by a marked suppression of the EEG reflected by the appearance of a low-voltage, high-frequency signal [386–388]. The low amplitude of the EEG recorded at seizure onset is thought to reflect desynchronization in the dipole layer.

Vertically oriented electrodes have also been used in animal models of epilepsy to identify the changes in the sinks and sources that occur in the gray matter during penicillin-induced tonic-clonic seizures [382]. At the very beginning, the source-sink pattern is very irregular and consists of a superimposition of sinks and sources of different intensities, durations, and locations in the cortex. During the tonic pattern, a regular "flip-flop" in the dipole layer occurs. During the clonic portion of the seizure, the dipole oscillations become less regular and variable in extent. In addition, during this period the flip-flops of the dipole seem to occur at shorter intervals than in the tonic phase.

"Flip-flops" of cortical dipoles in relationship to seizures have been observed in a number of animal models (for a review see [4]). Perhaps the most elegant work is that by Gloor and co-workers on penicillin-induced seizures in the hippocampus [389]. Of particular interest to this text is the observation that the phase of the oscillating dipoles in this preparation could be shifted by a single afferent discharge to the hippocampus.

These observations might lead the reader to the impression that the changes measured by the EEG electrode are merely a reflection of "flip-flops" of underlying cortical dipoles as portrayed in Fig. 5.6. However, it is important

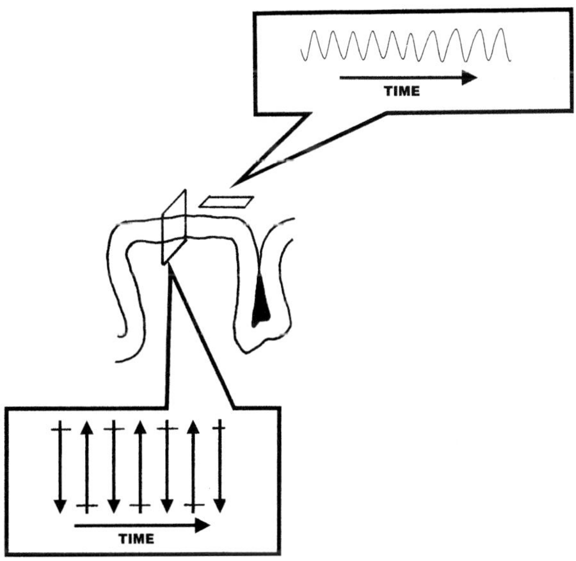

Fig. 5.6. Schematic relationship between "flips-flops" in the cortical dipoles (*upper right block*) and rhythmic changes in the EEG (*bottom left block*). As discussed in the text and Chap. 16 this picture is probably too simplistic

to keep in mind the conclusions of Sect. 5.4, i.e. the degree of synchrony in the underlying neural population necessary to produce the observed changes in the EEG is quite small. Thus at any one point in time the probability that a given neuron fires, let us say, at the peak of the EEG oscillation is very small. A perhaps more correct statement may be that if the neuron fires then it has the greatest probability of firing at the peak in the EEG. The importance of this discussion is to remind the reader that the EEG reflects the statistical properties of the underlying neural population. The possibility that the rhythmicity in the EEG reflects statistical periodicity, i.e. a statistical property of a stochastic dynamical system with retarded variables, is discussed in Chap. 16.

5.7 Ictal EEG Patterns

The pattern of scalp EEG ictal discharge of seizures that originate in the temporal lobe complex have been extensively studied [390–395]. Three distinct ictal patterns have been described [390]. Type I seizure patterns are characterized by a temporal and/or subtemporal rhythm of 5–6 Hz (Fig. 5.7A). A feature of these ictal patterns is that for a given frequency the waveform tends to remain regular and well modulated for time intervals longer than 5 s. Type II seizure patterns are characterized by slower rhythms (2–5 Hz) (Fig. 5.7B).

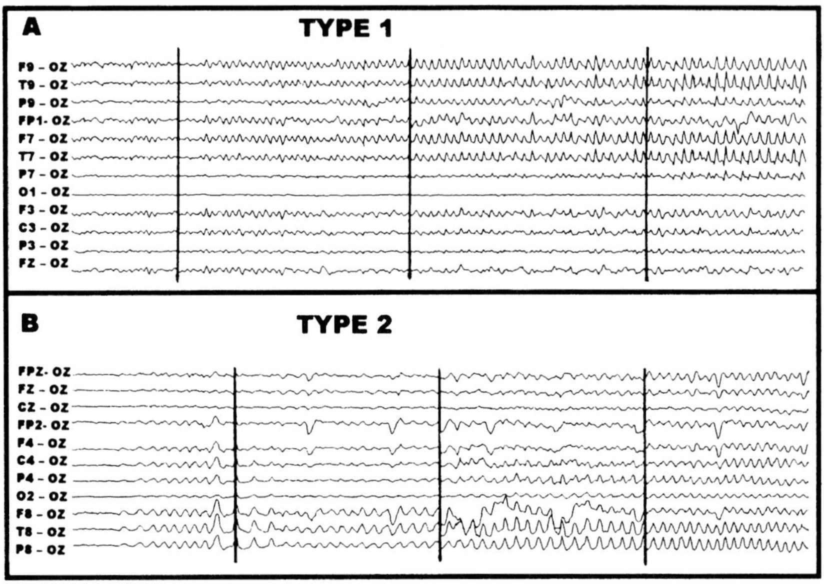

Fig. 5.7. Ictal EEG recordings for a (**A**) Type I seizure arising from the left hippocampus and (**B**) a Type II seizure arising from the right temporal neocortex. *Vertical bars* delineate a time interval of 5s. Figures adapted from [390]

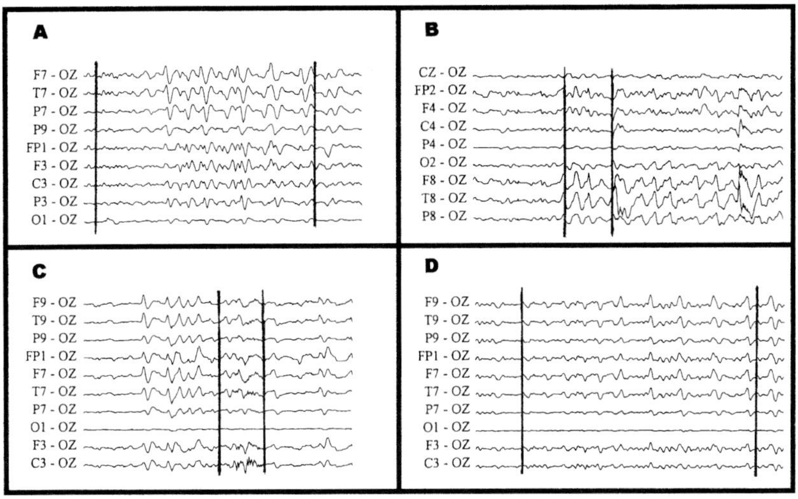

Fig. 5.8. Variability in Type II seizure patterns. Time intervals (*vertical lines*) are 1s in (**B**) and (**C**) and 5 s in (**A**) and (**D**). Figure adapted from [390]

In contrast to Type I seizures, the waveforms of Type II seizures show variability both between different seizures and during the seizure (Fig. 5.8). Moreover, the temporal development of these ictal patterns was much less stereotyped than observed for Type I ictal patterns. Finally, some seizures, referred to as Type III, were characterized by diffuse slowing or arhythmic background changes (example not shown).

Clinical associations and intracranial electrode recordings indicate that the above ictal patterns are strongly correlated to the site of origin of the seizure [388,396]. Type I ictal EEG patterns are almost exclusively associated with hippocampal onset seizures, whereas, Type II ictal EEG patterns point to a temporal neocortical seizure onset. The least specific ictal EEG pattern is Type III that may be seen in both temporal neocortical and hippocampal onset seizures.

Measurements with intracranial electrodes indicate that these three ictal EEG patterns are not a reflection of cortical activity at seizure onset. Rather the EEG patterns arise from the subsequent spatiotemporal dynamics and development of synchrony in the neuronal population as the seizure evolves. It is quite likely that these different EEG ictal patterns have at their basis the different wave phenomena observed in excitable media (see Chaps. 9–11). Indeed the difference in rhythmicity between Type I and II EEG ictal patterns is quite reminiscent of the fast ($\sim$ 9Hz) and slow ($\sim$ 3 Hz) bulk population oscillations observed in computer simulations of randomly connected networks of integrate-and-fire neurons [66]. Recently the wave properties of spatially extended media with retarded variables has begun to attract attention [397]. It has been conjectured that increasing the spatio-temporal delay in any spatially extended biological system modeled by a hyperbolic delayed partial differential equation will result in spatial complexity and irregular wave propagation, including multistability and spatial chaos. These observations emphasize the need for more careful characterization of the clinically observed EEG ictal patterns.

5.8 Limitations of the EEG

Current emphasis in EEG has been directed at determining the location and orientation of a hypothetical dipole generator that produces changes observed in the EEG, i.e. epileptic spikes and seizures (for reviews see [398,399]). Since a spatially localized dipole layer can be approximated by a single dipole this has led to the development of powerful methods to localize dipole sources and hence identify epileptic foci. Moreover since the regions of synchronization grow as seizure generalize, coherence methods are useful for monitoring the progression of seizures (see Chap. 6).

It is typically assumed that this dipole is a static entity and is localized at a fixed depth from the cortical surface. It is well known that even with this simplifying assumption this inverse problem in electrostatics does not, in general, possess a unique solution. However, with the addition of constraints such as realistic head and skull models, it has been possible to develop algorithms to localize dipoles to an accuracy that is smaller than the size of typically surgical resections for epilepsy (i.e. the anterior 5.5cm of the temporal lobe). Here we address the limitations of the EEG for exploring the spatio-temporal dynamics of ictal patterns.

The magnitude of the potential measured by an EEG electrode depends on 1) the size of the dipole layer generating the source; 2) the distance of the electrode from the dipole layer; and 3) the solid angle, Ω, subtended by the dipole surface to the electrode; and 4) the dielectric constant of the medium.

The amount of surface area occupied by a dipole layer generator is the major factor that enables it to be detected by an EEG electrode. Measurements using implanted subdural electrodes demonstrate that underlying a clinically evident epileptogenic EEG focus at the scalp there are multiple, independent spike generators that occupy areas of cortical surface $< 2\,\text{mm}$ in diameter. Despite being capable of generating high amplitude spikes, the independent generators are not detected by the scalp electrode. It is only when sufficient numbers of these spike generators fire synchronously can this epileptic activity be measured by the scalp electrode.

The surface area of the human cortex is $\sim 2200\,\text{cm}^2$ [153,400]. This surface is highly convoluted: the crests of the convolutions are referred to as the gyri, the infoldings that separate the gyri as the sulci. Nearly two-thirds of the cortical surface is located in the depths of the sulci. Thus if the generators of epileptic spikes were distributed randomly over the cortical surface, then most would be located too deep to be recorded by a scalp electrode. It is important

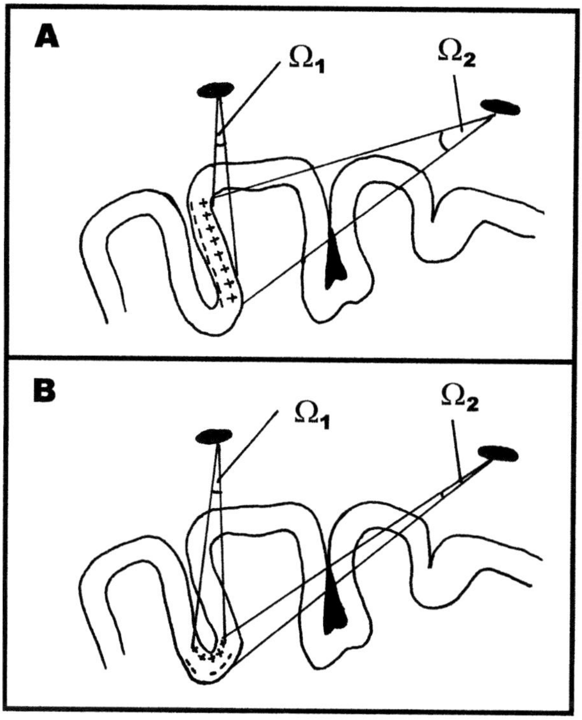

Fig. 5.9. Influence of the solid angle, Ω, on the potential "seen" by an EEG electrode for a source placed (**A**) on the lateral wall of a sulcus and (**B**) at the bottom of a sulcus

to realize that the EEG is much more sensitive to changes measured for neurons located on the gyri than from those located in the sulci.

First, the magnitude of the potential measured by an EEG electrode depends on the distance, r, the electrode is from the dipole layer (specifically the potential falls as r^{-2} provided that our measurements are taken sufficiently far from it). Indeed it is thought that one reason why epileptic foci arising in the frontal lobes are typically poorly located by the EEG is because many are probably located deep in the sulci. The fact that the EEG is relatively insensitive to deeply located generators becomes of major concern in view of the fact that rapid conduction of nerve impulses via myelinated fibers occurs preferentially in the vertical plane away from the surface (Chap. 2).

Second, the potential P at every point in space (except in the immediate vicinity of the dipole), is proportional to the solid angle subtended by the dipole surface to that point [401]. In other words,

$$P = \pm \frac{e}{4\pi} \Omega \,. \tag{5.1}$$

Whenever Ω is large the measured potential is large, and vice versa. The situation becomes more complex in the case of the curved surfaces overlying a sulcus since even an epileptic focus located directly under an electrode may not be detected, but the signal will be falsely located at an electrode some distance away from the true location (Fig. 5.9). When multiple electrodes are employed (e.g. bipolar montage) more difficulties arise because the solid angles subtended by different electrodes overlap, thus decreasing the effective solid angle seen by each electrode (see, for example, [389]).

Considerations of Ω limit the usefulness of depth electrodes. On the one hand, the insertion of a depth electrode enables EEG activity to be measured from a deeply situated dipole generator. On the other hand, the Ω is typically much smaller for a depth electrode than a scalp electrode and thus the area of the brain from which activity can be recorded is very small. Thus depth electrodes must be very precisely placed to be useful. The dilemma faced by the clinician is deciding whether the potential benefit of a depth electrode to the patient is worth the risk of implantation (the mortality rate is $\sim$ 1–2%) given that even a small error in the placement of the electrode may render it useless.

Finally, the potential gradient in a field set up by an electric charge is affected by the dielectric constant of the medium. Within the body, there are many media with different dielectric constants. The electric fields cross the boundary surfaces between these media and become distorted if the media have different dielectric constants. As far as the EEG is concerned, the most important boundary surface is that between the skin and surrounding air. Consequently the electrical fields set up by a dipole layer in the brain will be distorted in the region of the scalp (Fig. 5.10). This distortion is more marked for dipole layers near the surface than for those further away from it. In the boundary zone near the scalp, the isopotential surfaces are more curved and closer together. This is so because the current intensity is increased near the surface due to the inability of the current to be dispersed into the space beyond the boundary (i.e. the air behaves as a perfect insulator). Thus potential drops of equal magnitude to those occurring in an infinite volume conductor, now take place over a shorter distance, hence the crowding together of the isopotential surfaces.

In view of these considerations a dipole layer located in the depth will show less distortion of its gradient on the scalp than one near the surface. A small discharging lesion near the surface sets up a narrowly confined EEG focus with steeply sloping gradients, whereas a deep-lying small discharging lesion will set up a field on the surface of the scalp that slopes more gently, tends to show a plateau at its center and covers a wide area in which it may be difficult to extract the signal from noise. The steeply contoured, narrow-based potential fields of the superficial dipole layer will correspond to sharper waves on the EEG that are easily detected by the eye of the EEG worker.

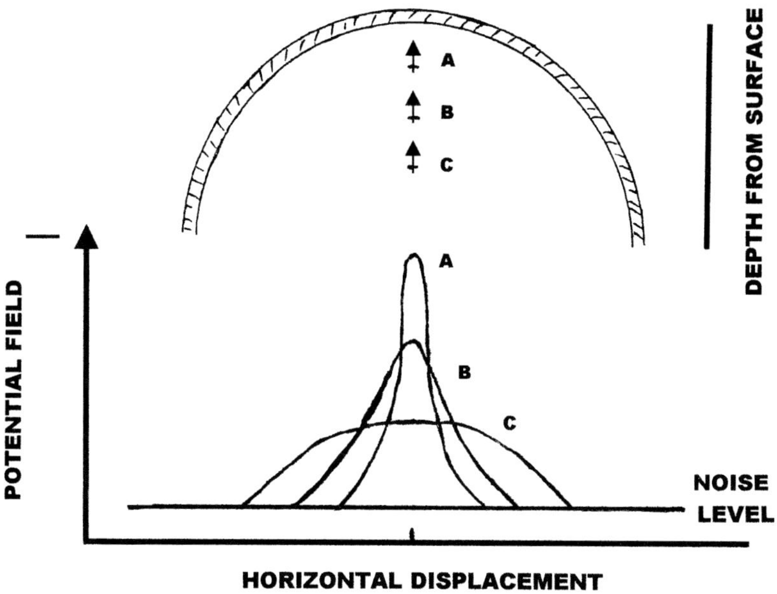

Fig. 5.10. Distortion of the isopotential fields produced by a vertically oriented dipole layer by the scalp-air interface. A, B, C correspond to dipoles located at different depths from the scalp (*top of figure*). The most superficial dipole arises above the noise level only within a narrowly confined zone, whereas the more deeply located dipoles rise above this level over a larger zone. Note that these considerations apply only to monopolar recordings

This is another mechanism that provides a further bias favoring the detection of more superficially situated dipole layers over deeply located ones.

5.9 Concluding Remarks

In recent years a great deal of attention has been given by neurobiologists to understanding the functioning of *single* neurons, the structure and function of their synapses, and the intracellular biochemical processes triggered when a neurotransmitter binds at its membrane receptor. Despite the great advances in each of these areas, the frustration is that it has not been possible to translate this progress into a deeper understanding of the EEG. However, a more accurate statement may be that EEG has not provided the investigators the message they wished to hear.

The EEG is not a phenomenon of a single neuron, but is a measure of the activity of a very large *population* of neurons. Moreover, it is not capable of monitoring the changing relationships between cortex and sub-cortex alluded to in Chaps. 1–3. The EEG is a measure of synchrony in selected cortical neural populations.

Thus the EEG points to the very essence of epilepsy and its treatment. Epilepsy is to be understood in terms of how neural oscillations arise and how oscillating neural populations become synchronized. It is clearly naive to think of treatment strategies for epilepsy based on simplistic concepts of restoring the balance between inhibition and excitation in the nervous system. Effective strategies must be aimed at preventing this synchronization from occurring or at desynchronizing populations that have become synchronized.

6 Electrocorticographic Coherence Patterns of Epileptic Seizures *

V.L. Towle, F. Ahmad, M. Kohrman, K. Hecox, and S. Chkhenkeli

One of the most difficult questions facing epileptologists involved in the surgical treatment of epilepsy is the identification of brain areas that are responsible for event initiation, or that participate in the propagation of seizures. Determining the initial location and extent of the epileptogenic focus and the spatio-temporal organization of the neural populations causing seizures are crucial for the successful surgical treatment of intractable epilepsy, and for the treatment of patients with lesional brain processes accompanied by epileptic seizures. The practical problem of epilepsy surgery – that of determining the limits of a cortical resection that will prevent the recurrence of epileptic seizures, is only partially resolved. The decision of whether to resect or leave a marginally active area of cortex intact elicits a wide range of opinions from clinicians. Advances in neuro-surgical and imaging techniques have increased the proportion of intractable epilepsy patients that can be successfully treated with surgery. Many of the expensive hospital days spent in monitoring units are spent waiting to record seizures so that these areas can be identified. If such areas could be more easily identified during the interictal period, the length and costs of epilepsy work-ups could be dramatically reduced. Further, if such areas could be identified more precisely, the efficacy of seizure surgery would be increased. Although it is expensive and manpower intensive, electrocorticography (ECoG) remains the cardinal method for solving this problem. By analyzing the lateral coherence patterns across the surface of cortex, it may be possible to identify a "cortical epileptic system" that is causing the seizures. It is desirable, but not yet possible, to characterize such a system using interictal recordings. If this becomes possible, analysis of spontaneous seizures and the spread of after-discharges from direct cortical stimulation may be relegated to a confirmatory role, rather than the primary means of functional mapping.

6.1 Lateral Coherence

One statistic that may be useful for understanding the spread of seizures is lateral coherence, which is a measure of the similarity and phase relationship between signals recorded at various locations across cortex. Coherence is like Pearson's correlation coefficient squared (r^2), and is scaled from 0 to 1. It

* This work was supported by The Brain Research Foundation.

differs from simple correlation in that it is calculated in the frequency domain, and separate values can be calculated for each frequency band observed in the signal. Like the cross-correlations calculated in the time domain, a phase lag is associated with each coherence value, suggesting the direction and relative timing of information flow. In this context coherence is a measure of the similarity of two ECoG waveforms. It can be defined as:

$$\gamma(f) = \frac{S_{xy}(f)}{\sqrt{S_{xx}(f)S_{yy}(f)}}, \tag{6.1}$$

where $S_{xx}(f)$ and $S_{yy}(f)$ are the autocorrelation power spectra of the two time series $x(t)$ and $y(t)$ in the frequency domain, and $S_{xy}(f)$ is their cross spectrum [402]. In spite of several interesting observations in the area of seizures and coherence measures, analysis of coherence patterns as a function of spatial location has not yet gained widespread use as a means to evaluate epileptic seizures. Different forms of coherence analysis of scalp potentials have been employed to investigate "functional coupling" [403], "information exchange" [404], "temporal coordination" [405], and "functional interaction" [406]. When recorded directly from cortex, the coherence statistic reflects the extent of synchrony of neuronal activity from discrete, and sometimes distant cortical areas. High coherence has been interpreted as evidence of structural and/or functional connections between cortical areas [407]. We have used the coherence method for the identification of the sensory/motor area and to reveal the localization of pathologic processes [408]. The instability of scalp coherence patterns has recently been attributed to the rapid switching of thought, and even formalized into a theory of consciousness, called the dynamic core hypothesis [409]. Synchronous rapid bursts of neuronal discharges are thought to result in graded activity in the 40–70Hz range (gamma band) and has been postulated to entrain neurons to mediate conscious perception [410] and selective attention [411].

Mapping of conventional ECoG slowing and spikes has been widely employed to identify epileptic foci. The temporal relationship between such phenomena in different areas of the brain is often complex, and has traditionally been determined by visual inspection of raw voltage traces. Development of objective indices of such relationships is desirable. Early studies of ECoG coherence examined the relationship between seizure activity recorded in the two hemispheres, and reported small inter-hemispheric delays that were suggestive of the lateralization of onset [151,152,261,299]. When seizure activity spreads within a hemisphere, coherence patterns are altered, and reflect "transference" of the seizure activity from one area to another [412]. There is a general trend for coherence to fall off with increasing inter-electrode distance [413–415], especially at higher frequencies [416]. However, during seizures ECoG coherence substantially increases in high-frequency bands [413,417,418]. Localized, low-amplitude, high-frequency activity has also been associated with the onset of some seizures [386,419], but coherence patterns have not been studied during this phenomenon. A pronounced increase in cortical lateral interdependency is evident when nonlinear analyses are performed [420]. Furthermore, the onset and spread of medial temporal lobe seizures have successfully been traced when a coherence-like statistic that emphasizes phase aspects of the signal is utilized [421]. Changes in the temporal pattern of epileptic activity recorded during perceptual discriminations in a single patient have been described by indicating that normal and pathological areas of the brain interact [420]. A better understanding of such interactions might create insight into how seizures spread, and possibly to new ways to suppress seizures. For example, multiple sub-pial transection can reduce the spread of seizures. After multiple sub-pial transection [160] one would expect coherence patterns to decrease and become oriented parallel to the axis of transection. Such analysis might provide an immediate index of the effectiveness of the procedure.

6.2 ECoG Recordings

Subdural ECoG recordings can be made in the epilepsy unit during the awake and sleeping states from 8×8 arrays chronically located over the frontal, temporal, parietal, or occipital lobes. A contralateral mastoid or forehead reference is used. Each ECoG channel is divided into contiguous 1–5 s segments and processed using the SCAN Edit Module (V. 4.1) (Neuroscan, Inc., Herndon, VA). Mean power spectra between 1–50 Hz are obtained [402] by averaging the spectra computed from each segment of ECoG. Analysis of the resulting spectra are divided into 5 frequency bands similar to those used to describe surface EEG recordings: 1–4 Hz (δ), 4–7 Hz (θ), 7–13 Hz (α), 13–30 Hz (β), and 30–10 Hz (γ).

The ECoG inter-electrode coherence is calculated for all possible electrode pairs for each frequency band using Neuroscan's Edit module. The number of different pairwise comparisons for N electrodes is $N(N-1)/2$. We have developed laboratory software that depicts selected inter-electrode coherences graphically as line segments or splines drawn between pairs of coherent electrodes. The width of the line is roughly proportional to the magnitude of the coherence, and the direction of the phase lag is indicated by the color or width of the tip of the line segment.

6.3 Ictal Coherence

The dynamic nature of epileptic seizures can be seen in the analysis of spontaneous seizures recorded with chronically indwelling subdural electrodes. We will present two spontaneous seizures that were recorded from grids similarly placed over the lateral posterior aspect of the temporal lobe recorded from two pediatric patients. The first seizure to be described was recorded from an

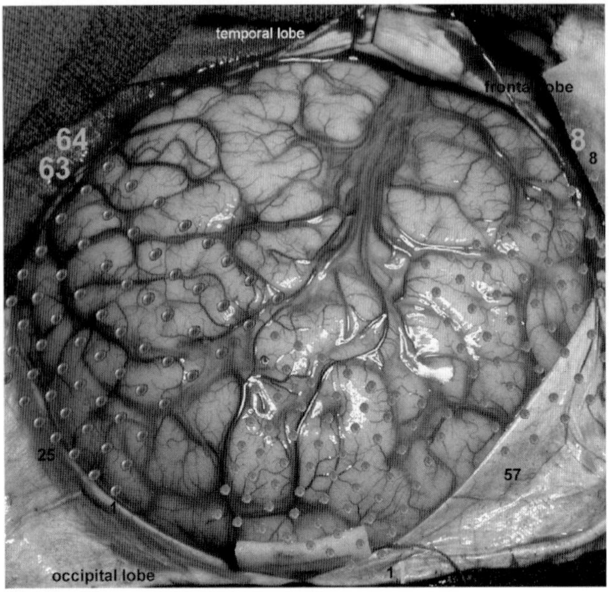

Fig. 6.1. A photograph of the exposed surface of the brain revealing the placement of the 8 × 8 grid relative to the abnormal gyral morphology. Electrode 64 is over the left inferior temporal lobe; prominent blood vessels indicate the left Sylvian fissure (The colour plate of this figure appears on page 355)

eight year-old male with left hemimegencephaly involving the left temporal-parietal-occipital region (Fig. 6.1). This child's seizure began in a malformed region of the lateral occipital lobe and spread to more superior and anterior regions of the posterior temporal lobe. The 5-Hz spike-and-wave activity exhibited the typical gradual slowing and then abruptly stopped in this 45 s duration spontaneous seizure (Fig. 6.2b,c). There were intervening areas in the middle of the grid that did not display seizure activity (Fig. 6.3). The frequency content was similar across electrodes located throughout the extent of the seizing area, which behaved as a unitary phenomenon in the time domain. This is reflected in the coherence patterns calculated from waveforms recorded during the late phase of the seizure, which are reminiscent of the distribution of increased power in the theta band. Areas several centimeters apart reveal coherent signals in this frequency band, suggesting that these regions were strongly linked to each other (Fig. 6.4a). The functional linkage could be either through cortico-cortical arcuate fibers, or from a common sub-cortical input. Interestingly, the seizing and non-seizing areas remained largely incoherent during the interictal phase before the seizure (Fig. 6.4b). Generally, normal patterns of coherence were observed over the areas that did not subsequently participate in the seizure, but the coherence in the seizing area was substantially lower, as if the seizure-prone area was functionally disconnected from the rest of the brain.

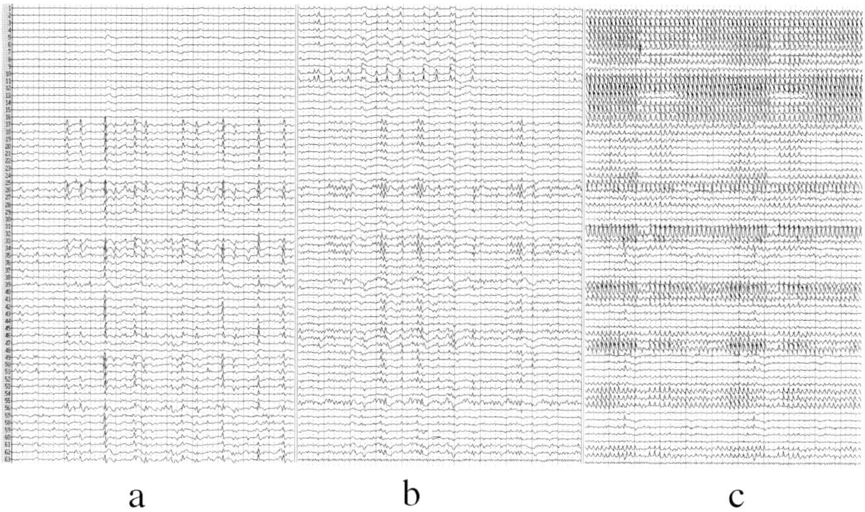

Fig. 6.2. Changes in the ECoG recorded from the occipital grid before (**a**) and during (**b,c**) the seizure. The seizure remains highly coherent as it spreads to more anterior areas of the brain. The successive epochs are labelled (**a**–**c**) and correspond, respectively, to the panels shown in Fig. 6.3

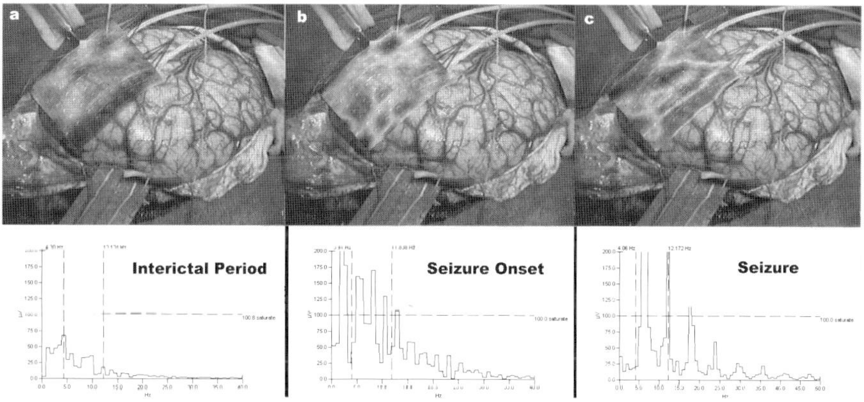

Fig. 6.3. Changes in ECoG power during the course of a seizure. The seizure (*red*) begins in the posterior aspect of the grid (**a**) and spreads along its superior edge and covers the anterior edge of the grid. The successive epochs are labelled (**a**–**c**) and correspond, respectively, to the ECoG segments shown in Fig. 6.2 (The colour plate of this figure appears on page 355)

A second patient is a seven year-old female with tuberous sclerosis of the right parietal-temporal area. Her spontaneous seizure lasted 96 seconds, and migrated across the recording grid as the seizure progressed (Figs. 6.5 and 6.6). It was preceded by a short electrodecromental period, followed by a localized polyphasic transient event recorded over the sensory hand

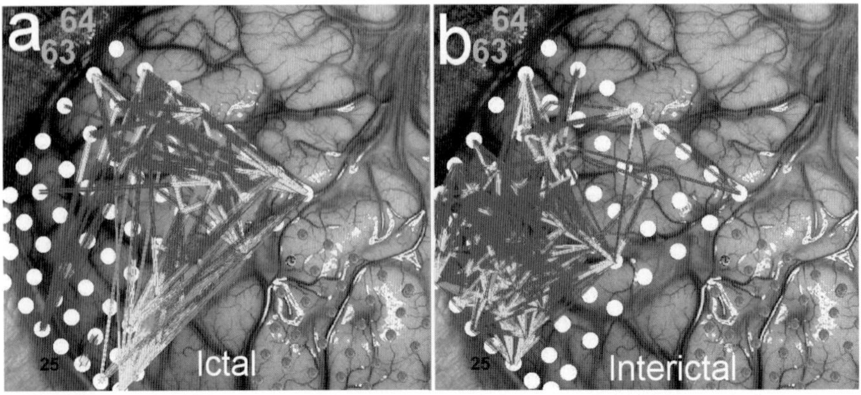

Fig. 6.4. Coherence patterns recorded during the ictal (**a**) and interictal (**b**) periods were largely complimentary, suggesting that during the interictal period the areas that would seize were disorganized and became synchronous during the seizure (The colour plate of this figure appears on page 356)

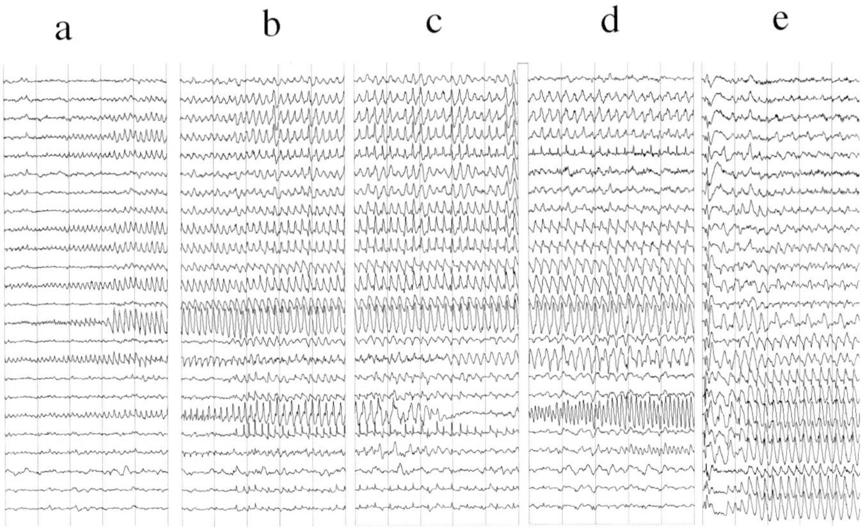

Fig. 6.5. Successive 5 s excerpts of selected channels recorded from an 8×8 grid, showing the migration of the seizure activity across the grid. The successive epochs are labelled (**a**–**e**) and correspond, respectively, to the panel shown in Fig. 6.6

area, from electrodes located in the posterior portion of the grid, from which frequent interictal spikes were recorded (Fig. 6.6a). This polyphasic transient was immediately followed by a continuous discharge at the anterior edge of the grid located over the morphologically abnormal middle and inferior temporal gyri (Fig. 6.6b). The original spike and wave activity was centered at 6 Hz, but as the seizure evolved, the power spectrums simultaneously revealed

a
b
c
d
e

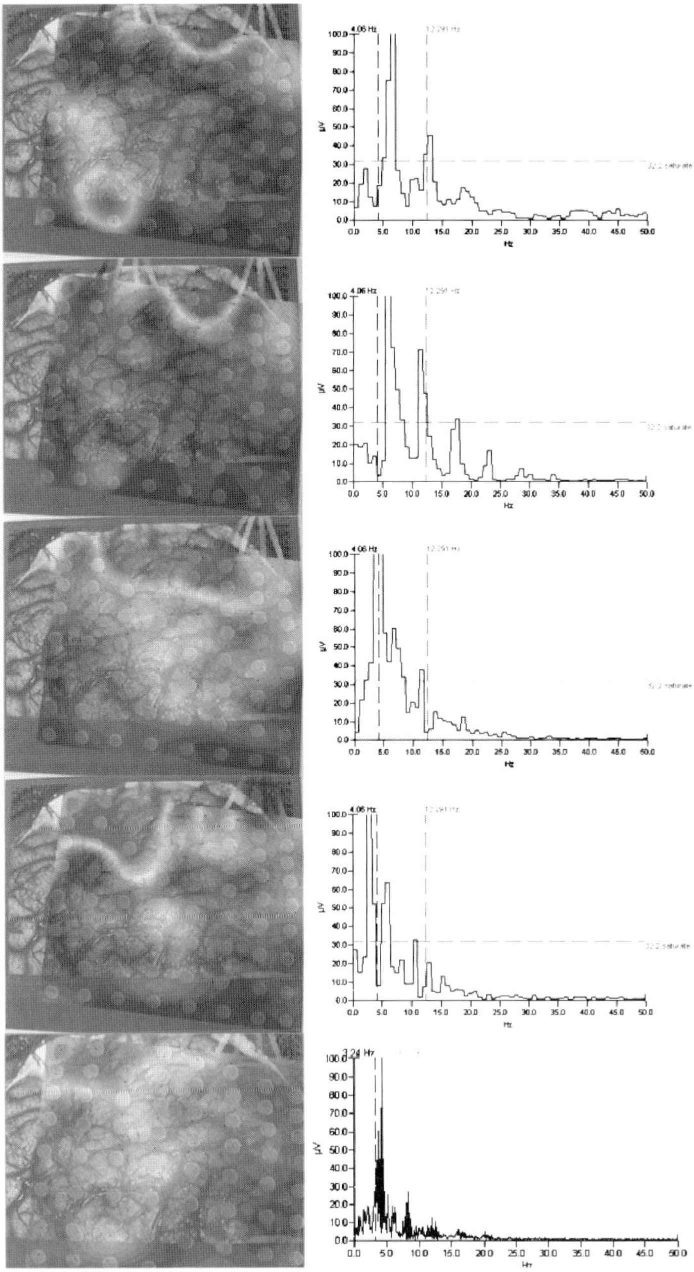

Fig. 6.6. The distribution of increased power (*red*) in the theta and alpha bands as a spontaneous seizure progresses. The seizure moves along the anterior edge of the grid (**a**–**c**) and plays out in the initial area while continuing to progress in the subsequently involved area (**d**–**e**). The successive epochs are labelled (**a**–**e**) and correspond, respectively, to the ECoG segments shown in Fig. 6.5 (The colour of this figure appears on page 357)

different peak activity at different electrodes along the anterior edge of the grid, with peak power between 4–12 Hz. Some electrodes revealed a waxing and waning of the seizure activity, while others simultaneously exhibited a steady 5 Hz activity. Gradually, the seizure slowed and then persisted in the inferior electrodes (Fig. 6.6d,e), while the later region continued to seize for several seconds, when it abruptly stopped and exhibited postictal slowing. Phase reversals of these potentials (not shown) suggested that the seizures were primarily in the depth of the sulci. As the seizure developed, it moved in a superior direction along the edge of the grid, where instantaneous phase reversals suggested it involved the superior surface of the temporal lobe. Although the hand area was active during the interictal period, it did not participate in the seizure.

This patient revealed a markedly different pattern of coherence from the previous patient. During the interictal period there was an increase in coherence in the temporal areas that would subsequently seize, but the progressive nature of the propagation of the seizure precluded a coherent pattern developing during the seizure (Fig. 6.7a). Unlike the previous patient, there was no noticeable increase in coherence in the postcentral hand area during the interictal period (Fig. 6.7b). A topectomy of the anterior lateral temporal seizing region was performed, and the postcentral hand area was spared.

6.4 Interictal Coherence

For some patients, as illustrated by the first case presented, the interictal recordings appear to herald the ictal patterns and provide insight into the symptomatology accompanying the seizure. Another case with a similar interictal pattern is a 33 year-old female with a history of intractable epilepsy and frequent generalized tonic-clonic seizures. The patient experienced focal motor phenomena involving the left half of her face, a turning of the head to the left or twitching in the left hand with the immediate generalization. Chronic subdural arrays of electrodes were placed over the right frontal, lateral temporal and inferomedial temporal surface of the brain. During the five-day monitoring of ECoG activity several clinical seizures that were typical for the patient occurred, which always started under the same electrode over the motor face area. Diagnostic cortical stimulation of that area provoked a typical clinical seizure with an aura, short partial motor phenomena and generalization. Similar diagnostic stimulation during surgery revealed the same area of cortex involved in the development of clinical seizures and epileptic after-discharges. The area that was involved in the seizures on the basis of conventional electrocorticography is marked with a thin black string in Fig. 6.8. There was an area that exhibited intermittent spikes and slowing over the anterior inferior cortex that was not resected. Analysis of the coherence patterns from interictal background activity revealed two areas of high coherence (> 0.7) over the frontal cortex of the right hemisphere. One

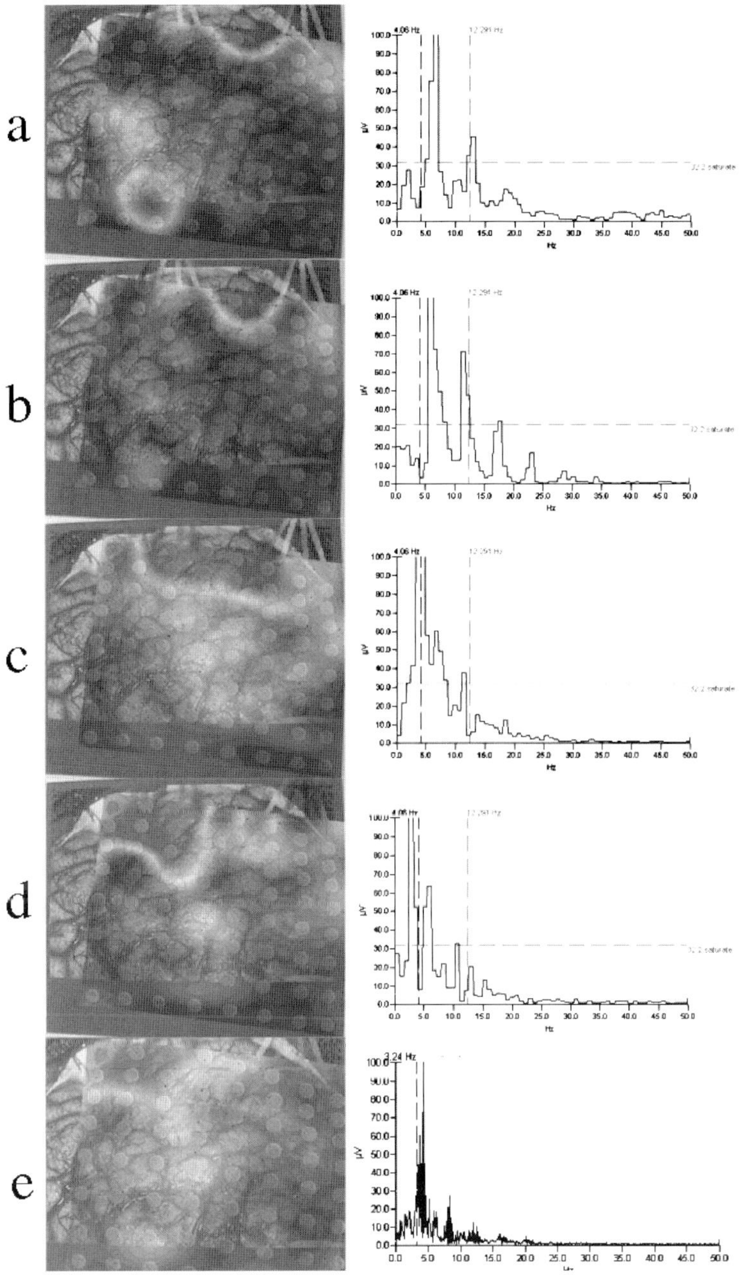

Fig. 6.7. Coherence patterns during the interictal (**a**) and ictal (**b**) periods. Because this seizure did not behave as a unitary phenomenon it does not reveal increased coherence during the seizure. However, during the interictal period before the seizure began, there is increased coherence in the region that would subsequently seize (The colour plate of this figure appears on page 358)

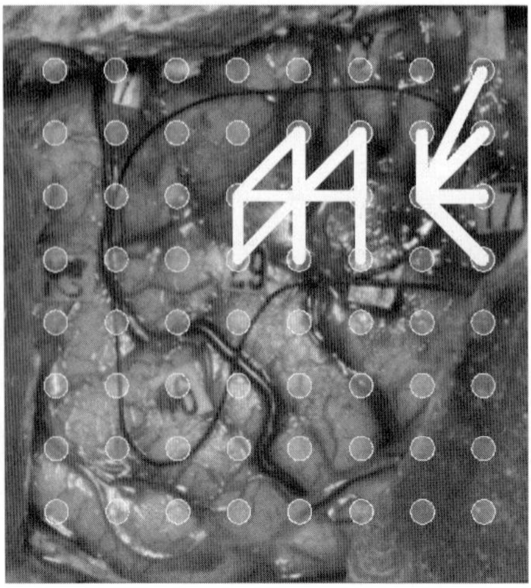

Fig. 6.8. A portion of the interictal coherence patterns calculated in the delta band recorded from an epileptic frontal lobe. The area to be resected is marked with a black string. Within that area, there was a region between electrodes 18 and 19 where the coherence was dramatically reduced. This region had a striking reduction in staining, suggesting a reduction in the number of viable neurons (The colour plate of this figure appears on page 356)

corresponded to the conventionally identified epileptic region, and the second corresponded to the anterior region that was not resected (not shown). Unlike the case of transient aphasia described above, long-distance coherences were not identified between the two areas of increased coherence, suggesting that the two areas were functioning independently. We interpret these results as indicative of a second independent focus with intermittent spiking that did not reach the clinical threshold for resection, but was evident in the coherence analysis. Histological studies of the resected tissue revealed a loss of cells in the regions associated with reduced coherence. In areas in which we observed increased lateral coherence abnormal inhibitory-like cells were identified (Fig. 6.9). The patient has experienced some reduction in her frequency of seizures, but still experiences frequent partial complex seizures and generalized seizures.

Coherence can establish functional connections between anatomical regions that can explain clinical symptomatology. For example, we studied a patient with transient aphasia as one of the symptoms of complex partial seizures. During periods of epileptic-like ECoG, unusually high coherences were observed between the subject's left mesial temporal lobe, where the tumor was located, and the left superior temporal gyrus, which had been associated with expressive language through direct cortical stimulation (Fig. 6.10). These unusual coherence patterns were not present during periods of normal ECoG patterns. The consistent phase of these long-distance coherences suggested that his speech area was being disrupted by activity arising from the

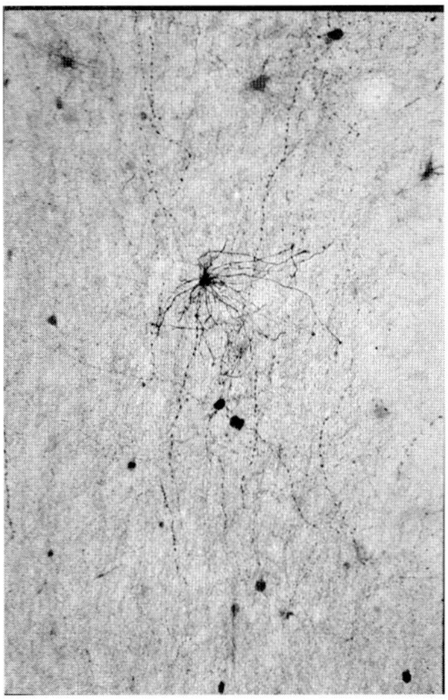

Fig. 6.9. A hypertrophic cortical cell from an epileptic region of frontal cortex that exhibited pathologically increased delta activity and increased lateral coherence (calbindin staining, calibration: 25 μm)

region near his mesial temporal mass. Although the superior temporal gyrus was not part of the epileptic system, it was occasionally being disrupted by the system during the interictal periods.

We have identified areas that appear normal in conventional EEG during the interictal period, but are found to be epileptic after stimulation or other provocative manipulations. This example is taken from recordings from the surface of the left temporal lobe of a patient with seizures due to an AVM (Fig. 6.11). The patient was a 52-year old female with a history of episodes of disorientation, short periods of speech disturbances, simple partial motor and sensori motor seizures (involving the right half of the face and the right hand) and complex partial seizures, all of which started almost two years ago. The MRI and cerebral angiogram revealed a vascular malformation in the left temporal lobe. A discolored region of cortex was observed during surgery in the left middle temporal gyrus (short black arrow). There was a clear area of spiking in the immediate region of the lesion with normal conventional recordings from the rest of the temporal lobe. However, analysis of interictal coherence patterns revealed high coherence between the spiking region and a more distant area near the anterior tip of the lobe. After the lesion and surrounding tissue was removed, a final recording revealed that the previously quiescent area near the tip began to spike vigorously. Therefore, further temporal cortex was resected in that area. The coherence patterns had revealed an association between this region and the lesional area that

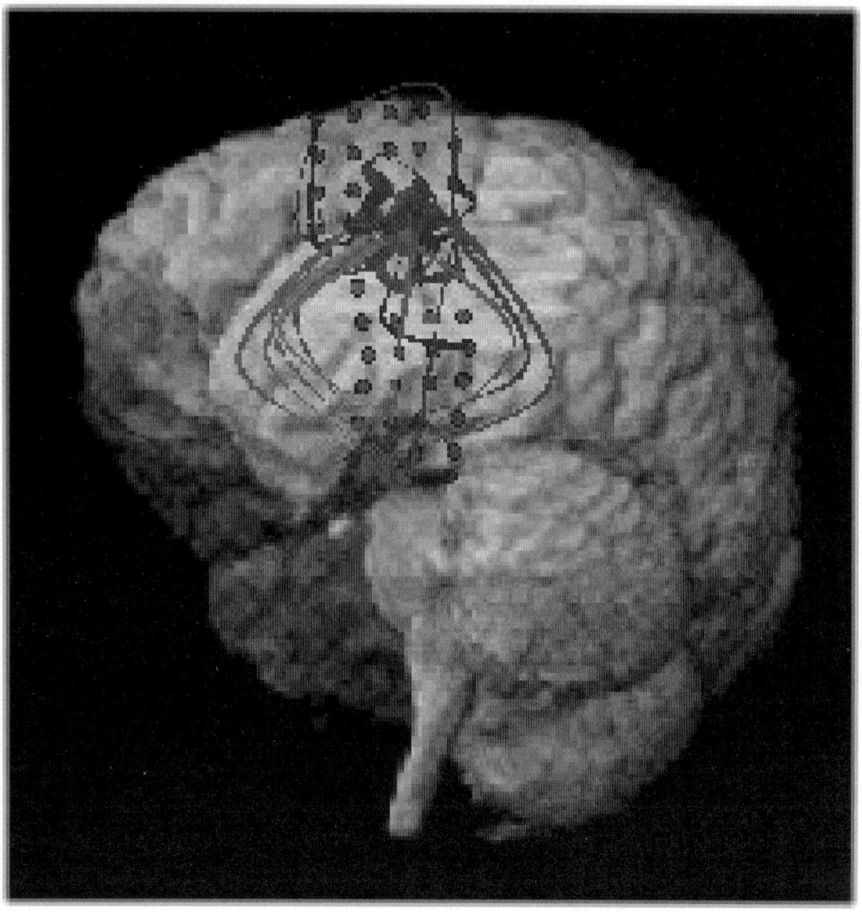

Fig. 6.10. Coherence patterns suggesting that an area near a mesial temporal lobe lesion was driving a distant area on the superior temporal gyrus that was related to expressive speech (The colour plate of this figure appears on page 359)

was not evident from visual inspection of conventional EEG recordings. The coherence revealed a hidden relationship between two areas that did not appear to be related using conventional recordings. Part of the boundaries of the epileptic system were revealed by this analysis that were not evident from conventional recordings, and the likelihood of postoperative seizures was reduced.

The application of coherence measures is driven by the concept that spatially separate anatomic sites can participate in an "epileptic system". The boundaries of such systems are difficult to define by morphological analysis, such as Brodman's areas, and are better identified by functional means. The linear coherence technique operationalizes these functional connections. As

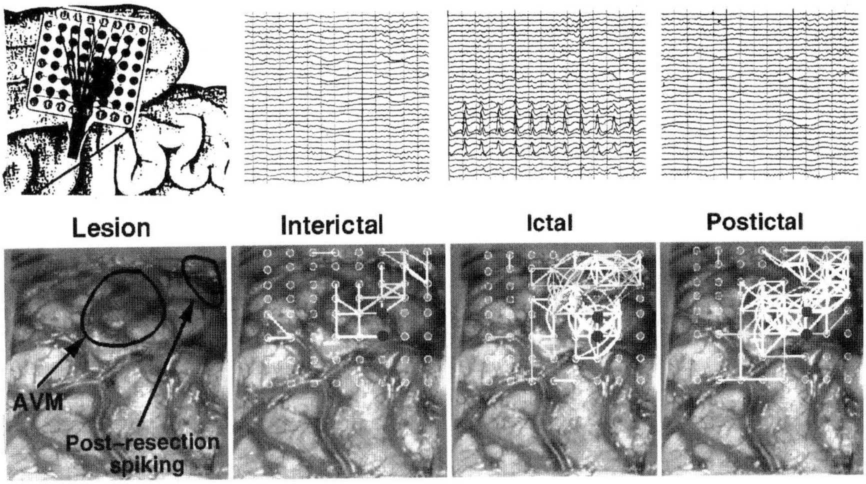

Fig. 6.11. Raw ECoG (*top*) and coherence patterns (*bottom*) observed before and during an after-discharge. The discharges in the conventional recordings were localized in the area of the AVM, but the coherence revealed functional connections between the site of the stimulation (*green dots*) and an area that began spiking only after the resection was completed. The interictal background coherence pattern heralded the appearance of the post-resection focus (The colour plate of this figure appears on page 359)

calculated here, however, coherence is sensitive to many variables. Changes in coherence can be secondary to the strength of the temporal relationship between two sites, but can also be diminished by increasing noise levels, or by an increase in the nonlinear component of the temporal relationship. This chapter has reviewed some of the evidence that the linear component of coherence is a good beginning descriptor of cortical spatio-temporal relationships. Despite substantial empirical evidence of fluctuating levels of nonlinear electrical activity during seizures, it is evident that linear coherence measures capture important activity levels of energy. The relative importance or quantification of the relative magnitude of linear and nonlinear contributors to these functional relationships is still largely unknown. Despite the caveats, the linear coherence measures reviewed here make clear the availability of important information about the functional and structural relationships involved in the production of seizures.

That stated, visual analysis of pathological waveforms and interictal spiking is still the most commonly employed basis for clinical decision making. Such analyses can be supplemented by examining the dynamics of early ictal ECoG power spectrums and lateral coherence patterns, which may yield greater insight into which areas of the brain initiate and participate in the spread of seizure activity and improve the outcome of the surgical treatment of epilepsy.

7 Synchronization of Synaptically Coupled Neural Oscillators

P.C. Bressloff and S. Coombes

The problem of epileptogenesis can be divided into two distinct but related components: the initiation of synchronous discharge, and its subsequent propagation. This division reflects clinical studies that suggest that a partial seizure originates in a localized area of cortex and either remains there or spreads to new areas. Moreover, an increasing number of experimental studies in vitro and in vivo suggest that the focus of seizure initiation involves a small fragment of a single cortical layer (possibly layer V) involving several thousand neurons that act as pacemaker cells (see [135] and references therein). This synchronous activity then rapidly spreads vertically to encompass all other layers as well as propagating horizontally to recruit neighboring cortical areas [126]. Neuronal rhythmic activity also manifests itself in various aspects of normal brain function. For example, recent experimental observations indicate that synchronization and waves of excitation can occur in cortex during sensory processing [422,423]. Other systems supporting oscillations include the reticular thalamic nucleus (RTN), which is thought to act as a pacemaker for synchronous spindle oscillations observed during sleep or anesthesia [424,425], the hippocampus [426], and central pattern generators in locomotion [427]. The characteristics of local cortical circuits are of fundamental importance in understanding the origins of seizures.[1] The major intracortical pathways have been summarized in the so-called "canonical model" of Douglas and Martin [369,428], which is shown in Fig. 7.1. Roughly speaking the neocortex has two major types of neuron. Excitatory (spiny) cells are mostly pyramidal cells, with a relatively large, polarized apical dendrite and a dense distribution of dendritic spines. The somas of pyramidal cells lie in layers 2 to 6 and have different output targets according to whether they are from layer 4 and above or from below layer 4. Their axons almost always leave the cortex to connect with various parts of the nervous system, but also send collaterals that make numerous connections within the cortex itself. Inhibitory (smooth) cells, on the other hand have a variety of dendritic morphologies, but have no apical dendrite and relatively few spines. They can be found in all cortical layers, and their axons almost never leave the cortex. The primary inhibitory transmitter is GABA. One of the most common methods of inducing experimental seizures is to compromise synaptic inhibition.

[1] Editor's note: Partial seizures of extratemporal onset are clinically quite different from psychomotor seizures. However, the basic principles of synchronization discussed in this chapter are applicable to all partial onset seizures.

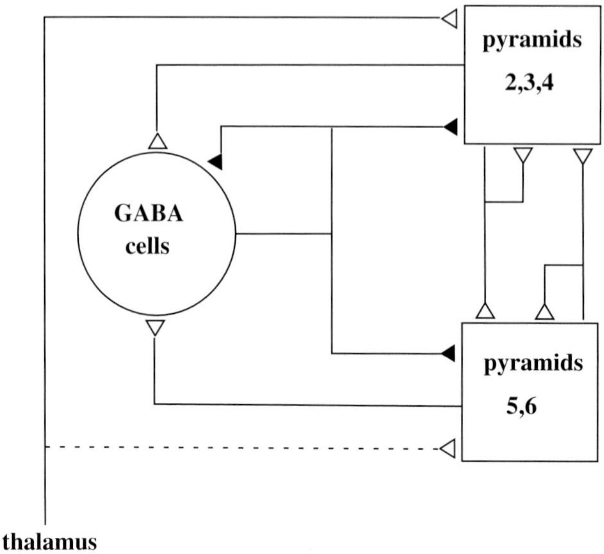

Fig. 7.1. The canonical intracortical circuit of Douglas and Martin [369,428]. The main features include reciprocal excitation among pyramidal cells both within and between layers, and both feedforward and feedback GABAergic inhibition to all cells. Open (*closed*) triangles denote excitatory (*inhibitory*) synapses

This is most easily accomplished pharmacologically using antagonists of the GABA_A receptor-channel complex such as bicuculline. Inhibitory responses on pyramidal cells are mediated by both GABA_A and GABA_B receptors. The former activate fast, large, chloride-selective conductances, and the latter open much slower, smaller, potassium-selective conductances [429,430]. High doses of bicuculline essentially eliminate the major form of synaptic inhibition. This greatly simplifies things from a modeling perspective since one need only consider populations of excitatory cells. There is also evidence that the GABA_A system is very sensitive to brain state and pathology, and even slight changes in its inhibitory efficacy can lead to profound modifications in cortical excitability [431]. However, certain caution should be exercised in considering such a simplified view of disinhibition, since the underlying pathology in various types of seizures is very likely to involve many cellular and molecular mechanisms. The principal cells thought to participate in epileptic seizures are excitatory pyramidal cells and these can display both regular spiking and bursting patterns of output activity [158], see Fig. 7.2. It has been proposed that intrinsically bursting pyramidal cells in layer 5 play a major role in epileptogenesis, since they have the ability to generate a relatively large excitatory output (burst) with a very small increase in stimulus intensity [135,136,157]. Although direct evidence for this hypothesis has not been obtained in the case of cortical oscillations, experiments in-vitro

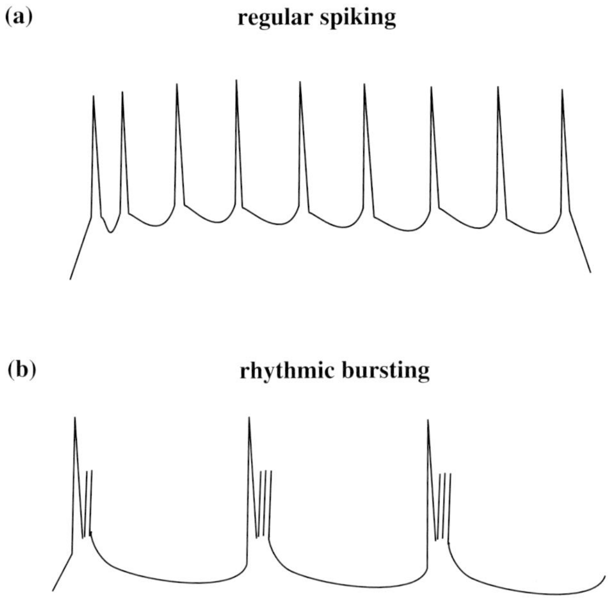

Fig. 7.2. Sketch of intrinsic firing patterns found in neocortical pyramidal cells: (**a**) regular spiking, (**b**) rhythmic bursting

have established that intrinsic bursters can trigger a full-blown, synchronized epileptiform event in the CA3 region of the hippocampus [426,432,433].

Many of the wide variety of burst phenomena observed in neural systems appear to be due to a similar underlying mechanism. First note that the various chemical and electrical dynamics of a neuron operate on many timescales, and for bursting neurons the dynamics can often be dissected into a fast system coupled to a slowly oscillating subsystem. Typically the fast system has a time-scale of milliseconds and models the membrane potential, and hence spike generation. The slow subsystem operates on a time-scale of tens of seconds and models transmembrane ionic currents. The fast system is modulated by the slow one, and has two parameter regimes: a stationary state or *resting potential*, and a periodic state during which action potentials are generated. Thus, for bursting to occur, the slow variable must parameterize bifurcations in the fast system. Bursting phenomena may be sub-classified according to the underlying wave form of the oscillatory component, e.g. triangular, parabolic and elliptic bursting (for a review see [434]). Unfortunately, from a modeling perspective, very little is known about how networks of synaptically coupled intrinsic bursters synchronize their activity. Much more is known concerning the synchronization properties of regular spiking neurons. Therefore, in this chapter we shall concentrate on the latter. We expect, however, that many of the results will generalize to the case of intrinsic bursters, particularly when the intra-burst interval is significantly shorter than the inter-burst interval as seen in Fig. 7.2b. It should also be noted that some forms of bursting phenomena may emerge at the network level rather than through intrinsic mechanisms at the single-cell level [435].

A significant feature of seizures from a clinical perspective is that the initial focus of abnormal activity can sometimes propagate and eventually usurp normal function over most, or even all, of the cortex. A number of recent experimental studies have revealed the propagation of traveling bursts of activity in slices of excitable neural tissue taken from the cortex [126,127], as well as the hippocampus [426,436] and the thalamus [424]. The underlying mechanism for propagation of these waves is thought to be synaptic in origin rather than diffusive as in the propagation of action potentials. When $GABA_A$ receptors are strongly blocked with bicuculline, synchronized discharges can be evoked by a weak electrical stimulus from any site on a cortical slice. Following rapid vertical propagation, each discharge propagates away from the stimulus in both horizontal directions at a mean velocity of about 6–9 cm/s [126]. The phenomenology of horizontal propagation is primarily determined by the spatial patterns of excitatory connections, which have a large variation between and within cortical areas. For example, spatial periodicity of waves in cat cortex appear to closely fit the periodicity of orientation columns [126]. Moreover, discharges occasionally jump across a region of cortex suggesting that there are feedforward excitatory connections with patchy discontinuities.

Computational models of synaptically generated waves have also been developed, in which neural tissue is treated as a one-dimensional continuum [127,436–438]. These various studies indicate that wave propagation in cortical (and hippocampal) slices only occurs if the synaptic strength of neuronal interactions exceeds some threshold; well beyond threshold the velocity is approximately linear in the coupling. Moreover, multiple forms of discharge patterns are observed to propagate at the same parameter values. Selection of a particular pattern occurs through the initial conditions such as the strength of the initial stimulus. Interestingly, there is not much variation in the speed as a function of the number of spikes that propagate. The mechanism of wave propagation in thalamic slices appears to be different from cortical and hippocampal tissue; thalamic waves propagate in a lurching manner resulting in an approximately logarithmic dependence of velocity on synaptic coupling strength and there is no minimal wave velocity [438].

Motivated by the above discussion, we shall address three particular theoretical issues related to synchronous oscillations in cortex:

1. *What are the intrinsic biophysical mechanisms underlying oscillatory activity in individual cells?* In Sect. 7.2 we describe how limit cycle oscillations (regular spiking) can arise in conductance-based models due to constant external current injection. We also discuss various aspects of the response properties of single neural oscillators that play an impor-

tant role at the network level. One such property is the *phase resetting curve* or PRC, which can be found by perturbing a neuron with a brief depolarizing stimulus at different points on its limit cycle and measuring the resulting phase-shift [188,439] (see also Chap. 16). For certain types of neuron such a stimulus always advances the onset of the next spike, that is, the PRC is always positive, whereas for others the stimulus may also delay the next spike. Differences in the nature of a neuron's phase resetting curve can have a major effect on phase dynamics at the network level [440]. Finally, the notions of frequency and phase locking are introduced by considering periodically forced oscillators. We illustrate this with the particular example of an integrate-and-fire neuron.

(ii) *How can a localized pool of neural oscillators synchronize their activity through excitatory synaptic interactions?* In Sect. 7.3 we investigate the important role of (a) intrinsic membrane properties and (b) synaptic/axonal delays in determining whether or not a pool of excitatory neural oscillators can synchronize. We also describe methods for reducing the complexity of single-neuron models in order to facilitate the analysis of network behavior. One such method is averaging theory [441], which reduces the network dynamics to a system in which the relative phase between oscillators is the relevant dynamical variable [440,442–444]. In neural models the effective phase interaction may be regarded as a convolution (over one period of oscillation) of the postsynaptic current and the phase resetting curve. The method of averaging can be used for sufficiently weak coupling such that, in an appropriate coordinate frame, relative phases evolve on a much slower time scale compared to the natural frequency of oscillation in the uncoupled limit. For higher levels of coupling, where the method of averaging breaks down, it is still possible to reduce certain classes of biophysical models to an effective phase model [442]. However, the interaction between neurons now involves the direct product of the PRC with the post-synaptic current such that the phases no longer occur as differences. This generally makes the analysis far more difficult. One way to proceed is to assume the following: (i) there exists a precise threshold condition (such as the phase variable passing through zero with positive velocity) that signals the firing of an action potential spike; (ii) the significant information transmitted between neurons is the sequence of spike firing times. Under these assumptions, it is possible to study synchronization and phase-locking outside the weak coupling regime of averaging theory in terms of the underlying firing time dynamics [445–448].

2. *How can activity propagate to other areas of cortex?* Recently Ermentrout (1998) has established that a simple integrate-and-fire (IF) model of a neuron captures much of the process underlying an excitation wave in cortical slices. This is due to the fact that the velocity of a wave in the case of strong synaptic coupling is essentially independent of the ionic details of cell membranes, depending mainly on the integration rise

time from the resting potential to threshold. In Sect. 7.4 we consider the propagation of a localized pulse of activity along a one-dimensional network of integrate-and-fire neurons with long-range excitatory synaptic coupling. Solitary wave solutions are found to come in pairs with the faster (slower) solution stable (unstable) in the case of zero axonal delays; for non-zero delays and fast synapses the stable wave can itself destabilize via a Hopf bifurcation [449]. Interestingly, this destabilization results in the generation of so-called lurching waves thought to play an important role in thalamocortical spindle waves [450].

7.1 Neural Oscillator Models

7.1.1 Conductance-based Models

The standard dynamical system for describing a neuron with spatially constant membrane potential V is based upon conservation of electric charge, so that

$$C\frac{\mathrm{d}V}{\mathrm{d}t} = -F + I_{\mathrm{s}} + I\,, \tag{7.1}$$

where C is the cell capacitance, F the membrane current, I_{s} the sum of external synaptic currents entering the cell and I describes any external injected currents. In the original Hodgkin–Huxley model [451] the membrane current arises mainly from the conduction of sodium and potassium ions through voltage-dependent channels in the membrane. The contribution from other ionic currents is assumed to obey Ohm's law. In fact F is considered to be a function of V and of three time- and voltage-dependent conductance variables m, n and h:

$$F(V,m,n,h) = g_{\mathrm{L}}(V - V_{\mathrm{L}}) + g_{\mathrm{K}}n^4(V - V_{\mathrm{K}}) + g_{\mathrm{Na}}hm^3(V - V_{\mathrm{Na}})\,, \tag{7.2}$$

where g_{L}, g_{K} and g_{Na} are constants and V_{L}, V_{K} and V_{Na} represent the constant membrane reversal potentials associated with the leakage, potassium and sodium channels, respectively. The conductance variables m, n and h take values between 0 and 1 and approach the asymptotic values $m_\infty(V)$, $n_\infty(V)$ and $h_\infty(V)$ with time constants $\tau_m(V)$, $\tau_n(V)$ and $\tau_h(V)$ respectively. Summarizing, we have that

$$\tau_X(V)\frac{\mathrm{d}X}{\mathrm{d}t} = X_\infty(V) - X, \qquad \text{with } X \in \{m,n,h\}\,. \tag{7.3}$$

The six functions $\tau_X(V)$ and $X_\infty(V)$, $X \in \{m,n,h\}$, are obtained from fits with experimental data. It is common practice to write

$$\tau_X(V) = \frac{1}{\alpha_X(V) + \beta_X(V)}\,, \qquad X_\infty(V) = \alpha_X(V)\tau_X(V)\,. \tag{7.4}$$

The parametric form for α_X and β_X is given in the Appendix.

Recently a systematic approach for reducing the number of degrees of freedom in the Hodgkin–Huxley model has been developed using the method of equivalent potentials [452,453]. First, it is well known that the timescale associated with changes in m and τ_m is much smaller than those associated with h and n. Thus m will reach its asymptotic value rapidly, suggesting the replacement of m by its asymptotic value $m_\infty(V)$. This instantaneous approximation reduces the number of dynamic variables from 4 to 3 at the expense of accuracy over very short timescales. A similar reduction for n and h is not appropriate since this would destroy the ability of the model to generate action potentials. A further reduction is possible, however, with the introduction of two auxiliary potentials U_n and U_h. The basic idea is to mimic the slower approach of n and h to their asymptotic values by the replacement $X = X_\infty(U_X)$, $X \in \{n, h\}$. This equation can always be solved exactly for U_X since the functions X_∞ are monotonic and hence invertible. In practice it is found that the auxiliary potentials U_h and U_n are remarkably similar. A further reduction is now possible with the replacement of the U_X by an average over the two potentials. In fact it is simpler still to assume that they are identical and let $U_h = U_n = U$. Hence, the reduced model is now two-dimensional with a membrane current $f(V, U) = F(V, m_\infty(V), n_\infty(U), h_\infty(U))$. By demanding that the time dependence of f in the reduced model mimic the time dependence of F in the full model at constant V we have that

$$C\frac{\mathrm{d}V}{\mathrm{d}t} = -f(V,U) + I_\mathrm{s} + I \tag{7.5}$$

$$\frac{\mathrm{d}U}{\mathrm{d}t} = g(V,U)\,, \tag{7.6}$$

where

$$g(V,U) = \frac{\dfrac{\partial F}{\partial h}\left[\dfrac{h_\infty(V) - h_\infty(U)}{\tau_h(V)}\right] + \dfrac{\partial F}{\partial n}\left[\dfrac{n_\infty(V) - n_\infty(U)}{\tau_n(V)}\right]}{\dfrac{\partial f}{\partial h_\infty}\dfrac{\mathrm{d}h_\infty(U)}{\mathrm{d}U} + \dfrac{\partial f}{\partial n_\infty}\dfrac{\mathrm{d}n_\infty(U)}{\mathrm{d}U}}\,, \tag{7.7}$$

and $\partial F/\partial h$ and $\partial F/\partial n$ are evaluated at $h = h_\infty(U)$ and $n = n_\infty(U)$. The variable V describes the capacitive nature of the cell whilst U describes the time dependence of the membrane conductance. In fact U may be regarded as a variable responsible for the refractory period of a neuron. The reduction to a two-dimensional system allows a direct visualization of the dynamics by plotting the flow in the (V, U) plane. A plot of the nullclines $\mathrm{d}V/\mathrm{d}t = 0$ and $\mathrm{d}U/\mathrm{d}t = 0$ (see Fig. 7.3) allows the visualization of both the fixed point and the flow. In the reduced model $g(V, V) = 0$ and the nullcline for $\mathrm{d}U/\mathrm{d}t$ is simply the straight line $U = V$. The fixed point is defined by the intersection of the two nullclines. For zero external input the fixed point is stable and the neuron is said to be *excitable*. When a positive external current is applied the low-voltage portion of the $\mathrm{d}V/\mathrm{d}t$ nullcline moves up whilst the high-voltage part remains relatively unchanged. For sufficiently large constant external

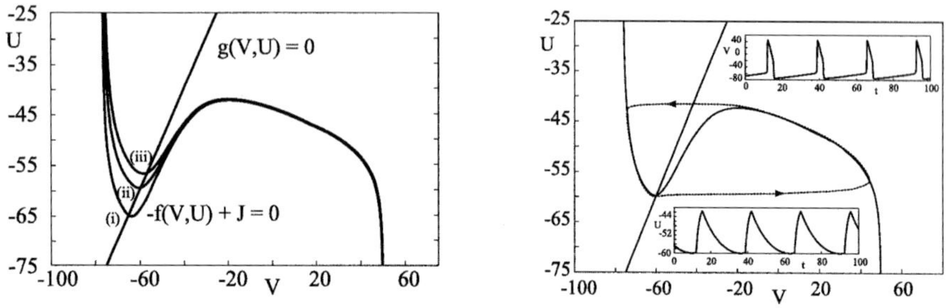

Fig. 7.3. (**a**) Nullclines of the reduced neuron model with (i) $I = 0$, (ii) $I = 10$ and (iii) $I = 20$. (**b**) The reduced neuron model in an oscillatory regime ($I = 10$) capable of generating a train of spikes inset for dynamics of V and the refractory variable U)

input the intersection of the two nullclines falls within the portion of the $\mathrm{d}V/\mathrm{d}t = 0$ nullcline with positive slope (see Fig. 7.3). In this case the fixed point is unstable and the system may support a limit cycle as shown in Fig. 7.3. The system is said to be oscillatory as it may produce a train of action potentials (see the inset for V in Fig. 7.3b). Action potentials may also be induced in the absence of an external current for synaptic stimuli of sufficient strength and duration. Another example of a planar system that behaves in much the same way as the Hodgkin–Huxley equations in a certain parameter regime is the Morris–Lecar model [454], which was originally introduced to describe how under constant-current injection barnacle muscle fibers respond with a host of oscillatory voltage waveforms.

$$\frac{\mathrm{d}V}{\mathrm{d}t} = I - g_{\mathrm{L}}(V - V_{\mathrm{L}}) - g_{\mathrm{K}} w (V - V_{\mathrm{K}}) - g_{\mathrm{Ca}} m_\infty(V)(V - V_{\mathrm{Ca}}), \tag{7.8}$$

$$\frac{\mathrm{d}w}{\mathrm{d}t} = \lambda(V)(w_\infty(V) - U), \tag{7.9}$$

with

$$m_\infty(V) = 0.5(1 + \tanh[(V - V_1)/V_2]), \tag{7.10}$$

$$w_\infty(V) = 0.5(1 + \tanh[(V - V_3)/V_4]), \tag{7.11}$$

$$\lambda(V) = \phi \cosh[(V - V_3)/(2V_4)]. \tag{7.12}$$

Here, g_{L} is the leakage conductance, $g_{\mathrm{K}}, g_{\mathrm{Ca}}$ are the potassium and calcium conductances, $V_{\mathrm{L}}, V_{\mathrm{K}}, V_{\mathrm{Ca}}$ are corresponding reversal potentials, $m_\infty(V)$, $w_\infty(V)$ are voltage-dependent gating functions and $\lambda(V)$ is a voltage-dependent rate.

7.1.2 Class I and Class II Excitability

It is useful to distinguish between two mechanisms whereby excitable membranes go from resting to repetitive firing as current is injected, which we

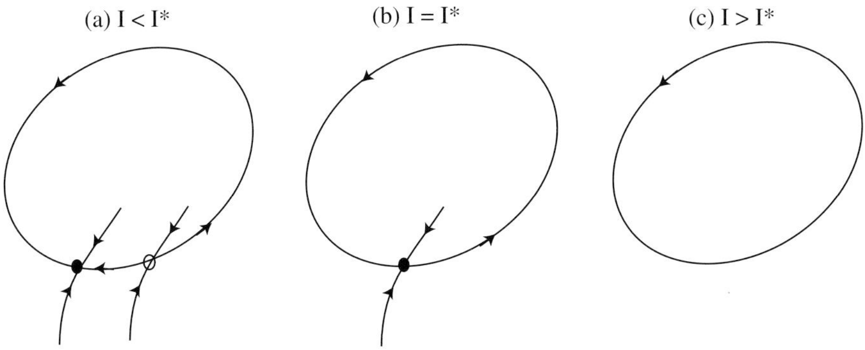

Fig. 7.4. Schematic diagram of a saddle-node bifurcation on a limit cycle occurring as the applied current I is increased. (**a**) For $I < I^*$ there is a unique asymptotically stable fixed point and a saddle-point with a one-dimensional unstable manifold whose branches form a closed loop. (**b**) At criticality, $I = I^*$, the node and saddle coalesce. (**c**) For $I > I^*$ all that remains is a closed limit cycle

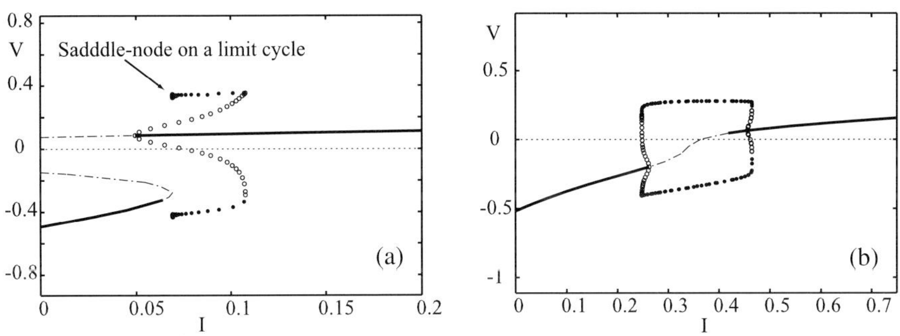

Fig. 7.5. Bifurcation diagram of (**a**) the Class I Morris–Lecar model and (**b**) the Class II Morris–Lecar model as a function of the external drive I. *Filled circles* show amplitude of stable limit cycle, *open circles* indicate unstable limit cycle behavior. *Solid line* shows stable fixed point, *thin line* shows unstable fixed point behavior

denote by Class I and Class II excitability respectively [455]. Class I excitability is characterized by the appearance of oscillations with arbitrary low frequency as current is injected, and typically occurs via a saddle-node on a limit cycle bifurcation as illustrated schematically in Fig. 7.4. On the other hand, for Class II excitability the onset of repetitive firing occurs at a non-zero frequency due to a stable equilibrium losing stability via a Hopf bifurcation. This is the mechanism underlying the formation of a limit cycle in Fig. 7.3 for the (reduced) Hodgkin–Huxley model. The Morris–Lecar model can exhibit both forms of excitability as illustrated by the bifurcation diagrams in Fig. 7.5. The particular parameter values used are given in the Appendix.

7.1.3 Phase Resetting Curves

One easily measurable property of a neural oscillator is its phase resetting curve (PRC), which we denote by the 2π-periodic function $R(\theta)$. The PRC is found experimentally (or numerically) by perturbing the oscillation with a brief depolarizing stimulus $\Delta I(t)$ at different times in its cycle and measuring the resulting phaseshift from the unperturbed system (see Chap. 16). Suppose that the unperturbed oscillator evolves on an attracting limit cycle, which is parameterized by a phase variable $\theta \in S^1$ such that $\theta(t) = \overline{\Omega} t$ with $\overline{\Omega}$ the natural frequency of oscillation. The perturbed dynamics then reduces to the following phase equation:

$$\frac{\mathrm{d}\theta(t)}{\mathrm{d}t} = \overline{\Omega} + \Delta I(t) R(\theta(t)) \,. \tag{7.13}$$

For certain types of neuron a depolarizing stimulus always advances the onset of the next spike, that is, the PRC is always positive, whereas for others the stimulus may also delay the next spike. We refer to models with a strictly positive PRC (such as Class I Morris–Lecar) as being of Type I and those for which the PRC has a negative regime (such as Hodgkin–Huxley and Class II Morris–Lecar) as being of Type II. Numerical examples illustrating both types of PRC are shown in Figs. 7.6–7.8. In certain cases it is possible to obtain analytical expressions for $R(\theta)$ by performing an appropriate change of coordinates [442,456]. Indeed, in the case of an oscillating neuron with Class I excitability close to the critical point (see Fig. 7.2) one has the universal result that $R(\theta) \sim 1 + \cos(\theta)$, which is a Type I PRC [457].

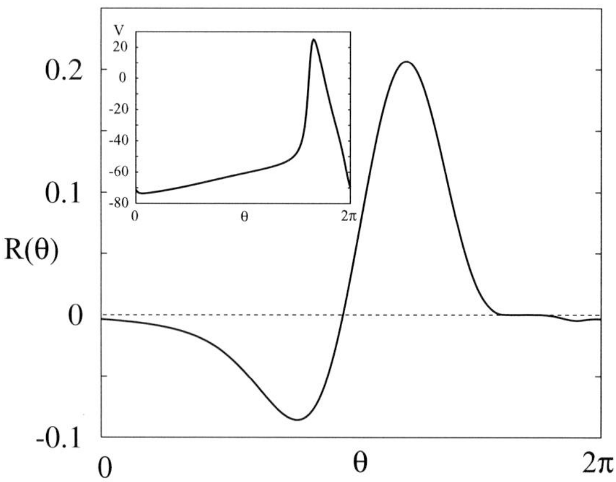

Fig. 7.6. Hodgkin–Huxley model with external drive $I = 10$ showing Type II phase resetting curve. Corresponding orbit over a single cycle is shown in the inset

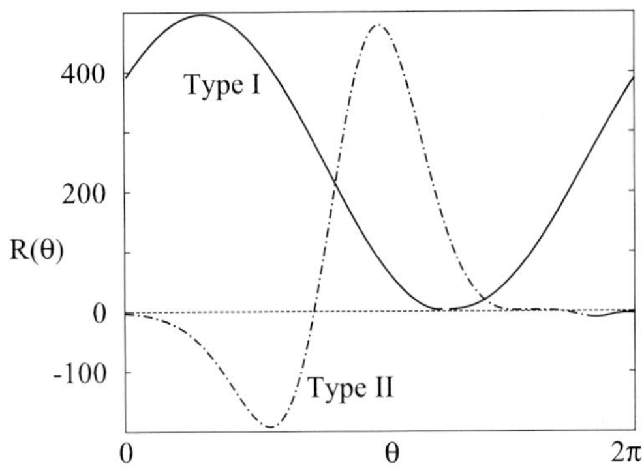

Fig. 7.7. Morris–Lecar model showing two different response types. Parameters given in the Appendix. Responses have been scaled to the same ranges

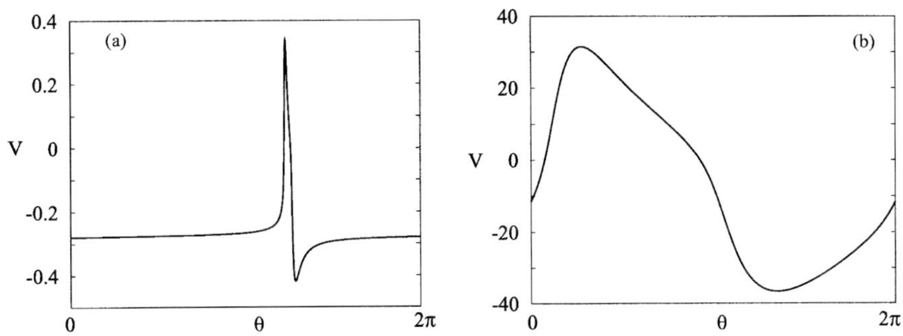

Fig. 7.8. Orbit of the (**a**) type I and (**b**) type II response curves shown in Fig. 7.7 for the Morris–Lecar model

7.1.4 Periodic Forcing: Frequency-locking, Phase-locking and Synchronization

It is known that nerve membranes can exhibit a variety of responses to periodic stimulation whether it is in the form of a smooth alternating current or the injection of current pulses [439]. Among the most commonly observed and studied behaviors are mode-locked states in which a repetitive regular output is entrained to some multiple of the forcing period (see also Chap. 8). In addition to mode-locked rhythms it is often possible to observe irregular or aperiodic rhythms in which entrainment is non-existent. Interestingly for many biological oscillators the stable behaviors that are most commonly observed correspond to low-order ratios between the number of cycles of forcing and the subsequently generated rhythm. Moreover, the stable rhythms can be organized in a regular fashion (so-called Arnold tongues) such that aperiodic motion is often generated in regions of parameter space separating rhythmic behavior [439]. For a weak periodic forcing we can use the phase equation (7.13), which we write in the convenient form

$$\frac{d\theta_1}{dt} = \Omega_1 + \epsilon R(\theta_1)F(\theta_2)\,, \tag{7.14}$$

$$\frac{d\theta_2}{dt} = \Omega_2\,, \tag{7.15}$$

where $\theta_1(t)$ is the phase of the oscillator, F is a prescribed forcing function with frequency Ω_2 and phase $\theta_2(t) = \Omega_2 t+$ constant, and ϵ determines the strength of forcing. It is useful at this stage to introduce some terminology (see also the discussion in Chap. 9 of [458]). The oscillator is said to be *frequency locked* when it has a stable periodic solution; such a solution is of type (p, q) if θ_1 makes q rotations while θ_2 makes p rotations. A $p\!:\!q$ frequency-locked solution is said to be (p, q) *phase-locked* if

$$q\theta_1(t) - p\theta_2(t) = \text{constant}\,.$$

In the square $[0, 2\pi] \times [0, 2\pi]$ trajectories consist of straight lines with slope q/p (see Fig. 7.9).

Of particular interest from the viewpoint of this chapter is the *synchronous* or in-phase state, which is a $(1, 1)$ phase-locked state with $\theta_1 = \theta_2$. Suppose that $\Delta\Omega \equiv \Omega_1 - \Omega_2 = \mathcal{O}(\epsilon)$ and ϵ is sufficiently small. Let $\Phi = \theta_1 - \theta_2$. Since $\mathrm{d}\Phi/\mathrm{dt} = \mathcal{O}(\epsilon)$, we can apply averaging theory [441] to obtain the approximation

$$\frac{\mathrm{d}\Phi}{\mathrm{d}t} = \Delta\Omega + \epsilon H(\Phi)\,, \tag{7.16}$$

where

$$H(\Phi) = \frac{1}{2\pi}\int_0^{2\pi} R(\theta)F(\theta - \Phi)\mathrm{d}\theta\,. \tag{7.17}$$

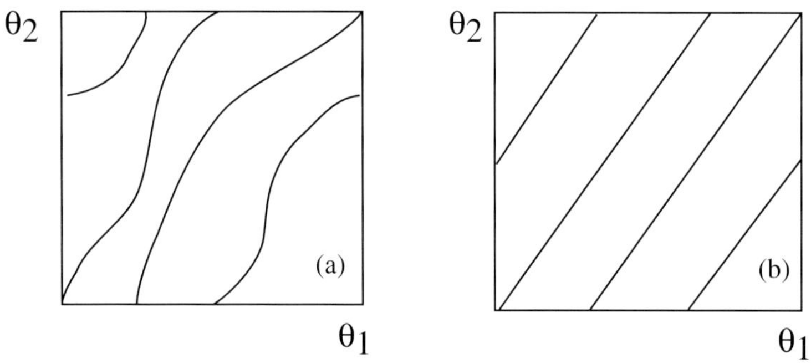

Fig. 7.9. Representation of a type $(2, 3)$ frequency-locked solution for (**a**) non-phase-locked and (**b**) phase-locked cases

The synchronous state exists if $\Delta\Omega+\epsilon H(0)=0$ and is stable if $H'(0)<0$. Note that the above ideas can be extended to the case of mutually coupled phase oscillators (see Sect. 7.3). Phase locking of a neural oscillator to a periodic stimulus can also be formulated without recourse to a phase description. Suppose that there exists a well-defined threshold κ signaling the onset of fast somatic membrane depolarization and the subsequent firing of an action potential spike. If T^n denotes the nth firing time (threshold-crossing time) of a neuron then

$$V(T^n)=\kappa, \quad \frac{dV}{dt}(T^n)>0\,. \tag{7.18}$$

Since the shape of an action potential is fixed for a given model, we can characterize the output of the neuron in terms of its sequence of firing times $\{T^n\}$. For example, in the case of a regular oscillator (with a constant input) as illustrated in Fig. 7.3b we have $T^n=n\overline{T}+\text{constant}$ where $\overline{T}$ is the period between spikes. More complex patterns of spikes can be generated in the presence of a periodic stimulus of period Δ, say. We define a phase-locked solution to be one for which the output firing times are of the form

$$T^n=\left([n/q]-\phi_{n(q)}\right)p\Delta\,, \quad n(q)=n \bmod q\,, \tag{7.19}$$

where [.] denotes the integer part and the $\phi_{n(q)}\in[0,1)$ denote a collection of firing phases. Such a solution corresponds to the firing of q spikes every p cycles of the input (see Fig. 7.10). If we define the mean firing period $\langle\Delta\rangle$ according to

$$\langle\Delta\rangle=\lim_{N\to\infty}\frac{1}{N}\sum_{n=1}^{\infty}\left(T^{n+1}-T^n\right)\,, \tag{7.20}$$

then the solution (7.19) satisfies $\langle\Delta\rangle=(q/p)\Delta$.

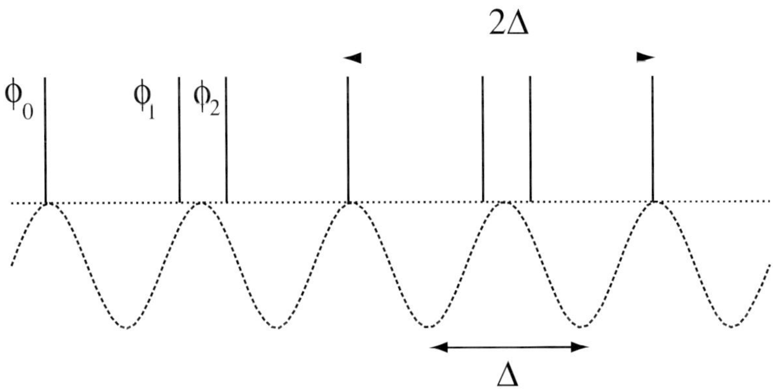

Fig. 7.10. An example of a 2:3 mode-locked solution in which a neuron fires 3 spikes (with phases ϕ_1,ϕ_2,ϕ_3) for every two cycles of the input

7.1.5 Integrate-and-fire Model

The integrate-and-fire (IF) model provides a caricature of the capacitative nature of cell membrane at the expense of a detailed model of the refractory process [459]. The IF model satisfies (7.1) with $F = F(V)$ together with the condition that whenever the neuron reaches threshold κ, it fires and V is immediately reset to the resting potential V_0. Using a curve-fitting procedure it is possible to approximate $F(V)$ by a cubic, $F(V) = a(V - V_0)(V - V_1)(V - \kappa)$ where the constants a, $V_{0,1}$ and κ can be determined explicitly from a reduction of the underlying Hodgkin–Huxley equations [452]. However, $F(V)$ is usually reduced further by considering the linear approximation $F(V) = (V - V_0)/\tau$. We thus obtain the equation (on setting $V_0 = 0$ and $C = 1$)

$$\frac{\mathrm{d}V}{\mathrm{d}t} = -\frac{V}{\tau} + I \,, \quad T^m < t < T^{m+1} \,, \tag{7.21}$$

supplemented by the reset condition (cf. (2.18))

$$V(T_+^m) = 0 \quad \text{whenever} \quad V(T_-^m) = \kappa \,. \tag{7.22}$$

Suppose that we decompose the external input as $I = I_0 + \Delta I(t)$ where I_0 is constant. In the absence of the perturbation $\Delta I(t)$, the IF neuron fires spikes at regular intervals $T^m = m\overline{T}$ (up to an arbitrary phase), where $\overline{T} = \tau \ln(\tau I_0/[\tau I_0 - \kappa])$. Under the nonlinear transformation

$$\theta(t) \bmod 2\pi = \frac{2\pi\tau}{\overline{T}} \ln\left(\frac{\tau I_0}{\tau I_0 - V(t)}\right) \,, \tag{7.23}$$

(7.21) naturally reduces to the phase equation (7.13) with

$$R(\theta) = \frac{2\pi}{\kappa\overline{T}}\left[1 - \mathrm{e}^{-\overline{T}/\tau}\right] \mathrm{e}^{\overline{T}\theta/2\pi\tau} \tag{7.24}$$

for $0 \leq \theta < 2\pi$ and $R(\theta + 2\pi k) = R(\theta)$ for all integers k.

Now suppose that the IF neuron has a sinusoidally modulated input of the form $I(t) = I_0 + \epsilon \sin(2\pi t)$. This case has been studied in some detail by Keener et al. [459], who showed that the sequence of firing times is generated by the first-order implicit mapping

$$F(T^{m+1}) = F(T^m) + \mathrm{e}^{T^m/\tau} \,, \tag{7.25}$$

with

$$F(t) = \mathrm{e}^{t/\tau}\left(I_0\tau - 1 - \frac{\epsilon\tau}{\sqrt{1 + 4\pi^2\tau^2}} \sin(\eta - 2\pi t)\right) \,, \quad \tan\eta = 2\pi\tau \,. \tag{7.26}$$

When $I_0 - \epsilon > \tau^{-1}$, the map F is non-invertible, and it is known that mode-locking can occur almost everywhere, that is, on the complement of a set of measure zero [459]. In Fig. 7.11a we plot $\langle\Delta\rangle^{-1}$ as a function of τ with $\epsilon = I_0 - \tau^{-1}$. The resulting devil's staircase structure shows that the preferred

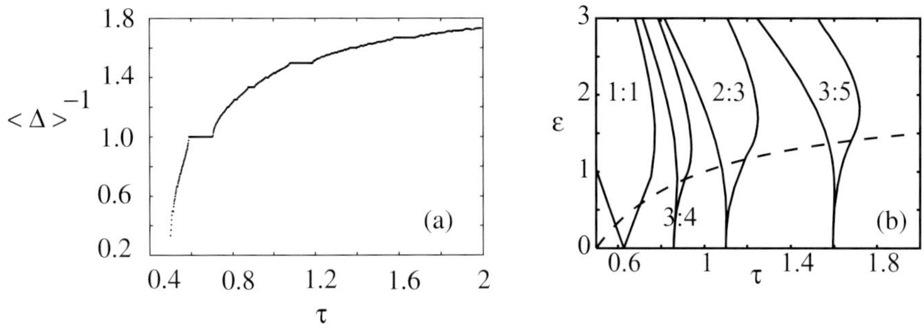

Fig. 7.11. (**a**) Plot of the inverse mean firing period as a function of the decay rate τ for $I_0 - \epsilon = \tau^{-1}$ with $I_0 = 2$ (i.e. along the border between invertible and non-invertible regimes of the firing time map). Note that the dominant solutions are $1:1$, $3:4$, $2:3$ and $3:5$ (with increasing τ). (**b**) The Arnold tongue structure for the dominant modes

mode-locked solutions are those with low ratios of q to p. In Fig. 7.11b we plot the borders of the regions (Arnold tongues) where these dominant modes become unstable (via a saddle-node bifurcation) in the (τ, ϵ)-plane. Note that in the standard IF model the Arnold tongues do not overlap even for large ϵ and thus do not support chaotic dynamics, in contrast to more complex conductance-based models with periodic forcing. (The explicit construction of Arnold tongues for IF neurons in [460] suggests that they do overlap). However, this analysis has to be supplemented by an additional condition that takes into account the presence of discontinuities in the underlying firing time map as highlighted in [459].

7.2 Mutual Synchronization in a Network of Synaptically Coupled Oscillators

7.2.1 Network Model with Synaptic Coupling

We now consider a network of oscillating neurons that interact via synapses by transmitting spike trains to one another. Let $V_i(t)$ denote the state of the ith neuron at time t, $i = 1, ..., N$, where N is the total number of neurons in the network. For pedagogical reasons we shall assume that the oscillators are identical, that is, they have the same natural frequencies, phase resetting curves etc.. Suppose that the variables $V_i(t)$ evolve according to the equations (cf. (7.1))

$$C\frac{\mathrm{d}V_i(t)}{\mathrm{d}t} = I + \sum_p g_p(V_i,t)(V_p - V_i(t)) + \epsilon X_i(t)[V_\mathrm{s} - V_i(t)]\,, \tag{7.27}$$

for $i = 1, \ldots, N$, where I is a constant applied current, and g_p are voltage and time-dependent conductances associated with the opening and closing of ionic channels with corresponding membrane reversal potentials V_p. These ionic conductances can be modeled in terms of one or more auxiliary variables along the lines of Hodgkin–Huxley, see Sect. 7.3. The final term on the right-hand side of (7.1) represents the total synaptic conductance X_i induced by incoming spike trains from other neurons in the network, with V_s the synaptic membrane reversal potential. In the case of excitatory (inhibitory) synapses $V_\mathrm{s} > 0$ ($V_\mathrm{s} < 0$). As a further simplification, we shall take $V_\mathrm{s} - V_i(t) \approx V_s$ and absorb V_s into the global coupling parameter ϵ that characterizes the strength of the interactions.

The synaptic current $X_i(t)$ is generated by the arrival of spikes from other neurons in the network, and is taken to have the explicit form

$$X_i(t) = \frac{1}{N-1} \sum_{j \neq i} \sum_{m=-\infty}^{\infty} J(t - T_j^m) , \tag{7.28}$$

where $J(t)$ determines the time-course of the postsynaptic response to a single spike with $J(t) = 0$ for $t < 0$, and T_j^m, integers m, denotes the sequence of firing times of the jth neuron such that

$$V_j(T_j^n) = \kappa, \qquad \frac{\mathrm{d}V_j}{\mathrm{d}t}(T_j^n) > 0 . \tag{7.29}$$

For the sake of illustration we consider the case of all-to-all coupling. A biologically motivated choice for $J(t)$ is the alpha-function

$$J(t) = s(t - \tau_\mathrm{a}), \quad s(t) = \alpha^2 t \mathrm{e}^{\alpha t} \Theta(t) , \tag{7.30}$$

where α is the inverse rise-time of synaptic response, τ_a is a discrete axonal propagation delay, and Θ is the unit step-function. Equation (7.30) is the limiting case of the more general synaptic response function $s(t) \sim (\mathrm{e}^{-\alpha_1 t} - \mathrm{e}^{-\alpha_2 t})\, \Theta(t)$ that has distinct rise and fall times. Note that the synaptic conductance could be modeled more comprehensively in terms of a system of differential equations based on the presynaptic potential [461]. However, we assume here that the important information transmitted between neurons concerns the firing times T_j^m and that the detailed shape of an action potential is not important.

In principle one could solve (7.27)–(7.29) to obtain a nonlinear iterative mapping for the firing times. However, in practice it is necessary to carry out some form of reduction. Suppose that in the absence of any coupling each neuron acts as an oscillator firing spikes at a constant rate $\overline{\Omega}$. In other words, $T_j^n = n\overline{T} + \text{constant}$ where $\overline{\Omega} = 2\pi/\overline{T}$. Treating the synaptic input as a perturbation along the lines of Sect. 7.2.3, we assume that (7.27) can be reduced to the following equation using an appropriate change of coordinates:

$$\frac{\mathrm{d}\theta_i(t)}{\mathrm{d}t} = \overline{\Omega} + \epsilon X_i(t) R(\theta_i(t)) , \tag{7.31}$$

where $\theta_i \in S^1$ and R is the phase resetting curve of the neural oscillators. (The degree of validity of (7.31) will usually depend on how strongly attracting the underlying limit cycle is relative to the strength of coupling). Taking $\theta_i = \overline{\theta}$ to be the point on the limit cycle corresponding to $V_i = \kappa$, the threshold condition given by (7.29) becomes

$$\theta_j(T_j^n) = \overline{\theta} + 2\pi n, \quad \frac{d\theta_j}{dt}(T_j^n) > 0\,. \tag{7.32}$$

7.2.2 Synchronization in the Weak Coupling Regime

Phase model and averaging. Let us introduce new phase variables $\psi_i(t)$ according to $\theta_i(t) = \psi_i(t) + \overline{\Omega}t$. In the absence of any coupling the phase variable $\psi_i(t)$ is constant in time and all oscillators fire with their natural periods $\overline{T}$. For weak coupling, each oscillator still approximately fires at the unperturbed rate but now the phases $\psi_i(t)$ slowly drift. To first order in ϵ, we can take the firing times to be $T_j^n = (n - \psi_j(t)/2\pi)T$. Under this approximation, (7.28) and (7.31) lead to the phase equation (with $\overline{\theta} = 0$)

$$\frac{d\theta_i}{dt} = \overline{\Omega} + \frac{\epsilon}{N-1}\sum_{j\neq i} R(\theta_i)P(\theta_j) + \mathcal{O}(\epsilon^2)\,, \tag{7.33}$$

where $P(\theta + 2\pi) = P(\theta)$ for all θ and

$$P(\theta) = \sum_{m=-\infty}^{\infty} J((m + \theta/2\pi)T), \quad 0 \le \theta < 1\,. \tag{7.34}$$

The summation over m in (7.34) is easily performed for $J(\tau)$ satisfying (7.30) and gives $P(\theta) = \widehat{J}(\theta - 2\pi\tau_a/T)$ where $\widehat{J}(\theta)$ is a periodic function of θ with

$$\widehat{J}(\theta) = \frac{\alpha^2 e^{-\alpha\theta T/2\pi}}{1 - e^{-\alpha T}}\left[\frac{\theta T}{2\pi} + \frac{T e^{-\alpha T}}{1 - e^{-\alpha T}}\right], \quad 0 \le \theta < 2\pi\,. \tag{7.35}$$

Equation (7.33) can be simplified further in the weak coupling regime by averaging over the natural period $\overline{T}$. This leads to a phase equation of the form

$$\frac{d\theta_i}{dt} = \overline{\Omega} + \frac{\epsilon}{N-1}\sum_{j\neq i} H(\theta_j - \theta_i) + \mathcal{O}(\epsilon^2) \tag{7.36}$$

with

$$H(\phi) = \frac{1}{2\pi}\int_0^{2\pi} R(\theta)P(\theta + \phi)d\theta\,. \tag{7.37}$$

(Note that certain caution has to be exercised in applying averaging theory [441]. In general, one can only establish that a solution of (7.36) is ϵ-close to a corresponding solution of (7.33) for times of $\mathcal{O}(\epsilon^{-1})$).

Following the discussion of Sect. 7.2.4, we define a (1:1) phase-locked solution of (7.36) to be of the form $\theta_i(t) = \phi_i + \Omega t$, where ϕ_i is a constant phase and $\Omega = 1/T + \mathcal{O}(\epsilon)$ is the collective frequency of the coupled oscillators. Substitution of this solution into (7.36) and working to $\mathcal{O}(\epsilon)$ leads to the fixed point equations

$$\Omega = \overline{\Omega} + \frac{\epsilon}{N-1}\sum_{j \neq i} H(\phi_j - \phi_i)\,. \tag{7.38}$$

After choosing some reference oscillator, the N (7.38) determine the collective period Ω and $N-1$ relative phases with the latter independent of ϵ. In order to analyze the local stability of a phase-locked solution $\Phi = (\phi_1, ..., \phi_N)$, we linearize (7.36) by setting $\theta_i(t) = \phi_i + \Omega t + \widetilde{\theta}_i(t)$ and expanding to first-order in $\widetilde{\theta}_i$:

$$\frac{\mathrm{d}\widetilde{\theta}_i}{\mathrm{d}t} = \frac{\epsilon}{N-1}\sum_{j=1}^{N} \widehat{\mathcal{H}}_{ij}(\Phi)\widetilde{\theta}_j\,, \tag{7.39}$$

where

$$\widehat{\mathcal{H}}_{ij}(\Phi) = H'(\phi_j - \phi_i) - \delta_{i,j}\,,\sum_{k \neq i} H'(\phi_k - \phi_i) \tag{7.40}$$

and $H'(\phi) = \mathrm{d}H(\phi)/\mathrm{d}\phi$. One of the eigenvalues of the Jacobian $\widehat{\mathcal{H}}$ is always zero, and the corresponding eigenvector points in the direction of the flow, that is $(1, 1, \ldots, 1)$. The phase-locked solution will be stable provided that all other eigenvalues have a negative real part [462].

Pair of identical oscillators. In the particular case of two oscillators ($N = 2$), a phase-locked solution satisfies $\theta_2(t) - \theta_1(t) = \phi$ where the constant phase difference ϕ is a zero of the function $H^-(\phi) = H(\phi) - H(-\phi)$. A given phase-locked state is stable provided that (see (7.38))

$$\epsilon\frac{\mathrm{d}H^-(\phi)}{\mathrm{d}\phi} > 0\,. \tag{7.41}$$

Note that by symmetry both the in-phase ($\phi = 0$) and anti-phase ($\phi = \pi$) states are guaranteed to exist. The stability of the in-phase state for a pair of excitatory neurons has been investigated in some detail in [440], where it was shown that there can be major differences between oscillators with Type I and Type II PRCs. We shall briefly discuss these results using the fact that $H^{-\prime}(0) = H'(0)$. First consider the case of a Type I PRC. In general there will be some refractory period T_r during which time the PRC will be approximately zero. The only contribution to $H'(0)$ will come from the rest of the limit cycle where $R(\theta) > 0$. Therefore

$$H'(0) = \frac{1}{2\pi}\int_{\theta_\mathrm{r}}^{2\pi} R(\theta)P'(\theta)\mathrm{d}\theta\,, \tag{7.42}$$

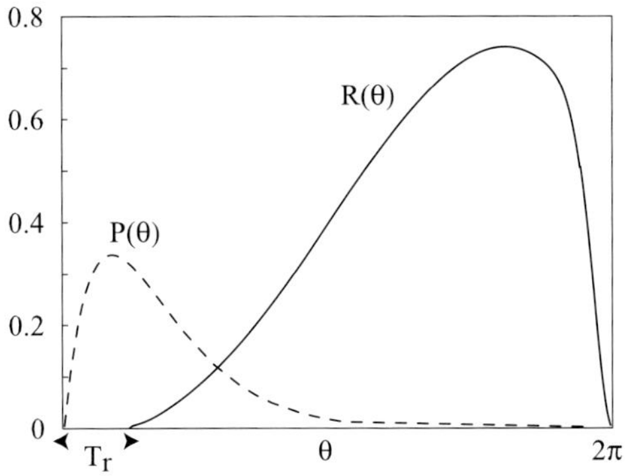

Fig. 7.12. Plot of a Type I phase resetting curve $R(\theta)$ and effective synaptic input function $P(\theta)$. The convolution of these two functions yields the effective phase interaction function $H(\phi)$

where $\theta_{\mathrm{r}} = 2\pi T_{\mathrm{r}}/T$ is the refractory period expressed as a phase. In the case of zero axonal delays the 2π-periodic synaptic function $P(\theta)$ is unimodal. Let us define $\theta^* = \max(\theta_{\mathrm{r}}, \theta_{\mathrm{p}})$, where θ_{p} is the phase at which $P(\theta)$ reaches its peak (see Fig. 7.12). We then have

$$H'(0) = \frac{1}{2\pi}\int_{\theta_{\mathrm{r}}}^{\theta^*} R(\theta)P'(\theta)\mathrm{d}\theta + \frac{1}{2\pi}\int_{\theta^*}^{2\pi} R(\theta)P'(\theta)\mathrm{d}\theta\,. \tag{7.43}$$

The first contribution to $H'(0)$ is positive and tends to stabilize the in-phase state whereas the second is negative and tends to destabilize it. Stability of the in-phase state thus depends on the balance between these two terms. If $\theta_{\mathrm{r}} > \theta_{\mathrm{p}}$ then the stabilizing term disappears and the in-phase state will be unstable. In other words, the in-phase state for a pair of Type I neural oscillators with excitatory synaptic coupling is unstable when the rise time is short relative to the refractory period. Furthermore, it can be shown that if $R(\theta)$ is a monotonically increasing function over almost all of its range, dropping rapidly to zero around $\theta = 2\pi$ (where the neuron fires), then the in phase state is unstable even when $\theta_{\mathrm{p}} > \theta_{\mathrm{r}}$. One such example is the integrate and-fire model discussed below.

Note that the above argument breaks down in the case of neurons with Type II response, since even when the maximum of the synaptic interaction occurs inside the refractory region, there is a synchronizing contribution to $H'(0)$ coming from the region of negative response. If this contribution is sufficiently large then the excitatory interactions can be synchronizing. The picture is also modified in the presence of discrete axonal delays, which shift the periodic function $P(\theta)$ in Fig. 7.10. The latter point will be illustrated below in the particular case of the integrate-and-fire model. The existence and stability of phase-locked solutions in the case of a symmetric pair of excitatory or inhibitory IF neurons with synaptic interactions has been studied in some

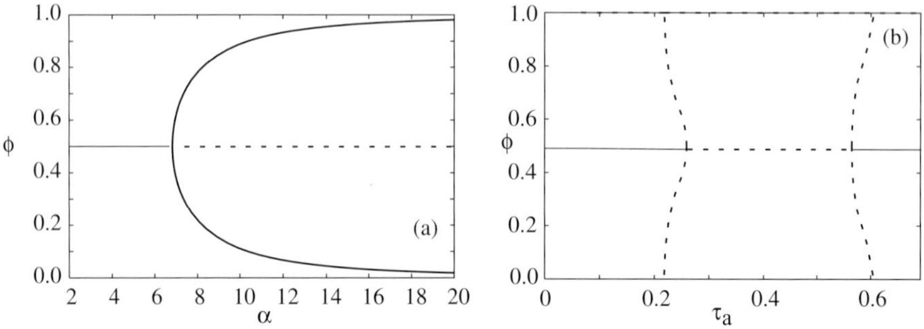

Fig. 7.13. (**a**) Relative phase $\phi = \phi_2 - \phi_1$ for a pair of IF oscillators with symmetric excitatory coupling as a function of α with $I_0 = 2$ and $\tau_a = 0$. In each case the anti-phase state undergoes a bifurcation at a critical value of $\alpha = \alpha_c$ where it becomes unstable and two additional stable solutions ϕ, $1 - \phi$ are created. The synchronous state remains unstable for all α. (**b**) Relative phase ϕ vs. discrete delay τ_a for a pair of pulse-coupled integrate-and-fire oscillators with $I_0 = 2$ and $\alpha = 2$

detail in [463]. First suppose that $\tau_a = 0$ and fix $\kappa = 1 = \tau$. For small α one finds that the in-phase and anti-phase solutions are the only phase-locked solutions. However, as α is increased, a critical value α_c is reached where there is a pitchfork bifurcation of the anti-phase solution leading to the creation of two partially synchronized states θ and $2\pi - \theta$, with $0 < \theta < \pi$ and $\theta \to 0$ as $\alpha \to \infty$ (see Fig. 7.13a). As shown by [464], variation of τ_a for fixed α produces a checkerboard pattern of alternating stable/unstable solution branches (see Fig. 7.13b) that can overlap to produce multistable solutions (see also [88]). Using (7.24), (7.34), (7.37) and (7.41) one finds that for zero axonal delays ($\tau_a = 0$) and excitatory coupling ($\epsilon > 0$) the synchronous state is unstable for all $0 < \alpha < \infty$. Moreover, the anti-phase solution $\phi = 1/2$ is stable for $\alpha < \alpha_c$ but it loses stability when $\alpha > \alpha_c$ with the creation of two stable partially synchronized states. In the limit $\alpha \to \infty$ these two states asymptotically approach the synchronous state so that the neurons are almost perfectly synchronized. This is consistent with the analysis of Mirollo and Strogatz [465] who proved rigorously that globally coupled IF oscillators almost always synchronize in the presence of instantaneous excitatory interactions. The stability properties of all solutions in Fig. 7.13 are reversed in the case of inhibitory coupling ($\epsilon < 0$) so that, in particular, the synchronous state is now stable for all α. If the discrete delay τ_a is increased from zero, then alternating bands of stability and instability are created. An example of such regions is displayed in Fig. 7.14. The corresponding stability diagram for the anti-synchronous state may be obtained by shifting the delay according to $\tau_a \to \tau_a + T/2$.

Globally coupled networks ($N > 2$). Synchrony in large networks of identical, globally coupled spiking neurons has recently been investigated by

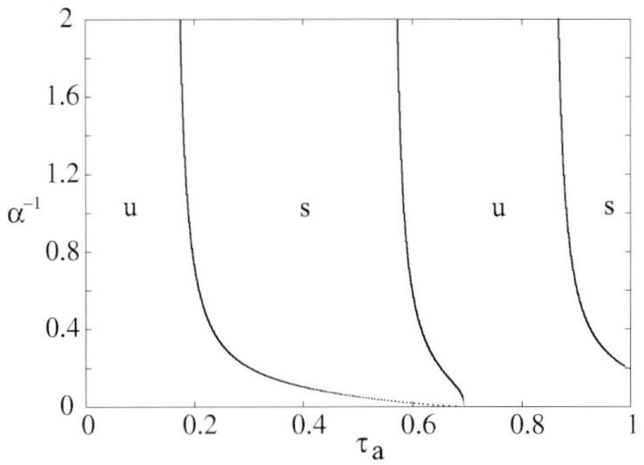

Fig. 7.14. Stability of the synchronous solution $\phi = 0$ as a function of α^{-1} and τ_a for weak excitatory coupling with $I_0 = 2$. Stable and unstable regions are denoted by s and u respectively. The stability diagrams are periodic in τ_a with period $\ln[I_0/(I_0 - 1)]$. (This periodicity would be distorted in the strong coupling regime since the collective period of oscillation depends upon the strength of coupling)

Gerstner et al. [466] using the spike response model, which is essentially an integrated version of the IF model. By carrying out a linear stability analysis of the underlying firing time map, they showed that in the thermodynamic limit ($N \to \infty$) the synchronous state of an IF network is stable if and only if $\epsilon H'(0) > 0$, and this holds for arbitrary coupling ϵ. However, the mode-locking theorem breaks down for finite networks and sufficiently strong coupling (see [447] and Sect. 7.3.3). Globally coupled identical neurons can exhibit a variety of other phase-locked states, which reflect the underlying permutation symmetry of the network. One of these is the so-called splay state $\phi_j = \pm j/N + \overline{\phi}$ for $j = 1, \ldots, N$. (In the thermodynamic limit this is often referred to as the asynchronous state in which phases are distributed uniformly around the circle, see [467,468]. Large networks can also break up into a small number of macroscopic clusters, each of which is fully synchronized [469,470].

Recently, it has been found that in the case of excitatory coupling ($\epsilon > 0$), the splay state of a globally coupled IF network is stable for small α (slow synapses) but undergoes a Hopf bifurcation at some critical α_c [471]. Beyond the bifurcation point the neurons are no longer phase-locked and their inter-spike intervals (ISIs) $\Delta_j^n = T_j^{n+1} - T_j^n$ vary either periodically or quasiperiodically on closed orbits. It was also observed that as one moves away from the bifurcation point ($\alpha > \alpha_c$) the degree of coherence or synchrony of the network increases. Elsewhere we have established that this picture is considerably more subtle due to the occurrence of a global heteroclinic bifurcation in the firing times, which is a natural consequence of the underlying permutation symmetry of the network [472]. That is, for sufficiently large values of α one finds that the quasiperiodic orbit has been destroyed in a global heteroclinic bifurcation due to a collision with invariant sub-manifolds of the state-space. Coincident with this global bifurcation is a saddle-node bifurcation that creates pairs of stable/unstable cluster states. Beyond the bifurcation point the system is in a periodic state close to synchrony (corresponding to one of the newly created stable solutions). We shall illustrate the above instabilities by considering how the degree of synchrony in the network varies with α. To quantify the degree of synchrony we use the firing time measure described in [473]. First, we introduce a set of phases, $\phi_k(j,m)$, associated with the firing times of the jth oscillator:

$$\phi_k(j,m) = \frac{T_k(j,m) - T_j^m}{T_j^{m+1} - T_j^m}, \quad T_j^m \leq T_k(j,m) < T_j^{m+1}, \tag{7.44}$$

where the $T_k(j,m)$ represent the set of firing times of neurons $i \neq j$ that occur on the interval $[T_j^m, T_j^{m+1})$. The number of such events will be denoted by $A(j,m)$ with $1 \leq k \leq A(j,m)$. For a set of phases $\Phi(j,m) = (\phi_1(j,m), \dots \phi_n(j,m))$ with $n = A(j,m)$ and fixed (j,m) we introduce the order parameter

$$r^2(\Phi) = \frac{1}{n^2} \sum_{k,l=1}^{n} \cos 2\pi(\phi_k - \phi_l). \tag{7.45}$$

An overall measure of synchrony, $\langle r^2 \rangle$, is defined by averaging $r^2(\Phi)$ over all oscillators and firing events in some time window. In Fig. 7.15, we show the

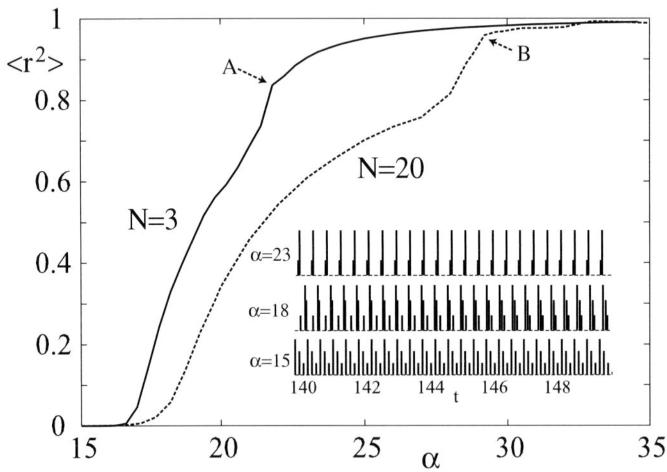

Fig. 7.15. Variation of the measure of synchrony $\langle r^2 \rangle$ as a function of α for $N = 3$ and $N = 20$ with $\epsilon = 0.4$, and $I_0 = 2$. Points A and B indicate where the dynamics has changed from a quasiperiodic to a periodic state due to a global heteroclinic bifurcation. Sample spike trains for $N = 3$ are also shown with different neurons distinguished by the height of their spikes

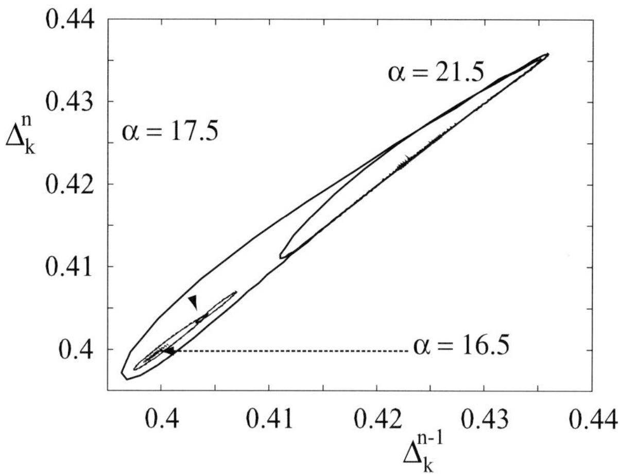

Fig. 7.16. Quasiperiodic motion of the ISIs on attracting invariant manifolds for $N = 3$ with $\epsilon = 0.4$ and $I_0 = 2$. This is an example of a weak instability in the firing times. Note that the temporal variation of the ISIs is relatively small in comparison to the mean ISI. For $\alpha < 16$ the splay state is stable and the map of the ISIs has a fixed point at around 0.4. For $\alpha > 22$ the *two-in-phase* state is stable, where two of the three oscillators fire in synchrony

variation of $\langle r^2 \rangle$ for $N = 3$ and $N = 20$. For relatively small α the network is in a stable asynchronous state for which $\langle r^2 \rangle = 0$. At $\alpha = \alpha_c(N)$ there occurs a discrete Hopf bifurcation to a partially synchronized state with $\langle r^2 \rangle$ a smoothly increasing function of α until the network undergoes a global bifurcation to a nearly synchronous state with $\langle r^2 \rangle \approx 1$. Typical orbits for the ISIs in the quasiperiodic regime are shown in Fig. 7.16.

7.2.3 Synchronization in the Strong Coupling Regime

So far we have discussed the issue of synchronization in networks of neural oscillators under the assumption of weak coupling so that the dynamics could be reduced to the simple phase model of (7.36). However, it is unclear that such an assumption is valid in real cortical circuits, and this raises the interesting question of what happens to the stability of in-phase states as the strength of coupling is increased. This issue has recently been addressed in a series of papers from our laboratory [445–448]. We briefly describe the results. If the synaptic interactions are sufficiently slow then the output of a neuron can be characterized reasonably well by a mean (time-averaged) firing rate function $f(X_j(t))$ (see for example [447,474,475]. Therefore, let us consider the case in which $J(\tau)$ is given by the alpha function (7.30) with a synaptic rise time α^{-1} significantly longer than all other time scales in the system. If the neuronal dynamics is fast relative to α^{-1} then the actual firing rate of a neuron will quickly relax to approximately its steady-state value. The latter

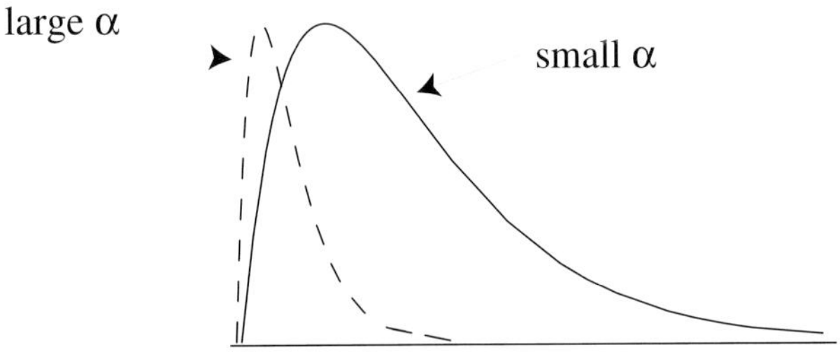

Fig. 7.17. Sampling of an input spike train by the delay kernel $J(t)$ for fast (large α) and slow (small α) synapses

is obtained by taking the total synaptic current X_j in (7.31) to be a constant X and integrating with respect to time t:

$$f(X)^{-1} = \int_0^{2\pi} \frac{d\Psi}{\overline{\Omega} + \epsilon X R(\Psi)} . \tag{7.46}$$

In the simple case of the IF model, (7.24) and (7.46) yield the firing-rate function (for $\kappa = \tau = 1$)

$$f(X) = \left\{ \ln \left[\frac{\epsilon X + I_0}{\epsilon X + I_0 - 1} \right] \right\}^{-1} \Theta(\epsilon X + I_0 - 1) . \tag{7.47}$$

We then introduce the approximation $\sum_{m \in \mathcal{Z}} J(t - T_j^m) \approx \int_0^\infty J(\tau) \cdot f(X_j(t-\tau)) d\tau$, which is valid for sufficiently small α. Heuristically speaking, for small α one expects a neuron to sample over a sufficiently long section of a spike train so that some form of time-averaging becomes reasonable (see Fig. 7.17). Under such an approximation, (7.28) can be rewritten as a second-order differential equation for the synaptic currentX_i:

$$\frac{1}{\alpha^2} \frac{d^2 X_i(t)}{dt^2} + \frac{1}{\alpha} \frac{dX_i(t)}{dt} + X_i(t) = \frac{1}{N-1} \sum_{j \neq i}^{N} f(X_j(t - \tau_a)) . \tag{7.48}$$

Equation (7.48) describes an analog or firing-rate model and is essentially a version of the Wilson–Cowan equation [70,71], except that here it is obtained using time-averaging under the assumption of slow synapses rather than using population averaging. Recall from Sect. 7.3.2 that networks of weakly coupled neural oscillators can have stable phase-locked solutions in which all the neurons have the same constant inter-spike interval (ISI). On the other hand, it is known that a strongly coupled network of analog neurons can bifurcate from a stable homogeneous state with identical time-independent firing rates to a state with inhomogeneous and/or time-varying firing rates (see [476]). The homogeneous state of the analog model is analogous to a phase-locked state of a spiking model, although all phase information is lost in the former. This suggests that there should exist some mechanism for

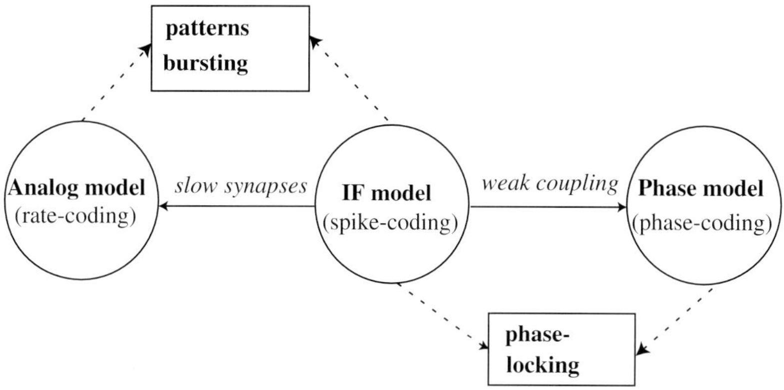

Fig. 7.18. Flow diagram indicating the relationship between the IF model and the two limiting cases of weak coupling (phase model) and slow synapses (analog model)

destabilizing phase-locked solutions of an oscillatory network in the strong coupling regime, at least for sufficiently slow synapses. Such a mechanism has recently been identified in the case of IF networks by carrying out a linear stability analysis of the firing time map. This establishes that desynchronization of integrate-and-fire (IF) neurons with strong synaptic coupling can occur via a discrete Hopf bifurcation in the firing times, and typically leads to non-phase-locked behavior characterized by periodic or quasiperiodic variations of the inter-spike intervals (ISIs) on closed orbits. In the case of slow synapses, the coarse-grained dynamics can be represented very well by the corresponding analog model (7.48) with firing-rate function (7.47), (see [447] for a review). The relationship between IF, analog and phase models is illustrated in Fig. 7.18. A number of particular results concerning strong coupling instabilities in IF networks have been obtained.

1. A homogeneous network of N IF neurons with weak excitatory coupling can synchronize in the presence of non zero axonal delays, see Fig. 7.14. However, in the strong coupling regime the synchronous state destabilizes resulting in a non-phase-locked state. The critical coupling for such an instability in the case $N = 2$ is shown in Fig. 7.19a, and the corresponding closed orbits for the ISIs and spike train plots are shown in Fig. 7.19b and Fig. 7.20. The behavior of the IF model reflects corresponding behavior in the analog model (7.48). Indeed, it is well known that the latter can undergo a Hopf bifurcation in the presence of discrete delays, leading to oscillations in the firing rate [477,478].
2. A homogeneous network of N synchronized IF neurons with global inhibitory coupling can destabilize in the strong coupling regime to a state in which some of the neurons became inactive, that is, oscillator death occurs [445]. There exists a critical value of the inverse rise time, $\alpha_0(N)$,

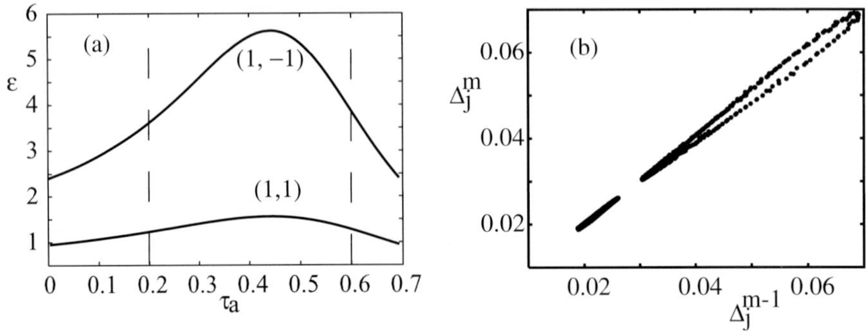

Fig. 7.19. (**a**) Two identical neurons with excitatory coupling. $I_0 = 2$, $\alpha = 2$. Between the vertical dashed lines the synchronous solution is stable in the weak-coupling regime. With increasing ϵ one first meets a Hopf instability of the firing times in the eigendirection $(1, 1)$. The higher curve represents the point at which the synchronous solution becomes unstable to a Hopf mode with eigenvector $(1, -1)$ and Hopf frequency $\omega = \pi/2$. (**b**) A plot of the inter-spike intervals $(\Delta_j^{m-1}, \Delta_j^m)$, $j = 1, 2$, for a non-phase-locked state following desynchronization illustrating how they lie on closed periodic orbits. Here $I_0 = 2$, $\alpha = 2$, $\tau_a = 0.5$, $\epsilon = 3$

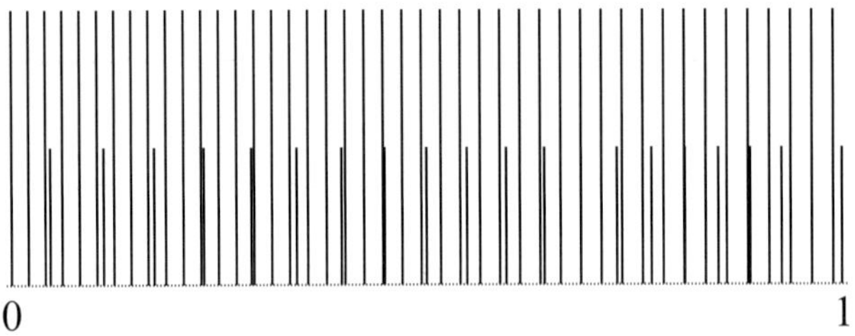

Fig. 7.20. Spike train dynamics for Fig. 7.19b over an interval of one cell time constant (neglecting transient behavior), showing clearly that the neurons are no longer synchronized

such that synchrony persists for arbitrarily large coupling when $\alpha > \alpha_0(N)$ (fast synapses). Moreover, $\alpha_0(N) \to 0$ as $N \to \infty$ (see Fig. 7.21). This establishes that the stability condition specified in the mode-locking theorem of Gerstner et al. [466] is valid in the thermodynamic limit, but no longer holds for finite networks with slow synapses. Note that oscillator death in a weakly heterogeneous network has been analyzed by Chow [479], and it has also been found in a recent study of mutually inhibitory Hodgkin–Huxley neurons [480].

3. Rhythmic bursting patterns can occur in asymmetric IF networks with a mixture of inhibitory and excitatory synaptic coupling [445]. The ISIs evolve on periodic orbits that are large relative to the mean ISI within a burst. There is good agreement between the temporal variation of the ISIs in the IF model and the variation of the firing rate in the corresponding analog model.

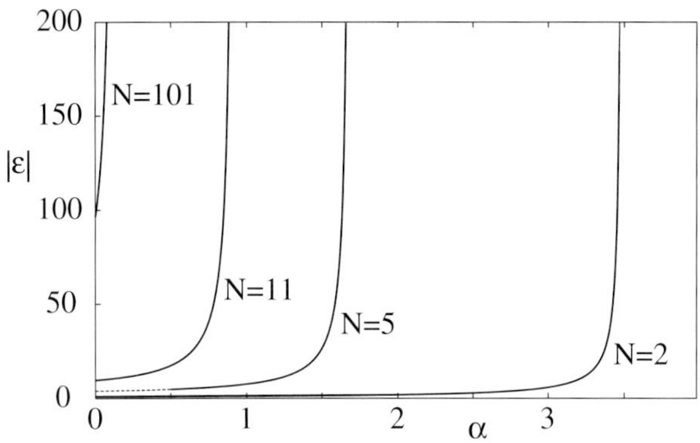

Fig. 7.21. Plot of critical coupling (*solid curves*) for desynchronization of an all-to-all network of identical IF neurons as a function of α for various network sizes N and $I_0 = 2$. Crossing the boundary from below signals excitation of the linear mode $(1, -1)$ leading to a stable state in which some of the neurons becomes quiescent (oscillator death). For $\alpha > \alpha_0(N)$ the synchronous state is stable for all ϵ. The critical inverse rise time $\alpha_0(N)$ is seen to be a decreasing function of N with $\alpha_0(N) \to 0$ as $N \to \infty$

4. A network of IF neurons with long-range interactions can desynchronize to a state with a regular spatial variation in the mean firing rate that is in good agreement with the corresponding analog model [446]. The ISIs evolve on closed orbits that are well separated in phase-space. The fine temporal structure of the IF model leads to deterministic fluctuations of the activity patterns whose coefficient of variation is an increasing function of the inverse rise time α. In other words, fast synapses lead to higher levels of deterministic noise.

It should be noted that the phase-locking instability of Figs. 7.15 and 7.16 is distinct from those described here. The former occurs in the weak coupling regime and is thus also found in the phase-averaged model (7.36). Indeed, a similar sequence of state transitions, namely a Hopf bifurcation from a splay state followed by a global bifurcation to a cluster state, are observed in the latter [472]. Despite of the fact that there are significant changes in the degree of synchrony, as highlighted in Fig. 7.15, the spatio-temporal variations in the ISIs characterized by the quasiperiodic orbits are relatively small (see Fig. 7.16), implying that $1 : 1$ frequency locking still holds approximately, which is consistent with the persistence of stable homogeneous states found in the corresponding analog model. Although the above results were obtained for the particular case of IF neurons, one expects that analogous results hold for conductance-based models. Indeed, a linear stability analysis of the firing time map associated with the reduced model (7.31) has established that strong coupling instabilities will occur as anticipated by the corresponding analog model (7.48), at least for sufficiently slow synapses [448].

7.3 Synaptically Generated Traveling Waves

Simulations of conductance-based models of propagation of activity in disinhibited cortical slices reveal that the propagation velocity is determined primarily by the response of the postsynaptic neuron to the first one or two spikes of the presynaptic neuron [127]. This motivates taking only the first spike of each neuron into account as a first approximation. Following Ermentrout [481] and Bressloff [449], we consider a one-dimensional network of neurons with excitatory coupling evolving according to the equation (see Fig. 7.22a)

$$\frac{\partial V(x,t)}{\partial t} = I_0 - \frac{V(x,t)}{\tau} + I(x,t)\,. \tag{7.49}$$

Here $V(x,t)$ denotes the membrane potential of the neuron at $x \in \mathcal{R}$ at time t, τ is the membrane time constant, I_0 is a constant external input, and $I(x,t)$ is the total synaptic input into the cell. We set $\tau = 1$ for convenience, which fixes the units of time to be of the order 1–10 ms. We also assume that the neurons are in the *excitable* regime $I_0 < \kappa$, where κ is the firing threshold, so that each neuron requires an additional stimulus before firing. One then supposes that a localized region of the network is stimulated (mimicking a localized, partial seizure) and considers whether or not this can result in the propagation of activity along the network in the form of a stable pulse. The synaptic current in Equation (7.49) is taken to be of the form

$$I(x,t) = g\int_{-\infty}^{\infty} W(x-x')E(x',t)dx'\,, \tag{7.50}$$

where $E(x,t)$ is the input at time t due to the firing of the presynaptic neuron at x, g is a positive coupling parameter, and $W(x)$ is taken to be positive, symmetric, and a monotonically decreasing function of $|x|$ with $\int_0^{\infty} W(x)dx < \infty$. Typical choices for such a weight function are $W(x) = (2\sigma)^{-1}\Theta(\sigma - |x|)$ (step) and $W(x) = (2\sigma)^{-1}\mathrm{e}^{-|x|/\sigma}$ (exponential). For concreteness, we choose the exponential function, although our basic results will not depend on the precise form of $W(x)$ (Fig. 7.22b).

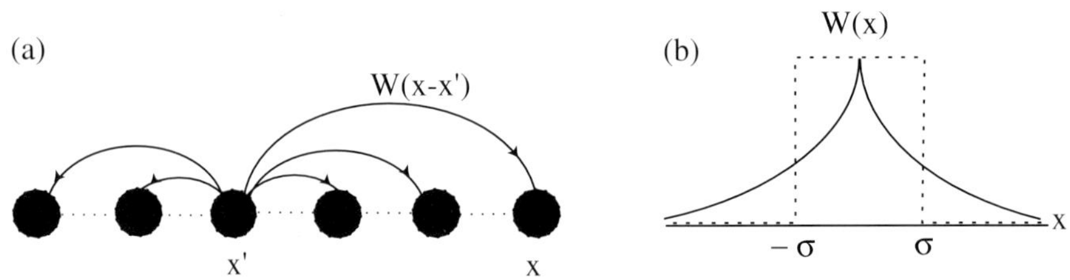

Fig. 7.22. (**a**) One-dimensional cortical network. (**b**) Exponential and step weight kernels $W(x)$

We define a solitary wave solution of (7.49) and (7.50) to be one for which $E(x,t) = J(t - x/c)$, where c is the velocity of the wave and $J(t)$ is the wave profile with $J(t) = 0$ for all $t \leq 0$. The physical interpretation of such a solution is that each neuron fires once with the neuron at x firing at time $T(x) = x/c$ (up to an arbitrary constant reflecting translation invariance). The function $J(t)$ represents the linear response of a neuron to an incoming spike idealized as a Dirac delta-function (see (7.30)). In order to determine the velocity c of a solitary pulse, we substitute (7.50) and (7.30) into (7.49), and then integrate with respect to t over the interval $-\infty < t \leq T(x) = x/c$. This generates a self-consistency condition for c of the form [481,449]

$$1 = g \int_0^{\infty} W(x) \mathrm{e}^{-x/c} \int_0^{x/c} \mathrm{e}^{t} J(t) \mathrm{d}t \mathrm{d}x \,. \tag{7.51}$$

We can explicitly integrate over x and t in (7.30) by Fourier expanding $J(t)$ and using the analyticity properties of $\widetilde{J}(\omega) = \int_{-\infty}^{\infty} \mathrm{e}^{-\mathrm{i}\omega t} J(t) \mathrm{d}t$, namely, that $\widetilde{J}(\omega)$ is analytic in the lower-half complex ω-plane since $J(t) = 0$ for $t \leq 0$. After performing a contour integral we arrive at the useful result

$$1 = \frac{gc}{2(1+c)} \widetilde{J}(-\mathrm{i}c) \,. \tag{7.52}$$

The (local) asymptotic stability of a solitary wave can be determined by considering perturbations of the firing times. Suppose that $T(x) = x/c + u(x)$ with $u(x) = \phi(x)$ for $x \leq 0$ and $\phi(x)$ a prescribed, bounded function on $(-\infty, 0]$. Asymptotic stability then corresponds to the condition $u(x) \to 0$ as $x \to \infty$ for arbitrary non-uniform initial data $\phi(x)$. Integrate (7.49) over the interval $(-\infty, T(x)]$ and expand the resulting nonlinear integral equation to first order in $u(x)$. Substitution of a solution of the form $u(x) = \mathrm{e}^{\lambda x}$, $\lambda \in \mathcal{C}$, into the linearized integral equation then yields the characteristic equation

$$\int_0^{\infty} \left[\mathrm{e}^{-\lambda x} - 1\right] W(x) f(x/c) \mathrm{d}x = 0 \,, \tag{7.53}$$

where

$$f(\tau) = \int_0^{\tau} \mathrm{e}^{-(\tau - t)} J'(t) \mathrm{d}t \,, \tag{7.54}$$

and J' denotes differentiation with respect to t. Asymptotic stability holds if all non-zero solutions λ of (7.53) have a negative real part. Such solutions depend indirectly on the coupling g via the speed c, which satisfies (7.52). Equation (7.53) can be evaluated using contour integrals to give $P(\lambda + c) = P(c)$ with

$$P(z) = \frac{z}{(1+z)} \widetilde{J}(-\mathrm{i}z) \,. \tag{7.55}$$

Note that $P(z)$ is real when z is real, and $P(z) \to 0$ as $|z| \to \infty$. We now investigate the existence and stability of solitary pulses as a function of α and τ_{a}. First, Fourier transform Equation (7.30) and substitute into

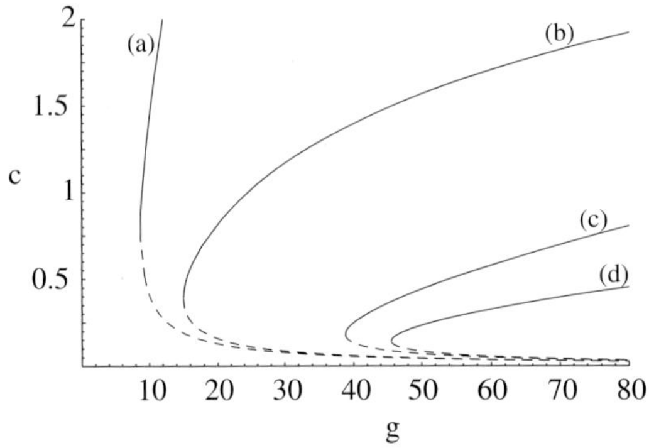

Fig. 7.23. Variation of the speed c of a traveling pulse as a function of the synaptic coupling g for different values of the inverse rise-time α and axonal delay τ_a. (**a**) $\alpha = 2, \tau_a = 0$, (**b**) $\alpha = 2, \tau_a = 1$, (**c**) $\alpha = 0.25, \tau_a = 0$, (**d**) $\alpha = 0.25, \tau_a = 1$. Solid (*dashed*) lines correspond to stable (*unstable*) branches. The external input is taken to be $I_0 = 0$ and $\sigma = 1$

Equation (7.52) to obtain the following equation for c [481]:

$$1 = \frac{g\alpha^2 c}{2(1+c)} \frac{e^{-c\tau_a}}{(\alpha+c)^2} . \tag{7.56}$$

In the case $\tau_a = 0$ the velocity scales according to a power law $c \sim \alpha\sigma\sqrt{g/2}$. This form of power law is consistent with the behavior observed in more detailed computational models [127]. The velocity c is plotted as a function of the coupling g in Fig. 7.23. It can be seen that for fixed α and τ_a there exists a critical coupling $g_s(\alpha, \tau_a)$ such that there are no traveling pulse solutions for $g < g_s$ and two solutions for $g > g_s$. The stability of the fast and slow solution branches displayed in Fig. 7.23 can be determined from the following theorem [449]:

Theorem 1. *Let C_+ and C_- denote, respectively, the fast and slow solution branches $c = c_\pm(g)$, $g \geq g_s$, of the self-consistency condition (4.8) for the velocity of a solitary pulse. Here g_s is the critical coupling for the existence of a solitary wave. The following stability results then hold: (i) The branch C_- is unstable for all α and τ_a. (ii) The branch C_+ is stable for all α in the case of zero axonal delays, $\tau_a = 0$ (iii) In the case of large axonal delays and fast synapses (small α), there exists a Hopf bifurcation point g_h such that $c_+(g)$ is stable (unstable) for $g > g_h$ ($g_s \leq g < g_h$).*

Note that for the parameter values used in Fig. 7.23 the fast branch is stable everywhere. The above theorem implies that for fixed α and g the continuous solitary pulse can become unstable at a critical value of the delay τ_{ac}. This is illustrated in Fig. 7.24, where the fast and slow solution branches are plotted

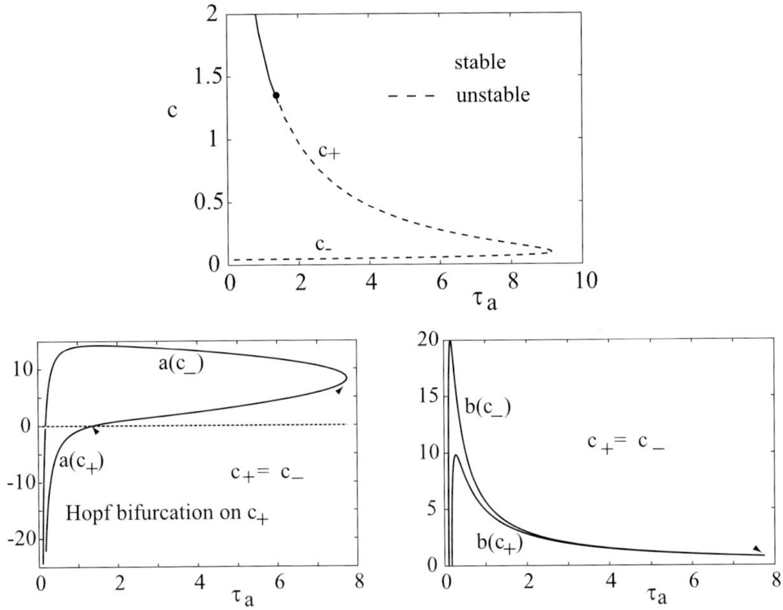

Fig. 7.24. *Upper figure* shows the fast and slow velocity branches as a function of the axonal delay τ_a for fixed α and g. The stable slow branch undergoes a Hopf instability at a critical value of axonal delay. The continuation of the real and imaginary parts of the eigenvalue pair associated with the Hopf instability are plotted in the *lower figures*

as a function of τ_a. The continuation of the real and imaginary parts of the pair of eigenvalue $\lambda_H = a \pm ib$ associated with the Hopf instability are also shown. Interestingly, Golomb and Ermentrout [450] have demonstrated numerically that in the regime where the continuous wave is unstable, $\tau_a > \tau_{ac}$, *lurching* pulses propagate with discontinuous, periodic spatio-temporal characteristics. Moreover, the lurching wave can persist in regions where the continuous wave does not exist and, for certain choices of weight kernel $W(x)$ such as the step, the two types of wave may co-exist (bistability).

The possible correlation between the nature of wave propagation and the effective size of discrete delays is suggestive of the observed differences between the spread of activity in cortical and thalamic slices. In computational and experimental studies of disinhibited neocortical slices, one finds that neuronal discharges propagate continuously at a velocity $c \sim 10\,\mathrm{cm/s}$ [127]. Axonal delays are relatively small. On the other hand, in models of thalamic slices, composed of excitatory thalamocortical neurons and inhibitory reticular thalamic neurons, waves propagate in a lurching manner at a velocity $c \sim 1\,\mathrm{cm/sec}$ [438]. This is thought to form the basic mechanism for the generation of 7- to 14 Hz spindle oscillations during the onset of sleep [425]. Each recruitment cycle of the lurching waves has two stages (see Fig. 7.24): (I) A

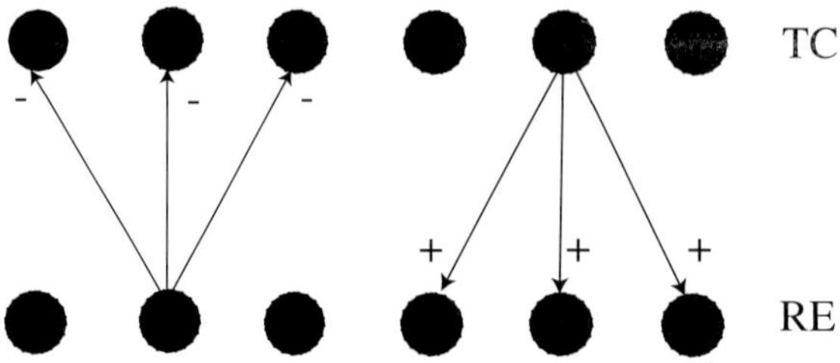

Fig. 7.25. Two-layer thalamic model. Inhibitory reticular (RE) cells inhibit excitatory thalamocortical (TC) cells, and TC cells excitep RE cells. Mutual inhibition between RE cells is neglected since it has a relatively small effect on discharge propagation

new group of inhibitory RE cells is excitepd by synapses from TC cells, and this RE group then inhibits a new group of TC cells. (II) The new recruited TC cells rebound from hyperpolarization and fire a burst of spikes, which further recruit more RE cells during next cycle. As discussed by Golomb and Ermentrout [450], one can reduce the two-layer thalamic model to a single-layer excitatory network with a large effective delay ($\tau_a \approx 100\,\text{ms}$) caused by the time needed for a TC cell to rebound from inhibition.

An interesting question concerns what happens in two dimensions, which reflects the more realistic case of horizontally propagating waves in cortex (Fig. 7.25). One expects to find several new types of wave beyond standard plane waves such as target patterns and spirals. Such waves have indeed been observed numerically by Fohlmeister et al. [482] and by Chu et al. [65].

Appendix

The Hodgkin–Huxley description of nerve tissue is completed with the set of equations

$$\alpha_m = \frac{0.1(V+40)}{1-\exp[-0.1(V+40)]}\,, \alpha_h = 0.07\exp[-0.05(V+65)]$$
$$\alpha_n = \frac{0.01(V+55)}{1-\exp[-0.1(V+55)]}\,, \beta_m = 4.0\exp[-0.556(V+65)]$$
$$\beta_h = \frac{1}{1+\exp[-0.1(V+35)]}\,, \beta_n = 0.125\exp[-0.125(V+65)]\,.$$

All potentials are measured in mV, all times in ms and all currents in μA per cm^2.

The parameter values used in the Morris–Lecar model are $V_K = -0.7$, $V_L = -0.5$, $V_{Ca} = 1$, $g_K = 2$, $g_L = 0.5$, $V_1 = -0.01$, $V_2 = 0.15$. For a type I response we use $g_{Ca} = 1.33$, $V_3 = 0.1$, $V_4 = 0.145$, $\phi = 1/3$ and $J = 0.0695$. For a type II response we use the same parameters with $g_{Ca} = 1.1$, $V_3 = 0$, $V_4 = 0.3$, $\phi = 0.2$ and $J = 0.25$.

8 Controlling Neural Synchrony with Periodic and Aperiodic Stimuli*

J.D. Hunter and J. Milton

A seizure involves the synchronization of the activity of a very large population of periodically firing neurons [4,483]. It is well known that periodic electrical or sensory stimulation can be used to trigger the onset of a seizure. Here we discuss the use of periodic stimuli to abort seizures once they are occurring.

There are a number of ways in which periodic electrical stimulation can be used to abort seizures. A periodic stimulus can be delivered directly to the epileptic focus. Since the coupling of the neurons to the external field differs depending on the distance of the neuron from the stimulating electrode, it is possible that such stimuli could dis-coordinate the neural firing to the point that population synchrony, and hence the seizure, is disrupted [484]. This approach is discussed in Chap. 18.

In this chapter we focus on strategies that emphasize the interplay between the frequency content of a periodic or aperiodic stimulus relative to the firing frequency of epileptic neurons. One possible scenario is that the periodic stimulus is applied to a site located some distance from the epileptic focus and exerts its effects either by augmenting the intrinsic brain inhibitory mechanisms or by influencing those mechanisms that regulate the spread of epileptic activity from its point of onset. Another possibility is that a coherent periodic stimulus to a population may push the population from a synchronized to an incoherent state.

There is an extensive literature that deals with the question of how a periodic stimulus influences a periodically firing neuron or population of neurons (for reviews see [439,485,486]). It would be expected that this theory would play important roles in the current uses of deep-brain stimulation [188]. Surprisingly, current use of vagal nerve stimulators and deep-brain stimulation is based on an empirical trial and error approach making essentially no use of this theoretical basis. Here we review some aspects of the interactions between periodic stimuli and neural populations that may be of some use for treating epileptic seizures. An important observation is that some of the results can be extended to aperiodic stimuli.

* This research was supported by grants from the National Institutes of Mental Health and the Brain Research Foundation. J. D. H. was supported by a scholarship from the National Science Foundation. We thank J. Crate, L. Fox and P. Lloyd for technical assistance.

8.1 Arnold Tongue Diagrams

From a theoretical point of view the effects of a periodic input with frequency f on a periodically firing neuron with frequency f_o depends on 1) the relative frequencies, f/f_o; and 2) the amplitude, A, of the periodic input. Figure 8.1 summarizes the salient features (herein we refer to diagrams of this type as Arnold tongue diagrams in honor of the mathematician who first discovered them). Stable $n : m$ phase locking zones are commonly observed for commensurate integer ratios, i.e. the frequency of the neural oscillator, m, is a harmonic or sub-harmonic of that of the stimulus, n. The stable phase locking zones are organized in the frequency-amplitude plane in an orderly fashion. Typically only the low-integer ratios are observed experimentally since the higher ratios become obscured as the intensity of the noise increases. Consequently, at very low stimulation amplitudes, it can be difficult to maintain stable phase locking. Modeling studies involving integrate-and-fire neurons suggest that the $1:1$ phase locking mode is the most robust [487]. Irregular dynamics can be seen in regions of the frequency-parameter space between the stable phase-locking zones. Arnold tongue diagrams have been measured for periodic stimulation for a variety of excitable cells [488–490], including neurons [491,492] and neural populations such as the epileptic hippocampus [493], and the respiratory rhythm generators [494,495].

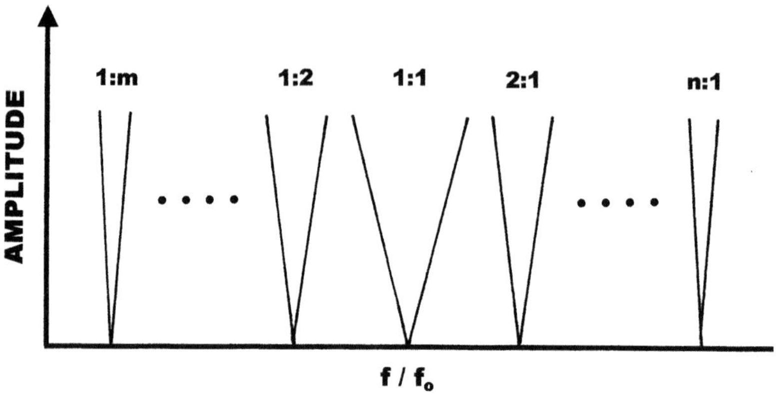

Fig. 8.1. Schematic of an g diagram. The ordinate represents the amplitude and the abscissa the relative period of the stimulus and the periodically firing neuron. The 1 : m Arnold tongues are to the left of $f/f_o = 1$; the n : 1 to the right. Between any two of these tongues, n : m tongues exist for all positive integers n, m. In addition, more complex dynamics such as quasi-periodicity and chaos exist in these regions

8.1.1 Entrainment Versus Synchrony

There are two ways in which the effects of periodic stimulation on a periodic oscillator can be described. First, we can measure the phases at which the periodic stimulus arrives relative to the cycle of the stimulated oscillator. If T_o is the period of the oscillator, then the phase, ϕ, at which the periodic inputs arrive is

$$\phi = \frac{t}{T_o} \pmod 1, \tag{8.1}$$

where t is the time that the input arrives. Note that ϕ has been defined to be between 0 and 1; to convert to radians it is necessary to multiple by 2π. Figure 8.1 is an example of an Arnold tongue phase diagram.

More relevant to our discussion of epilepsy is to measure synchrony. Entrainment does not necessarily imply a high degree of synchrony. This is because for $n : m$ patterns where $n \neq 1$ there are different time solutions for different initial conditions, even though the phase solutions are equivalent. Thus entrainment is necessary, but not sufficient, for synchrony in uncoupled populations of oscillators. In general, the $1 : m$ Arnold tongues generate the highest degree of synchronization in a population of uncoupled neurons receiving a common periodic input.

Epileptic seizures involve the synchronization of large populations of neurons. Thus measurements of synchrony are more relevant than phase. To compute the degree of synchrony, we have developed a statistic S that can be used to compute in-phase synchrony between N spike trains. For a spike at time t_{ij}, where $i = 1, \cdots, N$ is the train number and j is the spike number, we identify the spike at time t_k in a second train $k \neq i$ that is closest to it in time and compute the value $S_{ij} = \exp(-\lambda|t_{ij} - t_k|)$. Performing this procedure for each of the $N-1$ other spike trains. Thus each t_{ij} is identified with $N-1$ other spikes which are its nearest neighbors. If N_i is the number of spikes in the i-th train, repeating this procedure for every spike in every train will yield a maximum value of the statistic $S_{\max} = (N-1)\sum_{i=1}^{N} N_i$. The sum of the S_{ij} is normalized to the maximum values to yield the synchrony statistic

$$S = \frac{1}{S_{\max}} \sum_{i=1}^{N} \sum_{j=1}^{N_i} S_{ij}, \tag{8.2}$$

where $\lambda = 5/d$, where d is half of the interspike interval.

Figure 8.2a shows the degree of synchrony, S, of a population of uncoupled Morris-Lecar neurons as a function of the frequency and amplitude of a periodic current input. The Morris-Lecar model neuron has been found to accurately describe many of the phase resetting properties of *Aplysia* motoneurons [90]. As expected S is highest for the $1 : m$ phase locking regions.

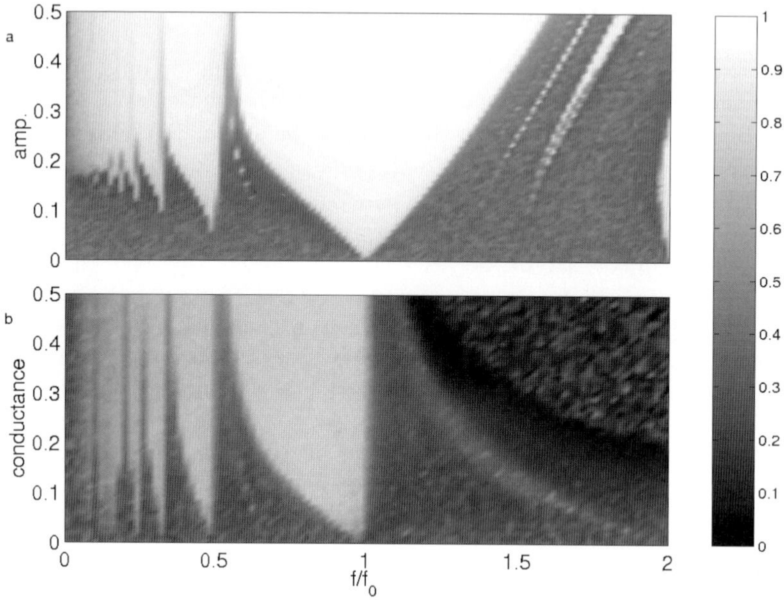

Fig. 8.2. Arnold tongue synchrony diagrams obtained for a periodically stimulated Morris-Lecar model neuron. This diagram was generated by measuring the spike train that resulted from periodic stimulation, repeating the same procedure with different initial conditions (chosen randomly) to generate 20 different spike trains, and then calculating S using (8.2). (**a**) Sinusoidal input. (**b**) Synaptic input with conductance parameter g. *White regions* show the amplitude frequency combinations which generate highly synchronous firing and *black* the desynchronized regions

8.1.2 Synaptic Inputs

Introducing a current to a neuron through a microelectrode is not physiological. Neural inputs are introduced across a synapse. Most attempts to control seizures with electrical stimulation in the brain will be mediated in part through synapses. For example, vagal nerve stimulation probably works in part through GABA-ergic synapses in the brainstem and cerebral cortex [496,497]. Thus an understanding of synaptic dynamics is essential to the development of rational control paradigms.

When the neural input is a synaptic input, rather than a tonic current, there are a number of additional considerations. Synapses are not static entities, but undergo use-dependent dynamics that must be taken into account [498–503]. Moreover the synaptic input will have its own entrainment dynamics [460,504,505]. Additionally, synaptic inputs enter via a change in conductance and hence the effects of reversal potential may be important [89,506]. In the special case of periodic inhibitory input, synaptic dynamics do not affect the timing of the inputs, but rather introduce a time-dependent fluctuation in the amplitude of the input. However, these amplitude effects are small when the input firing frequency is less than 20 Hz and thus we do not consider them further here [507].

Figure 8.2b shows the effect on synchrony of a population of uncoupled Morris-Lecar neurons receiving periodic synaptic input as a function of the relative synaptic frequency f_{I} and conductance amplitude g. Figure 8.2b is grossly similar to that of Fig. 8.2a for $f_{\mathrm{I}} < 1$; but completely different for $f_{\mathrm{I}} > 1$. The target neurons are not able to follow the pace of the IPSCs when they arrive at $f_{\mathrm{I}} > 1$ because the mean component of the IPSCs slows the firing rate. Moreover, for high-frequency large-amplitude periodic inputs, the inhibition silenced the model neuron's firing altogether (upper right-hand corner of Fig. 8.2b). For excitatory inputs this effect is symmetric about the axis $f_{\mathrm{I}} = 1$, with the synchrony region below 1 abolished (not shown).

8.1.3 Epileptic Hippocampal Slices

The effects of periodic stimulation of the mossy fiber bundle on epileptic activity generated by a hippocampal slice has been extensively studied [493,508]. The mossy fiber bundle is one of the main excitatory inputs to pyramidal neurons located in the CA3 region of the hippocampus. A schematic representation of the typical experimental paradigm is shown in Fig. 8.3a. Although the hippocampus in the brain spontaneously produces rhythmical activity, isolated slices of the hippocampus tend to be more quiescent. However, in perfusion media containing a high concentration of potassium ions the slices generate regular epileptiform bursts (plus penicillin; [493]) and even electrographic seizures (minus penicillin; [508]).

Hayashi and Ishizuka [493] studied the Arnold tongue structure for spontaneous epileptiform activity generated by the CA3 region of rat hippocampal slices in response to periodic stimulation of the mossy fiber bundle. Figure 8.3b summarizes the Arnold tongue phase diagram for the effects of the frequency, f, of periodic mossy fiber stimulation on epileptiform activity, measured by the field potential, generated by the CA3 region of the hippocampal slice. n is the number of current pulses delivered by Mossy fiber stimulation per m hippocampal epileptiform bursts. Stable $n:m$ phase locking zones are commonly observed for commensurate integer ratios, e.g. $1:1$, $2:1$, and so on. The effects of mossy fiber stimulation at frequencies less than the frequency of the epileptic bursts were not studied. In the regions between the $n:1$ phase locking tongues, complex dynamics including quasiperiodicity and possibly chaotic patterns were observed.

An important aspect of the work by Hayashi and Ishizuka [493] is that the epileptiform bursts in CA3 had duration < 50 ms. The CA3 neurons contribute a major excitatory input to the CA1 region via the Schaffer collaterals. Jerger and Schiff [508] compared the effects of mossy fiber bundle and Schaffer collateral stimulation on the generation of epileptiform bursts of duration > 1 s in the CA1 region of the hippocampus (they referred to these bursts as seizures). Figure 8.4 shows that stimulation of the mossy

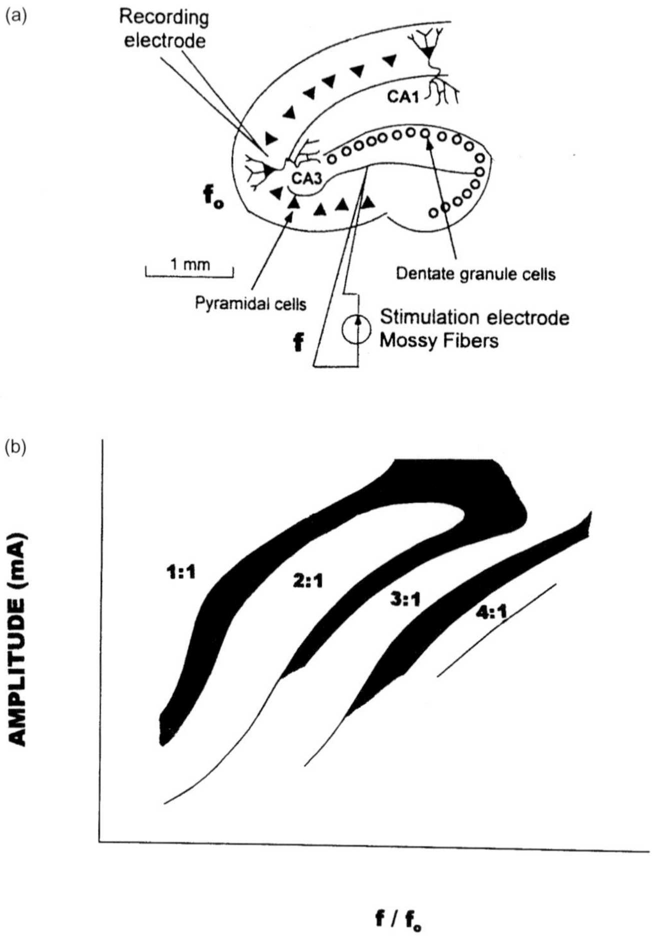

Fig. 8.3. (**a**) Schematic representation of the experimental set-up that enables the effects of periodic stimulation of the mossy fiber bundle on hippocampal slice epileptic activity to be studied. (**b**) Arnold tongue synchrony diagram for the field potential response of the CA3 region of a hippocampal slice to periodic mossy fiber stimulation. Figure adapted from [493]

fiber bundle at ~ 1 Hz reduced the number of seizures generated by the CA1 region.

The observations in Fig. 8.4 indicate that seizure suppression is not simply due to orthodromic stimulation of the CA1 pyramidal neurons, but that the frequency of the CA3 neurons is also a critical component of the seizure suppression mechanism. Thus, it is tempting to speculate that the difference between the studies of Hayashi and Ishizuka [493] and Jerger and Schiff [508] is the same as the difference between, respectively, the entrainment and syn-

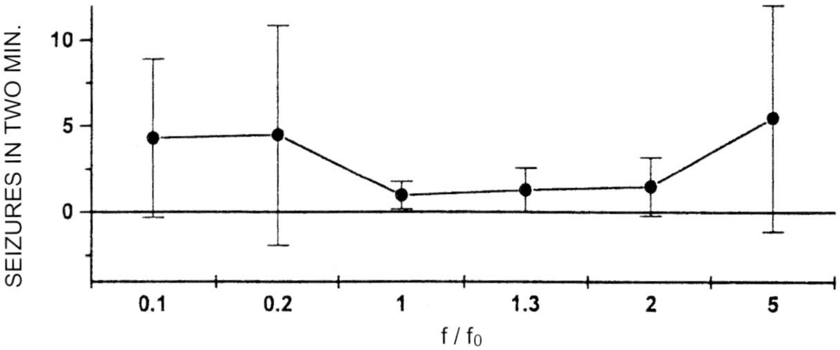

Fig. 8.4. Effect of periodic stimulation of the mossy fiber bundle on the generation of seizures in the CA1 region of the rat hippocampal slice. A seizure is defined as an epileptiform burst of duration ≥ 1 s. The seizure frequency is the number of seizures recorded in a 2 min interval. More complete seizure suppression was obtained for Schaffer collateral stimulation; however, the frequency dependence of the seizure suppression was similar. Figure adapted from [508]

chrony Arnold tongue diagrams alluded to above. The spontaneous frequency of epileptiform bursts in isolated hippocampal slices ranges from 0.3 to 1.0 Hz. In both studies $f/f_o > 1$. An epileptiform burst > 1 s probably involves a larger population of neurons than are involved when the burst is < 50 ms. Thus the decrease in seizure frequency could be simply a reflection of the fact that the synchronizing drive will be reduced when $f/f_o > 1$. However, Jerger and Schiff [508] suggested that the results in Fig. 8.4 reflected a frequency-dependent increase in inhibition in CA1 that suppressed seizure generation. This possibility is discussed in Sect. 8.5.

8.2 Aperiodic Current Inputs

It is now recognized that many of the features of the Arnold tongue diagram carry over to the situation in which the input to the periodically firing neuron is an aperiodic signal [509,510]. In this case the relevant parameters are the frequency content of the aperiodic signal and the stability of the $n:m$ phase-locking regimes to the random fluctuations. The importance of these observations is that they raise the possibility that neural networks could be desynchronized using appropriately designed noisy signals.

The first experimental observation to draw attention to this possibility was the phenomenon of spike-timing reliability [511,512]. Spike-timing reliability refers to the precision of spike timing when a neuron is repeatedly given the same time-varying input. Precisely timed spike trains are produced by neurons in response to aperiodic input signals in which current fluctuations resemble synaptic activity, in contrast to those produced by a constant-current stimulus [512].

In interpreting these results an important consideration is the magnitude of the fluctuations in the input relative to the current necessary to cause neural spiking. It is not difficult to appreciate that spike timing will be reliable when the relative magnitude of the input fluctuations is high. In the presence of noise, the width of inter-spike interval (ISI) distributions is inversely proportional to the slope of the membrane potential ($\mathrm{d}V/\mathrm{d}t$) at threshold [513,514]. A current with a large-amplitude fluctuating component will cause threshold crossings with a steeper slope than will a constant current and thus generate more reliably timed spikes in the presence of noise. An analogous slope condition exists for the synchronization of a network of neurons to a coherent input [466,515].

However, when the fluctuating component of the input is relatively small it is also possible for spike-timing reliability to arise from a resonance phenomena. An input signal may be divided into two components, a DC component and a time varying component. The neuron's firing rate (f_{DC}) in response to the DC component places constraints on the types of signals that may be encoded with reliably timed spikes in the presence of noise. Suppose at time 0 a spike occurs on the rising phase of an input signal and the membrane potential is reset to rest. If the time-varying component of the input signal is small compared with the DC component, then the next spike will not occur until some time approximately equal to T_{DC}, the inverse of f_{DC}. If the time-varying component of the input signal is increasing around time T_{DC}, the spike time distribution will be narrower than if the time-varying component is decreasing at this time because of slope considerations. Consequently small-amplitude periodic signals that are modulated by a frequency of f_{DC}, or its harmonics, will generate more reliably timed spike trains than other frequencies.

It follows from these considerations that spike timing-reliability is in fact a consequence of the stability of the $1:1$ Arnold tongue. Modeling studies of integrate-and-fire type neurons demonstrate that the $1:1$ phase-locking mode is the most stable one; the modes involving harmonics and subharmonics are much less stable in the presence of aperiodic fluctuations [487,509]. This conclusion is demonstrated in Fig. 8.5. This figure shows spike timing reliability of an *Aplysia* motoneuron to three different aperiodic signals: 1) broadband, low-pass filtered noise (Fig. 8.5A); 2) broadband, low-pass filtered noise in which the frequencies around and including f_{DC} were removed by digital filtering (Fig. 8.5B); and 3) a broadband, low-pass filtered noise in which a frequency band below f_{DC} was removed (Fig. 8.5C). As can be seen, spike timing reliability is highest provided that the input contains frequencies around f_{DC}. Qualitatively similar results to those shown in Fig. 8.5 were obtained simultaneously and independently in an analytic treatment of the phase model with *phase noise* and in numerical investigations of the van der Pol oscillator [510].

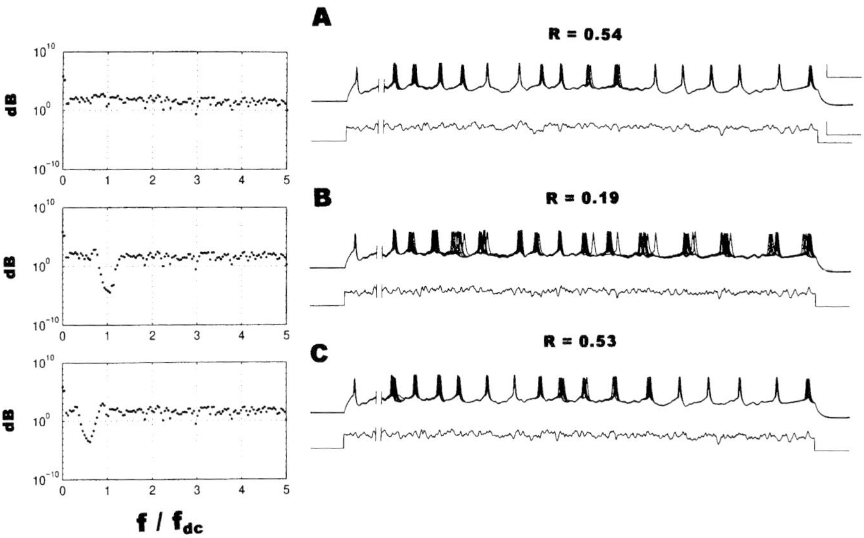

Fig. 8.5. Spike-time reliability in *Aplysia* motoneuron with aperiodic inputs. Superimposed voltage traces from 10 different trials recorded from a buccal motoneuron for four different input signals: (**A**) broad-band aperiodic input, (**B**) band-stop input lacking frequencies around $f_{\rm DC}$, and (**C**) band-stop control input lacking frequencies around $0.55 f_{\rm DC}$ but containing frequencies around $f_{\rm DC}$. The means and SDs of the aperiodic current inputs are equal: $\mu = 10.9\,\text{nA}$, $\sigma_s = 1.6\,\text{nA}$. While all aperiodic input signals generated more reliable firing times than did the DC trial (not shown), signals containing power in the frequency bands around $f_{\rm DC}$ (**A**,**C**) generated significantly more reliable firing times than those that did not (**B**). The spike-time reliability for each series is given above each voltage trace. The hatch marks denote a region of 0.68 s in length that is not plotted because the neuron is accommodating and the firing frequency is outside the analysis region. Scale bar above voltage trace in a) is 20 mV by 100 ms and above current trace is 10 nA by 100 ms; axes in **B**,**C** are scaled similarly. The power spectra of the three traces are shown to the left

8.3 Rate-dependent Control of Neural Synchrony

A simple picture of an epileptic seizure assumes that a large population of neurons becomes synchronized to a common periodic input, frequency f, generated by the epileptic focus. The results in Fig. 8.6 suggest that there are two ways in which this synchronization could be altered and hopefully the seizure aborted. First, it might be possible to alter the intrinsic firing rate, f_o, of the neurons. By moving f/f_o horizontally (Fig. 8.6) the population of neurons could be shifted from, for example, an Arnold tongue having $1:1$ entrainment to a more complex entrainment pattern. An alternative strategy would be to change the amplitude, or strength, A, of the periodic input. In this case changes in synchronization occur because of a vertical change in A

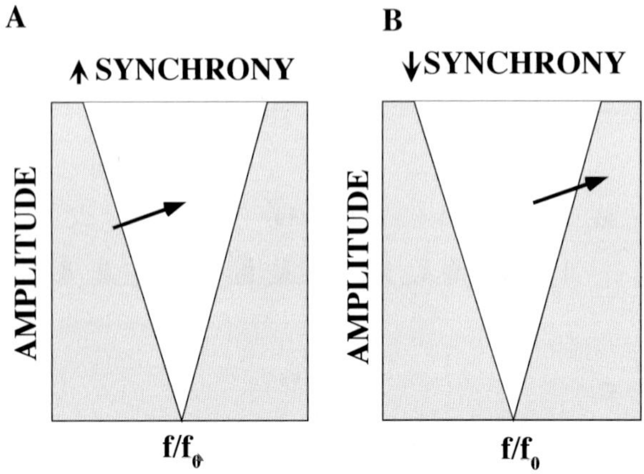

Fig. 8.6. Schematic of rate-control approach. A small tonic hyperpolarizing current increases the relative firing rate (f/f_o) and relative amplitude (amp/DC). This corresponds to a rightward and upward movement in the Arnold tongue space. (**A**) If the population is not on a tongue, the shift can synchronize firing. (**B**) If the population is on a tongue, the shift can desynchronize firing

(example not shown). At present the only tools we have at our disposal to affect these changes are pharmacological agents and electrical stimuli. In the discussion that follows we consider the control of synchrony by the introduction of electrical stimuli.

Figure 8.7 illustrates the rate control of synchrony of a population of uncoupled *Aplysia* motoneurons using an injected hyperpolarizing current. In this experiment the motoneurons are caused to spike periodically at f_o by the injecting of a depolarizing DC current. A periodic sinusoidal current, frequency f_S, is then injected. By varying f_S and f_o the initial synchrony of the motoneuron to the sinusoidal input can be adjusted. Once the synchrony (or desynchrony) has been established, a hyper-polarizing current is injected. The effect of this current is to reduce f_o, and hence increase f/f_o. This hyperpolarizing current can cause the population to become synchronized (Fig. 8.7A) or to become desynchronized (Fig. 8.7B).

8.4 Inhibitory Control of Neural Synchrony

In order to understand the use of an implanted brain stimulator to control seizures it becomes necessary to understand the interactions between at least three oscillators: 1) the intrinsic oscillation of the neuron; 2) the synchronizing oscillation of the epileptic focus; and 3) the controlling oscillation of the inhibitory interneurons due to the action of the implanted brain stimulator. Experimentally this can be readily studied in the *Aplysia* buccal motor ganglion where use is made of the large, identified interneurons that

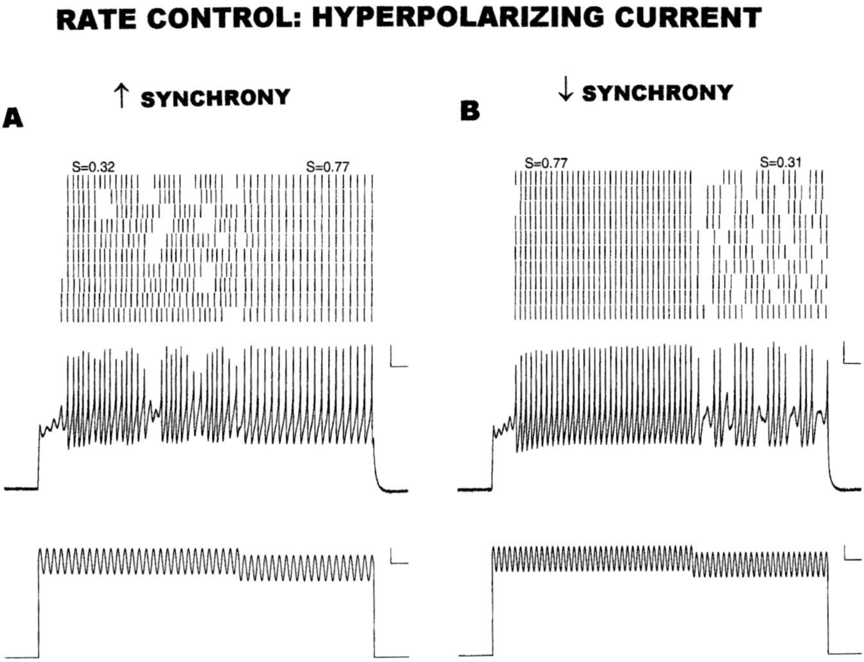

Fig. 8.7. Rate control of synchrony in the *Aplysia* buccal ganglion. (**A**) A hyperpolarizing current can increase synchrony (see Fig. 8.6A). The motoneuron (*middle*) is stimulated with a sine wave with a frequency too low to generate synchronous firing. A hyperpolarizing current step (*bottom*) slows the neuron's firing rate and synchronizes the neural firing, as seen in the raster and synchrony statistic (*top*). (**B**) A hyperpolarizing current can decrease synchrony (see Fig. 8.6). Experimental conditions are same as (**A**) except that the sine wave frequency is (1 : 1) and the hyperpolarizing current desynchronizes firing

are presynaptic to a population of identified motoneurons. The experimental paradigm is similar to that shown in Fig. 8.7 except that the control comes from a periodically firing inhibitory interneuron. Depending on the initial choice of f_{o} and f_{S} the effect of the inhibitory synaptic input is either to increase synchrony (Fig. 8.8A) or to desynchronize (Fig. 8.8B).

These experiments illustrate that synaptic inputs can either be used to synchronize or desynchronize neural firing in a rate-dependent manner. A major effect of the third oscillator is to introduce a distortion in the synchrony diagram. This distortion arises from the interplay between the epileptic focus's synchronizing input and the inhibitory controller's input. Figure 8.9 illustrates this observation for a population of uncoupled Morris-Lecar neurons. Here the input of the epileptic focus is represented by a sinusoidal input (f_{S}) and that of the periodic inhibitory interneuron by f_{I}. In each case the

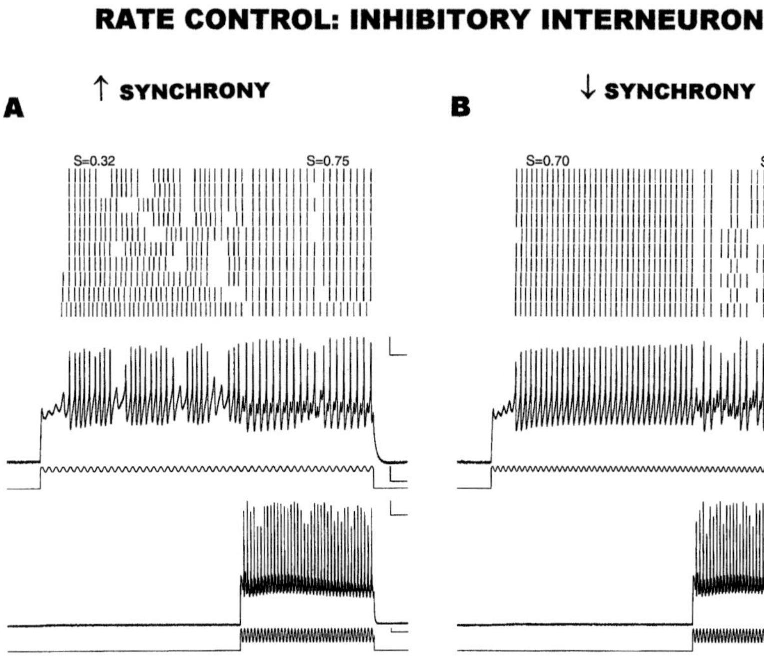

Fig. 8.8. Rate control of synchrony in the *Aplysia* buccal ganglion. This experiment is similar to the one in Fig. 8.7 except that an inhibitory interneuron is used as the controller. (**A**) Periodic inhibitory input synchronizes neural firing. (**B**) The same input desynchronizes neural firing when the sinusoidal frequency input to the motoneuron is 1 : 1

frequencies have been normalized to the spiking frequency of the Morris-Lecar neuron.

In the relative frequency ranges of 0.5 to 1.5 the synchrony is primarily related to the 1 : 1 phase locking region. The "no sine" and "no synaptic input" are plotted in the first column and row. There is a vertical bar of synchrony that widens as the amplitude of the sinusoidal input increases across columns. This is the tongue generated by the sinusoidal input. A horizontal bar of synchrony is generated by the synaptic input that widens one moves down the rows. This widening corresponds to the widening of the tongues as the amplitude increases, as shown in Fig. 8.2a (vertical bars) and Fig. 8.2b (horizontal bars).

The interaction between sinusoidal and inhibitory inputs is shown in the remaining panels. When the amplitude of both inputs is small, the synchrony regions form a cross in the synchrony diagram. As the amplitude of the inhibitory input increases, the vertical band of synchrony is progressively distorted to the left. This is caused by the slowing of the firing of the neuron. At sufficiently high amplitudes of the two inputs, the widening and bending be-

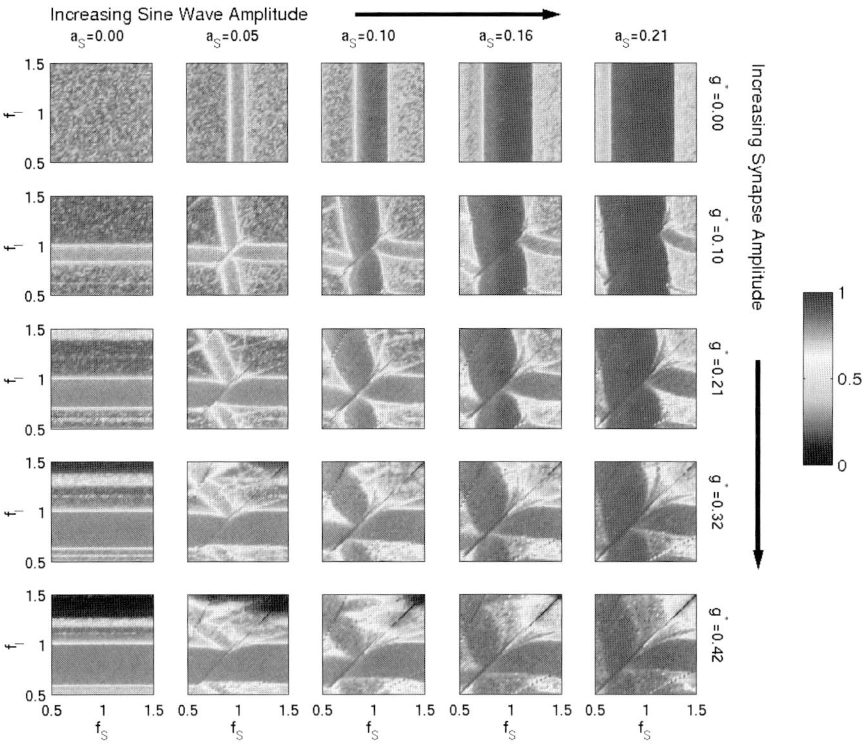

Fig. 8.9. Investigation of the synaptic control of synchrony in the Morris–Lecar model. A periodic inhibitory synaptic input is used to control synchrony in a population of sinusoidally driven neurons. The two experiments in Fig. 8.8 correspond to two points in the parameter space investigated here. Colorbar to the right indicates the population synchrony. *Parameters*: synaptic frequency f_I, conductance g, sine frequency f_S and amplitude a_S (The colour plate of this figure appears on page 360)

comes so pronounced that the horizontal and vertical synchrony regions fuse and there is synchrony over practically the entire parameter range. Qualitatively similar results are obtained when a dynamic synapse [499], with both facilitation and depression, is used as a controller, provided that the synaptic input frequency is not too large (not shown).

8.5 Periodically Stimulated Inhibitory Networks

The axon collaterals of the CA3 pyramidal neurons ramify extensively within the CA3 region and make excitatory connections on other pyramidal neurons, including those in CA1, as well as on inhibitory interneurons. The inhibitory interneurons, in turn, project to a large number of pyramidal neurons to form recurrent inhibitory loops. Normally the effects of the excitatory connections are damped out by the stronger inhibitory inter-neural influences. Thus the hippocampus, and the CA3 region in particular, is only reluctantly recruited into "ictal-like" seizure activity, even during the course of a partial complex seizure [95]. These observations suggest that, at least to a first approximation, the hippocampus can be thought of as a laterally inhibited neural network.

The effects of periodic stimulation on laterally inhibited networks of neurons were well characterized over 40 years ago by the classical work of Hartline, Ratliff and their co-workers [516]. We briefly review this work here because of its obvious relevance to this discussion.

These investigators studied the effects of sinusoidal illumination on the output of the retina of the horseshoe crab, *Limulus* [516,517]. They choose to study the *Limulus* retina since it is known to contain only inhibitory interactions. They showed that the spatial and temporal characteristics of the retinal output could be well described by a model of the form

$$r_{\mathrm{m}} = e_{\mathrm{m}} - \sum_{n \neq m} k_{\mathrm{mn}}(r_{\mathrm{n}} - r^{o}_{\mathrm{mn}}) , \tag{8.3}$$

where r_m is the rate of discharge of the m-th ommatidium, e_m is the input (light) to the m-th ommatidium, and r^{o}_{mn} is a threshold.

In order to discuss the effects of periodic stimulation we first pass to the continuum limit

$$r(x,t) = e(x,t) - \int_{-\infty}^{\infty} \int_{-\infty}^{t} \mathrm{d}t' k(x - x', t - t') r(x', t') , \tag{8.4}$$

where the notation (x,t) takes into account the spatial as well as the temporal dependencies in the retina. If we assume that the input stimuli are uniform in space, i.e. $r(x,t) = r(t)$ and $e(x,t) = e(t)$, we obtain

$$r(t) = e(t) - \int_{-\infty}^{t} k(t - t') r(t') \mathrm{d}t' . \tag{8.5}$$

Taking Fourier transforms of both sides of (8.5) and re-arranging we obtain the transfer function for the *Limulus* retina

$$\frac{\overline{r(\omega)}}{\overline{e(\omega)}} = \frac{1}{1 + \overline{k(\omega)}} , \tag{8.6}$$

where the overline designates the Fourier transform. Since the integral of $k(x)$ over all x is less than unity, the condition for resonance (i.e., 1 : 1 phase locking) is $\overline{k(\omega)} < 1$.

Numerical calculations showed that if τ_{p} represents the time for peak inhibition, then the amplification occurs at $\sim (2\tau_{\mathrm{p}})^{-1}$ [517] (Fig. 8.10). The time to peak inhibition is equal to the neural delay, or dead time, plus the time required for inhibition to build up in the network (technically, this is a lag). In the *Limulus* retina, the time delay is 0.1 s, the lag is 0.5–1.0 s, and hence $\tau_{\mathrm{p}} \sim 0.15 - -0.2$ ms. The resonance peak is predicted to occur

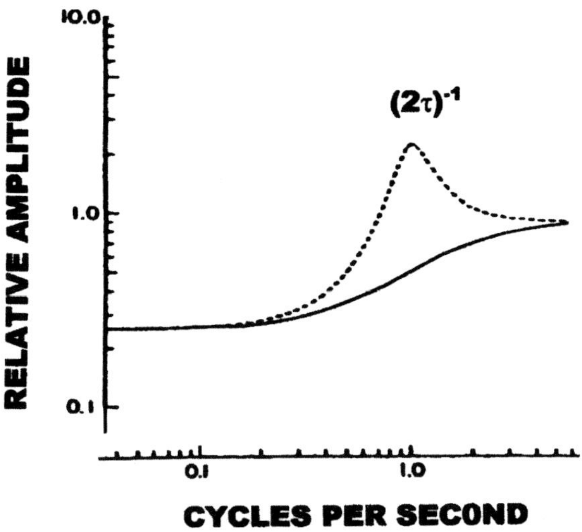

Fig. 8.10. Effect of the time to peak inhibition, τ_p, on the predicted Bode plot for a laterally inhibited neural network. The solid line corresponds to $\tau_p = 0$ and the dotted line to $\tau_p = 0.5$ s

at 2.5–3.3 Hz. This is in excellent agreement to the experimentally observed resonance ~ 3 Hz [517].

The Jerger and Schiff results demonstrate that there is a frequency dependent suppression of seizures in the CA1 region in the hippocampal slice, with ideal suppression frequencies between 1–2 Hz. This suggests that $\tau_p \sim 250$–500 ms. Stimulation of the CA1 pyramidal cells via the Schaffer collateral pathway show that the maximal inhibitory current flows around 200–250 ms after stimulus onset. This inhibitory cuurent is thought to represent the effects of slow acting $GABA_B$ receptors [518]. Numerical simulation of recurrent inhibitory neural networks suggest that τ_p increases with network size and that τ_p equal 500 ms or larger can easily be obtained (C. Hauptmann, personal communication). Thus the prediction by [508] is consistent with the known behavior of a laterally inhibited neural network [517].

8.6 Discussion

The results presented in this chapter emphasize that the effects of electric stimulation depend on both the frequency and strength of the stimuli and the intrinsic firing frequency of the neurons or neural population. For example, in the hippocampus, dramatic changes in neural synchrony and entrainment can be seen for stimulus frequencies (~ 1 Hz) which are close to the natural frequency of the neural population (see Fig. 8.3 and 8.4). In contrast, the present-day use of deep-brain stimulators favors the use of high frequency stimulation (e.g. > 100 Hz) to cause a reversible blockade of the stimulated area [188]. An exception is the vagal nerve stimulator that typically employs a stimulus frequency ≤ 60 Hz [497,519]. It is important to note that stimulating a peripheral nerve in order to stimulate a brain region must involve a polysynaptic pathway. The integrate-and-fire properties of neurons, and the relatively small magnitude of EPSPs relative to a neuron's firing threshold, imply that a frequency reduction will occur at each synapse [68]. For example, the firing frequency of an α-motoneuron in the spinal cord is about one-tenth of the frequency of its input [520]. Thus it is possible that, indirectly, vagal nerve stimulation may be taking advantage of the sensitivity of the synchrony of a neural population to low-frequency inputs.

The success of an implanted brain stimulator to control seizures obviously is dependent on the proper choice of the stimulating parameters, such as frequency and amplitude. The observations in this chapter emphasize how difficult choosing the appropriate parameters is likely to be. However, there seems to be a simple strategy available to decrease synchrony when it becomes too large. If $f_{\rm S}$ is within the vertical synchrony bar, the best choice for desynchrony is $f_{\rm I} > 1$. If $f_{\rm S}$ is outside the bar, the best strategy is to do nothing, since there are many regions in the parameter space where the addition of $f_{\rm I}$ will increase synchrony.

The Arnold tongue structure that underlies neural synchrony to periodic stimuli underscores the difficulties inherent in the use of pharmacological agents to suppress seizures. On the one hand, pharmacological agents that affect synaptic plasticity can act to decrease A, the strength of the entraining stimulus. On the other hand, it is likely that these same agents would also affect the intrinsic firing frequency of the neural population. It is possible that the effects of pharmacological agents on both A and $f/f_{\rm o}$ could work against each other. A potential advantage of the use of electrical stimulators is that it may be possible to design them to ensure that only one of the entrainment parameters is changing, which would be a simpler therapeutic strategy to implement.

In terms of the future design of an implantable seizure-aborting device, the observation that low-frequency stimuli could be used has the advantage of prolonging battery life. However, this advantage may be largely offset by the need to monitor the intrinsic firing properties of the neural population to be stimulated and very precisely regulate stimulator frequency accordingly. The success and reproducibility of implantable seizure aborting devices probably hinges on the extent to which this latter problem can be solved.

9 Modeling Pattern Formation in Excitable Media: The Legacy of Norbert Wiener

A.S. Mikhailov

The pattern of rotating spiral waves is typical for excitable media. Such waves are observed in chemical reactions in aqueous solutions [521,522], in catalytic chemical reactions on metal surfaces [523], in liquid crystals [524], in optical systems [525], in populations of the social amoebae *Dictyostelium discoideum* [526,527], in the heart [528], in the retina [115], and even inside single biological cells [529]. Remarkably, the properties of spiral waves in all these diverse systems are similar. This suggests that the emergence and the dynamics of spiral waves are based on general mechanisms, not sensitive to the details of a particular system.

The mathematical theory of pattern formation in excitable media originates from the pioneering work by Norbert Wiener. The great mathematician of the 20th century, the founder of cybernetics, was striving to understand the processes of self-regulation in living matter. During a stay of Wiener in Mexico, he had many discussions with his friend Arturo Rosenblueth from the Mexican Institute of Cardiology. They talked about patterns of circulating electrical excitations in the heart, their similarity to the processes in the nervous tissue and mathematical models able to reproduce this behavior. These discussions and subsequent studies resulted in their article "The mathematical formulation of the problem of conduction of impulses in a network of connected excitable elements, specifically in cardiac muscle" [530] which appeared in 1946 in *Archivos del Instituto de Cardiologia de Mexico*. In this large paper Wiener and Rosenblueth formulated the basic mathematical concept of an excitable medium, showed the existence of rotating spiral waves, analyzed their properties and considered the role of such patterns in cardiac disorders and fibrillation.

The paper was actively discussed in the seminars at the Massachusetts Institute of Technology (B. Hess, personal communication). It was followed by publications [531,532] where its generalizations and more detailed studies were performed. My own copy of this article stems from A.T. Winfree and contains numerous comments made by him on the margins, as he was reading the paper in 1970 at the University of Chicago. In the 1960s this rare publication was discovered by the participants of the Seminar on Mathematical Biology, held by I.M. Gel'fand at the Mathematical Department of the Moscow University, and its Russian translation has been published.

Modern theoretical studies of pattern formation in excitable media are much influenced by the subsequent discovery of partial differential equations that describe reaction-diffusion systems with excitable properties. The highly increased computation speed permits fast numerical simulations of such systems. However, in many biological applications such detailed equations are not known and their very existence is questionable. In contrast to this, the more abstract approach, advocated by N. Wiener, has wider applicability. In its framework, an excitable medium is defined by a number of its characteristic parameters, such as the wave propagation velocity, the excitation and recovery times, etc. These parameters can often be directly deduced from the experiments. Once they are known, the theory predicts the properties of spiral waves and their dynamics.

The aim of this article is to review the impact of the work by Wiener and Rosenblueth on modeling of pattern formation in excitable media. First I summarize the contents of their fundamental paper and then look at how this approach has been extended to describe various phenomena related to spiral waves in excitable media.

9.1 Excitable Media

The Wiener–Rosenblueth (WR) theory of excitable media was based on the experiments with cardiac and nervous tissues performed in the beginning of the 20th century [533–536]. In these early publications, such concepts as an excitation wave or the refractory period were introduced. However the description of excitable media by Wiener and Rosenblueth was much more general and also included references to physical aspects of these phenomena, as can be seen from the following citations:

"Conduction in nervous tissue resembles that in somatic striated in cardiac muscle. The laws which apply to the muscle fibers are also applicable to the nerve fibers. In both instances the propagation is active, with energy supplied locally. In both cases an impulse travels with a nearly uniform velocity. In both cases the excitation and transmission are all-or-none and do not allow for impulses of varying degrees of strength. In both cases activity is followed by an unexcitable period of definite duration, the absolutely refractory period; and this stage is followed in turn by a relatively refractory period, during which the tissues have subnormal excitability" ([530], p. 205–206).

The structure of the cardiac muscle is synticial. Its fibers interconnect abundantly and excitation can easily pass through any connection. Therefore, the cardiac muscle can be treated as a pseudo-continuous conductive medium. In contrast to this, conduction in nervous structures is more complicated (see, for example, Chap. 2). When an impulse reaches a synapse, it does not necessarily stimulate the following neuron. Transsynaptic stimulation usually requires a repetition of impulses in time at the same synapses (temporal summation) or the simultaneous arrival of impulses at a sufficient number of adjacent synapses. Moreover, the arrival of impulses may have other consequences than the stimulation of the postsynaptic element. It may have an inhibitory effect, rendering the elements less excitable to the stimuli. Nonetheless, some of the responses mediated by the nervous system bear a close analogy to those of the heart. "Stimulation at any point on the surface of the cerebral cortex will usually elicit a wave which spreads in all directions for some distance, the counterpart of the [heart] beat. This wave usually degenerates within a short distance of the stimulated point; but under certain experimental conditions, say during the clonic stage of a widespread tonic-clonic response (experimental epileptic seizure) it is usual to find many cortical elements coupled physiologically so that stimulation of one of them activates the rest in a relatively ordered fashion" ([530], p. 206).

The normal beat is accompanied by repeated generation and subsequent concentric spreading of an electrical excitation wave over the heart muscle, that causes its contractions. Besides its physiological beat, the heart can exhibit two other quite different types of contraction.

"If a turtle' s ventricle is overdistended or stimulated with rapidly repeated electrical shocks, a continuous cyclical wave appears, running along a definite closed path. In this condition, unlike physiological beating, the whole muscle is never at rest (in diastole). Similar activity has been observed in human auricles. The conditions that initiate this activity are not known. The phenomenon is called flutter.

A type of nervous conduction quite like cardiac flutter is well known in the umbrella disc of the coelenterates. Appropriate stimuli to the quasi-circular nerve net of a medusa or of an anemone will start a circulating wave that may keep going around for many hours. Although neither the anatomical structure nor the physiological organization of this network is adequately known, the analogy to flutter in cardiac muscle is striking.

The third type of cardiac contraction and conduction is known as fibrillation. Rapidly repeated stimulation of the mammalian ventricle will again set up circulating waves, as in flutter. These waves, however, do not seem to progress in an orderly fashion but rather to follow random paths. Several waves may progress simultaneously. The ventricle looks like a quivering mass of worms ([530], p. 207).

At the middle of the century, the detailed mechanisms of pulse conduction in nervous or muscle fibers were not known. Therefore, Wiener and Rosenblueth chose a phenomenological approach that led them to the formulation of an abstract concept of an excitable medium. According to them, the excitable medium consists of locally connected excitable elements. Each of these elements can be found in one of three different functional states of *rest, excitation* and *recovery* (the refractory state). The arrival of an external perturbation at an element in the rest state triggers its transition into the state of excitation. The element stays a short time in the excited state and then goes into the refractory state of a fixed duration. In this state, the element is not responding to any external perturbations. Only when the refractory state has expired and the element has returned to the original rest state, it can again undergo excitation.

An element that is currently occupying the excited state induces excitation of neighboring elements that are in the state of rest. Thus, application of a local perturbation to the array of such elements would produce a propagating excitation wave. Though this system actually consists of individual excitable elements, similar to the cardiac tissue it can be approximately treated as a continuous medium. When such continuous formulation is chosen, the propagating excitation waves represent traveling fronts of excitation that are followed by a refractory zone. Wiener and Rosenblueth assumed that each small element of the curve, specifying the excitation front, moves with time in its normal direction at a constant velocity V. In their original model, the duration τ_{e} of the excited state was assumed to be vanishingly short compared with the duration τ_{r} of the subsequent refractory state. Thus, the width of the wave practically coincides with the width of its refractory zone and is equal to $V\tau_{\mathrm{r}}$. To complete the definition of the excitable medium, the law of motion of excitation waves near the medium borders should be specified. It has been assumed in [530] that the wave could not separate itself from a border and moved orthogonally to the boundary.

This model reproduces the basic properties of the physiological heart beat. If a certain element of the medium is periodically excited, it emits concentric excitation waves that spread across the medium. Moreover, when two excitation waves collide, they should annihilate. Indeed, the excited elements in the collision zone are then surrounded from both sides by the elements in the refractory state and therefore they cannot forward excitation to other elements of the medium. In order to explain the appearance of circulating waves, responsible for flutter in the cardiac tissues, motion of excitation waves around obstacles must be further considered.

9.2 Spiral Waves Around Obstacles

The excitable medium can contain obstacles. These may be produced by cutting a hole or creating a region not accessible for excitation. The obstacles for propagation of excitation waves in the heart muscle are typically formed by blood vessels spanning the muscle. As a consequence of an infarct, small regions of the cardiac tissue can also die out, forming regions where propagation of excitation is impossible. Wiener and Rosenblueth considered motion of excitation waves around obstacles with different shapes. Since in their theory the wave could not separate itself from the boundary of an obstacle, this implied that the wave must circulate indefinitely around the obstacle. The shape of the rotating wave is then found from simple geometrical considerations:

Suppose that the wave is rotating around a circular hole of radius R (Fig. 9.1). Obviously, the wave cannot circulate as a straight ray (in that case

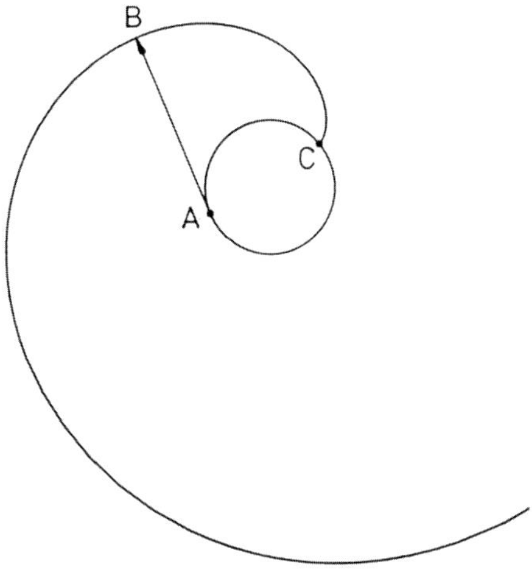

Fig. 9.1. Geometrical construction of the spiral wave in the WR model

the linear propagation velocity of the wave should have indefinitely increased with the distances from the center, but it must remain constant). Therefore, the wave coils into a spiral. To determine the form of this spiral, it suffices to note that, since all wave elements move at the same velocity V, the length of the line segment AB should be equal to the length of the arc AC in Fig. 9.1. As known in analytical geometry, the spiral possessing this property represents an *involute* of the obstacle. An involute is easily produced if we take a cord and wind it several times around the obstacle. If we now attach a pencil to the end of the cord, and start to unwind it by keeping the cord strained, the pencil would draw on the sheet of the paper a trajectory that represents the involute of the obstacle. The involute of a circular hole is a spiral that approaches far from its center an *Archimedian spiral* of a constant pitch. This pitch, i.e. the spatial period of the rotating spiral wave, is equal to the perimeter $L = 2\pi R$ of the hole. Respectively, the rotation period of the spiral wave is given by the time $T = 2\pi R/V$ that is needed for the excitation to make a full round of the hole boundary.

These were the main conclusions of Wiener and Rosenblueth. They also constructed the wave shapes for the media with obstacles of different forms. Figure 9.2, reproduced from the original paper, shows an example of two spiral waves rotating in opposite directions around two holes of equal sizes. The subsequent wave positions are indicated and annihilation of colliding wave parts is taken into account here.

Looking again at the above construction of the spiral wave, we can notice that not every obstacle can support a rotating spiral wave. Indeed, if the obstacle is too small, so that its perimeter L is shorter than the width $L_{\mathrm{min}} = V\tau_{\mathrm{e}}$ of the excitation wave, circulation of excitation around it is not possible (when the excitation front would return after a full rotation it would find

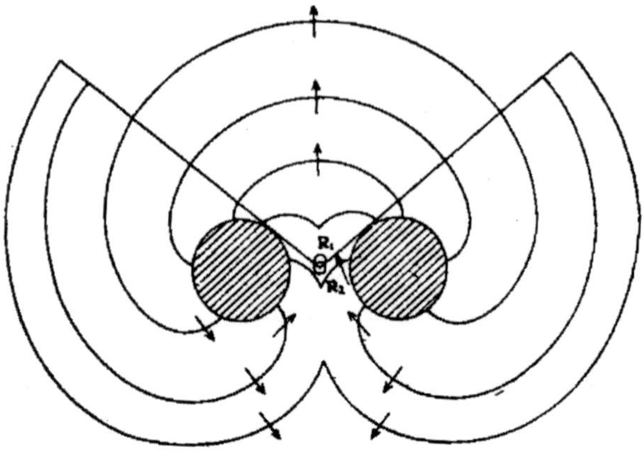

Fig. 9.2. Spiral waves around two obstacles. Figure taken from [530], p. 221

the medium still in the refractory state and thus die out). This implies the existence of the critical radius $R_c = V\tau/2\pi$ of a hole. The smaller holes cannot support flutter in this theory.

Circulation of spiral waves around obstacles allows us to explain the phenomenon of flutter in the cardiac muscle. Wiener and Rosenblueth analyzed the examples of experimentally observed flutter [535], and concluded that they were consistent with this explanation. It was also clear that similar arguments would explain persistent circulation of excitation in the rings of an excitable medium and hence circulating excitation regimes in the nervous net of a medusa. Though Garey [534] earlier reported an observation of flutter in the cardiac tissue in the absence of a hole, Wiener and Rosenblueth believed that these particular experiments were not carried quite accurately and therefore they did not violate their firm belief that the circulation of spiral waves was possible only around obstacles of a sufficiently large size.

9.3 Free Spiral Waves

However, just two years later Selfridge [532] showed that freely rotating spiral waves were also possible in the WR theory. His argument, independently formulated in 1965 by Balakhovsky [537], was very simple. Let us take a minimal hole of radius R_c that is still able to support a circulating spiral wave and start to deform it, while keeping the wave rotating around this hole. The performed deformation represents stretching the hole in the vertical direction and squeezing it horizontally, so that the perimeter of this obstacle always remains constant. Continuing this distortion, we arrive eventually at a situation when the obstacle would represent just a linear cut in the medium with the length πR_c. Going up the right side of this cut and down along its left side would make the distance $2\pi R_c = V\tau_r$. Clearly, the wave can

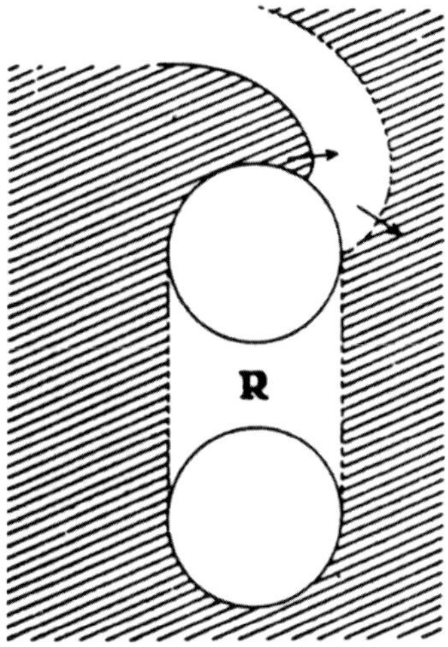

Fig. 9.3. Geometrical construction of the spiral wave in the WR model

still circulate around this special linear obstacle. If we look closer at the circulation regime, we notice however that while, for instance, the excitation wave is coming up along the right side of the cut, the elements on its left side are in the refractory state and hence would not become excited even if the cut were suddenly removed. The same observation can be made if we watch the wave going down the left side of the linear obstacle. Accordingly, Selfridge and Balakhovsky suggested that, once steady circulation of an excitation wave along a linear cut has been established, the cut is completely removed so that a continuous excitable medium is produced. The spiral wave, obtained by squashing flat the minimal hole and removing the cut, will thus circulate in the medium in the absence of any holes. Its tip then moves up and down a linear segment of the length equal to half of the perimeter of the minimal hole. The temporal period of such a free spiral wave is equal to the recovery time of the medium.

It is interesting that Wiener and Rosenblueth were very close to getting this result. In their paper they considered a spiral wave rotating around *two* holes of equal radius r (Fig. 9.3). "The solid left-hand line with the arrow in Fig. 9.3 represents parts of the wave-front and its direction of progression; the dotted right-hand line represents the corresponding position of the wave back. All the shaded region is refractory at the instant represented in the diagram; the unshaded region between the two obstacles (R) may be partly or wholly refractory or not refractory at all. When the wave front reaches that region it either finds it refractory and keeps on toward the other object, or else it penetrates. But when it penetrates, it cannot go beyond R to the left because that part will still be refractory. The left boundary of R can only be reached around the lower obstacle. When the wave has reached this left

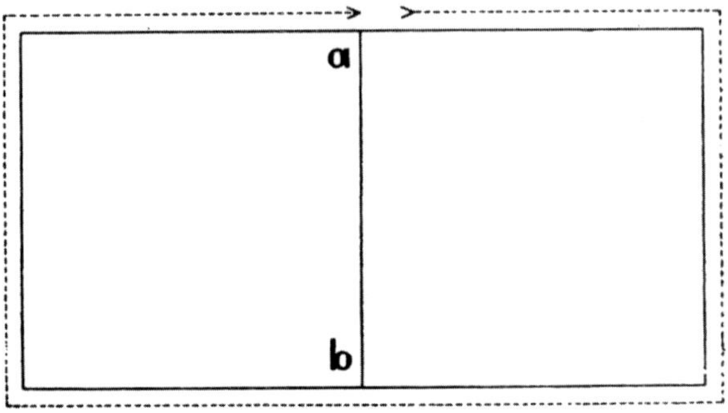

Fig. 9.4. A spiral wave rotating around two small obstacles. Figure taken from [530], p. 224

boundary, the conditions will then hold for the right part of the boundary of R. The wave will therefore repeat along a path surrounding the two obstacles. Every part of R will fire once in every cycle. Essentially the same situation as that of Fig. 9.3 is more simply illustrated by the line diagram of Fig. 9.4. The solid lines represent transmitting fibers and the dotted arrow spans the region between the wave front and the wave back. The state of the conducting line *ab* has no effect on the rest of the system." ([530] pp. 224–225).

Simple geometrical analysis shows that such a circulation regime is possible if the distance L between the centers of two holes satisfies inequalities $2L + \pi r > L_{\rm min} > 2L + \pi r/2$. The dimensions of the holes are unimportant otherwise. But, if this is indeed so, we may make the holes as small as we wish and eventually completely eliminate them! In this way we again come to the solution, described by Selfridge and Balakhovsky.

This theoretical analysis has prepared the ground for experimental observation of freely rotating spiral waves. Though the original motivation for studies of excitable media had come from cardiology and neuro-physiology, the cardiac and neural tissues did not actually represent good experimental systems for observation of spiral waves. They were strongly non-uniform and anisotropic. Moreover, the visualization of excitation patterns in cardiac and neural tissues was very difficult. A simpler experimental system was therefore needed.

Such a system was indeed discovered in 1950 by Belousov [538]. Experimenting with a reaction, intended to mimic the biochemical Krebs cycle, he found that this reaction shows persistent oscillations. All his attempts to report this observation in Soviet chemistry journals failed. Finally, much later, a short description of the reaction was published [539]. However, the reaction became widely known and Zhabotinsky started experimental studies of this chemical system at the University of Moscow. Uniform oscillations, propagating waves and "target patterns" of concentric spreading waves were observed [522,540,541].

The WR model of excitable media was discussed at the seminars on mathematical biology held by I. M. Gel'fand at the University of Moscow. Several extensions of this model were proposed [542–545]. Since this theory was formulated in general terms, it was clear that its applications would not be limited to cardiac or neural tissues. Similar phenomena of pattern formation were expected for other systems consisting of locally coupled active elements. The reaction, discovered by Belousov, provided an example of an active medium where individual volume elements were coupled through diffusion of reactants. In 1971, Zhabotinsky and Zaikin published the first photographs of spiral wave patterns in the oscillating reagent [546] (see also [522,547]).

The reaction was also studied by Winfree, who got its recipe in 1968 from Zhabotinsky during his visit to Prague. His original motivations stemmed from theoretical studies of synchronization phenomena in populations of coupled oscillators, though he was also familiar with the paper by Wiener and Rosenblueth. Winfree modified the chemical composition of the solution and thus created an excitable chemical medium. Photographs of spiral waves in this system first appeared in 1971 in a Swedish popular science magazine [548]; the detailed account of the experimental results together with their theoretical interpretation was published one year later [549]. It should be noted that at about the same time free spiral waves were observed in populations of *Dictyostelium discoideum* [527] and were actively studied in the heart [528].

The first model simulations of spiral waves were made in 1972 by Gul'ko and Petrov [550] for a system of partial differential equations that represented a reduced version of the Hodgkin–Huxley model. These were followed by other numerical studies of such patterns in reaction-diffusion systems. Surprisingly, the experimental and numerical studies of freely rotating spiral waves did not completely agree with the theoretical predictions by Selfridge and Balakhovsky, based on the WR theory. The rotating waves indeed had the form of Archimedian spirals. However, the phenomena taking place in the central region of such patterns were different. Apparently, the tip was rotating around a circle of a definite radius rather than moving up and down along two sides of a line segment.

Figure 9.5 displays the results of one such simulation [551]. Initially (frames a–e) the spiral wave rotates here around a relatively large hole that is seen in the center. Subsequently, the size of the hole is decreased (frame f). We see that the wave does not disappear. Instead its tip detaches from the actual hole and starts to rotate around a circle of a *larger* radius (frames g–k). This circle forms the core of the spiral wave. Now, the small hole lying in the center of the core can be completely removed without influencing in any way the rotation regime of the spiral wave. The numerical investigations have revealed that the elements in the core region never became excited and remained in the state of rest. Moreover, the rotation period of free spiral waves could be many times larger than the recovery time of the medium.

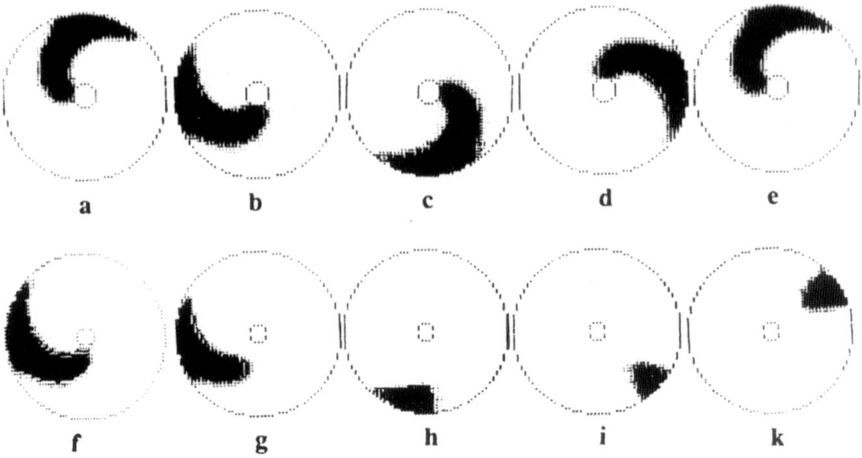

Fig. 9.5. Circulation of the wave around a large obstacle (**a**–**e**) and transition (**f**–**k**) to the free rotation after the size of the obstacle was decreased. Figure taken from [551] with permission

The numerical simulations of interactions of excitation waves with obstacles have also been performed. As an example, Fig. 9.6 shows passing of a flat excitation wave through a small obstacle located in the center of the integrated system [552]. When the excitation threshold was relatively low, the wave recovered its integrity after being temporally broken by the obstacle (Fig. 9.6a). If, however, the excitation threshold was increased, a different effect was found. The two ends of the broken wave started to retreat until the wave completely disappeared (Fig. 9.6b).

The system shown in Fig. 9.7 has a long thin obstacle, extending from the left to approximately the center of the medium. Initially, an excitation wave is sent along the thin channel in the upper left corner of the medium. This wave

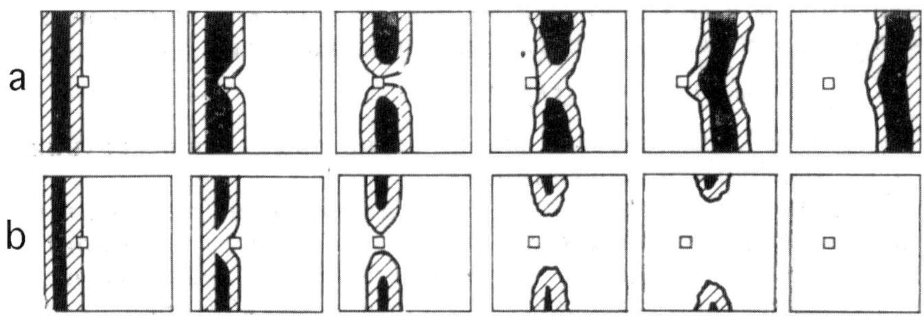

Fig. 9.6. Wave propagation through a medium with an obstacle (the *small square* in the center) at higher (**a**) and lower (**b**) excitability of the medium. Single frames are separated by equal time intervals. Figure taken from [552] with permission

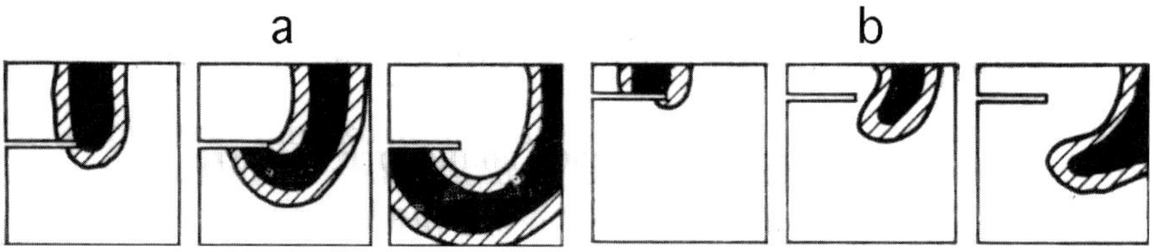

Fig. 9.7. Wave motion near a long thin obstacle (the *narrow rectangle*) for two different excitabilities of the medium: (**a**) the wave follows the boundary of the obstacle, (**b**) the wave becomes separated from the boundary and a free tip is formed. Figure taken from [553] with permission

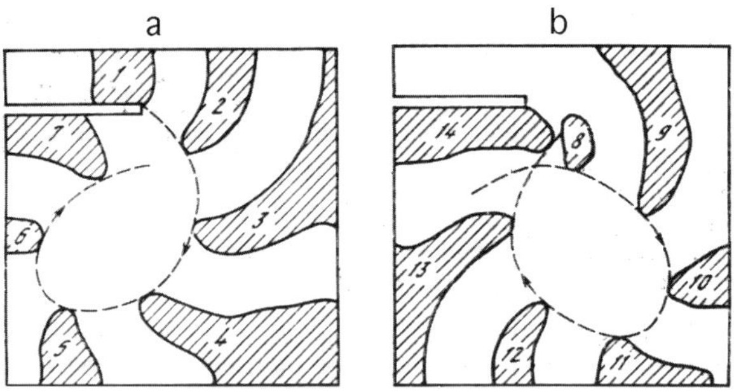

Fig. 9.8. Formation of a spiral wave; the first two rotations (**a,b**) are shown. Figure taken from [553] with permission

travels through the channel and suddenly comes to a point where the obstacle is ended. The subsequent evolution of the wave again depends on the excitation threshold in the considered system [553]. When it was higher, the wave remained stuck to the obstacle. Following its perimeter, the wave turned and started to move in the reverse right direction along the obstacle (Fig. 9.7a), in accordance with the predictions by Wiener and Rosenblueth. When, however, the threshold was increased, the wave tip *separated* itself from the obstacle (Fig. 9.7b). The subsequent evolution is schematically shown in Fig. 9.8. We see that the wave tip starts to rotate and a free spiral wave is thus generated.

The process leading to the formation of a free rotating spiral wave is clearly seen in Fig. 9.9. The initial condition in these simulations has been chosen in the form of a broken flat excitation wave. Figure 9.9a–d show subsequent wave shapes and positions for four different values of excitability [554]. If the excitability is low (Fig. 9.9a), the wave contracts and disappears (cf. also Fig. 9.6b). On increasing the excitability of the medium, a marginal situation can be achieved (Fig. 9.9b) where the wave steadily moves in the right direction, neither growing nor contracting with time. Further increase in the medium's excitability gives rise to a growing wave (Fig. 9.9c). We see that the tip moves a little slower here than the rest of the wave, as a consequence of this effect, the wave bends already tending to form a spiral wave with the

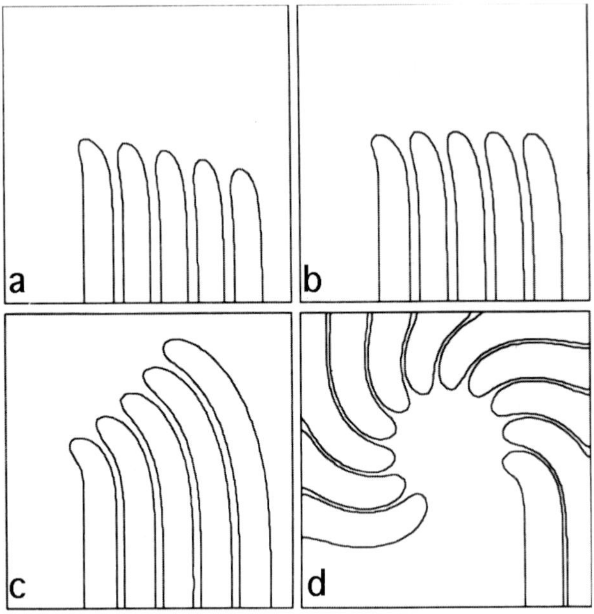

Fig. 9.9. Evolution of a broken flat wave in a reaction-diffusion model of an excitable medium. Contours of the constant activator concentration are superposed at subsequent time moments for four different (**a** – **d**) excitabilities of the medium. Figure taken from [554] with permission

core of a large radius. If the excitability is made even higher, the growth and curling is stronger, so that a rotating free spiral wave is clearly seen in Fig. 9.9d.

Formation of free-rotating spirals from broken waves can also be clearly seen in cellular automata models of excitable medium [555,556].

9.4 Kinematics of Curved Waves with Open Ends

The experimental data and the numerical simulations, described in the previous section, show that the development of free spiral waves is intrinsically related to the motion of wave tips. These situations, where the excitation waves were broken and had freely moving open ends (tips), were not considered in the original WR theory that assumed that the waves could only move sticking to the medium boundaries. Thus, if we want to describe such spiral waves in the framework of the WR approach, the theory should be modified to specify the laws of motion of the waves with open ends.

The tip can grow or retract. Therefore, besides the propagation velocity V of the front, we should introduce another important parameter, i.e. the growth velocity G of the tip. We can assume that, if G is negative, this corresponds to a retracting tip.

However, it may easily be seen that this modification alone is not sufficient to describe the formation of a free spiral wave from a wave break. Indeed, at the initial stage of this process the tip grows. But, as the wave curls and approaches the spiral shape, the growth slows down. In the final state of a steadily rotating spiral wave, the growth velocity of the tip is zero since it moves along a circular trajectory and remains orthogonal to this circle. Apparently, the tip growth velocity should therefore gradually change in time during such a process.

In the WR model, an excitation wave is pictured as an infinitely thin oriented curve. Hence, the only local property of the wave can only be its local curvature K. Evidently, the curvatures of all wave elements, including that one which is next to the tip, increase with time until a steady rotating spiral with a constant shape is formed. Therefore, the model can be improved by assuming that the tip growth velocity is a function of the front curvature near the tip, i.e. $G = G(K)$. Note that the tip itself represents in this model simply a point and has therefore no curvature of its own. It is convenient to choose the curvature of a wave segment as positive if the wave is locally convex in the propagation direction. When the curvatures are small, a linear dependence can be expected so that we have

$$G = G_0 - \gamma K \,, \tag{9.1}$$

where G_0 is the tip growth velocity for a flat wave and γ is a coefficient that specifies the sensitivity of the tip to variations of the curvature. Once the curvature dependence of the growth velocity is incorporated into the model, we must, to be consistent, also take into account the curvature dependence of the wave propagation velocity V. For small curvatures, a linear law should hold, i.e.

$$V = V_0 - DK \,, \tag{9.2}$$

where V_0 is the propagation velocity and D is a constant coefficient.

The modification of the WR theory, including motion of broken waves with open ends and employing the curvature dependences (9.1) and (9.2), has been proposed by Davydov and Mikhailov. It was based on the earlier studies by V.S. Zykov [557,558] who considered only effects of the curvature dependence (9.2) on the wave propagation in excitable media (see, for example, [559]). The linear curvature dependence (9.2) of the propagation velocity has been introduced for combustion waves in 1951 by Markstein [560]. In the problems of crystal growth, such dependence was employed in 1951 by Burton et al. [561].

The coefficient D must be positive in order to guarantee the stability of propagating waves with respect to their spontaneous deformation. If D is negative, any small protrusion of a flat front will have a positive curvature and thus will have moved faster than the main flat front eventually destroying it.

Using (9.1) and (9.2), the evolution leading to the formation of a freely rotating spiral wave can be qualitatively understood. Curling of a broken wave takes place because its curvature is higher near the tip and hence the wave propagation velocity is smaller there. As the wave curls, its curvature at the open end is increasing and this diminishes, according to equation (9.1), the tip growth velocity. Eventually, the tip growth is terminated and the steady rotation is established when the curvature at the wave end reaches its critical value $K_c = G_0/\gamma$.

This conclusion is confirmed by direct numerical simulations of the extended WR model with (9.1) and (9.2). Figure 9.10a,b shows the formation of a steady rotating spiral wave from a flat wave for two different initial perturbations near its free end [554]. Though the transients are different, the final spiral wave is the same in both cases. The open wave end moves orthogonally to a circle representing the core boundary of the spiral wave. It does not grow or act in the radial direction. The wave in Fig. 9.10 is freely rotating and the tip does not penetrate into the core region not because it is occupied by an obstacle but because of its own dynamics, controlled by the local curvature of the wave.

Numerical simulations and analytical studies of the extended WR model are based on the evolution equation for the local wave curvature. The derivation and detailed discussion of this nonlinear partial differential equation can be found in [555,562]. One of the important results of this investigation is

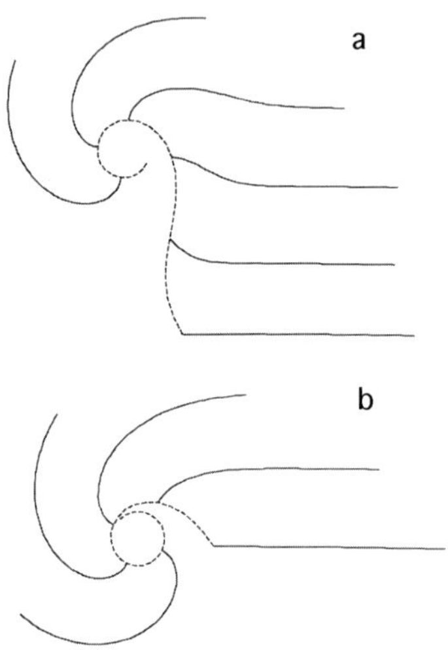

Fig. 9.10. Formation of a steadily rotating spiral wave from an initial broken flat wave with two different perturbations (**a,b**) near its end. The extended WR model with parameters $V_0 = 1.5$, $G_0 = 0.96$, $D = 1.0$, $\gamma = 1.5$. Figure taken from [554]

that the rotation frequency of free spiral waves can be estimated in the limit $DG_0 \ll \gamma V_0$ as

$$\omega_0 = \zeta \left(\frac{DG_0}{\gamma V_0} \right)^{3/2} \frac{V_0^2}{D} , \qquad (9.3)$$

where the numerical constant is $\zeta = 0.685$.

It can be noted that spiral waves are possible in this model only when G_0 is positive, i.e. the broken flat wave grows. The rotation period $T_0 = 2\pi/\omega_0$ of free spiral waves goes to infinity as $T_0 \approx G_0^{-3/2}$ when G_0 approaches zero. This is accompanied by a fast increase of the core radius $R_0 = V_0/\omega_0$. The model parameter G_0 is controlled through the effective excitability of the medium. Hence, one can reproduce the basic regimes illustrated in Fig. 9.9.

Remarkably, the extended WR model is able to explain several interesting effects that have been recently found in experiments with excitable media and obtained in numerical simulations of reaction-diffusion models. We discuss some of these applications in the next three sections.

9.5 Complex Anisotropy

Formulating their original model, Wiener and Rosenblueth noted that "the definition of homogeneity adopted here implies isotropy, although the fibrillar structure of cardiac muscle, with some orientation of fiber bundles in certain regions, makes absolute isotropy unlikely" ([530], p. 208). For a long time, the role of anisotropy in excitable media remained only little investigated (because the chemical excitable media based on aqueous solutions, that have been actively studied in the last decades, did not show any anisotropy at all). Only recently has this problem been thoroughly analyzed, motivated by the experimental studies of chemical reactions on catalytic metal surfaces where strong diffusion anisotropy of adsorbed molecules is typical. Generally, the propagation anisotropy may significantly influence the wave propagation properties and even lead to the appearance of a new kind of wave pattern in such systems.

The simple anisotropy is characteristic, for instance, for activator-inhibitor systems with one diffusing species whose diffusion coefficient is constant. Such anisotropy can easily be eliminated by rescaling the coordinates. If, however, both the activator and the inhibitor are diffusing and their diffusion anisotropies are different, the situation with the *complex* propagation anisotropy is realized. Similar behavior takes place in the models with a single diffusing activator species when the anisotropy is state–dependent.

The original WR model can be generalized to describe the phenomena in planar anisotropic media by assuming that the wave propagation velocity is a function $V(\alpha)$ of the angle α between the wave tangent vector and the anisotropy axis of the medium. When the propagation anisotropy is simple, this function has a universal form

$$V(\alpha) = V_{\min}(q \sin^2 \alpha + \cos^2 \alpha)^{1/2}, \tag{9.4}$$

where the constant parameter q is fixed by the ratio of the maximal (at $\alpha = \pi/2$) and minimal (at $\alpha = 0$) propagation velocities, i.e. $q = (V_{\max}/V_{\min})^2$. In systems with complex anisotropy, there is no universal angular dependence of the propagation velocity. Hence, if we write $V(\alpha)$ in the form (9.4), the parameter q will then depend on the propagation direction, i.e. we would have $q = q(\alpha)$.

Using the angular dependence $V(\alpha)$ of the propagation velocity, evolution of initially circular spreading wave has been considered [563]. If it is described by (9.4), the wave approaches an elliptic shape with time. In the excitable media where the anisotropy is complex and (9.4) does not hold, the wave shapes are not elliptic. Figure 9.11a displays a non-elliptic target pattern in the medium with complex anisotropy, characterized by the dependence $q = a + b \sin^2 \alpha$ for $a = 2$ and $b = 0.5$.

However, the effects of the complex anisotropy may be even stronger. Spreading convex waves in such media may spontaneously develop *edges*. This is possible if the function

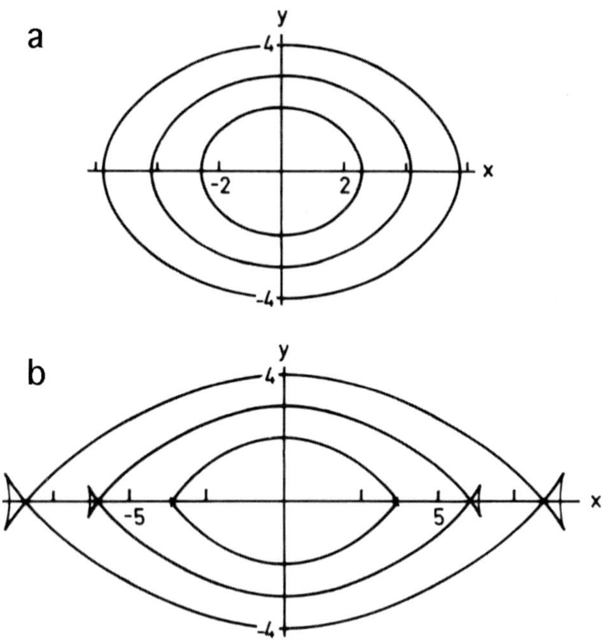

Fig. 9.11. Target patterns (expanding concentric waves) in the medium with the complex anisotropy for (**a**) $b = 0.5$ and (**b**) $b = 5$; other parameters are $a = 2$ and $V_{\min} = 1$. The outer loops ("swallow tails") of the waves are unphysical and should be cut off. Figure taken from [563]

$$Q'(\alpha) = V(\alpha) + \frac{d^2V}{d\alpha^2} \tag{9.5}$$

is negative in an interval of the angles α. Figure 9.11b shows an eye-shaped target pattern for the model with $q = a + b\sin^2\alpha$ for $a = 2$ and $b = 5$. From the mathematical viewpoint, this phenomenon is related to the formation of caustics on convex fronts in such anisotropic media (see [563]).

The spontaneous development of sharp edges, where the local wave curvature reaches high values, may have important practical consequences. Indeed, if the medium is not uniform and the excitability varies across it, breaking of the waves would be greatly facilitated at their edges. This may lead to the generation of multiple spiral waves and eventually to turbulent regimes, or to fibrillation. Square-shaped target patterns with sharp edges have been observed in the experiments with catalytic surface chemical reactions [564].

Another interesting effect of complex anisotropy is related to the motion of excitation waves with open ends. In isotropic media the motion of a broken wave is characterized by its growth velocity G (when G is negative, the wave is contracting). The curvature dependence of the tip growth velocity is responsible in such systems for the development of steadily rotating spiral waves. In anisotropic excitable media, the tip growth velocity depends on its direction of motion, so that $G = G(\alpha)$.

When the anisotropy is simple and can be eliminated by rescaling coordinates, the function $G(\alpha)$ is either positive or negative for all angles α. This can easily be understood, if one takes into account that a scaling transformation may change the magnitude of motion, but will not influence its direction. When, however, the anisotropy is complex, the velocity $G(\alpha)$ can change its sign at a certain angle α_0. Thus, depending on their direction of motion, the waves may grow or contract in the same medium.

Suppose that $G > 0$ for the motion in the vertical direction, but $G < 0$ for the motion in the horizontal direction. If we initially take a short flat wave segment moving in the vertical direction, it begins to grow and to curl at its both open ends, tending to form a double spiral. But at the same time the growth velocity $G(\alpha)$ is decreasing, because the local propagation direction near the tips turns closer to the horizontal axis. When the critical angle α_0, such that $G(\alpha_0) = 0$, is reached, the growth is stopped before a full tip rotation is made. Hence, instead of spiral waves, *traveling wave fragments* are produced [565,566]. The development of such a traveling fragment, computed using the extended WR model with complex propagation anisotropy, is shown in Fig. 9.12b.

In a previous section we noted that, as the effective excitability of a medium is increased (e.g. the excitation threshold is lowered), a transition from shrinking to growing broken waves, which form spirals, takes place (Fig. 9.9). In the extended WR model, this corresponds to a transition from the negative to the positive tip growth velocity G_0 of a flat wave. In isotropic media, this transition occurs at the same critical excitability for all propagation directions. If an excitable medium is characterized by complex propaga-

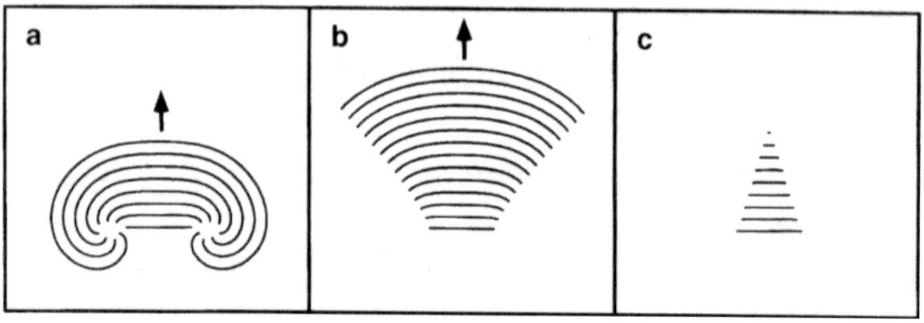

Fig. 9.12. Evolution of an initially flat small wave fragment in the medium with complex anisotropy in the extended WR model. The angular dependence of the tip growth velocity is $G_0 = A - B(\sin\alpha)^2$. The different model parameters are (**a**) $A = 1.5$, (**b**) $A = 0.5$, and (**c**) $A = -0.5$; the common parameters are $V_0 = 1.5, D = 1.0, \gamma = 1.5$, and $B = 1.0$. Figure taken from [565]

tion anisotropy, this is no longer the case. As the excitability is increased, G_0 will first change its sign and become positive only for the motion along a certain preferred direction. As shown above, steadily traveling wave fragments are formed under these conditions.

Hence, a transition from shrinking wave fragments to rotating spiral waves will generally be separated in the excitable media with complex anisotropy by the regimes where curved wave fragments, steadily traveling along a certain preferred direction, are found (Fig. 9.12). This behavior has indeed been seen in numerical simulations of a FitzHugh–Nagumo model with state-dependent anisotropy [565]. It was observed in the experiments with surface chemical reactions, providing a good example of an excitable medium with the complex anisotropy (Fig. 9.13).

9.6 Resonances

An advantage of the extended WR model, including motion of open ends, is that it describes not only steadily rotating free spiral waves, but also processes leading to their formation and relaxation of a spiral wave to its stable form after a perturbation has been applied [563]. Furthermore, this simple theory can be used to consider the effects caused by spatial or temporal modulation of the medium's properties.

A steady spiral wave has a fixed rotation period and therefore it is natural to ask whether *resonance* effects would take place if the excitability is periodically modulated with the same time period. This problem was the first application of the extended WR model. To incorporate time dependence of the medium excitability into the model, we assumed that the growth velocity

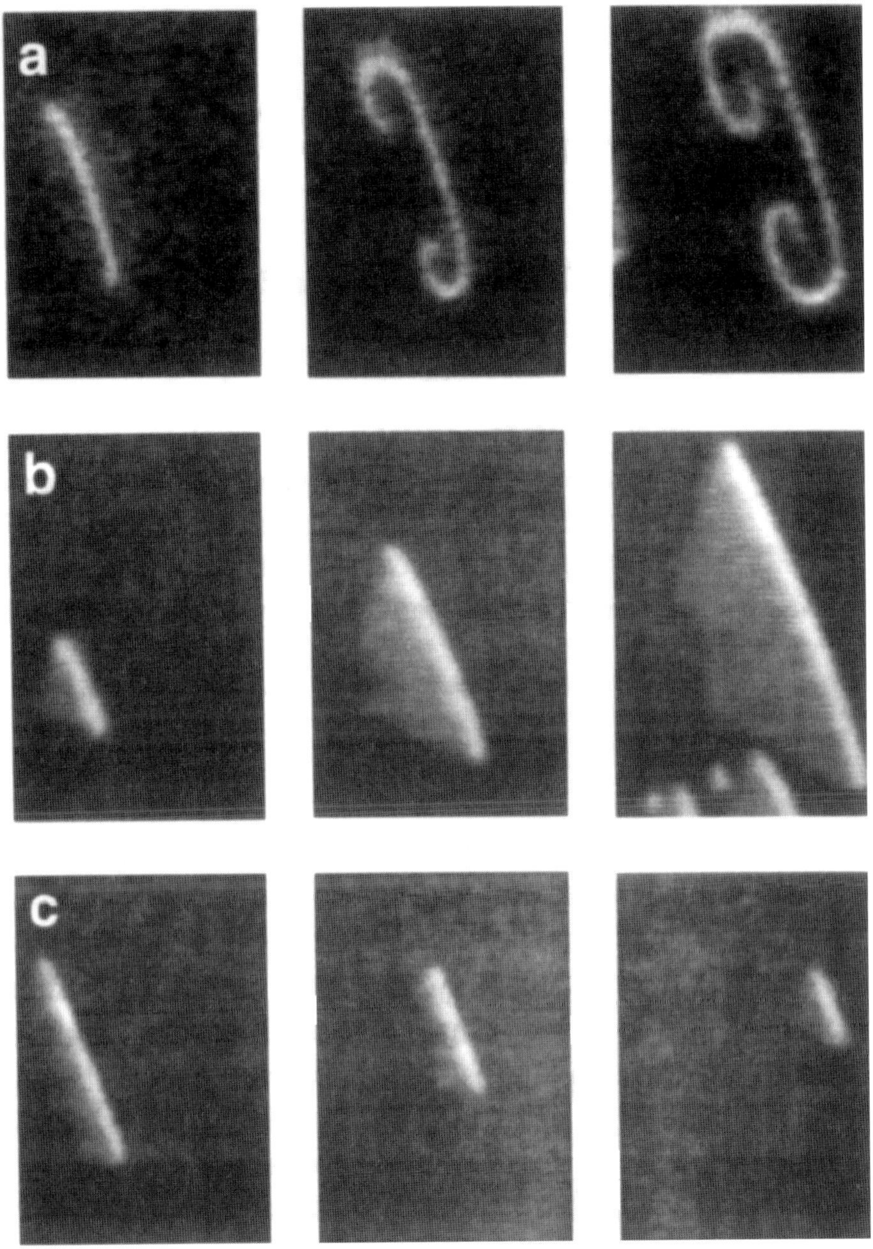

Fig. 9.13. Evolution of a single wave fragment for three different experimental conditions (**a,b,c**) in the excitable, medium formed by a surface chemical reaction. Subsequent frames are shifted to accommodate the entire traveling objects. The size of the frames is $75 \times 100\,\mu\mathrm{m}$. Figure taken from [565]

G_0 of the open flat wave that is most sensitive to variations in the excitability was periodically modulated as

$$G_0(t) = \widetilde{G}_0 + G_1 \cos(\omega_1 t + \phi_1)\,. \tag{9.6}$$

The modulation amplitude was small ($G_1 \ll \widetilde{G}_0$) and the modulation frequency ω_1 was close to the frequency ω_0 of the spiral wave.

The analysis reveals [555,562,567] that in the presence of such periodic modulation the steady rotation of a spiral wave is destabilized and replaced by a complex motion. Introducing the complex coordinate $Z(t) = X(t)+\mathrm{i}Y(t)$ of the end point, this motion is described (for a spiral wave rotating in the clockwise direction) as

$$Z(t) = R_0 \exp(-\mathrm{i}\omega_0 t - \mathrm{i}\phi_0) \pm R_1 \exp\left[-\mathrm{i}(\omega_0 - \omega_1)t - \mathrm{i}(\delta + \phi_0 - \phi_1)\right], \tag{9.7}$$

where

$$R_1 = \frac{3\gamma G_1 V_0}{4\widetilde{G}_0 \left|\omega_0 - \omega_1\right| \sqrt{\gamma^2 + D^2}}, \tag{9.8}$$

and $\delta = -\arctan(\gamma/D)$. The plus sign is taken here if $\omega_1 > \omega_0$, the minus sign corresponds to the case $\omega_1 < \omega_0$. The rotation frequency ω_0 of the spiral wave is determined by (9.3); $R_0 = V_0/\omega_0$ is the core radius of the spiral wave.

Figure 9.14 shows the tip trajectories, described by (9.7) for three different choices of the modulation frequency ω_1. We see that they represent a superposition of two circular motions with radii R_0 and R_1 and angular velocities ω_0 and $\omega_0 - \omega_1$. The tip moves along the circular core of radius R_0, while the core center itself moves along a circle of radius R_1. If the modulation frequency is smaller than the frequency of the spiral wave ($\omega_1 < \omega_0$), the directions of rotation around these two circles are the same. In this case the tip trajectory is a cycloid ("a flower") with inside-looking petals. When

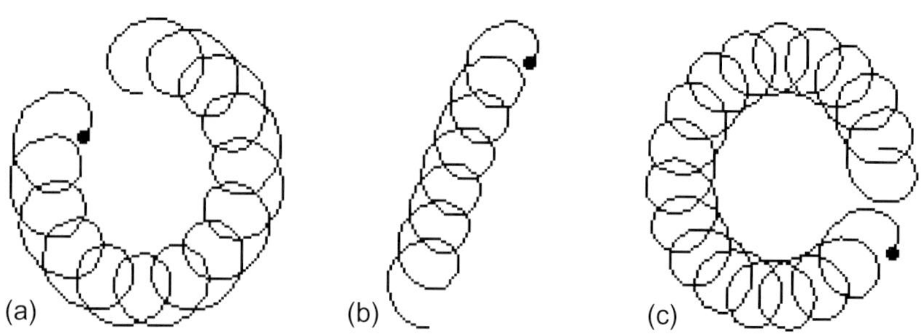

Fig. 9.14. Resonance of spiral waves. The trajectory of the wave tip is shown for three different frequencies of the external forcing: (**a**) $\omega_1 = 0.0205$, (**b**) $\omega_1 = 0.0217$, and (**c**) $\omega_1 = 0.023$. Other parameters of the extended WR model are $V_0 = 1, D = 1, G_0 = 0.15, \gamma = 1.5$ and $k_1 = 0.025$. The natural rotation frequency of spiral waves is $\omega_0 = 0.2166$

$\omega_1 > \omega_0$, the core center rotates in the direction that is opposite to the rotation direction of the tip along the core boundary. Then the trajectory is again a cycloid, but with outward–looking petals. Remarkably, the rotation radius R_1 of the core center is inversely proportional to the difference $\omega_0 - \omega_1$ between the frequency of the spiral wave and the modulation frequency. Hence, when the resonance $\omega_1 = \omega_0$ is approached, this radius is indefinitely increasing. Under the condition $\omega_1 = \omega_0$ of the complete resonance the rotation center of the spiral wave moves along a straight line, so that its tip performs the motion described by

$$Z(t) = R_0 \exp(\pm i\omega_0 t - i\phi_0) + vt \exp\left[\pm i(\delta + \phi_0 - \phi_1)\right] , \qquad (9.9)$$

where the drift velocity v is given by

$$v = \frac{3\gamma G_1 V_0}{4\widetilde{G}_0 \sqrt{\gamma^2 + D^2}} . \qquad (9.10)$$

The plus sign in (9.9) corresponds to the spiral wave, rotating in the counterclockwise direction; the minus sign is for the clockwise rotating wave.

When we derived these results, we realized that they provided a nice way of testing the extended WR model. By that time it was known that by modifying the chemical composition of the Belousov–Zhabotinsky reaction it could be made light-sensitive. Hence, by varying the illumination intensity, it was possible to change the effective excitability of this system. In our experiment [568] we first initiated a spiral wave in a thin layer of the BZ solution. After measuring the rotation period of this wave, we started to periodically and uniformly illuminate the solution with an intense light source. The wave pattern was photographed against the background of fixed reference lines.

Figure 9.15 shows three subsequent frames taken at time intervals of 10 min when the modulation period was chosen equal to the natural rotation period of the spiral wave. In agreement with the theoretical predictions, the wave center drifts at a constant velocity along a straight line (the total displacement in Fig. 9.15 is 2 cm).

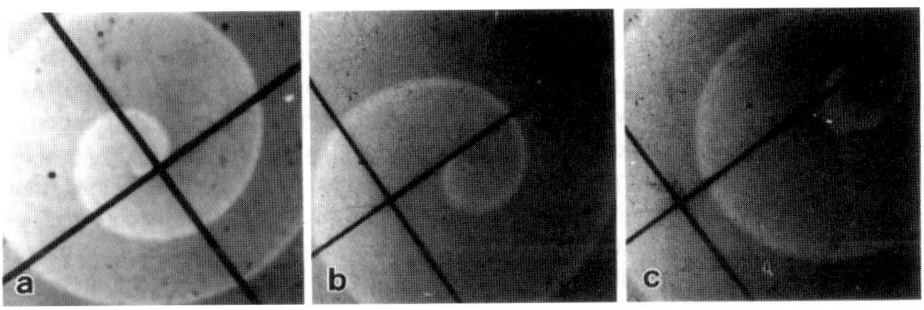

Fig. 9.15a–c. Drift of the spiral wave in the BZ reaction induced by external periodic forcing. Figure taken from [568]

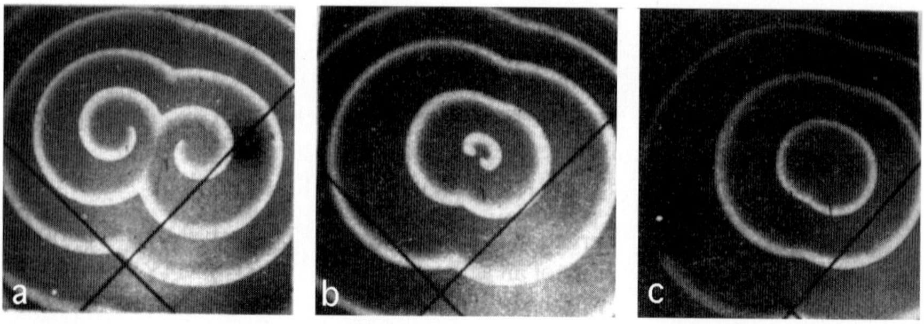

Fig. 9.16a–c. Annihilation of two spiral waves in the BZ reaction, induced by external periodic forcing. Figure taken from [568]

Another experimental verification of the theory involved the action of a periodic light modulation on a pair of spiral waves rotating in opposite directions. As follows from (9.10), under the resonant conditions such waves should drift in different directions. Therefore, they may collide and annihilate each other. This behavior was indeed observed in the experiment (Fig. 9.16).

Subsequently Zykov has also numerically tested these predictions of the extended WR model. He integrated the differential equations of the FitzHugh–Nagumo system with a periodically varying rate parameter and was able to reproduce both possible cycloidal regimes, depending on the relationship between the modulation frequency and the natural rotation frequency of the spiral wave [567].

The resonances of spiral waves play a fundamental role in such phenomena as the drift of spiral waves in the presence of spatial gradients or external fields. Moreover, they provide a simple explanation for the migration of spiral waves along physical boundaries of the medium and the formation of bound states of spiral waves.

Though a spiral wave is an extended pattern that may fill the whole medium, its properties and the response to external perturbations are essentially determined by what happens only in a narrow region close to its tip. If a perturbation is created at a certain distance from the tip, it is rapidly convected away and cannot influence the tip motion.

Let us consider, as an example, motion of a spiral wave in a non-uniform excitable medium whose properties slowly change along a certain direction, so that the spatial gradient is constant. When the gradient is absent, the tip moves along a circle representing the core boundary of the spiral waves. Introducing a weak gradient, we find that, while moving along the core boundary and thus repeatedly visiting regions with different properties, the tip effectively experiences a periodic modulation of the excitability. Moreover, since this modulation is caused by the tip rotation, its period coincides with the rotation period of the spiral wave. But this is exactly the situation that we have under external resonant forcing of the spiral wave that has been considered

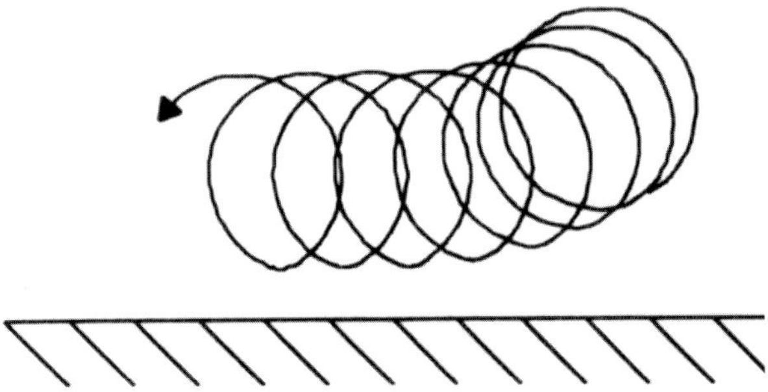

Fig. 9.17. Drift of the spiral wave along a flat boundary ($V_0 = 1.0, G_0 = 3.0$, $D = 1.0, \gamma = 2.0$). Only the tip trajectory is shown. Figure taken from [562]

above. Hence, the spiral wave should then drift with a constant velocity along a straight line forming a certain angle with the gradient. Indeed, drift of spiral waves in the presence of spatial gradients has been reported already in early numerical simulations of reaction-diffusion systems and is often observed in experiments with the BZ reaction.

A related effect is the drift of a spiral wave induced by application of external (e.g. electrical) fields [569,570]. In such systems the propagation velocity and the tip growth velocity vary, depending on whether the wave moves along or against the applied field. Since the rotating tip of the spiral wave moves in different directions, depending on its angular position, it again experiences an effective periodic modulation of the medium's properties that takes place with a period equal to the natural rotation period of the spiral wave. Therefore, the spiral wave should drift with a constant velocity along a straight line. In the framework of the extended WR model, the drift velocity and the drift direction with respect to the external field have been theoretically determined [566].

Numerical simulations of excitable reaction-diffusion systems [571] have shown that, if a spiral wave is placed close enough to a border of the medium, it may start to drift parallel to it. This migration of spiral waves proceeds at a fixed distance from the border, determined only by the medium properties. If initially the spiral wave is a little farther away from the border, it moves closer to it until this characteristic distance is restored. If the spiral wave is pushed slightly nearer to the border, it is repelled by it.

It is interesting that the migration phenomenon is contained in the extended WR model [562]. Figure 9.17 shows the tip trajectory for a spiral wave drifting along a flat boundary. The subsequent wave positions within a single rotation during this process are displayed in Fig. 9.18. A qualitative explanation of migration in the WR model is obtained if we notice that, as the tip rotates, it periodically comes closer to the medium's boundary. When this occurs, the curvature of the wave decreases which, in turn, influences the tip growth velocity G. Hence, we again have a situation when the tip

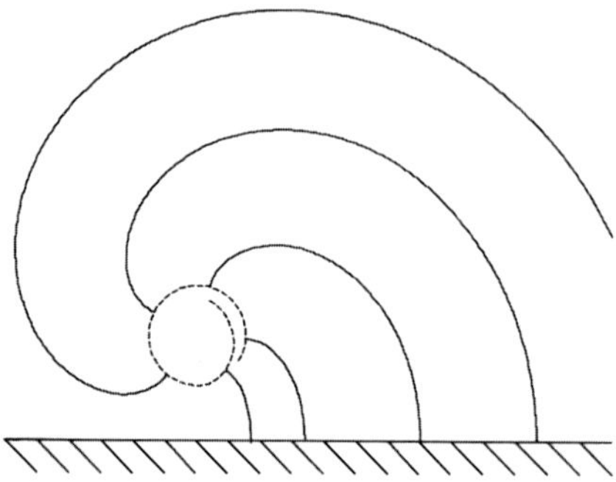

Fig. 9.18. Subsequent positions of the wave rotating near a flat boundary (time intervals are $\Delta t = 15$, other parameters are the same as in Fig. 9.17. Figure taken from [562]

feels a resonant modulation that would tend to produce its drift through the medium.

The waves always move orthogonally to the medium's boundary. Therefore, by making the mirror reflection of the wave pattern shown in Fig. 9.18 and pasting it with the pattern in the upper half-plane, we obtain a pattern that represents two spiral waves rotating in the opposite directions, or a *bound state* of two spiral waves. Such bound states have indeed been observed in numerical simulations of reaction-diffusion systems [572]. Migration of the spiral wave along the flat boundary implies that the spiral waves, forming the bound state, will slowly drift along the symmetry axis of this pattern. Moreover, a certain fixed distance between their centers is maintained during this process.

9.7 Meandering

When free spiral waves in the Belousov–Zhabotinsky reaction were discovered, it was found that the rotation of these waves is often not quite regular, i.e. their centers could wander through the medium [549]. Initially this *meandering* behavior was attributed to possible imperfections of excitable media. When numerical simulations of excitable media, described by reaction-diffusion models were later performed, it was, however, established that meandering could also take place in perfectly uniform systems.

It is natural to ask whether meandering of spiral waves can be understood in the framework of the WR theory and its extensions. In the above theoretical analysis of free spiral waves we have neglected the refractory properties of excitable media. This was possible if the estimated rotation period $T_0 = 2\pi/\omega_0$ of spiral waves was longer than the recovery time τ_r of the medium. In the original WR model, new excitation of elements was strictly forbidden before time τ_r has elapsed after they were last excited. However,

Wiener and Rosenblueth emphasized that this was only a simplification and remarked that "at the end of the [refractory] period they can be stimulated by a sufficiently strong electric pulse; in other words, the refractory period considered is that which is functional from the standpoint of conduction" ([530], p. 209). Hence, a medium element recovers some excitability near the end of the refractory period (i.e. in the "relative refractory phase") though the excitation threshold is then relatively high. This threshold gradually decreases during the relative refractory phase until it reaches its value characteristic for the rest state of the element when the recovery is completed.

A higher excitation threshold implies a lower propagation velocity. This means that if a wave moves through the medium whose elements are in the relative refractory phase its propagation velocity is decreased. The pulse propagation is not possible if the interval is shorter than a certain minimal time, representing the absolute refractory period. For larger intervals, pulse propagation is already possible, though the pulse velocity is smaller. The relative refractory phase can be approximately defined as corresponding to the time intervals during which the propagation velocity is significantly reduced.

In its original formulation, the WR model had no relative refractory phase. One possible extension of the WR model consists of the introduction of the relative refractory phase into this model. This can be done assuming that the propagation velocity is a function of the time elapsed after the last excitation of a given element of the medium. If the curvature dependence of the wave propagation velocity is also taken into account, this yields

$$V = V_0(T) - DK\,, \tag{9.11}$$

where $V_0(T)$ is the propagation velocity of a pulse or a flat wave through the medium whose elements have experienced previous excitation time T ago.

The dependence given by (9.11) has been used by Keener and Tyson [573] in their theory of spiral waves. For spiral waves around obstacles, this dependence yields corrections to the rotation period caused by the fact that the wave is curved at the obstacle's boundary and the medium elements near this boundary may have not completely recovered when an excitation waves again runs through them. Free spiral waves were treated by these authors as rotating around an effective obstacle whose radius should be empirically determined. The disadvantage of the Keener–Tyson theory was that it did not explicitly consider the motion of the wave tip. Therefore, it provided no way to describe transient processes, leading to steady rotation of free spiral waves. The response of such waves to external perturbations could not be considered and therefore the stability of steadily rotating free spiral waves could not be discussed.

Zykov [574] has suggested to further modify the WR model and take into account that not only the wave propagation velocity, but also the growth velocity G of the tip should depend on the time T elapsed after the last excitation:

$$G = G_0(T) - \gamma K\,, \tag{9.12}$$

where K is the curvature at the open end and $G_0(T)$ is a function that decreases for smaller times T.

Once this last extension is made, it becomes possible to consider arbitrary motions of broken waves, taking into account the recovery effects in the medium. The actual numerical simulation is performed by defining for each point (x, y) of the medium the variable

$$T(x,y,t) = t - T^*(x,y)\,, \tag{9.13}$$

where $T^*(x, y)$ is the last time moment when the excitation wave has passed through this point. The form of the functions $V_0(T)$ and $G_0(T)$ is not universal and should be determined for each particular excitable medium (see [575]). As an example, the functions

$$V_0(T) = V_0\left(1 - \frac{T_{\mathrm{r}}}{T}\right), \tag{9.14}$$

$$G_0(T) = G_0\left(1 - \frac{T_{\mathrm{t}}}{T}\right), \tag{9.15}$$

can be chosen. By varying the parameter T_{r}, one can then control the effective recovery time of the medium.

Figure 9.19 shows the evolution of free spiral waves in the extended WR model under an increase of the recovery time. Only the trajectory of motion of the wave tip is shown; the initial positions of the tip are indicated by arrows. If T_{r} is small, the recovery effects are negligible and stable steady rotation of the spiral wave is established (Fig. 9.19a). This freely rotating spiral wave has the period T_0 which can be estimated as $T_0 \approx 14.6$ using (9.3). As T_{r} increases, steady rotation becomes unstable and a small-amplitude modulation develops (Fig. 9.19b). At $T_{\mathrm{r}} = 10$ the amplitude increases and the two-period motion is clearly seen (Fig. 9.19c). If T_{r} is further increased, the trajectory approaches a rosette form (Fig. 9.19d,e) and cycloidal orbiting is observed (Fig. 9.19f). At a certain value of the parameter T_{r}, the spiral wave drifts along a straight line (Fig. 9.19g); this behaviour is similar to that observed for free spiral waves under resonant external forcing. At still larger values of T_{r}, the tip trajectory represents a flower with outward-looking petals (Fig. 9.19h).

These meandering regimes have also been found in numerical simulations of reaction-diffusion models described by partial differential equations [576,577].

To illustrate the mechanism of meandering, Fig. 9.20 shows a superposition of curves, representing the excitation wave, that are computed at subsequent time moments in the extended WR model [562]. We see that the tip starts to grow and performs rapid rotation around a circle of a smaller radius. This rotation is ended when the curve runs into the refractory tail left in the medium (this will actually occur at the next moment for the last

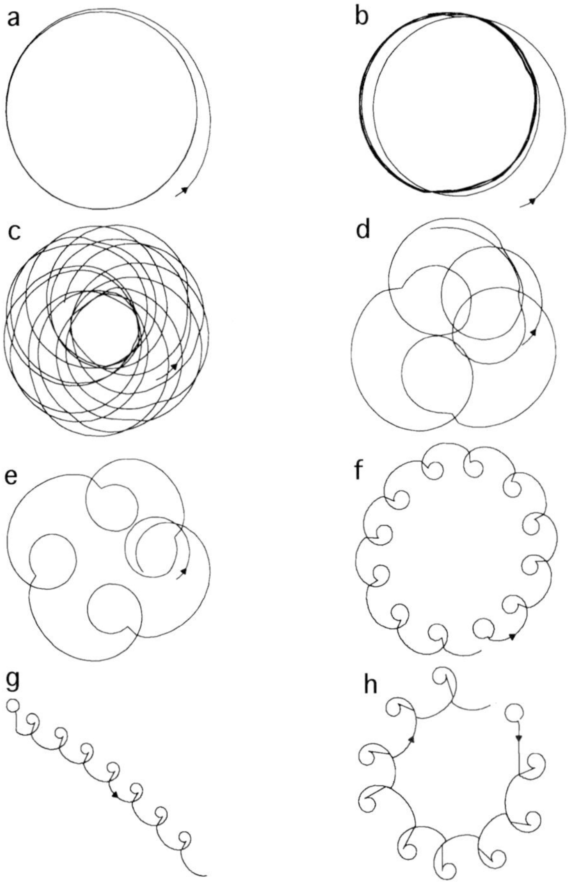

Fig. 9.19. Meandering of spiral waves. The trajectories shown correspond to the tip motion calculated in the extended WR model with the parameters $V_0 = 1.50, G_0 = 3.2, D = 1.0, \gamma = 5.0$ for different recovery times $T_r = 1$ (**a**), 5 (**b**), 10 (**c**), 20 (**d**), 30 (**e**), 50 (**f**), 70 (**g**) and 100 (**h**). Figure taken from [562]

of the curves shown in Fig. 9.20). After that a much slower motion along a trajectory of a larger radius will take place. For comparison, Fig. 9.21 shows a similar diagram for the meandering spiral wave in the BZ reaction [578]. It is constructed by digital overlay of images at subsequent moments separated by equal time intervals.

Another way of understanding meandering in the framework of the WR model is to recall the results of Selfridge and Balakhovsky. As we have seen, they argued that free spiral waves were already possible in the original WR model. The core of such waves should then represent an effective linear cut of the length $L = V\tau_r/2$. While moving along the cut, the wave could not pene-

Fig. 9.20. Subsequent wave positions for the meandering regime shown in Fig. 9.19h. Figure taken from [562]

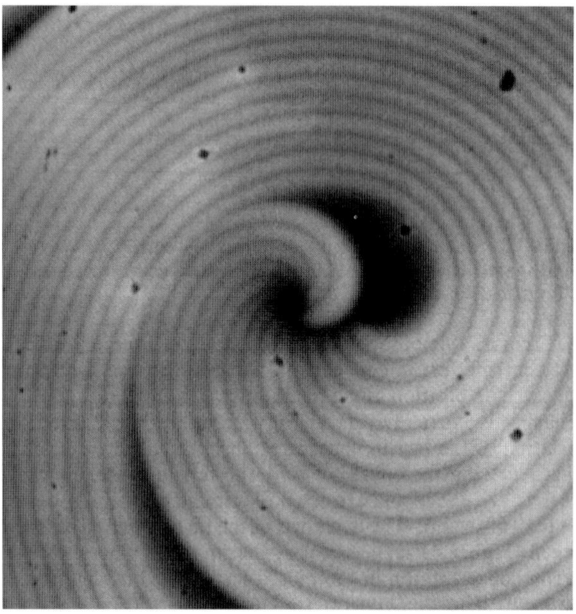

Fig. 9.21. Digital overlay of 14 subsequent images of a meandering spiral wave in the BZ reaction. Figure taken from [578] with permission

Fig. 9.22. The spiral wave of spreading depression in the retina. Overlay of 13 subsequent images separated by time intervals of 12.6 s. Figure taken from [579] with permission

trate into the region on the other side of it because it was in the (absolutely) refractory phase.

In this picture, the tip should rotate by 2π and reverse its direction of motion, whenever it reaches an end point of the linear core. But in the extended WR model, where the curvature effects on the propagation and growth velocity are taken into account, such instantaneous rotation is not possible. Instead, at the end of the linear core the tip will move to make a rotation along a circle of a small radius and then will move along a relatively straight trajectory. The same behavior would then be repeated at the opposite end. As a result, the core is not linear, but Z-shaped. Since the spatial orientation of the Z-shaped core is not fixed, it would slowly rotate. Hence, if the tip trajectory is followed over long times, it would look like a rosette. Spiral waves with Z-shaped cores can therefore be viewed as a special case of meandering spiral waves. Figure 9.22 shows the rotating spiral wave of spreading depression in retina [579]. To display the Z-shaped path of the wave tip within one rotation period, thirteen subsequent images of the spiral wave, separated by equal intervals of 12.6 s, are overlaid here.

9.8 Fibrillation and Turbulence

The motivation of Wiener and Rosenblueth was to mathematically explain flutter and fibrillation. The phenomenon of flutter was attributed by them to the circulation of spiral waves around obstacles. Concerning fibrillation, they noticed that this phenomenon apparently involved propagation of the same type of circulating waves, as in flutter. Their definition of fibrillation was that it "consists of continuous activity over randomly varying paths in a network of connected conducting elements" ([530], p. 254). In contrast to this, flutter was defined as consisting "of a wave or waves of activity in a conducting system with a regular cyclic recurrence of paths, and therefore a well-defined regular wave front and period" ([530], p. 254).

Because Wiener and Rosenblueth believed that the waves could circulate only along obstacles, the appearance of such random paths could be explained by them only assuming that the excitable medium contained a large number of randomly distributed holes of varying sizes. If the fraction of the medium occupied by the nonconductive tissue (i.e. obstacles or holes) was large, the remaining conductive part could be pictured as a mesh formed by the conductive tissue. Alternatively, one could consider "a tangle of fibers distributed at random and anastomosing with one another" ([530], p. 235). An excitable network may support many waves circulating along various paths. Such waves in a regular cubic network are shown in Fig. 9.23.

Wiener and Rosenblueth expected that excitation waves, propagating over a large random network of conducting fibers, would follow complicated paths. Moreover, these paths would not be fixed, but would irregularly change with time. In modern scientific language, this activity regime can thus be described as *spatio-temporal chaos* or *turbulence.*

The theoretical description of thus–defined fibrillation regimes might only be given in statistical terms. Wiener tried to construct a mathematical formulation of this statistical problem. He introduced joint probability densities, specifying distributions over firing moments for the entire system of the network nodes, and derived some general conditions that should be satisfied by these probability functions. This was of course still very far from the solution of such extremely difficult statistical problem. So far as I know, it remains unsolved even now.

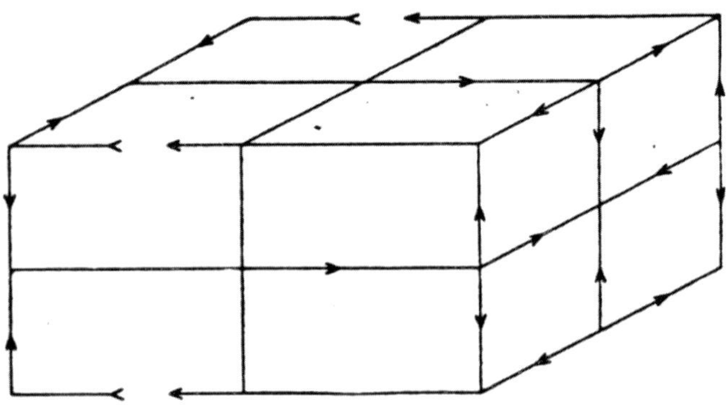

Fig. 9.23. Line diagram of the excitation circulation in a three-dimensional network. Figure taken from [530], p. 231

The ideas of Wiener and Rosenblueth have played in important role in the subsequent research of fibrillation and turbulence in excitable media. When the existence of free spiral waves was shown by Selfridge and later by Balakhovsky, this immediately raised the question, whether fibrillation was also possible *in the absence of obstacles.* Indeed, the function of obstacles in the explanation of fibrillation, proposed by Wiener and Rosenblueth, was that they did not allow excitation to pass through certain regions. But some regions of the excitable medium may also become temporally inaccessible because the elements inside such regions are in the absolute refractory phase. Thus, if freely rotating single spiral waves exist, one cannot exclude the possibility that the same system may support many interacting free spiral waves or traveling wave fragments displaying complex spatio-temporal behavior.

The main problem here was to explain how such complex wave regimes may naturally develop. An important step has been made by Krinsky [543,544] who has shown how spiral waves could spontaneously emerge and reproduce in a non-uniformly refractory medium. In our popular science book [580] this creation mechanism of spiral waves was described as follows: "If a medium is not uniform, the recovery time (i.e. the duration of the refractory state) is not the same in different spatial regions. As seen from Fig. 9.24, when two excitation waves, separated by a sufficiently short time interval, propagate through a medium with varying refractoriness, the front of the second wave is ruptured. Creation of a spiral wave on an inhomogeneity is illustrated in Fig. 9.25. The initial positions of two waves are shown in Fig. 9.25a. Excitation cannot penetrate into region D with an increased duration of the refractory state. As a result, the front of the second wave becomes broken (Fig. 9.25b). Subsequently, the broken wave turns, due to diffraction, over the region D (Fig. 9.25c). As the elements inside the region D go back from the refractory state to the state of rest, the excitation wave starts to penetrate inside the region D and curls into a spiral (Fig. 9.25d)".

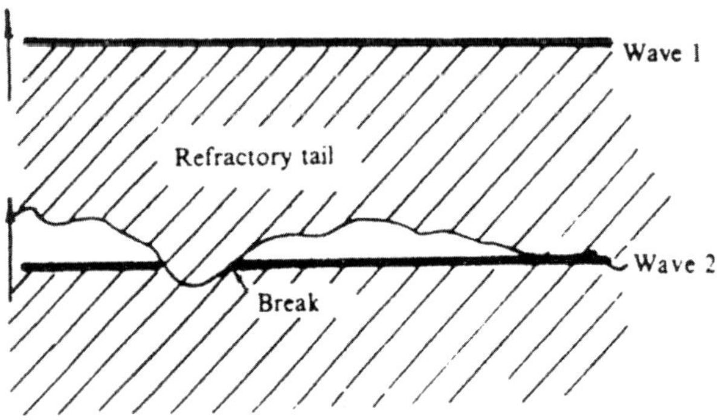

Fig. 9.24. Destruction of the excitation wave in a medium with the non-uniform refractoriness. Figure taken from [580]

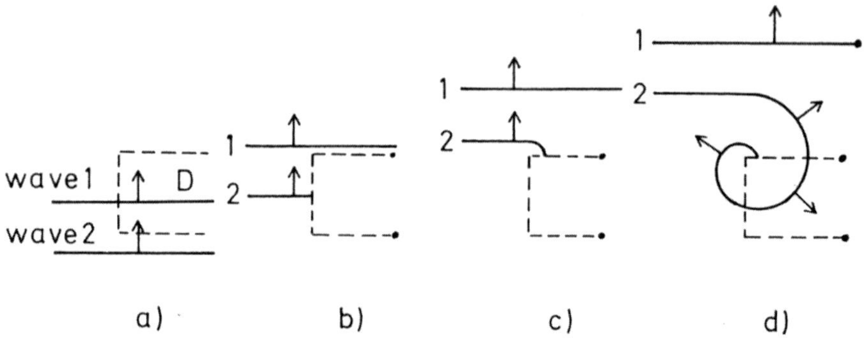

Fig. 9.25. Creation of a spiral wave. The *dashed line* marks the boundary of the region D with an increased refractory time. The *solid lines* show the wave-fronts (**a**) Initial positions of two waves, arrows indicate the direction of motion; (**b**) the second wave does not enter the region D, and the wave breaks up; (**c**) the right end of the broken wave starts to curl into a spiral; (**d**) a spiral wave is created. Figure taken from [580]

If a high-frequency periodic source of waves is operating in the medium, the waves will again and again break as they pass through the part of the medium, where inhomogeneity is located. Thus, a large number of spiral waves will be produced. Moreover, each spiral is itself a wave source. "Secondary breakups of the spiral may start to develop when its excitation front (i.e. the arm of the spiral wave) passes through regions with an increased refractory time". As a result, *reproduction* of spiral waves occurs. If the reproduction rate of spiral waves exceeds the rate of their disappearance at the boundaries, the number of spiral waves in the medium grows. The growth continues until the whole medium is filled by short "fragments" of spiral waves and its elements perform irregular, or chaotic, oscillations. The chaotic wave regime can be interpreted as turbulence in an active medium".

Such transition to spatio-temporal chaos has indeed been observed in the experiments with the BZ reaction [581]; it was attributed to breakups of excitation waves on the boundaries of convection cells due to thermal gradients in the solution. Subsequently it was systematically studied in the BZ reaction in the presence of artificially introduced random heterogeneities [582].

It should be emphasized that the role of heterogeneities in the phenomena of wave turbulence consists only in triggering a transition to the turbulent state. The asymptotic statistical properties of such turbulence may be almost independent of inhomogeneities that have led to its appearance.

Spontaneous breaking of waves and a transition to turbulence have also been observed in numerical simulations of uniform excitable media [583–587]. Various mechanisms could be responsible for this. In some cases, the breakup is preceded by meandering of spiral waves that can create fluctuations in the local refractory properties of the medium that would be sufficient to rupture a wave. From a general point of view, any dynamical instability in an excitable medium that gives rise to small-amplitude chaotic variations may also lead to this effect, since it is equivalent to the action of a noise source that creates spatial inhomogeneities in the medium.

It should be noted that since the time of Wiener and Rosenblueth much progress has been made in experimental and theoretical investigations of cardiac disfunctions, such as arrhythmia (flutter) and fibrillation. Today we know that these phenomena are complex and have diverse origins. Three-dimensional effects, such as the appearance of scroll waves, may also play an important role in the fibrillation.

9.9 Conclusions

The study by Wiener and Rosenblueth was motivated by cardiological research. However, they understood that similar approaches would be applicable in many systems that involve "continuous self-sustained transmission of impulses in physiologically connected non-spontaneously-active networks". Therefore, they concluded that "the method of approach used here is equally applicable, with the proper modification of details, to other networks, such as nerve-nets" ([530], p. 260).

The aim of this article was to trace the developments in the theory of excitable media initiated by the pioneering WR paper. The model they proposed was revolutionary in many aspects. Though the main properties of excitable media, such as propagating excitation and refractory time, were introduced in the beginning of the 20th century in experimental cardiology, Wiener and Rosenblueth were the first to construct a mathematical framework for these concepts and have made possible quantitative analysis of excitable media. Even when their results were not accurate, as with their false conclusion that free spiral waves were not possible, this analysis played a highly stimulative role for the subsequent investigations. Taken together, the WR model and its extensions provide good background for theoretical interpretation and understanding of pattern formation in excitable media.

These models can be used to describe a wide class of media, irrelevant of their nature and detailed mechanisms. However, several conditions should be satisfied. The duration of the excited state should be, for instance, small when compared with the recovery time of the medium. Therefore, the entire excited zone in front of the propagating wave can be approximately represented by a single curve of zero thickness. In some excitable media, such a condition is not satisfied. When relaxation of the excited state is slow, new effects become possible. Back-firing of propagating waves, i.e. emission of pulses that propagate in the rear direction, may take place. Moreover, attraction between propagating pulses can develop. These special properties provide additional mechanisms for meandering and breakup of spiral waves.

Furthermore, an intrinsic feature of the WR models is that the recovery process is strictly local and, therefore, the effective excitability of any given element depends only on the interval of time that has passed after its last excitation. For the cardiac tissue, this condition is always satisfied. However, when chemical excitable systems are considered, they will be typically described in terms of activator and inhibitor species (the excitable and refractory states corresponding, respectively, to the dominance of the activator or the inhibitor species). If the inhibitor is not diffusing or its diffusion constant is significantly less than that of the activator, the WR models are applicable. Strong inhibitor diffusion leads to instabilities of excitation waves and a different kind of turbulence (see, for example [588,589]).

10 Are Cardiac Waves Relevant to Epileptic Wave Propagation?*

A.T. Winfree

10.1 Introduction

The mechanisms of cardiac fibrillation and of the epilepsies differ markedly. Also, there are many kinds of fibrillation and many epilepsies. Yet some collective parallels stand out. Their tissue substrates (heart muscle, brain neocortex and sub-cortical nuclei) both involve electrically active cells connected to neighbors, and both aberrations are electrical modes that seem alternative to the normal modes in the same tissue without change of parameters. The onset and offset of each is an abrupt switch from or back to the normal mode, both can be evoked by electrical stimulation at 10–100 Hz and current density around $20\,\mathrm{mA/cm^2}$, and both can be terminated by a brief stimulus. In the case of fibrillation a single harsh direct-current shock of several milliseconds' duration is in fact the only known acute therapy. In the case of epilepsy, Gluckman et al. [590] report prompt suppression of epileptiform activity in hippocampal slices using much smaller 60 Hz AC fields, and it is well known that a single brief sensory stimulus given close to seizure onset often appears to abort it [110]. Both fibrillation and epilepsy look quite irregular electrically, yet both have dominant periodicities in the 3 to 10 Hz range, and both processes seem to be determined at least in part by propagation geometry on the scale of centimeters. But they differ in that fibrillation seems to be a geometrical mode of normal propagation between cells of just one kind through only excitatory electrical connections, whereas epilepsies involve feedback loops between distinctive cell populations through both excitatory and inhibitory synapses. The dispersion curves of cardiac propagation (the dependence of speed on periodicity) do not much resemble those proffered for waves contributing to the EEG [144,147].

That said, the aim of this chapter is to present what is known of cardiac fibrillation in a way that may assist theorists of the epilepsies in drawing their own conclusions about helpful or misleading analogies between the two phenomena.

* My efforts have been supported by the US National Science Foundation since 1967, currently from the Applied Mathematics program. I thank John Milton and Ivan Osorio for thoughtful reading and critique which improved the 11 June 1999 manuscript in several places, and Jeff Reznik for direction to recent literature.

10.2 Different Molecular Mechanisms of Propagation

Do propagating activations in any CNS tissue resemble those in heart muscle? This seems a silly question. The answer might seem obviously "No, not at all: heart muscle is a relatively uniform electrical syncytium like a 3-dimensional squid axon, but brain tissues are composed of neurons connected through excitatory or inhibitory chemical synapses." Yet among the diverse non-synaptic communication modes referred to as "volume transmission" in the CNS [591], there are some rather like those encountered in heart muscle. Some of these slow communications involve calcium-induced calcium release, and the like for potassium. Hippocampal glial networks are thought to be well coupled by gap junctions [592], as heart cells are. Non-synaptic propagation seems implicated in some forms of epilepsy [592–594]. And the CNS abounds with commissures, fasiculi, and association tracts of diverse lengths and speeds; along these fast-lane projections action potentials propagate in much the same way as they do in normally beating heart muscle.

There are basically two kinds of propagating signal in the CNS:

- The fastest propagations are mediated by *electrotonic currents*. These traverse single-cell arborizations that may bypass several cells, or long fiber tracts that bypass many thousands to end at a remote synapse. Whether myelinated or not, the axon's intracellular mechanism is Hodgkin-Huxley-like: local "reactions" (the voltage–dependence of ionic channels affecting local membrane voltage) electrotonically connected by "diffusion" of electrical potential through a conductive medium. Then final passage to the target cell is synaptically mediated; and
- The slowest propagations, e.g. calcium waves or laterally spreading depression, are mediated by local *diffusion of ions* or molecules through the gap junctions linking adjacent cells, or through the extracellular medium. Such chemically mediated impulse propagation is familiar to physical chemists in much simpler systems, e.g., the Belousov–Zhabotinsky reagent.

A mixture of these two modalities is familiar in the heart. Excitable cells in heart muscle connect to several immediate neighbors through electrically conductive gap junctions; there is nothing like arborization to contact remote neighbors, nothing like chemical synapses, and no postsynaptic membrane to respond by inhibition or excitation: these connections are only excitatory. Adjacent cells are also relatively similar in structure and function in heart muscle. Excitation conducts through a volume of such connected cells much as it does across a uniform sheet of excitable membrane. Transmembrane potential in the various tissues of the heart plays exactly the role of chemical concentration, Hodgkin–Huxley-like channel kinetics plays exactly the role of chemical reactions, and electrotonus plays exactly the role of molecular diffusion. From the viewpoint of dynamical equations, these are exact correspondences, not loose analogies.

In all these contexts (chemicals with diffusion in glial networks, or voltage with synapses in nerve networks, or their hybrid, voltage with diffusion in heart muscle) similar phenomena occur on different scales of time and space. The differences of scale are mainly due to electric diffusion coefficients being 10^5 to 10^8 times greater than those for molecules and ions, and ionic channel responses to voltage being 10^3 to 10^4 times faster than the pertinent chemical reactions. Derivation of the consequences for large-scale activity has been conducted in terms of distinctively different approximations:

- Naturally, the best models are founded explicitly on quantitative cellular neuro-physiology. This tradition goes back at least to Beurle's analytical mathematics [63] and to the micro-anatomical computer simulations of Farley and Clark [67] (which, incidentally, first discovered the rotating spiral waves that thirty years later became so prominent in the study of fibrillation in heart muscle.) If anyone is unfamiliar with the overwhelming fruitfulness of modern nerve network dynamics, excellent presentations from quite different perspectives are available [68,75,144,426,595,596], among many others cited throughout this volume. My job in this chapter is to draw attention to some less-familiar alternative perspectives.
- The activation front in an excitable medium need not span as many cells as it does in heart muscle: it can also hop discretely, almost one cell at a time. Propagation in such discrete media is witnessed, for example, in the c-AMP waves that organize behavior and development in the cellular slime mold. In this case a cell fires in response to the aggregate influence of nearby cells, conveyed to it by molecular diffusion through an extracellular medium. The response is mediated by membrane receptors and intracellular cascades constituting, in that respect, a sort of excitatory "synapse". Supposing a sufficient stimulus and response, the impacted cell then fires, disgorging "neurotransmitter" (c-AMP) that mingles with the earlier and later contributions from other cells, impacting nearby non-refractory cells, some of which may similarly fire if local concentrations exceed their threshold. This slime-mold analogy may seem in many respects fitter to CNS neurons than the analogy of myocardium, but let us also examine two other options:
- An alternative approach to propagation in excitable media arises from the opposite extreme case of continuous excitation spanning so many cells that we may as well think of the medium as a continuum. Continuous media may be conceived of in which an excitation front moves under a few axiomatic rules. These rules may be made as elaborate as needed in an ad hoc way. This is the approach initiated by Wiener and Rosenblueth (1946) for heart muscle, successfully generalized by Mikhailov and Zykov [566], and elaborated here in Chap. 10. One of its clear merits is that it depends on no specific mechanism, so it is freely adaptable. Given thoughtful attention to the essentials, it may well provide an elegant

description of propagation in such complex media as found in brainstem nuclei, allocortex, and neocortex. One more analogy remains:

- In this chapter I try to succinctly outline the main features of propagation in 3-dimensional normal, healthy, electrically continuous and uniform myocardium by starting from the actual mechanism. The mechanism boils down to the "cable equation" augmented with detailed kinetics of the specific ionic channels involved, in the style made famous by Hodgkin and Huxley. This model is thus a reaction-diffusion partial differential equation. The "reaction" part describes local ionic channel responses to changing membrane potential; the "diffusion" part is the cable equation, which can be viewed as electrotonus, i.e. diffusion of electric potential, scaled by a coefficient compounded of passive capacitance and resistivity. It turns out to have dimensions of length^2/time, the same as chemical diffusion coefficients. This is the mechanism also of chemically propagated excitations such as calcium-induced calcium-release waves, and potassium-induced potassium-release waves. Although these various propagating excitations have completely different molecular mechanisms, they share common principles. All involve local membrane events that propagate by diffusion: of electric potential in myocardial activation, of ions and neurotransmitters in the other cases. The local membrane depolarization constitutes excitability, whether of postsynaptic or non-synaptic membrane, and the recovery process may or may not involve active transport. The pertinent equations are remarkably similar in form, and so are their solutions, so analogies are expected between phenomena in the several corresponding situations. This chapter focuses on the best understood case, the cardiac electrical activation wave.

10.3 Electrical Propagation Between Adjacent Cells

There remains some debate over the quantitative extent to which healthy heart muscle resembles an electrical continuum in all three dimensions, on scales exceeding the passive electrical space constant of sub-threshold responses (about 1 mm). It clearly differs from skeletal muscle in that its fibers are electrically well coupled not only along the length of the fibers but also laterally, in the two dimensions transverse to the sarcomere length. This coupling in skeletal muscle is almost nil, so the space constant and the would-be activation front thickness in transverse directions are less than the cell spacing, so propagation in those directions is at least saltatory, and in fact fails. In heart muscle during "active" super-threshold responses, membrane nonlinearity precludes use of exponential linear models characterized by the space constant. But the thickness of the activation front serves as a comparable standard of comparison against cell spacing: it exceeds cell spacing several-fold in all directions. So I adopt the approximation that, while transverse coupling is about nine-fold weaker transversely than longitudinally in heart muscle (and conduction is therefore about three-fold slower transversely), its continuity is still perfect. CNS projection tracts are more like skeletal muscle in having essentially no transverse connectivity or propagation. They have a wide range of propagation speeds because they have both fast (myelinated) and slow (non-myelinated) fibers, not because coupling is direction dependent. More refined treatment of myocardium also takes into account the anisotropy of the interstitial fluid medium. Here, too, conduction is easier in the fiber direction, but not so much so as intracellularly. This double anisotropy has remarkable consequences for stimulation by extracellular current [597,598], as described below.

Supported by this continuous mechanism (whatever its molecular underpinnings), flat activation fronts propagate at speed $f\sqrt{D/T}$. This "fudge-factor", f, is a positive dimensionless multiplier that depends on underpinning details, but seldom exceeds about 2 in mathematical models. We might expect natural selection to keep it close to that in contexts, such as heart activation, where speed is important. It falls toward 0 as the period of repeated excitation shortens. In this rule-of-thumb T is the nominal rise time of space-clamped excitation. It takes longer and so propagation gets slower as the period of re-excitation gets shorter. At a critically short period re-excitation fails, T becomes very long, and conduction blocks. At periods longer than a few hundred milliseconds this T is about 1/3 ms in normal myocardium. D is the passive diffusion coefficient for electrical potential, about $1\,\mathrm{cm}^2/\mathrm{s}$ along fibers. As mentioned above, this is the reciprocal product of specific membrane capacitance, a surface/volume ratio, and a resistivity that includes cytoplasm and the resistive junctions coupling cells. Thus in myocardium, solitary action potential speed is expected to be about $55\,f\,\mathrm{cm/s}$ and is observed to be about 60 cm/s, suggesting a dimensionless f around 1.1. The corresponding thickness of the solitary excitation front is speed × rise time, thus $f\sqrt{DT}$ is around 0.02 cm or 200 µ. As noted above, the rising activation thus spans more than a full cell length: this tissue is not too grainy to treat as a continuum. (For a treatment of propagation speed spanning the range from continuum to the discrete-cell case of saltatory conduction, like myelinated axons or slime molds, see [599,600].) If this is correct then the product of activation front thickness by speed should be close to D, and their ratio should be close to T. Both are so in normal myocardium, but not in ischemic myocardium after the gap junctions between cells have largely shut down, markedly decreasing D, so that propagation shifts into the saltatory regime.

If the solitary activation front is slightly non-flat, its speed is augmented in proportion to D times the concave curvature, or diminished in proportion to D times the convex curvature. The notion of "curvature" is a somewhat tricky in 3D anisotropic media like myocardium (see [601] for examples), but in isotropic 2-dimensional cases, it is just the reciprocal radius of the local best-fitting circle. This approximate speed rule is good for huge radii but fails as that radius shortens towards the front thickess. It is good for a solitary pulse in fully recovered medium but needs amendment in the case of periodic stimulation [602].

The activation front thickness is about the same as the radius of the minimum ball of excited cells capable of nucleating a spreading activation. In less excitable tissues (e.g. those affected by prolonged rapid pacing or by lidocaine) rise time, T, is longer, so the activation front's thickness and the critical nucleation radius embrace more cells, not less. In contrast, if mutual coupling is attenuated by disabling the gap junctions (with octanol, or with connexin antibodies), then the critical nucleus correspondingly shrinks, ultimately to the length of a few cells, beyond which point we don't want to be caught using continuum approximations. In the limiting case of no coupling, current injected though a micro electrode would directly excite only the one cell impaled. In normal myocardium, however, current injected into one cell passes into its neighbors, so that many cells necessarily fire together. Even with many cells activated, this local activation will not propagate if the boundary of the activated volume, the initial activation front, is too convex (i.e. if it bulges outward in the direction of would-be propagation, with too small a radius of curvature). It takes quite a lot of current (like 30 μA for 2 ms) to excite the hundreds of cells needed to make up a ball of more than critical radius for propagation. In an anisotropic medium like normal myocardium, this mechanism of nucleation has more subtle consequences, but masses of quite a few cells are involved in any case, so continuum approximations remain workable (see [601]).

In 2-dimensional heart muscle, excitation propagates in an expanding disk from a point-like source. In a 3-dimensional setting we see an expanding sphere (or at least some closed potato-shaped volume) of activation. Except for the slight modulation of speed by curvature of the boundary mentioned above, the mechanism and dynamics along any radius are the same as in a 1-dimensional setting. But most excitable media are also capable of supporting an alternative and distinctly different 2- or 3-dimensional mode of excitation. This distinction is *not* a matter of cellular mechanism nor of parameter values, but simply of initial conditions. If the *alternative* mode is excited, it takes over the whole medium forever; if it is not, we see only one-shot propagation from one stimulus. These qualitatively distinct modes of activity in one and the same medium were first recognized in pioneering numerical simulations of a neural network supposed to resemble mammalian cerebral cortex [63,64,66,67]. This alternative mode is a *vortex* rotating with a certain characteristic period not much longer than the nominal refractory period of the cells involved. For more, see [65,69,72,89,482,603].

To observe these (cortical) vortices or the wave trains they emit one would need resolution better than 1 mm in space and better than 1 ms in time.

Such resolution in space and in time together has been achieved in myocardium [604] but I think not yet in the cortex (see Figs. 10.1–10.2 of [144], for space-time resolutions of all current observational techniques). Anyway, in myocardium this vortex mode is much coarser and does not even need such resolution: it rotates at about 10 Hz, and waves are spaced centimeters apart. The myocardial vortex was the first to be examined carefully in the laboratory because it is lethal: its period is too short for effective mechanical response, so blood pressure plummets. Then not only the heart, but more importantly, the brain served by the heart goes irreversibly ischemic within a few minutes. Instigation of these "rotors" is one mode of onset of sudden cardiac death, a mode that afflicts even perfectly healthy hearts by total surprise. Depending on factors we need not examine now, rotors may persist stably or they may evolve into a more complicated mode called "fibrillation". So we have at least three distinct modes, as do other excitable continua and as do nerve networks coupled by excitatory and inhibitory synapses:

- quiescence except in immediate response to a stimulus,
- rotors keeping the medium active perpetually at their characteristic period, and
- a poorly described, imperfectly periodic, spatio-temporally complicated mode called fibrillation in myocardium.

There are doubtless other modes, too, especially in neural networks, e.g. see [89]).

From another perspective let's now re-examine this vortex-like mode, as it is the most conspicuous pathology of almost every kind of excitable tissue. Setting aside "fibrillation" as still something of a mystery, the other two modes of continuing activity in excitable membranes correspond to two familiar alternative solutions of the electro–physiologist's nonlinear cable equation. One occurs in much the same way in 1-, 2-, or 3-dimensional contexts. This is impulse propagation along a fiber tract, or activation-front propagation in 2- or 3-dimensional settings. It looks the same with passing time but for a progressive change of *rectilinear distance*. This kind of action potential translates in space at fixed speed. The other is the localized vortex or rotor. It looks the same with passing time but for a progressive change of *angular orientation*. This kind of action potential can rigidly rotate in place rather than moving. This spinning source excites familiar rectilinear propagation radially, but it seems viable even in such a confined space that no such action potential appears. The rotor has a certain characteristic *period* just as the rectilinear propagation has a certain characteristic *speed*, in both cases regardless of how the thing was instigated.

In myocardium the rotor turns in time $\tau_o \sim 100\,\mathrm{ms}$. This necessarily exceeds any absolute refractory period, of course, but that is a slippery sort of statement since the refractory period (nominally 200–300 ms when tissue is paced at a physiologically normal rate) shortens as the excitation interval shortens. How much, in percentage, does the rotor period exceed the absolute refractory period the rotor's period? How much does it exceed the (shorter) minimum possible period of regularly repetitive response? Answers vary enormously from one membrane or tissue or model to the next, but they *are* known, together with their dependence on diverse pertinent electrophysiological parameters.

Anatomically, the rotor in myocardium is a disk whose diameter in any direction (in an anisotropic tissue) is nominally $1/\pi$ of the distance a flat activation front propagates in that direction during one rotation period. Within this disk a rotating gradient of membrane potential spans the range from almost full depolarization to almost complete repolarization, corresponding mostly to the open-ness of sodium channels. Transverse to this gradient, the channels involved with refractoriness (Ca^{2+}, K^{+}) exhibit a similar gradient of open-ness. This electrical tornado drifts freely (and slowly relative to propagation speed) through the 2-dimensional myocardium until it either encounters a boundary and there snuffs out, or as a second option (if the period is short enough? if the medium is grainy enough?) it breaks up into multiple vortices, or as a third option, drives various patches of tissue at various lower rational-fraction average frequencies (because they are unable to respond to every impulse at this frantic pace) and so produces one sort of ventricular fibrillation. There may be another sort in real 3-dimensional myocardium.

There is a fourth option: the wandering rotor can snag on a small patch of relatively inexcitable tissue. In this case an action potential rotates around that electrical inhomogeneity at some longer period, producing a stable ventricular tachycardia. In the extreme case that the snag is an absolutely inexcitable patch, this situation is little different from a familiar action potential propagating along the rim of the functional hole.

So far as observed to date, fibrillation arises promptly after a rotor is started in normal 3-dimensional heart walls, but this transition does not occur in artificially thinned, functionally 2-dimensional preparations of the same myocardium. A threshold thickness was predicted theoretically and found experimentally: it is close to the vortex' diameter in that direction [605]. The same has now been achieved by a different theoretical mechanism in an electro-physiologically more realistic model [606]. This is speculative theory; it is not known for sure whether the 2-dimensional preparations fail to support multiplication of rotors simply because their membranes have inadvertently been made abnormal, or because thin layers lack some cell type found only at depth, or because their area is insufficient, rather than because they fall short of the proffered minimum thickness.

The effective radius, r, of the rotor is also roughly the diffusion distance of the propagator species (membrane potential, in this case) during one rotation period, $\tau_o : r \sim \sqrt{2D\tau_o}$. In isotropic tissue, excitation propagates around the rotor's perimeter a distance $2\pi r \sim 2\pi\sqrt{2D\tau_o}$ in time τ_o at speed $< f\sqrt{D/T}$. This f was 1.1 for the solitary activation front in normal ventricular muscle, but here we deal with the lesser speed of activations at short period moving into tissue that still remains relatively refractory and excites with longer rise time $T \sim 1$ ms. This factor f shrinks with shortening period: $0 < f(\tau_o) < 1.1$. Taking the ratio of distance to time to get the speed, then squaring it,

$$8\pi^2 D\tau_o/\tau_o^2 \sim f^2 D/T, \text{so}$$

$$8\pi^2/\tau_o \sim f^2/T, \text{so}$$

$$\tau_o \sim 8\pi^2 T/f^2 > 79\text{ms}/1.2 \sim 66\text{ms}$$

Accordingly *radius*, $r \sim \sqrt{2 \times 1\,\text{cm}^2/\text{s} \times \tau_o} > 0.36\,\text{cm}$. Rotor **diameter** in the direction along fibers is twice that (0.72 cm).

How much greater are τ_o and r above these low estimates? That has to do with refractory periods and other features of the membrane which become algebraically complicated [558]. In normal myocardium τ_o turns out to be closer to 100 ms and r correspondingly turns out to be closer to 1 cm in the longitudinal direction (along fibers) or about 1/3 of that in the transverse direction. More importantly, it turns out that the vortex rotation in normal myocardium is so far from steady that quantitative estimates of "the" period command little interest, and its nominal radius is immensely enlarged by its rapid, seemingly random wanderings.

Notice that $8\pi^2 D\tau_o/\tau_o^2 \sim c_o^2$, the speed at rotor period, so $8\pi^2 \sim c_o^2\tau_o/D$, a dimensionless "quality factor" (called Q) independent of the units chosen for time and space, and measurable on a vortex (i.e. rotor or any other mechanism) in any medium. It turns out that this estimate, about 79, is way off base in plenty of cases, but those cases have certain features in common. If the medium is only marginally excitable, Q can be arbitrarily large. It becomes, in fact, infinite at the parameter-space border of rotor viability, where speed c_o approaches that of the solitary action potential while period τ_o grows boundlessly. In spontaneously oscillatory media, Q seems commonly near its theoretical lower bound, $6\pi \sim 19$. Other cases (the ones we care about) show $Q \sim 100$, give or take a factor of 3 or so, even though such cases cover a range of physical, chemical, and biological media in which all other pertinent quantities range over a dozen orders of magnitude. In particular, Ca^{2+} wave rotors in hippocampal glial syncytia and rotors in heart muscle show $Q \sim 100$–200.

I do not know how to evaluate Q in the neural network models mentioned above, because I do not know what, if anything, corresponds to "D". But $c_o^2\tau_o = QD$ is about $10\,\text{cm}^2/\text{s}$ in the vortices of Chu et al. [65]. If D were evaluated as it is in myocardium, but from the larger surface/volume ratios of CNS dendrites, it would be maybe $1/5\,\text{cm}^2/\text{s}$, so Q would be around 50, again in the range observed for rotors in wildly diverse systems.

The foregoing algebra can be put into more useful form for media with propagation speed only slightly dependent on period. In such cases (myocardium included) c_o^2 above can be replaced very approximately by $f^2 D/T$ (see above) without acknowledging f as a weak unknown function of the period. So we find $8\pi^2 D\tau_o/\tau_o^2 \sim f^2 D/T$, so $QT \sim f^2\tau_o$. In the case of myocardium, this should be around 100–200 ms, so with $f \sim 1.1$ we would expect $\tau_o \sim 90$–180 ms, which happens to be about right. For neural network models with risetime T near 1 ms and Q near 50, the observed 21 ms rotation period would require fudge-factor f to be around 1.6.

In a sufficiently spacious setting, the depolarized end of the rotor's electrical gradient kicks off an outward excitation that propagates away in the form of a spiral, like the stream of water radiating from a rotating garden sprinkler. The radial spacing between turns of this spiral is the period times propagation speed at that period, which is several centimeters, about the same in most mammalian myocardia. So a full turn of the spiral is not evident in hearts smaller than a large dog's. Since the rotor itself has a characteristic size, there is no room for fibrillation (interpreted as a proliferation of rotors) in hearts much smaller than a rabbit's. (Exception: in the mouse heart, the action potential is dramatically shorter. It lacks the long refractory plateau phase that characterizes all larger species. So its rotor is correspondingly smaller, does fit within that small heart, and spins about twice as quickly as is usual in larger mammals [607].)

One way to initiate this rotary mode is to transiently block propagation along some arc of activation front wider than two rotor diameters. Mirror-image rotors develop, one at each of the dangling edges of the interrupted front. (If those edges were closer together, healing would occur instead.) Another way is to excite a disk of tissue that overlaps the refractory wake of a propagating front. The rim of the directly activated disk would be a closed ring of outward activation except that the part in refractory tissue suffers block, and so we have again the foregoing case. Another way is to stimulate twice or more in sufficiently rapid succession at a point (a familiar procedure in cardiac electro-physiology), creating a quatrefoil of mirror image rotors in doubly anisotropic myocardium [608]. In a cortical network model something similar is shown in [65].

Rotors occur in the widest variety of excitable media. Except in special cases of marginal excitability, their period is on the order of Q/f^2 (on the order of 50–100) times the excitation rise time of that medium, or eliminating the elusive f, about $QD/(\text{propagation speed})^2$ (this being the speed at rotor period, grossly estimated by the higher speed of the flat solitary pulse). This observation pertains to uniform continuous media. In cellular media with such rapid excitation and/or weak coupling that the moving activation front is no thicker than the cell spacing, conduction is instead saltatory, as noted above, and easy estimates are hard to come by, even for propagation speed [599,600]; no one has attempted to predict rotor period. For example, in ischemic myocardium the coupling between cells is attenuated by closure of gap junctions, then propagation becomes saltatory. Continuum rules can mislead badly in this situation, as clearly illustrated in [609]. Might theory elaborated for this case apply also to propagation between cortical neurons? Maybe not. First, in the cerebral cortex overlapping neural arborizations extend across n cell spacings, suggesting continuity if n much exceeds 1, which it probably does. Nonetheless prediction from the heart-motivated continuum models may fail because brain tissues are commonly composed of diverse neuron types connected though excitatory and inhibitory synapses, commonly at great distances, not composed of similar cells all connected to immediate neighbors by excitation as in heart muscle. Lord Sherrington is said to have marveled a half century ago at "how much easier it would be to understand the brain if neurons would light-up when they talk to one another." We have since achieved exactly that in cardiology, using image-amplified CCD cameras with 1000 Hz frame rates watching voltage-sensitive dyes in the epicardium, but only because "talking to one another" is implicit in adjacency. The unique challenge of the CNS is that its topology is so much more intricate. This may be an essential factor in epilepsy.

10.4 Chemical Propagation Between Adjacent Cells

Turning to mechanisms coupled by *chemical* diffusion rather than *electrical* diffusion (electrotonus), we find rotors of spreading depression in the chicken retina [115], in the pigeon forebrain [610], and in the hippocampus [611]. Largo et al. [611] declare, contrary to all prior assumptions, that spreading depression propagates by diffusion through the gap junctions, not primarily through the extracellular medium, and this spread seems independent of functioning glia. Spreading depression seems to play no role in epilepsy, but has been implicated in migraine with aura [612–614]; see [615] for a more exotic possibility.

At about the same speed as spreading depression, Ca^{2+} waves propagate intercellularly among astrocytes cultured from the hippocampus [123] or from the suprachiasmatic nucleus [616–618]. In some cases the channel of cell-cell coupling is the gap junction passing Ca^{2+} ions and/or inositol (1,4,5)-triphosphate [619]; in others the intercellular medium is the channel, with ATP serving as the agent [620]). Because glial Ca^{2+} waves evoke electrical activity in neurons and vice versa [620–623] these waves may have functional importance for neural communications. Their rotors, presumed to be the source of observed spiral waves, have roughly the period anticipated above from propagation speed and diffusion coefficient or from observed rise time of excitation: about half a minute in astrocyte networks [624].

Calcium rotors have also been observed in the Xenopus egg cortex [625] and (with Ca^{2+} rise time an order of magnitude shorter and rotor period only 1 s) inside a single cardiac myocyte during Ca^{2+} overload [600,626]. The corresponding periodic pulse sequences are so fine (10 μ spacing) that the usual continuum descriptions fail. Such saltatory conduction results whenever the units of excitation (Ca^{2+}-sequestering sub-cellular organelles in this case) are farther apart than the thickness (where transition occurs from rest to full excitation) of the wavefront. As noted above, that thickness is roughly $f\sqrt{DT}$, whether the diffusing quantity be electrotonic potential (D on the order of 0.1 to 1 cm^2/s) or small molecules or ions (D on the order of 10^{-6}

to 10^{-8} cm^2/s). Also in astrocytes there seems little reason to imagine that Ca^{2+} waves propagate in a continuous rather than saltatory way: the rising activation does not span multiple cells.

Saltatory or continuous, the velocities, periods, and wavelengths seem roughly compatible with the estimates given above in terms of observed rise times and diffusion coefficients in a continuum. Bazhenov et al [627] report waves of excitation propagating at fixed speed in the thalamic reticular nucleus. The speed is only a fraction of a cm/s, being some scores of cells/s, passing from one cell to the next in an ostensibly saltatory way in a time comparable to the membrane's passive time constant. In a model basically identical to those used for myocardium in the pathological case of attenuated intercellular coupling, propagating spiral waves appear, with speeds and periods compatible with the estimates used above.

The waves synchronizing calcium oscillations in mammalian liver suggest a locally oscillatory process coupled across space by a few times 10^{-5} cm^2/s extracellular molecular diffusion. So far as I know, vortices have not been sought. If they exist and if propagation is continuous and Q is around 100, as seems common for the widest variety of rotors, then by the foregoing estimates ($QD = c^2T$, which predicts T, given the diffusion coefficient, D, and the observed propagation speed, c) their periods would be around 200–600 s.

10.5 Electrical Propagation in the Long-Range Fast-lane

In addition to both electrical and chemical diffusion through gap junctions, the human heart contains a long-range projection, the Bundle of His, for firing the entire ventricular muscle synchronously upon command from the atrioventricular node. The brain diversely elaborates such high-speed projections. The modularity of cortical architecture in all three dimensions of cerebrum and cerebellum starkly contrasts with the relative uniformity of myocardium. I presume that sub-cortical nuclei are comparably intricate. Cabling between nuclei (projections), and between nuclei and cortices, and between patches of cortex (association tracts or fasiculi and commissures) presents still more idiosyncrasies. Such long-range bypasses of lateral connectivity sometimes occur in heart muscle, but there they are regarded as pathologies. They are commonly detected first by their malignant effects on cardiac rhythm: impulses sometimes circulate along them at dangerously short periods, as for example in Wolff–Parkinson–White syndrome. The solution is simply to locate and cut them. The consequences of such intervention in the brain are dramatically evident, for example in split-brain surgeries for epilepsy (corpus callosum, anterior commissure, etc.) and in prefrontal lobotomy.

But continuing the loose comparison with the heart, are there any pathologies of long-range connectivity? Certainly so, for example, secondary generalization of epilepsy across the corpus callosum [40]. But could long-range projections be involved in a primary way, in the genesis of seizures? A number of pathways exist for seizure propagation (Chap. 2). Since the conduction velocities of central myelinated and unmyelinated axons differ, time delays arise and probably play a role in propagation dynamics. Recent investigations into the dynamics of spatially extended media with retarded variables emphasize that increases in time delays increase spatial complexity and can result in irregular wave propagation, including multistability and chaos ([397] and Chap. 16).

An example of the effect of a time delay on neural network dynamics is the observation that stuttering might come from excessively delayed auditory feedback from Wernicke's to Broca's speech area, especially if they happen to be in opposite hemispheres. Stuttering is often alleviated by talking more slowly. Presumably time delays in general become functionally more conspicuous during more frequent activation. What about periodic stimulation, as in music, strobe lighting, and video watching? On 17 December 1997 about 685 Japanese children and adults experienced seizures while watching Pokemon episode 38 cartoon flickering at 12 Hz on television [628,629]. About one person in 200 is susceptible to epileptic seizures, and of these one in 20 is triggered by such flickering. The effect is enhanced by luminance contrast (spatial patterns) and by red light (and according to another report, by green light). It requires bright light in both foveae. The most effective period is about 60 ms (range roughly 33 to 100 ms, though isolated extreme cases have been reported at 12 ms and at the 1/3 s period associated with absence seizures). Could this have anything to do with propagation delays? A literature search turned up scores of papers on this subject (xeroxed, underlined, filed but not cited here). None mention propagation delays. Nor do any mention epileptogenesis from alternative modalities of periodic stimulation, such as one might expect were there any merit to such speculations about visual periodicity: neither acoustic vibration at the short-period end of this range, nor mechanical vibration (unless vibrator-induced orgasm is a sort of ictal episode).

What sort of delays could be functionally significant in the human brain? According to Chap. 2 and cited sources, 20–60 ms delays are not uncommonly observed along the corpus callosum and perhaps other commissures, and along association tracts (fasiculi) between ipsilateral cortical regions. Including thalamic processing (LGN), retinal illumination registers with higher visual areas only after 50–100 ms, depending on intensity. Freeman [630] also notes that the 3 Hz frequency of seizures in the cat's olfactory bulb may be determined by round-trip propagation delays along the lateral olfactory tract connecting to the pyriform cortex. There are also lots of slowly conducting fiber tracts in the cerebellum. Remembering the many sources of major bias in favor of short delays (e.g. predominant use of small mammals and the relative smallness of action potentials on thin, slow fibers), Miller [143] supposes, contrary to usual opinion, that delays of 100 ms and more may be common along sub-populations of fibers within projection tracts in the human brain. Could cortical processing, routed through many visual areas and between hemispheres, result in failures of synchrony or in unstable resonances?

If we think of "the brain" as a network of visual processing centers (nuclei, cortical areas) reciprocally interconnected by delay lines, then we find many multi-synaptic closed loops, each with a certain circulation time. Most of those times are probably shorter than 60 ms, but some could be longer. What happens as we feed in stimuli periodically at gradually increasing rate? Suppose the duty cycle is 1/2 (i.e. 50% pattern A, 50% pattern B). We neglect the relatively short refractory period of cells/fibers excited during either pattern. First of all, what prevents a single stimulus from circulating indefinitely in any such loops? Crick and Koch [631] suggest that loops with connections that strong simply do not occur, at least intracortically and between cortex and parts of the thalamus. But there must be loops of some sort for integration of cerebral processing; this "reentry" (the same word as in the literature of cardiac arrhythmias) is, after all, the main theme of Edelman's vision [632]. Perhaps circulation of the usual kinds of stimuli is highly damped along all loops. But what about such evolutionarily rare stimuli as precisely periodic discrete impulses? In myocardium, in a narrow range around 10 Hz and rational multiples, these evoke fibrillation basically because they can propagate in loops of any size in a 2- or 3-dimensional syncytium, and some of those loops prove self-sustaining (rotors) while shorter periods are unstable.

Consideration of the electro-encephalogram (EEG) suggests that at the "large" population level, what "propagates" during a seizure is a condition of synchrony rather than an isolated activation front (Chap. 5). If this is true, then any trading of wares between aficionados of propagation in cardiac syncytia and neurologists seeking mechanisms for the spread of epilepsy seems to end almost before it begins. However, it is quite possible that at the level of neuron to neuron propagation (e.g. Pathway 1 in Chap. 2) closer similarities exist between propagation in heart and brain. Indeed, recent optical imaging studies of epileptic brain and brain slices have pointed to the wavelike properties of epileptic propagation at the level of "small" neuronal populations [130,189,633–636].

10.6 An Alternative Description of Propagation in Complex Media

Analytical thinking about fibrillation originated before it was known that myocardium is an electrical syncytium and before many of the peculiarities of its ionic mechanism had been quantified. Wiener and Rosenblueth (1946) accordingly adopted a flexible framework for describing propagation. Their vision turned out to need fundamental revisions, both because myocardium turned out to be a syncytium and because of some mathematical slip-ups. But the overall framework has been much elaborated and generalized during a half century of theoretical work, and may well prove more suitable to synaptically coupled tissues.

As an undergraduate in 1963 I moon-lighted in Frank Rosenblatt's Perceptron lab at Cornell and took his course. Wondering what "brain waves" were, I found Norbert Wiener's Nonlinear Problems in Random Theory (1958) a major inspiration. Studying it led to my first technical publication, about mutual synchronization among neural oscillators connected by gap junctions or purely electrical synapses, such as occurs in the pacemaker nucleus of electric fish [637,638] and in the olivary nucleus of the brain stem [639]. My introduction to propagation disturbances in the heart came about by tracking this book back to its origins, partly in Wiener and Rosenblueth's comparably inspiring analysis of mutually coupled non-oscillatory excitable units. They extended the contemporary analysis of one-dimensional propagation in terms of piecewise linear partial differential equations, e.g. in Rashevsky [640] and Rushton [641], to a two-dimensional context by simplifying: instead of deriving the quantitative attributes of propagation from a laboratory analysis of ionic mechanisms, they invoked a simpler, purely descriptive principle. This is Christian Huygens' (circa 1673) notion that each point on a mathematically thin wave front is a source from which excitation radiates at the same speed in all directions, regardless of front curvature or the temporal period of repetition (he was thinking of light and sound).

In the Wiener and Rosenblueth model, the amplitude of such propagation was declared to be 1 into all forward directions and 0 into the refractory wake. The local membrane state was allowed to be discontinuous in space, as in Rashevsky [640] and Rushton [641], who analyzed propagation mathematically in terms of the cable equation, involving continuous space and continuous time, but allowed membrane polarization to flip-flop discontinuously between "resting" and "activated" potentials. Wiener and Rosenblueth's vision of 1946 is sometimes described in terms of 3 such discrete states: a long-lasting quiescent state, an instantaneous excited state, and an absolutely refractory state with a continuous timer built in. The actual electrotonic mechanism of conduction from one cell to the next was then well understood but the 1946 model just postulated that propagation somehow happens, at "1 space unit per time unit". In terms of mechanistic continuous concepts not present in this model, conduction speed in fact depends on the quickness of local excitation ($1/T$, above, which depends on timing in the relatively refractory phase) and on the magnitude and range of spatial coupling (D, above), and on the wavefront's curvature. But the Wiener and Rosenblueth description made that quickness infinite, and there was no relatively refractory phase nor any coupling rule capable of scaling the relation of time to space. They apparently felt that in order to infer anything rigorously these simplifications were necessary and that they probably had no important qualitative consequences.

That turned out to be wrong: the nature of relative refractoriness and the actual manner of coupling (electrotonus, alias diffusion of electric potential) do have consequences quite different from the all-or-none, fixed-speed coupling analyzed in Wiener and Rosenblueth [558,642,643]. Equally important, several later authors felt compelled to comment that Wiener and Rosenblueth's "theorems" 3–7 were not proved and are in fact false even of the model as they axiomatized it. For example, their Section V, "Flutter in a 2-Dimensional System", analyzes a stimulation protocol from Mines [644] and Lewis [535] that is still much used in laboratories today under the new names "method of successive adjacent stimuli", "S1-S2 stimulation", "cross-field stimulation", etc. [645]. This protocol induces what are now recognized as rotors: two successive electrical stimuli at sufficiently separated places, if rightly timed for that separation, evoke reverberations at a characteristic period. As understood by Wiener and Rosenblueth, these persist only because excitation is circulating around a hole in the otherwise-continuous medium, whose perimeter exceeds speed × refractory period. The speed being 60 cm/s independent of period or curvature, and refractory period being a fixed quantity at least 0.2 s, this was > 12 cm, larger than the entire ventricle of a guinea pig or small dog. This should have served as a flashing red light that the simplifications adopted, while gaining for us some geometrical transparency, had also lost something physiologically important.

This analysis desperately needed refinement. It makes sense only if you pass over certain of the theoretical diagrams inattentively and it corresponds to the experiments only if you imagine that laboratory investigators somehow overlooked such a large "obstacle" of inexcitable tissue. Theorem 4 asserts there can be no rotors, even though Garrey [533] had long earlier demonstrated something like rotors in heart muscle; Wiener and Rosenblueth thought Garrey must have made a mistake. First reading Theorem 4 in late 1970, soon after I had discovered rotors in a chemically excitable medium, that medium seemed (to me, wrongly) an even more perfect realization of the Wiener and Rosenblueth model. So I found their argument incomprehensible, and then found that others had the same experience:

- Selfridge [532] noted that were any excessively big hole flattened to a slit-like discontinuity in the medium, then circulation at the same period would still persist stably. If the slit were next healed so we could say there are no holes (in a sense: but the local state is still spatially discontinuous along the suture line) then the line would shorten to Wiener and Rosenblueth's minimum perimeter. Circulation around it immediately goes unstable in favor of extinction unless the refractory period and propagation speed exhibit no noise fluctuations whatever.
- In the former USSR, Balakhovsky [537] discovered this slit argument independently, but overlooked the radical instability of propagation that Selfridge had emphasized.

- Krinsky [543] showed that by introducing a finite duration for the excited state and not only spatial discontinuities of state, but also spatial discontinuities of a fixed parameter governing the refractory period, reverberation could be made to persist in the absence of holes, and in fact around slits of short enough perimeter to be plausible in real hearts. None of these supplementary requirements bothered cardiac electro-physiologists, who believed anyway that *diseased* heart muscle, at least, (the only kind they encounter) is in fact dominated by parametric heterogeneity and discontinuities.;
- Abakumov et al. [646] rediscovered the internal contradictions in Theorem 4 and the corresponding diagrams. They showed by their own analysis of the original axioms something like Balakhovsky's slit arising from two-shock stimulation. This was closer to a believable rationalization of the known facts, but still the modeled circulation proved unstable in two dimensions. Comparable circulation in normal heart muscle is perniciously persistent, but this was not yet known.
- Observing my discovery of chemical rotors in continuous time, space, and state, in the adjacent lab Durston [647] rediscovered the same inconsistencies in Wiener and Rosenblueth's argument.

Escape from this hall of mirrors came through simply computing from the nonlinear cable equation, using principles of electrotonus accepted by all electro-physiologists since at least 1937: circulation then proves stable without a hole, at a unique period, in a plausibly small space, even without imputing dramatic irregularities to the tissue, much as observed in normal heart muscle.

The sort of analysis pioneered by Wiener and Rosenblueth has much appeal, especially for its geometrical simplicity and independence of cellular details. If we consider this model for potential insight into brain waves, an essential first step in understanding it must consist in appreciating exactly why it does not serve for heart waves. Is the whole problem simply that it is unable to describe some small region around the pivot of rotations? Is this problem adequately evaded by cutting out that region and replacing it by a hole along whose border excitation propagates much as it does at the same distance from the center of a rotor? In the same spirit contemporary physics abandons any attempt to describe what goes on "inside" the Schwartzchild radius around the space-time singularity that John Wheeler called a"black hole". To implement such a dodge, however, we must accommodate three difficulties:

- As noted above, the hole required turns out to be as large as the entire heart in familiar laboratory animals, reinforcing Wiener and Rosenblueth's conviction that a heart without holes (e.g. normal left ventricle) cannot harbor such a vortex. So if vortices are observed anyway, then it would seem to follow ipso facto that functional holes (maybe dead tissue)

evidently have been detected. The trouble is that the observed vortex wanders freely while the putative region of deadness cannot.
- The period of circulation around the putative hole turns out to be an order of magnitude longer than that of observed rotors and of clinical arrhythmias associated with fibrillation.
- It has also since turned out that myocardium is characterized by a dual anisotropy, one consequence of which is that two successive stimuli at one and the same point elicit a quadrupole of mirror-image rotors, something entirely impossible in the class of models deployed in 1946 [648–651].

Nit-picking here is at least ungenerous. The work of 1946 led the way. Presumably some tinkering with the axioms would recover the familiar phenomena at some expense in simplicity, if we only want a tidy conceptual description of what is already known empirically. It is here that the road splits five ways, into distinct lines of logically consistent approximation, and it is here that we might look for one path along which to understand the spread of epilepsy (to the extent that its lateral spread within a block of gray matter is important in comparison with spread along white matter):

1. The middle tongues of Fig. 10.1 extend the previous road straight ahead: continuous-time, continuous or discrete space, continuous or discrete state models. The "kinematic" branch of "axiomatic" descriptive models such as Mikhailov presents in Chap. 9 are very much in the spirit of Wiener and Rosenblueth. This branch may lead to useful description of propagating disturbances in the brain if their multiple synaptic substrates and interpenetrating arborizations prove too difficult to analyze in terms of explicit mechanism;
2. Along the left tongue a more extreme caricature of reality leads to discrete-time, discrete-space, discrete-state "cellular automaton" models such as Moe et al. [652] implemented for heart muscle; and
3. The right tongue leads us to continuous-time, continuous-space, continuous-state partial differential equations. They are harder to solve, but closer to real mechanisms in syncytia (e.g. in heart muscle, but not in skeletal muscle nor in cerebral cortex.) This is the tradition pioneered by Rashevsky [640], Rushton [641], and Cole [653]. It ramifies into two of the five final branches: the "monodomain" form of Hodgkin and Huxley [451] and the "bidomain" form required for analysis of extracellular electrical stimulation (a need noted in Freeman [596]), starting with Muler and Markin [654]. Both apply only to syncytia, which arise only in heart muscle, in networks of glia, and in special circumstances in networks of brain neurons (e.g. the inferior olive and the electric fish's pacemaker, mentioned above).

Figures 10.1 and 10.2 give a visual impression of the variety of modeling implemented in cardiology. The citation landmarks given on Fig. 10.1 do not include all those just mentioned above, but only a sampling among quantitative accounts only of spiral wave propagation in myocardium, but within this

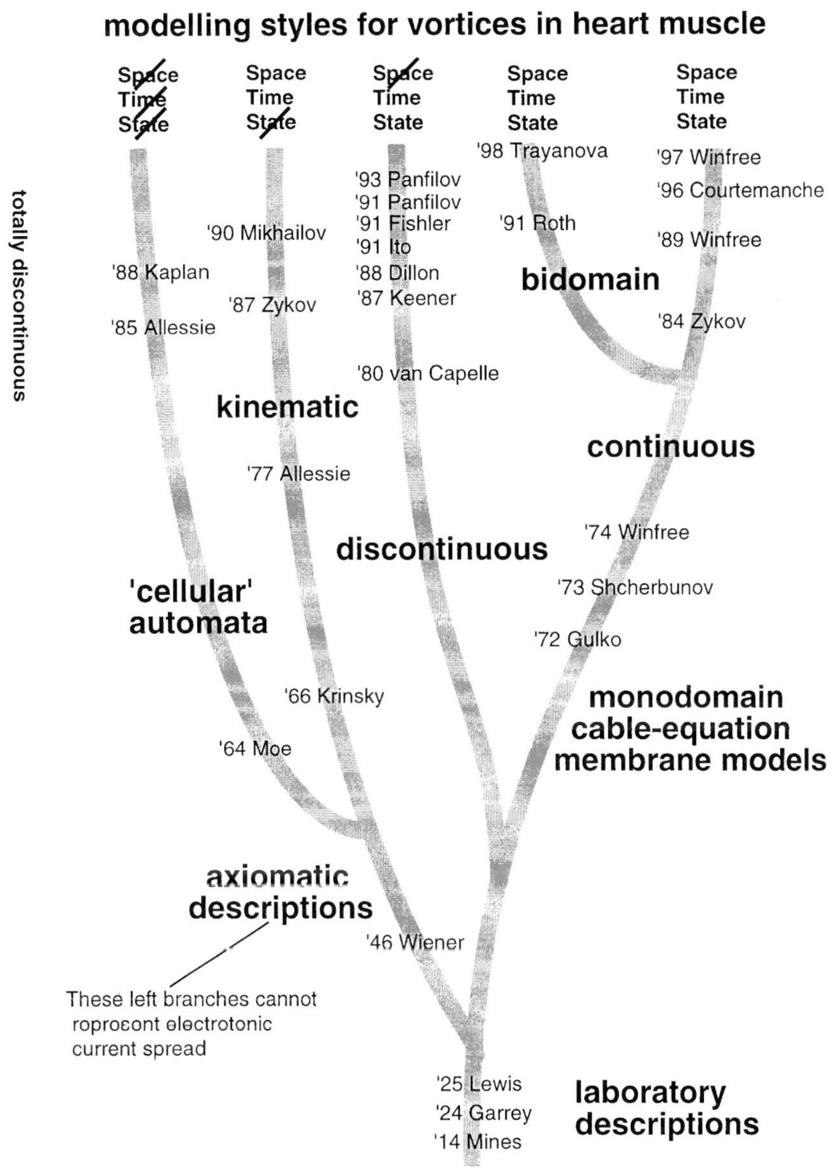

Fig. 10.1. A variety of quantitative styles have evolved to emphasize different features of wave propagation in myocardium. Which, if any, would comparably illuminate brain waves of diverse kinds remains to be seen. This figure locates some papers in the reference list according to their choices to give time, space, and/or membrane state a continuous description. The studies sampled here are all 2-dimensional, concerning vortex action potentials

10.7 Electrical Stimulation of Excitable Tissues

So much for propagation. What about electrical stimulation? This matter cannot be quantitatively addressed in myocardium without strict attention to the "bidomain" architecture of fibrous tissue. Because it is technically inconvenient to impale enough cells (hundreds) with separate intracellular stimulating electrodes, current from an extracellular electrode must be applied to the intermingled intracellular and extracellular compartments of the tissue. These have different anisotropies of electrical conductivity. A result is that a single anode or cathode produces a 4-fold symmetric arrangement of depolarized and hyperpolarized tissue around it [608,648–651]. There are also distinctions to make between anodal and cathodal make-and-break excitations [598,660]. When the local stimulus is thus quantified, we can address the question, "If a little volume should be successfully excited, will activation propagate or will it die out on the spot?" The answer in myocardium has to do with activation front thickness compared to the radius of curvature of the hopeful front, "curvature" being understood in the correct subtle way for an anisotropic monodomain continuum [601] or the more subtle way appropriate in a doubly anisotropic bidomain. I presume that the anisotropies of cortical tissues are more complicated, if the concept even has any use in that context. I do not now see how experience with myocardium is likely to illuminate electrical stimulation of the brain by point-like electrodes.

A conspicuous related mystery is the efficacy of defibrillation by a single DC shock applied through large or remote electrodes. Applied through 6-cm paddles on the heart surface or even the chest wall, or through an intracardiac catheter and a grounded defibrillator case under the pectoral muscle, a few milliseconds of about $20\,\mathrm{mA/cm^2}$ extracellular current suffices. It is not easy to achieve this uniformly, i.e. without wasting energy and damaging tissue in some volumes or leaving others under-electrified. The conceptual problem is that electric current is not effective unless it is crossing cell membranes, and theorists are not sure why or how it would do that in uniform tissue more than a few millimeter away from the electrodes. The best guess at present is that abrupt irregularities of fiber direction and/or of conductivity create "virtual electrodes" at short intervals throughout the heart: a heart made of perfectly combed uniform fibers probably could not be defibrillated electrically by a single shock [661].

If one had to predict the effect of gross electrification on cerebral tissues, it might be easiest in the fiber bundles of white matter. To get started, each cell might first be considered in isolation, insofar as no electrical junctions akin to those in heart muscle connect adjacent fibers. Given a fiber of diameter $10\,\mu\mathrm{m}$ ($1/1000\,\mathrm{cm}$) in an extracellular field of $10\,\mathrm{V/cm}$, then roughly $10\,\mathrm{mV}$ of potential drops across the cell, most of it split between cell membranes on entry and exit sides. The intervening cytoplasm may be nearly isopotential, since little current penetrates the very resistive membrane to flow through the less-resistive interior. This $5\,\mathrm{mV}$ transmembrane drop on the depolarizing side might approach excitation threshold in that portion of the membrane. It is then necessary to ask whether this would suffice to depolarize the whole cell [662,663]. Probably not, but if the impressed voltage oscillates then a) membrane impedance is less, and b) the potential drop can have another effect: synchronizing a cellular electrical rhythm of nearly the same period. Such oscillating fields are sometimes used for major surgery [664].

Following such a fiber to the soma and its dendritic arborizations spanning perhaps 100 to 1000 μm, the same argument would indicate comparable sensitivity at 1/10 to 1/100 the extracellular field in gray matter, i.e. 1 to 0.1 V/cm. So there might be no point in refining the argument for white matter: the gray would excite first anyway. But is such a loose argument meaningful? One might expect enormous complications from the elaborate architecture of dendritic branching. However, according to Tranchina and Nicholson [665] a realistic approximation depends mainly on the orientation of the extracellular field to the cell, and 0.1 to 0.2 V/cm, optimally oriented, suffices at least to so modulate membrane potential at the soma that the cell's rate of firing is noticeably affected. This seems to be the case in a wide variety of cells. This is in the high end of the range 0.04 to 0.20 V/cm believed to induce seizures [592,666]. Gluckman et al. [590] used 60 Hz AC fields in this range to suppress epileptiform activity in hippocampal slices.

It is hard to see much analogy from the case of heart muscle, though the outcome is similar in that several tenths of a V/cm at optimal orientation suffices to fire quiescent myocardium. Perhaps cells in the hippocampus or inferior olive, having electrically transparent gap junctions between neighbors, would present a more favorable inroad from the direction of cardiac analogies.

Refinements of such thinking about the single cell come from attention to details of cell shape and the influence of adjacent cells on the local field. In contrast, the refinements for heart muscle come more from quantitative attention to cell-cell connectivity and its anisotropies, both internal and external. A recent summary of such considerations and their consequences in heart muscle is given in [660]. With a wave of the hand, it might be argued that gray matter accentuates to the conceptual limit such inhomogeneities and anisotropies as cardiac electro-physiologists currently imagine to provide the "virtual electrodes" needed for defibrillation more than a centimeters or so away from metal electrodes.

It would be of interest to quantify the effects of electroshock therapy used for intractable depression. So far as I can estimate from data in Abrams [667], the effective range of current density is comparable to that used for defibrillating the heart: perhaps 20 to 200 mA/cm^2, quite heterogeneously distributed with some regions sub-threshold while others are over-electrified and permanently damaged.

It is clear that I am out of my depth in trying to draw out analogies from myocardium to the much more intricately structured tissues that support epileptic “fibrillation”, about which I know so little. But it has been fun to learn a little while preparing this chapter and I hope that by sticking my neck out here I might elicit some instructive feedback.

11 Pattern Formation in the Microbial World: *Dictyostelium Discoideum*

H. Levine

Over the past few decades, there has been significant progress made towards understanding how non-equilibrium systems can generate interesting spatial patterns. These patterns range from the dendritic, sometimes fractal, morphologies seen in diffusively unstable interface motion [668] to the self-organized nonlinear waves in excitable dynamics [669] to the interacting localized structures which occur in Turing-unstable media [670]. In each of these examples, we have in place reasonably accurate theoretical models that allow us, by combining numerical simulation with mathematical analysis, to assess the importance of different parameters and make predictions for future experiments. While much work needs to be done and new surprises undoubtedly lie in wait, we are confident that the framework being used to study these processes is ultimately capable of resolving any issues that arise in this field.

Based on this success, some researchers have turned to applications of pattern-formation principles to biological systems. Most advanced to date is work on microbial systems. These include the colony patterns formed during the nutrient-limited growth of a variety of bacterial species [671,672], structures due to the chemotactic collapse of *E. Coli* [673,674] and, the subject of this review, the spatial coordination of amoebae behavior in *Dictyostelium* [675]. We will focus on the first stages of the genetically controlled developmental program that takes the initially individualistic amoebae and transforms them into the motile multicellular slug and the fruiting body thereafter [676]. Our aim is to introduce the relevant phenomenology, describe the modeling ideas (and results therefrom) that have been applied to this process and finally to point out some directions for future research.

11.1 Spiral Waves and Aggregation

Dictyostelium discoideum is a species of soil-living eukaryotic (i.e. nucleated) amoebae that undergoes an extensively studied change in behavior as the cells cooperate in the face of starvation. The standard laboratory protocol for studying this behavior is quite simple. One first grows cells in a flask with sufficient nutrient. Then, one removes the nutrient and spreads the amoebae on an agar surface. After a few hours, they begin to signal to each other and thereby establish a dynamic wavefield that subsequently provides guidance

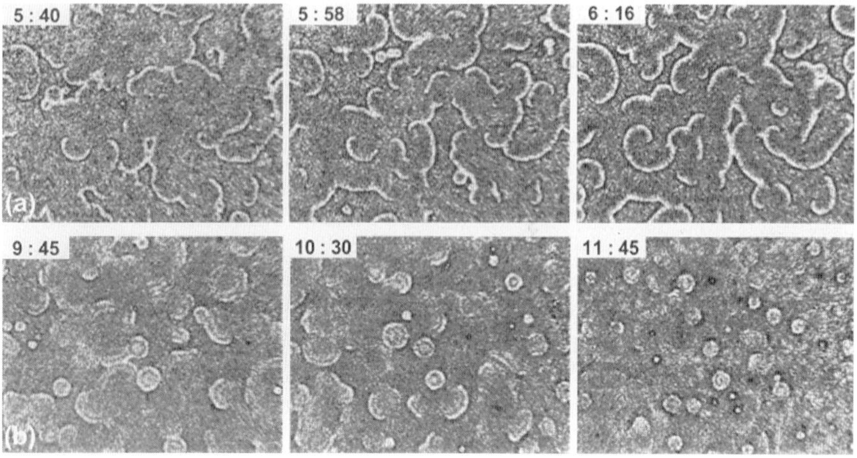

Fig. 11.1. Experimentally observed darkfield waves, from [681]

to aggregation centers. In this section, we discuss what is known about these early developmental steps.

It has been known since the work of Bonner [677] that the the cells utilize the small organic molecule cyclic AMP (cAMP) for signaling. Upon binding of cAMP to a receptor, a signal propagates internally and results in the manufacture of more cAMP, part of which is then secreted. Much of the detailed biochemistry has been worked out for this process [678,679], but from the point of view of the pattern-forming process only a few facts are relevant. First, there is a concentration threshold below which binding is too rare to lead to cAMP emission; a typical number for this threshold for fully aggregation-competent cells might be a few nanomolar. Next, binding of cAMP to the receptor leads to receptor inactivation. So, the cell will not continue to produce cAMP in the presence of a constant external cAMP concentration, but instead is tuned to respond to cAMP concentration jumps. This inactivation leads to a refractory period during which the cell cannot respond to cAMP. These facts can be directly tested by studying spontaneously oscillating cells in a test tube [680]; pulsing them with external cAMP will retard the oscillator phase in the refractory state and advance the phase as the system starts to recover. On the agar surface, cAMP can spread diffusively, allowing waves to propagate across the system. These waves can be seen experimentally by using darkfield microscopy [682]. In this methodology, the only light that reaches the objective is light which has been scattered by a significant angle by the object being imaged. As the cAMP wave passes over the cell, the cell undergoes what has been called "cringing" – it undergoes some change in its internal cytoskeleton (its structural backbone) that causes it to round up and detach from the substrate [683]. This shape change causes enough of a difference in the scattered light for bands to appear wherever the cAMP wave is near its peak. That the observed waves are directly correlated with cAMP concentration was established by Tomchik and Devreotes [684]. In Fig. 11.1, we show a typical set of darkfield pictures taken from the recent work of Goldstein et al. [681].

Based on this description, each region of cells behaves as an element of an excitable medium. Excitable media have been extensively studied, with particular emphasis on using simple models to make predictions of the generic properties of such processes as the Belusov–Zhabotinskii reaction [576,685], the oxidation of CO on platinum surfaces [523] and electrical waves in cardiac tissue [521]. The suggestion that the observed waves in *Dictyostelium* should be thought of as arising from excitable dynamics seems to have been first put forth by Durston [647]. After this realization, several efforts were made to quantitatively model the underlying biochemical kinetics; the most widely used model[1] is due to Martiel and Goldbeter (MG) [687]. After reviewing this formulation, we present a simple automaton approach [688,689] that has been able to capture the essence of the cAMP response and can subsequently be used in computational studies of the aggregation process.

The essential facts that must be explained in any signaling model is the nonlinear threshold response and the refractory period. Starting with the latter, MG assumed that the cAMP receptor could exist in two states, active and inactive. The active state binds to cAMP much more strongly than the inactive one; also, binding of cAMP shifts the kinetic balance away from the active state to the inactive one. This allows for the possibility of having the cycle unbound receptor → bound receptor → bound inactive receptor → unbound inactive receptor → unbound receptor. To get the desired excitability and therefore provide the impetus for the above binding cycle to occur, one needs a way to couple cAMP binding with an auto-catalytic increase in cAMP, thereby forcing all receptors to the bound inactive state once the process gets started with an above-threshold perturbation. MG postulated that the receptor acts as a dimer when it excites adenlyl cyclase to produce more cAMP. Using this ad hoc assumption and measured rate constants, they were able, with some fudging, to fit a variety of experimental data.

Since its inception, the MG model has proven to be the most useful semi-quantitative approach to this system. In its simplest two-component reduced form, the governing equations are

$$\begin{aligned}\dot{\psi} &= \epsilon \nabla^2 \psi + \frac{1}{\epsilon} f(\psi, r)\,,\\ \dot{r} &= g(\psi, r)\,,\end{aligned}$$

where ψ is the (extracellular) cAMP concentration and r is the fraction of activated receptors. The kinetic laws are given by $f = s\Phi - \psi$ with

[1] For a different model of *Dictyostelium* signaling kinetics, see [686]; there is a general consensus that this model is less realistic than the MG approach.

$$\Phi(\psi, r) = \frac{\lambda_1 + \Lambda^2(\psi, r)}{\lambda_2 + \Lambda^2(\psi, r)} \qquad \Lambda = \frac{r\psi}{1+\psi}$$

and $g = -r\, f_1(\psi) + f_2(\psi)(1-r)$ with

$$f_1 = \frac{1+\sigma\psi}{1+\psi} \qquad f_2 = \frac{L_1 + \sigma\, L_2\, h\, \psi}{1 + h\, \psi} .$$

The various parameters typically have values listed in Table 11.1; it is important to remember that all of these depend on the cell density. This set of equations embodies the "receptor loop" mechanism described earlier and can be understood in detail by studying the original paper by Goldbeter and Martiel [687]. In this formulation, f represents both creation of cAMP by the auto-catalytic process that is dependent on the product of cAMP concentration ψ and activated bound receptor $r/(1+\psi)$, and depletion via breakdown. Similarly, g embodies the activation of deactivated receptors (the 1-r term) as well as the de-activation of activated receptors. One can check in a straightforward manner that the nullclines of these functions give rise to either a monostable excitable medium or to a self-oscillatory one (if the parameters are varied slightly). Since the receptors are attached to (as yet stationary) cells, there is no diffusion for the second field.

Let us return to studies of these darkfield waves. Recent work by Lee et al. [681] as well as earlier work [682,690] in the biological community have established many important facts regarding those waves.

- The waves behave exactly as expected for excitable systems. This includes the effect of curvature on wave speed [691], the wave dispersion whereby speed depends on frequency and the fact that wavefronts annihilate upon impingements. These processes can be directly simulated via the Martiel-Goldbeter equations, with good agreement with experiment.
- The waves start out as small isolated pulses emerging from some randomly positioned "pacemaker" sites. At sufficiently large cell density,

Table 11.1. Parameters for the MG kinetic equations

L_1	10
L_2	0.005
h	18.5
σ	10
λ_1	10^{-3}
λ_2	2.4
s	47
ϵ	0.01
Time scale	28 min
Space scale	8.2 mm

spiral waves form spontaneously and rapidly suppress the pacemaker activity. At low density, this does not happen and the wavefield remains dominated by outgoing circular wavefronts from the emitting locations.

If one starts with a suitably arranged set of concentration and receptor fields in the MG model a stable spiral will quickly emerge. This approach was used by Tyson and Murray [692] in their pioneering effort to use the MG approach to model *Dictyostelium* waves. This does not, of course, explain why spirals would form dynamically in the actual experiment, merely that they are possible.

It is useful to get some quantitative sense of these dynamical events. The refractory period is of the order of a few minutes. Typically, the waves initially have period of 5–10 min, and these decrease as the wavefield evolves. The waves propagate at around 300 μm/min, which means that they pass over a 10 mm cell in a few seconds. The wavefield takes 4–6 h to establish itself after starvation and after a few hours, cells start to move and the waves themselves become more complex. We will return to the effects of cell motion due to chemotaxis in the next section.

As mentioned already, details are often less relevant than the generic features that this system shares with other excitable systems. Appealing to this idea, Kessler and Levine [688,689] introduced a simple model of the cellular response to cAMP stimulation. A cell becomes excitepd if the cAMP concentration at its spatial position is above some threshold C_{T}. After excitation, a cell emits cAMP for some fixed length of time after which the cells enter a refractory period when re-excitation is impossible (the absolute refractory period). Subsequently, the cells can be re-excitepd. The threshold is taken to vary continuously from its maximum value $C_{\max}$ at the end of the absolute refractory period to a fully recovered value $C_{\min}$ after some fixed relative refractory time. In detail, we have

$$C_{\mathrm{T}} = \left[C_{\max} - A \frac{x}{x + T_{\mathrm{ARP}}} \right] , \tag{11.1}$$

where the time variable x measures time from the end of the absolute refractory period T_{ARP}. The constant A is chosen to make $C_T = C_{\min}$ at the end of the relative refractory period $x = T_{\mathrm{RRP}}$. The typical parameters used for this model are absolute refractory period $T_{\mathrm{ARP}} = 8$, relative refractory period $T_{\mathrm{RRP}} = 2$, concentration release $= 300$ over an emission time $= 1$, $\Gamma \equiv$ cAMP decay rate $= 8$, $C_{\max} = 100$, $C_{\min} = 4$. These have been determined phenomenologically by attempting to get our simplified model to produce waves that are qualitatively in accord with observation.

The cell dynamics is then supplemented with a simple equation for the cAMP concentration

$$\frac{\mathrm{d}C}{\mathrm{d}t} = D\nabla^2 C - \Gamma C + \text{sources} , \tag{11.2}$$

where the sources are just the emitting cells. Placing cells randomly in a two-dimensional plane creates an excitable system. One can directly check that this model reproduces all the generic features of such a system. This is hardly surprising, and there is a long history of attempts to use automata models for this class of dynamics [693]. The great advantage is the simplicity; instead of having to solve a complicated set of kinetic equations, one merely has to program an internal clock for each cell to determine at any moment the state of the cell's responsiveness to cAMP. This model will be used in what follows to simulate aggregation.

11.2 Wavefield Evolution

Immediately after starvation sets in, there is no wave activity among the *Dictyostelium* cells. After a few hours, a few bursts of darkfield waves can be visualized; these waves propagate short distances and then die away. Subsequently, this activity becomes more pronounced and, depending on the cell density, one of two patterns emerge. At low density, some of the sites responsible for the initial bursts become periodic pacemakers, sending out circular waves and leading to target patterns that govern the eventual chemotactic motion. If the cell density is higher, the initial bursts lead to rotating spiral waves that thereafter dominate the wavefield. In this section, we review recent efforts at modeling this self-organization process.

First, the fact that targets dominate at low density is easy to understand. The general theory of excitable waves predicts that the fastest oscillators win. At weak excitability, spirals have low frequencies and should be suppressed relative to the aforementioned pacemakers. These pacemakers are presumably cells (or clumps of cells) that happen to have their signaling parameters fall within the oscillatory regime of the cAMP response dynamics. That is, there is enough cAMP left over after the re-activation of the receptors so as to directly re-excitep the system without needing any signal coming from some other region. Their frequencies remain fixed as we lower the overall density of the system, thereby weakening the excitability and slowing the spirals; eventually, the spirals will not occur at all even though the system can still support waves. Somewhere in this progression, targets will dominate. This pattern selection method is also apparent in other cases where the excitability is low. For example, one can delete the gene for PDI [694] (phosphodiesterase inhibitor) that then leads to over-secretion of phosphodiesterase, the enzyme that degrades cAMP; this too leads to targets rather than spirals [695]. So, the real issue is the mechanism that causes spirals to form. Spiral formation requires wave-breaking that in turn requires some type of sufficiently strong heterogeneity in the medium. One possibility, which we view as unlikely, is that there is always enough cell density variation caused by the experimental protocol to accomplish wave-breaking. Attempts to get spirals via small density fluctuations in the aforementioned automaton model

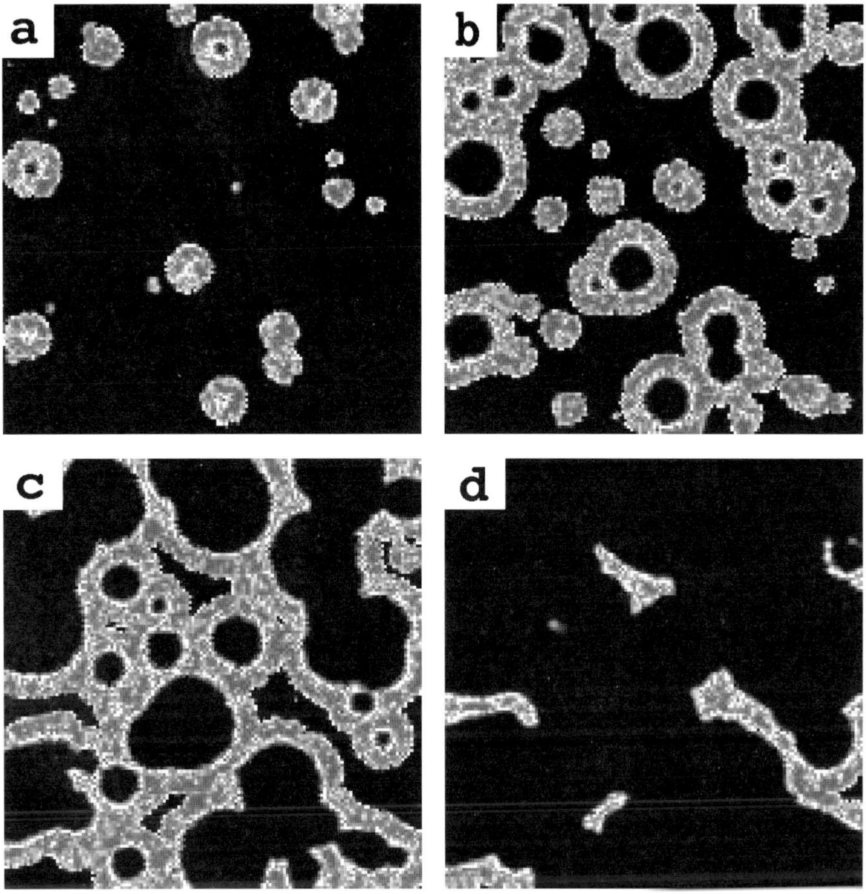

Fig. 11.2a–d. Time evolution of a concentric wave pattern due to random firing of cells in a system with time-independent excitability. No spirals are formed in the absence of sufficient heterogeneity to break the outgoing waves. Taken from [605,688]

were unsuccessful (see Fig. 11.2). The reason is that small variations in an excitable system will not break waves. In fact, the density explanation is inconsistent with a recent experiment [681] showing that one could eliminate spirals (even at high density) by quenching the fully developed wavefield (accomplished by spraying enough cAMP so as to saturate all the receptors) and then allowing it to re-form. Density inhomogeneities should always increase in time (see next section), but they were clearly inadequate even at this later times to generate spirals. In the work of Palsson and Cox (PC) [696], there was postulated an initial variation in the PDI distribution, due to different starting states for the cells. One can make this inhomogeneity large enough by hand to produce spirals, but this may be to some extent cheating.

The most promising answer to date has to do with the developmental dynamics of the cells. Specifically, it is incorrect to think of the *Dictyostelium* system as a static excitable medium. Over the several-hour time scale of the wavefield evolution, the cells undergo dramatic changes in their response capabilities [697]. Biochemical experiments have established that initially, the cells produce only a small amount of the cAMP signaling apparatus. As they begin to receive signals, they up-regulate [698] the expression of the genes controlling the cAMP system and thereby go from inert to excitable and eventually to oscillatory. In fact, the isolated pacemakers seen in the experiments are probably just cells that for some reason or another are temporally advanced on this developmental trajectory. Note too that any model that focuses on cell development as the necessary ingredient for spiral formation will automatically be consistent with the quenching experiment. At this stage, there is a distinction to be made between two classes of models that have appealed to developmental dynamics. One set of models works with some initial inhomogeneity in some aspect of the cells. For example, Goldbeter et al. [700] introduced a model in which the cell excitability increases autonomously in time starting from some initally heterogeneous "age" for the cells. In models such as this one in which excitability is the critical variable, one can use the fact that the system will automatically pass from being inert to being wave-supporting, as the typical cell increases its excitability. Small inhomogeneities are much more important near the onset of medium excitability, since a small fluctuation can easily drive a region sub-threshold and thereby break waves. An example of this as seen in the automaton model is shown in Fig. 11.3; here we modify the above equation for the threshold to

$$C_{\mathrm{T}} = \left[C_{\max} - A \frac{x}{x + T_{\mathrm{ARP}}} \right] (1 - E) ,$$

where we now have a new degree of freedom for each cell, the excitability E.

Fig. 11.3. Spiral pattern produced by automata model without genetic feedback, but with an autonomous increase in excitability – for details see [699]

We assume that E obeys the kinetic equation

$$\frac{dE}{dt} = -\alpha E + \beta\,, \tag{11.3}$$

and the excitability uniformly increases up to its maximum value. From the biology point of view, however, the up-regulation is not autonomous but instead depends on cAMP signals. This leads to a new class of models that incorporate what we have called genetic feedback [699].

Again using the automata model, we take

$$\frac{\mathrm{d}E}{\mathrm{d}t} = -\alpha E + \beta C\,, \tag{11.4}$$

where C is now the cAMP concentration and again there is a maximum excitability past which E is no longer increased. In addition to being more in sync with biological fact, this framework has several advantages. Most significant in our opinion is that the system generates its own inhomogeneity. As mentioned above, there is a phase of small wave bursts that in this model imprints a large degree of inhomogeneity on the cells even when there were initially only small differences. It is therefore straightforward to get wavebreaking with almost any reasonable assumption for the initial distribution. A picture of the spiral state for one realization of this model is shown in Fig. 11.4. We can combine this idea with the MG framework to make a semi-realistic model of the entire wavefield evolution [701]. In this extension of the MG approach, we augment the basic kinetic equations with the introduction of a new variable which governs the changing amount of signal transduction components. In the absence of detailed quantitative data regarding how different genes are separately regulated, we assume that this single variable controls all components, namely the receptor, the adenyl cyclase level, and the amount of both intracellular and extracellular phosphodiesterase. For

Fig. 11.4. Spiral pattern produced by automata model with genetic feedback – for details see [699]

simplicity, the dynamics of this new genetic variable is taken to depend on the amount of cAMP-bound active receptor, such that at levels typical of pulse peaks, it is increased. To fix the overall scale of how changes in this new variable affect the signaling, we require that our results match the experimentally observed increase of about a factor of 10–30 [702,703] in the spatially averaged amount of cAMP bound to receptors over the aggregation time scale. The details are given in a recent paper [701]. A simulation of the wave-quenching experiment is shown in Fig. 11.5, comparing favorably to the observations in [681].

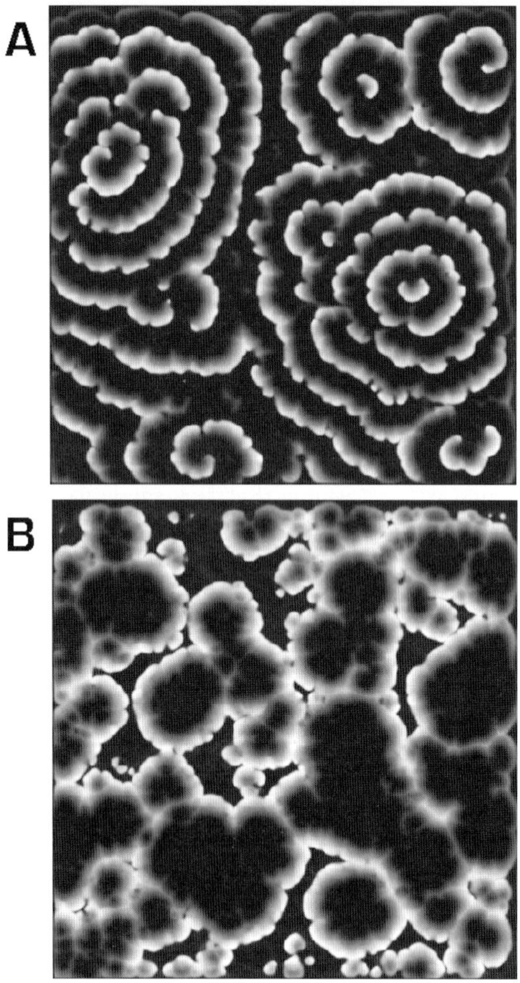

Fig. 11.5. Simulation of the wave-resetting experiments. Shown is the concentration of extracellular cAMP for reset times of 120 min. (**A**) and 240 min, (**B**), 360 min, and 347 min respectively, after re-setting – for details, see [701]

Fig. 11.6. The final state of the formation of a single spiral domain from a counter-rotating spiral pair, due to an infinitesimal initial gradient in the gene expression variable (from [701])

Finally, we note that this new class of excitable systems exhibits a novel spiral symmetry-breaking instability [704]. In standard excitable media, nearby counter-rotating spirals almost always peacefully co-exist. Here, however, any perturbation that leads to one spiral rotating more quickly will be amplified by the positive feedback effect on the cells underlying the faster rotator. A picture of the final stage of this process, from the aforementioned model, is shown in Fig. 11.6. There is some evidence from experiment that these competitive events are common in the wavefield dynamics. The real proof of the feedback idea, though, would require that the spiral competition take place in a self-amplifying manner (and thereby grow exponentially) as opposed to the winning of one spiral merely as a result of it being by chance in a region of higher but static underlying excitability.

11.2.1 Cell Streaming and Mound Formation

The entire purpose of the cAMP wavefield is to provide guidance for the cells as they move towards aggregation centers. Observationally, the cells start to move chemotactically (i.e. towards the nearest wave source) at around the time that the wavefield becomes fully developed. Once the cells begin the move, the darkfield pattern becomes disorganized and one no longer sees unbroken spiral wavefronts. If one directly images the cell density (for example by making the cells fluorescent), one discovers that this disorganization is due to the cell density becoming highly non-uniform. After a short while, a pattern of dense one-dimensional streams forms [705] and remains in place as the cells progress towards the nascent mound. This can be seen in Fig. 11.7.

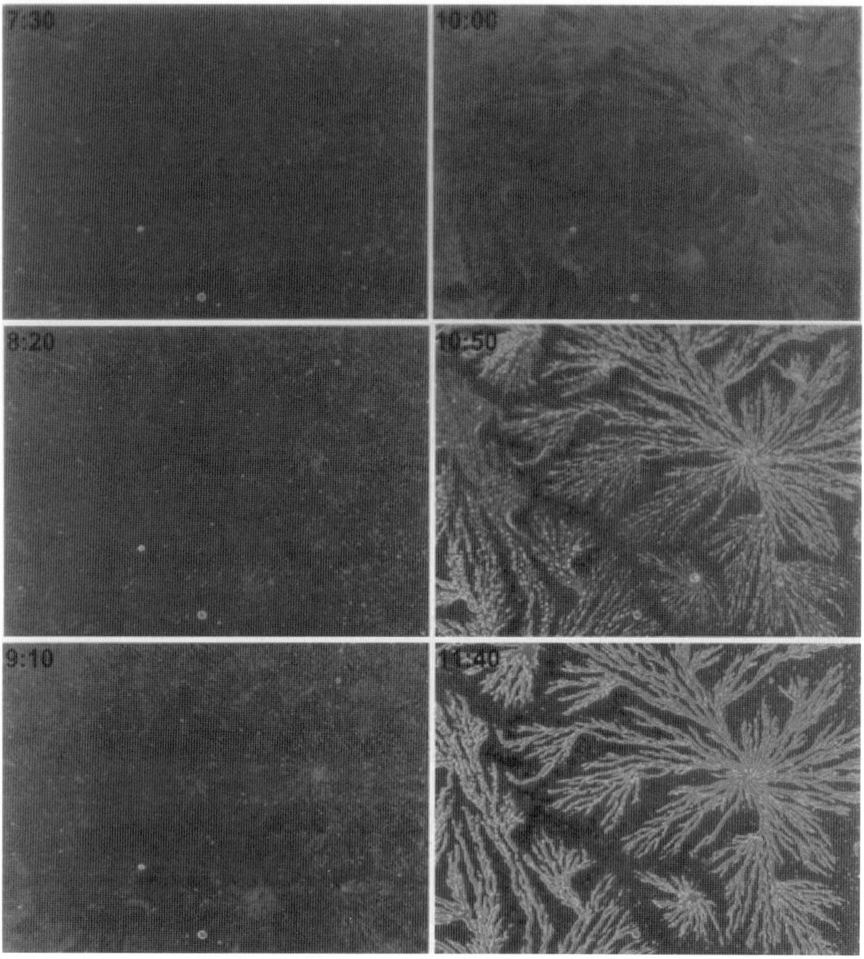

Fig. 11.7. Series of cell density images showing the formation of streams starting from an initially homogeneous density (courtesy of R. Kessin) (The colour plate of this figure appears on page 361)

Sometimes, one can see in the streaming structure the memory of the spiral wave that gave rise to it, as the streams curl as they approach the mound and as there is sometimes a low density "hole" at the mound center where the core of the spiral is located [706]. One such hole is shown in Fig. 11.8. When faced with a new behavior such as stream formation, one must consider different possible causes. One alternative is that the behavior is directly caused by a specific set of genes; put operationally, one asks whether mutant with specific gene knockouts could exhibit signaling and chemotaxis, but have no streaming. The other extreme is that the behavior follows directly from the collective physics of the cells with their pre-existing responses and there

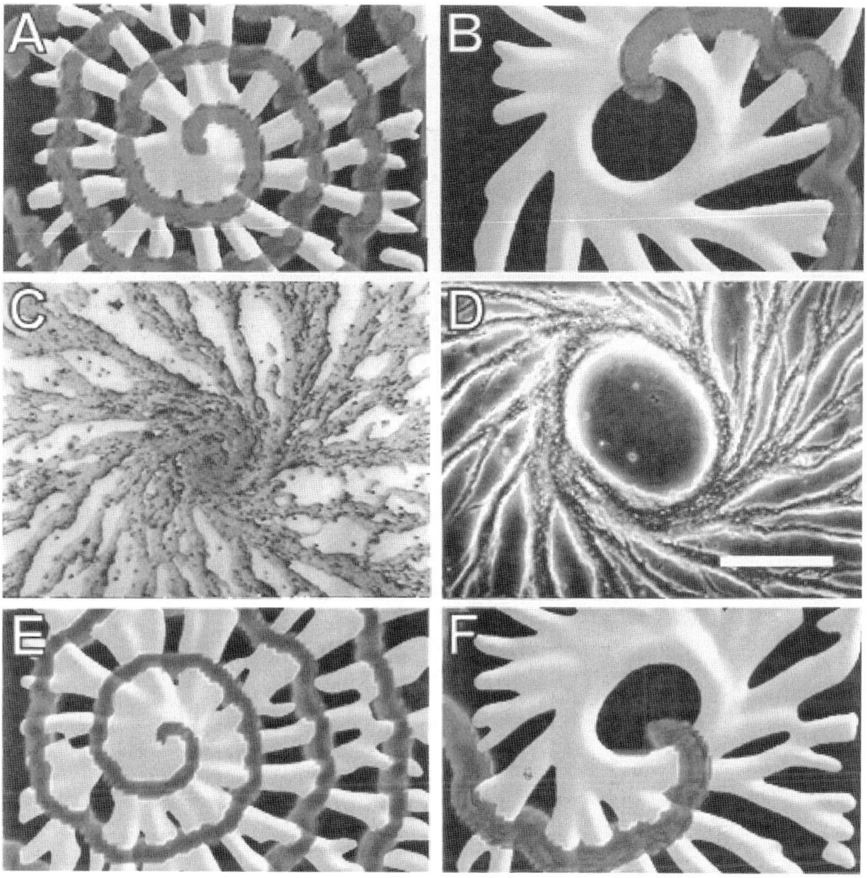

Fig. 11.8. Close-up view of the aggregation center in experiments (**C,D**) and a model (**A,B**; also **E,F**) due to Vasiev et al. [707,708]. The hole has been enhanced in going from **A** to **B** and **C** to **D** by lowering the excitability of the system, accomplished experimentally by adding caffeine. (Courtesy of C. Weijer) (The colour plate of this figure appears on page 362)

is no required change in single cell response. Reality is often somewhere in between. For streaming, the cells do at this stage become more adhesive via expressing some specific genes and this stickiness clearly influences the integrity of the streams. Nevertheless, it has been established that the basic mechanism responsible is traceable to a physical process, that of an instability in the coupled density-signaling dynamical system. Let us first discuss simulations. In our automata model [688,689], we imagine that an individual cell can detect a cAMP gradient across its own diameter. If the cell is in the quiescent phase, this gradient causes it to translocate forward (towards the source) by one cell diameter (assuming no other cell is blocking its path). Once the cell becomes refractory, it cannot respond to gradients; this provides

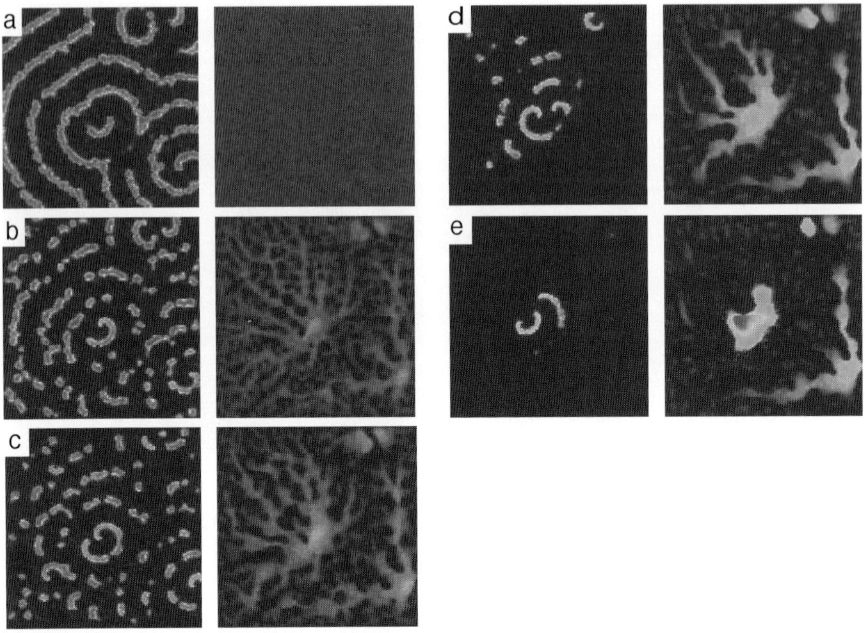

Fig. 11.9a–e. Stream formation in the simple automata model – taken from [699] (The colour plate of this figure appears on page 362)

a direct solution for the rectification needed to respond chemotactically to traveling waves [709]. The results of using this motion rule together with the previously discussed chemical dynamics is shown in Fig. 11.9. The streams that form in the simulation are quite realistic in terms of spacing, thickness, etc. We note in passing that one can also combine this simple motion rule with more complex signaling models [710], with similar results.

What causes streams to form in this simple model? The key insight was provided in [711] (see also [712]) who analyzed an augmented MG kinetics model with the cell density acting as a new dynamic field. The density obeyed the advection equation

$$\frac{\partial \rho}{\partial t} + \boldsymbol{\nabla} \cdot (\rho\, \boldsymbol{v}) = 0\,, \tag{11.5}$$

with a velocity determined by

$$\frac{\partial \boldsymbol{v}}{\partial t} = -\Gamma\, \boldsymbol{v} + k(\psi)\, \boldsymbol{\nabla}\, \psi\,, \tag{11.6}$$

with Γ equal to a relaxation time. The rate $k(\psi)$ is chosen to be a (rapidly) decreasing function of the cAMP concentration ψ. Because of the nature of the cAMP pulse, this motion rule corresponds to giving the cells a positive kick every time a wavefront (but not a waveback) passes. Levine and Reynolds (LR) [711] showed that this model exhibits a linear instability to density fluctuations, a sort of excitable medium version of the famous Keller–Segel [713,714] mechanism that is responsible for the chemotactic collapse seen in *E. Coli* colonies [673,674,715,716]. That is, regions of high cell density are more excitable and the waves move more quickly through them. This deforms the wavefront so that cells tend to detect waves coming primarily from high-density locations and move accordingly. The motion leads to even higher density, closing the feedback loop. This instability is strong enough that within a few wave cycles, noticeable inhomogeneities can be seen in the density. Note that this mechanism relies on the dependence of wave speed on cell density. This dependence has been directly measured by Van Oss et al. [717] and the results were in agreement with predictions based on the MG framework and utilized in LR; namely, speed increases with density.

Past onset, one can carry out numerical simulations of continuum models to study the nonlinear streaming state. A variety of groups have done this with fairly consistent findings. To take one example, Hofer et al. [718] utilized the equation

$$\frac{\partial \rho}{\partial t} = \boldsymbol{\nabla}\left(\mu(\rho)\boldsymbol{\nabla}\rho\right) - \boldsymbol{\nabla}\left(\chi(r)\rho\boldsymbol{\nabla}\psi\right) , \tag{11.7}$$

where $\mu(\rho)$ is a non-linear diffusivity to take into account random cell motion and χ, the chemotaxis coefficient, depends explicitly on the state of the receptor variable r. The fact that χ decreases at large r will accomplish the required rectification of the traveling wave. These authors showed in a series of papers [718–720] that this model has the streaming instability and that simulations can produce realistic streaming structures. We cannot help but point out that once one turns to simulation, the discrete model is simpler and more flexible, while leading to practically identical results.

To get beyond stream formation and make predictions regarding the structure of the mound, one must allow cells to climb atop each other and extend the aggregate in the vertical direction. There have been several attempts to deal with mound formation, with modest results to date. We will first discuss the automaton model. Levine et al. [721] generalized the motion rule to allow climbing whenever the chemotactic driving was sufficiently strong. This does allow the aggregate to expand vertically. To get realistic shapes, two additional mechanisms proved crucial. First, the effects of cell-cell adhesion [722] were included by adding a relaxational step to the allowed cell motion. This adhesion energy gives rise to an effective surface tension for the mound without which there would be no smoothness, in direct disagreement with observations. However, this approach did not prove sufficient since one also needed to reduce the role of chemical signaling in the mound. If the cells continued to move inward due to chemotaxis (say for a circular wave), the mound would have a sharp central peak instead of the observed flat structure. This can easily be seen in a simple mean-field description of the stochastic evolution of mound shape. In this model of the simulations, the shape is determined by a balance between chemotactic flux and surface-energy-driven relaxation.

tained from discrete molecular-dynamics simulations, and we feel that that is a lesson worth taking to heart for *Dictyostelium*.

11.3 Cell Sorting and Tip Formation

As long as the *Dictyostelium* amoeba are communicating solely or mainly by the cAMP chemical signal, the dynamics remains relatively simple. As cells enter the mound and begin life as components of a rudimentary multicellular organism, complexity proliferates. Now, one has to take into account cell-cell adhesion, interactions between the cells and the extracellular matrix (a set of secreted protein that can form some type of gel external to the cells) including the slime sheath, possible signaling due to various chemicals, etc. On the positive side, progress here should directly translate to progress in understanding similar developmental processes in more complex eukaryotes such as ourselves.

As cells enter the mound, they differentiate into two major cell types, pre-spore and pre-stalk. Most evidence to date [697] suggests that this differentiation happens in a spatially uniform manner, leading to a mixed population of 80% pre-spore and 20% pre-stalk. Exactly which signals are used to coordinate differentiation and to ensure proper proportioning is at present unknown. Over a period of roughly 2–4 h, the cells sort such as to place most of the pre-stalk cells at the apex of the mound. These pre-stalk cells then push their way upwards and form the protruding tip, distending the covering sheath as they do so; a picture of this transformation in progress is shown in Fig. 11.11. Mutants that cannot form (a specialized sub-class of) pre-stalk cells do not show tip formation [727]. Almost nothing is known for certain about how these events come about, but that has not prevented attempts at simulating these processes using (hopefully) reasonable assumptions about cell motion. In this section we discuss some recent ideas regarding the modeling of this stage of the developmental process.

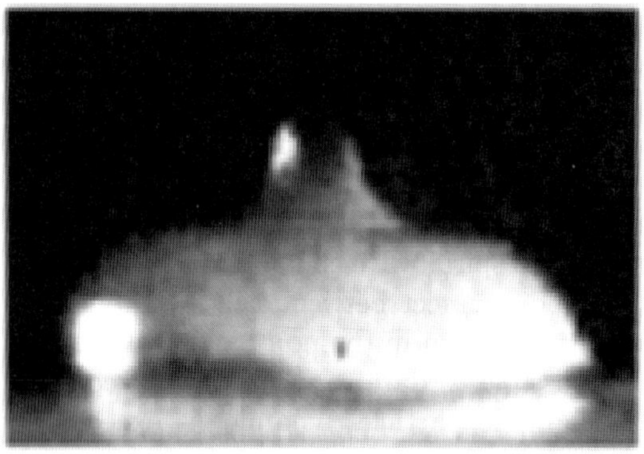

Fig. 11.11. Photograph of a tipped aggregate at 12 h

It is known that cells can sort due to differential adhesion, a biological version of phase separation. In this idea, originally due to Steinberg [728], the bonding strength between two cells depends on the types of cells involved, where the affinity of cell adhesion molecules on the cell membranes determines the adhesion energy. This energetics biases the otherwise random movement of cells in aggregates, causing them to re-arrange into patterns with minimal surface energy. Examples of such patterns in 2D can be seen in Glazier and Graner [725]. Research based on sorting in chicken embryo cells shows that the cells perform a biased random walk, with diffusion caused by cytoskeletally driven membrane fluctuations [729]. Such differential adhesion driven aggregates are usually convex in shape, since a quasi-ellipsoidal shape minimizes total surface area, and thus total surface energy. In our opinion, it is unlikely that differential adhesion provides the entire impetus for sorting or for cell motion in general. One reason is the repeated observation of swirling motions in the mound ([730], A. Nicol, private communication) (see Fig. 11.12), a collective effect that cannot be explained as a set of random shape changes exploring the adhesiveness landscape. Another is that it is very difficult to understand why tips would form under purely adhesive forces. Historically, one has always appealed to cAMP wave chemotaxis to get tips and to organize cell motions [731]. As we have seen, this can lead to tips and might help sorting as well if the response to cAMP is taken to be different in the different cell lines. Unfortunately, if we take at face value the recent results discussed at the end of the last section regarding sorting (and tip formation) in an adenyly cyclase null mutant [726], these models are wrong. It is therefore fair to say that no convincing answer has emerged as yet for this process.

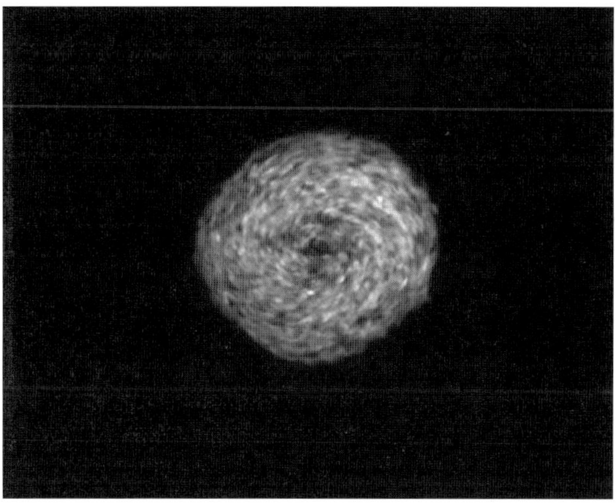

Fig. 11.12. Snapshot of a mound with swirling cell motion – courtesy of A. Nicol

We now briefly describe a computational approach that can investigate these issues [732]. First, we assume that all cells have differentiated to either pre-stalk cells (20% by volume) or pre-spore cells (80%). They have fixed surface properties that do not change during sorting and their cell volumes are constrained to be more or less constant. Second, the surface properties of cells are isotropic. Hence, cells adhere with an energy per unit contact area which depends on the cell types involved. Third, as we do not explicitly consider extracellular matrix in the mound, cells interact with their neighbors through direct contact. We assume a layer of sheath, which may have different contact energy with different cell types, has formed around the mound before we start our simulations.

In our lattice model, each lattice site contains a number, σ, corresponding to the unique index of a cell. The domain of sites with the same index belongs to the same cell that has type τ and volume v, and the membrane surfaces are described by the differences of indices. Adhesive energy resides on the membrane surfaces only, while cells have no internal boundary energy. The sheath around the mound and the substrate supporting the mound are also represented as "cells", but with particular properties and different surface tensions. To keep the aggregate from falling apart, we choose the surface tensions such that cells are more adhesive to each other than to substrate or to sheath. As the cells do not grow or shrink during sorting, we set a target volume V_σ for both pre-spore and pre-stalk cells. Deviation from the target volumes increases the total energy and is not favored. As mentioned above, this method was used in a model of mound morphology and slug formation, but these authors did not consider cell sorting.

Once one has an expression for the adhesive energy, sites are updated by a Monte Carlo procedure; the probability of accepting a re-assignment is determined by

$$\mathcal{P} = \begin{cases} 1 & \Delta H \leq 0 \\ \exp(-\Delta H/T) & \Delta H > 0 \end{cases} , \tag{11.10}$$

where ΔH is the change in effective energy caused by the trial modification and T is a parameter that controls the amplitude of cytoskeletally driven membrane fluctuations. Chemotaxis can be added by including in the energy a chemical potential term reflecting the "desire" of a cell to be at a higher value of the concentration. There are many open issues regarding the appropriateness of this type of equilibrium sampling dynamics for cell motions, but this approach does get us started in the right direction towards a better understanding of cell motions in dense multicellular structures.

Results to date show that indeed chemotaxis can cause tips to form and these tips can be composed primarily of pre-stalk cells if differential adhesion drives the pre-stalk cells to the mound surface (see Fig. 11.13). While this picture agrees reasonably well with experiment, the lack of any suitable chemotactic agent to provide the inward driving might mean that this agreement is fortuitous. Work is continuing on both the foundations of this

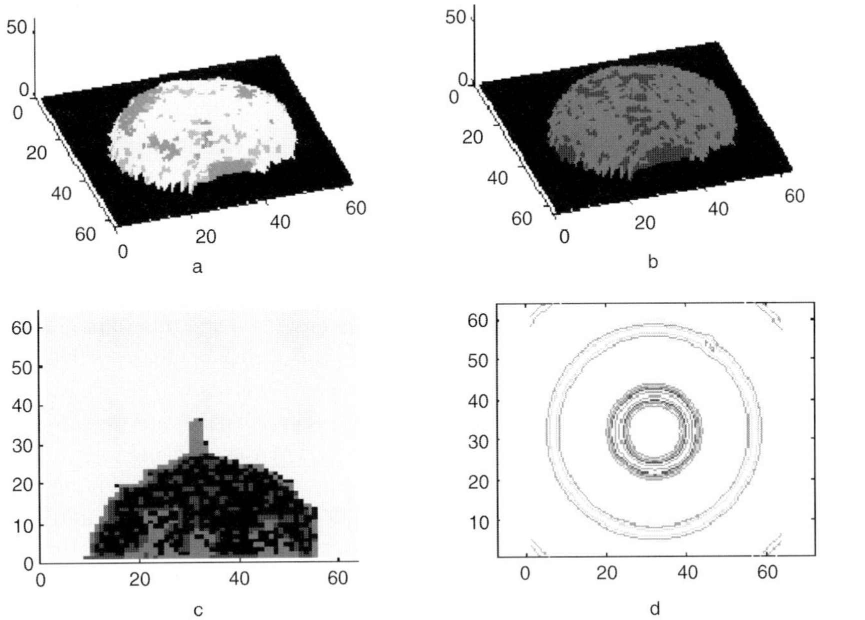

Fig. 11.13. Snapshot of a simulation with the Potts model framework, showing sorting and tip-formation due to a combination of differential adhesion and chemotaxis. Part (**a**) shows the three dimensional surface plot of the mound, color codes indicate the different states of cells: *green* for quiescent, *purple* for active and *yellow* for refractory. (**b**) shows the cell type distribution on the surface of the mound: *blue* represent pre-spore cells and *red*, pre-stalk cells. (**c**) is a vertical cross section through the mound; we can see *red* (pre-stalk) cells forming the protruding tip. (**d**) shows the chemical concentration (a view from the top) at the surface of the substrate (The colour plate of this figure appears on page 363)

model and on applying it to other hypotheses. One idea under study concerns repulsive chemotaxis. Ben-Jacob et al. [733] have proposed, based on pattern formation in Bacillus, that cells that go inactive in general might emit a repulsive chemical as a sort of warning beacon to the colony. Might the pre-spore cells in *Dictyostelium* adopt a similar stratagem?

As we do not as yet understand the transformation from a mound to the finger stage, it is probably premature to focus much attention on the slug that is formed from the finger. It is fair to say, though, that suggestions [734] regarding cAMP chemical waves (in the form of scrolls) for governing slug motility are probably incorrect in the light of recent biological work on mutant strains. Other researchers have focused on the forces generated by the cells on each other and on the substrate [735]; while this is of course a necessary component of any picture of how the slug moves, these models seem overall simplistic in the light of the complex repertoire of responses that a cell can utilize so as to accomplish coordinated motion.

Finally, it is worth noting that formation of fruiting bodies as an evolutionarily selected solution to starvation occurs in the bacterial world as well, namely in the myxobacteria [736]. It will certainly be worthwhile to compare *Dictyostelium* dynamics with those in these other organisms as a way of seeing whether there has been convergence in the types of physical processes that underlie morphogenesis of this type.

11.4 Reprise

The developmental dynamics of *Dictyostelium* provides a rich phenomenology upon which one can test notions of how to apply ideas from the field of non-equilibrium pattern-formation physics to processes involving biological complexity. It is fair to ask about the extent to which this framework has been successful. In other words, have we shown that one can make testable predictions for *Dictyostelium* that do not depend on all the ignored biological specificity but instead arise as a result of the generic features of the dynamical system formed by the interacting cells? The answer seems to be a resounding yes. Predictions regarding the selection of spirals and the formation of streams are indeed possible and have in general been borne out by experiment. One can hope that this state of affairs will be extended to sorting and tip formation in the near future. One can, with suitable attention to the biological details, unravel the physics from the biology.

Thus, in addition to understanding *Dictyostelium* our work can serve as a paradigm for using non-equilibrium physics in the service of developmental biology. The lessons learned are

- Pick systems with a critical mass of biologists so that the relevant information at the cellular and sub-cellular scales is either available already or at worst are under intense investigation. Systems like the *E. coli* spot patterns, while interesting to think about and model, rapidly reach a state where the models have to make arbitrary assumptions about the biology with little chance of rapid verification or discredidation.
- Pick issues that focus on collective phenomena that cannot in any obvious way be encoded directly by specific genes. For example, there cannot be a gene that tells the cell to make a spiral, since the spiral is a dynamic consequence of the whole system. There can, of course, be genes that bias the system towards spirals (as for example the PDI gene that increases excitability of the cAMP signaling system), but this will not obviate the need for an understanding of the collective nature of the spiral problem.
- Pick models that are as simple (and as easy to simulate) as possible for studying any specific issue. Since the models are not likely to be fully quantitative anyway, the gain in simplicity is much greater than the loss of detail. As a corollary, any finding that relies on some very precise

details of a complicated model is almost assuredly going to be irrelevant for the actual biological system.

Armed with these principles, one can go forth and start work on a variety of the fascinating processes that make up development. Hopefully, some of you readers will be motivated to do just that. Good luck!

12 Predicting Epileptic Seizures*

Robert Savit

In this chapter we will be interested in methods for anticipating seizures. It would, of course, be best if one could predict seizure occurrence with absolute accuracy and indefinitely into the future. But that is not likely to happen. One could reasonably hope to make such predictions in a fairly simple, non-chaotic, autonomous system. But the brain is not such a system. However, a more modest goal may be achievable. There is reason to believe that seizure prediction with reasonable confidence minutes to tens of minutes ahead may be possible, at least in some patients. Such an achievement would be a great boon to treatment and to the quality of life of patients with epilepsy. First, it would open up the possibility of interventive measures that might abort the seizure. But even if that were not possible, a few minutes warning could be invaluable. It would allow a patient with epilepsy to stop a potentially risky activity in which they were engaged, and find a safe situation in which to be when the seizure occurred.

Before beginning the discussion of seizure prediction several important caveats are in order. First, we will not be concerned with the precise definition of what a seizure is or how it is recognized. In particular, we will not be concerned with the difference between clinical, electrographic, or electroclinical seizures. For our purposes, we will assume that all seizures have both clinical and electrographic manifestations, and that the onset of clinical and electrographic manifestations occur within a time which is short compared to the horizon over which we seek to predict the seizure. For most of our discussion, we will consider seizure prediction of about 2 to 30 min prior to seizure onset, so a typical difference of a few seconds between electrographic and clinical onset counts as a short time relative to the prediction horizon. A corollary to this is that we will not discuss predictions that can anticipate seizures only a few seconds before electrographic or clinical manifestations (see Chap. 13). Because of the vagaries of defining precisely when seizure onset occurs, it is not even clear that such very short predictions are always really predictions: One mans prediction may be anothers detection.

Second, we will not discuss precise electrographic signatures of a seizure. This is a complicated subject that is discussed at length elsewhere and that

* I am grateful to Radu Manuca and Ivo Drury for illuminating discussions and collaborations on this topic and Dingzhou Li and W. Zhou for help with the figures and references. This work was supported in part by the US National Institutes of Health under grant No. 1 RO1 NS36803-01A1.

involves subjective as well as objective elements. We will assume that most reasonable encephalographers will agree on the occurrence of most seizures, at least within a few seconds.

Third, we will not systematically distinguish among different types of seizures. Most of the work that has been done on seizure prediction (and all of the work we will discuss in detail in this chapter) has been in the context of medically intractable temporal lobe epilepsy (TLE), largely because intracranial recordings, which are taken primarily from this patient population, present a very clean data source. We do not yet know the extent to which the tantalizing results obtained in the context of TLE may apply to other epilepsies. Thus, the reader should bear in mind the nature of the data behind the analyses we are discussing, but should also keep in mind the possibility that these techniques may be more generally applicable. Later in this chapter, we will also discuss some results on scalp recording from patients with TLE.

Fourth, we will not extensively discuss the significant literature that uses ideas from nonlinear dynamics to help localize the seizure focus or to characterize or otherwise explain other features of EEG. Our interest in this chapter is on prediction. There are, however, important theoretical implications of the work on seizure prediction, and we will briefly allude to those in the discussion section.

Before reviewing the results of work on seizure prediction, it is worthwhile discussing briefly the nature of modeling, measurement and prediction for nonlinear systems, and the dynamical and theoretical context in which these results should be understood. Epistemological questions that are merely bothersome for simple systems become central for complex nonlinear systems. Following this discussion, we will turn to a more specific discussion of prediction of epileptic seizures. We will set the stage by briefly summarizing the commonly accepted, qualitative picture of seizure generation. We will then review results on seizure prediction. Here we will first discuss some work that seeks to use nonlinear empirical indicators of signal complexity as a predictor of seizure onset. We will then turn to some work that seeks to exploit the intrinsic non-autonomy and non-stationarity in EEG signals to indicate the impending onset of a seizure. The chapter ends with a summary, a discussion of open questions and a prospectus for future research.

12.1 Empiricism and Theory

Our premise, consistent with the premise of this book, is that seizures are events whose origin is best understood in the context of nonlinear dynamics. By nonlinear dynamics we mean to include also systems that are non-autonomous, and possibly even adaptive. Biological systems are obviously adaptive, and since they are open systems they are clearly subject to non-autonomous effects. Thus, nonlinear dynamics, including non-autonomous effects, seems to be a reasonable theoretical framework in which to place theories of many neurological diseases. However, placing our considerations in this context requires careful attention to the precise nature of our statements and their theoretical and empirical meaning.

Unlike simple linear systems, nonlinear systems can often have a wide variety of different parsimonious descriptions, depending on the nature of the questions being asked. An important consideration in defining the question is the spatial and temporal scale over which the system is being viewed. So, for example, a description of phenomena lasting a few seconds in a neural system comprising 10^5 neurons may be qualitatively different from a description of phenomena lasting milliseconds in a system of only a few neurons. The fundamental reason that different spatial and temporal scales can give rise to qualitatively different descriptions in nonlinear or adaptive systems, has to do with the property of emergence. Collections of large numbers of elements (say, neurons) can lead to phenomena that are qualitatively and fundamentally different from those observed in smaller collections of elements. The simplest example of an emergent property is the phenomenon of water freezing into ice at 0°C. The notion of a phase, a liquid or a solid, cannot even be applied to a single water molecule. For the concept of phase to be meaningful requires that we consider a collection of many water molecules. Similar considerations apply to any nonlinear or adaptive system in which there are large numbers of interacting elements.

In the context of the brain and the phenomenon of epilepsy, these issues play a central role. First, although the firing mechanisms of a single neuron may be fairly well understood, it may be that the best theory of epilepsy will involve descriptive dynamics that are not at all similar to those of a single neuron, or even to those of a collection of a small number of neurons, just as the detailed structure of water molecules is largely (although not entirely) irrelevant to the description of the phase transitions of water. Second, it may also be that the best theory of epilepsy will involve qualitatively different descriptions depending on the time and length scales over which phenomena are considered. For instance, the dynamics appropriate to the prediction of seizures several minutes ahead may be markedly different from the description of the dynamics associated with the prediction of seizures seconds ahead, or hours ahead.[1] These realizations require that we keep a very open mind about the nature of models or theories when dealing with complex phenomena such as epilepsy. There may not be one right theory. It may be that the best we can hope for is a sequence of theories or descriptions that are effective

[1] Although different time and length scales may give rise to qualitatively different effective dynamics, there may be, in principle, calculable relations among the effective dynamics over different length and time scales. One approach to computing these relationships is that of the renormalization group and related computational schemes. There is a large literature on this subject, but its discussion is beyond the scope of this chapter.

and context dependent. I stress that this possibility may not be due to any inadequacy on our part, but may be intrinsic to the nature of the complex phenomena we are investigating.

As we have discussed, the phenomenon of emergence gives rises to serious theoretical and epistemological issues, and to the possibility of limitations on models and the necessity of effective descriptions. This is a potential concern for nonlinear systems in general, including neurological ones. But for problems that are, in addition, non-autonomous and data limited, such as the study of epilepsy, there is a further complication.

A common way to characterize nonlinear dynamical systems is with values of certain quantities based on correlation integrals and that are topologically invariant, such as Lyapunov exponents or Hausdorff or other dimensions. If the dynamical system under consideration satisfies certain conditions, including autonomy, then, in the limit of infinite precision, these quantities have a clear and specific theoretical meaning. However, these quantities lose their theoretical interpretation if conditions such as autonomy are not met. Nevertheless, it may be possible to formally compute such quantities and use them as strictly empirical indicators for various purposes. It is almost certainly the case that, in any sensible use of the word, epilepsy cannot be described as a nonlinear autonomous process. Nevertheless, quantities that, strictly speaking, have a simple theoretical meaning only in the context of autonomous nonlinear systems, may have some empirical or predictive value. The reason behind this possibility is that these quantities are sensitive to the nonlinear dynamics in the system. In a non-autonomous system, the effective dynamics may change over time, and these changes may, in some cases, be related to seizure onset. Consequently, variations in quantities based on correlation integrals may be useful indicators of relevant changes in the dynamics of the system, even though their theoretical interpretation is problematic.

In our discussion of results on seizure prediction, we will try to be clear about whether the results under discussion have clear theoretical significance, or are just intended as empirical statements.

12.2 Predicting Seizures

Attempts to predict epileptic seizures have a long history [15]. Prior to the recognition that ideas from nonlinear dynamics may be applicable to brain function, linear methods, such as spectral analysis were used to try to predict seizures. These attempts were largely unsatisfactory, although there was some success in very short-term predictability a few seconds before seizure onset (see also Chap. 13). With the realization that brain function is intrinsically nonlinear, and that traditional statistical methods generally fail to capture the nonlinear structure inherent in EEG new empirical methods based on intuitions garnered from the study of nonlinear dynamics were applied to the problem of seizure prediction.

A common view of the general nature of seizure onset in focal epilepsy is that a collection of neurons in a small region of the brain begin to fire in a pathologically synchronous fashion, and somehow entrain nearby neurons to a similar firing pattern. When a sufficiently large area of the brain has been so entrained, a seizure occurs with its attendant clinical and visual electrographic consequences. In the seizure state, therefore, neurons in regions of the brain near the seizure focus (and possibly those more distant), fire in a way that is more regular and synchronized, and consequently less complex (in some sense) than normal. This suggests that some measure of complexity of neuronal firing may decrease during a seizure. Moreover, if the process of entrainment takes place over an extended period, one may be able to observe this hypothesized decrease in complexity some time prior to the onset of the seizure. This general picture of entrainment appears to be consistent with much of the work to be reviewed below, although the precise notion of decreased complexity is somewhat more problematic. However, even if the idea of entrainment is in a general sense correct, it still leaves open a number of important questions both about the specific process of entrainment and about parsimonious descriptions thereof, and these may in turn have consequences for predictive strategies. We will return to this issue below.

Attempts to analyze EEG recordings, whether using linear or nonlinear methods face a number of obstacles. First, scalp recordings, by far the easiest to obtain, are dense with artifact, noise and contamination. Muscle artifact and attenuation of the signal through the surrounding bone and other tissue make a detailed, precise analysis difficult. In addition, the fact that the electrodes are relatively far from the brain means that one is observing signals that are averages over very many neurons. Since brain signals are inhomogeneous, much of the information associated with that inhomogeneity is lost.

Recordings from intracranial electrodes are a much cleaner data source. Muscle artifact is absent. Moreover, the electrodes, which are in direct contact with brain tissue record signals from a much smaller number of neurons, typically 10^5 or 10^6. Consequently, these data are considerably more sensitive to the spatial inhomogeneities of the brain. Because one may be interested in the propagation of seizure activity, sensitivity to spatial variations in the brain may be of great significance (see Chap. 6).

Of course, one limitation of intracranial data is that the population from which such data can be drawn is restricted. These data are available only from patients who are candidates for surgical resection of a suspected seizure focus. This typically limits the population to patients suffering from medically intractable localization-related (partial) epilepsy, most commonly arising from the temporal lobe. In particular, therefore, no direct comparison between intracranial EEG from these patients and from healthy subjects is possible. It is, however, possible to compare scalp recordings of persons with epilepsy and healthy subjects if sufficiently sensitive statistical methods are used.

We will present a brief review of some of the work on seizure prediction as defined above. Following this summary we will present some comments and caveats about these results. A number of groups have published reports suggesting that nonlinear methods can be used to predict seizures several minutes ahead [737,738]. At least one other group has published a methodological piece on problems associated with nonstationarity and nonlinearity in physiological time series [739]. In addition, our group, at the University of Michigan and the Henry Ford Hospital has some preliminary data on the issue. The review in this chapter will not be exhaustive. Rather, I will review the work of two groups that are representative in their general approach of some of the studies. These reviews will serve to illustrate some of the promise of nonlinear methods for seizure prediction, as well as some of the weaknesses and limitations of current work. I will then discuss the role of non-autonomy in epileptic EEG and refer to the preliminary results of our group and one other on this topic.

12.3 Simple Nonlinear Measures

From the point of view of nonlinear dynamics, the simplest approach one can take to the problem of seizure prediction is to define some quantity that may be sensitive to changes in the dynamics of the system, and examine the time dependence of that quantity. In the papers discussed in this subsection, the authors use simple quantities related to the correlation integral and seek to discover changes in those quantities that precede a seizure. Formally, these quantities have a well defined meaning in an autonomous dynamical system. In the papers discussed here, they are given only empirical, not theoretical meaning.

Lehnertz and Elger [741,740] have published several papers in which they report systematic changes in a quantity they call $D_2^{\rm eff}$, which is an empirically defined quantity inspired by (but not to be identified with) the definition of the correlation dimension. These analyses are done on data obtained from intracranial electrodes. In their work, Lehnertz and Elger are careful to state that this quantity, while formally an estimate for the correlation dimension, should be regarded only as an empirical quantity and not necessarily one with theoretical significance. The authors compute $D_2^{\rm eff}$ by calculating the correlation integral, $C(\epsilon)$ over epochs of 30 s and then computing

$$D^* = \frac{1}{N_\epsilon} \sum_{\epsilon=\epsilon_l}^{\epsilon_u} C'(\epsilon) \,, \tag{12.1}$$

where

$$C'(\epsilon) = \frac{\mathrm{d}\log C(\epsilon)}{\mathrm{d}\log \epsilon} \,, \tag{12.2}$$

and N_ϵ is the number of ϵ-values in the range $[\epsilon_l, \epsilon_u]$.

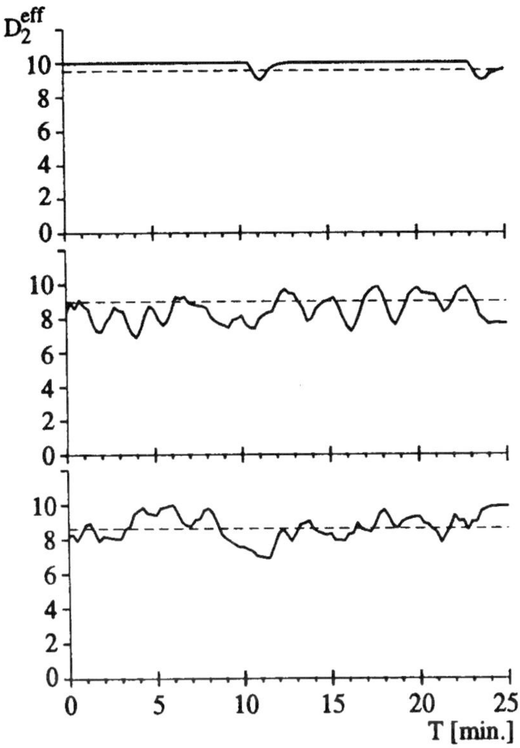

Fig. 12.1. Bottom to top: D_2^{eff} (*solid lines*) as a function of time taken from (*bottom to top*) within the primary epileptogenic area, an adjacent area, and a remote area during a state-1 (seizure free) epoch. The *dotted line* is a base-line value of D_2^{eff}. Figure taken from [740] with permission

D^* is regarded as an estimate of D_2^{eff}. The authors define two temporal regions: 1) at least 24 h away from any seizure; and 2) a period immediately preceding a seizure of duration between 10 and 30 min. In addition, the authors report values of D^* for data from electrodes in the epileptogenic focus, regions of brain adjacent to the seizure focus, and regions distant from the focus. In all, the authors report on results from 16 patients with medically intractable temporal lobe epilepsy. One epoch of type 2 (pre-ictal), and up to several epochs of type 1 (interictal) were analyzed from each patient. Results that the authors state are typical are shown in Fig. 12.1 and 12.2. Note that in Fig. 12.2 D^* falls below its "baseline" value and stays low about 8 min prior to seizure onset in recordings from the epileptic focus and from an area of the brain adjacent to the focus. This is not the case in the recording from a remote area of the brain, nor from any of the recordings during the type 1 (interictal) epoch. (Although there are decreases in D^* during type 1 pre-ictal epochs, the authors state that these decreases are typically not as intense,

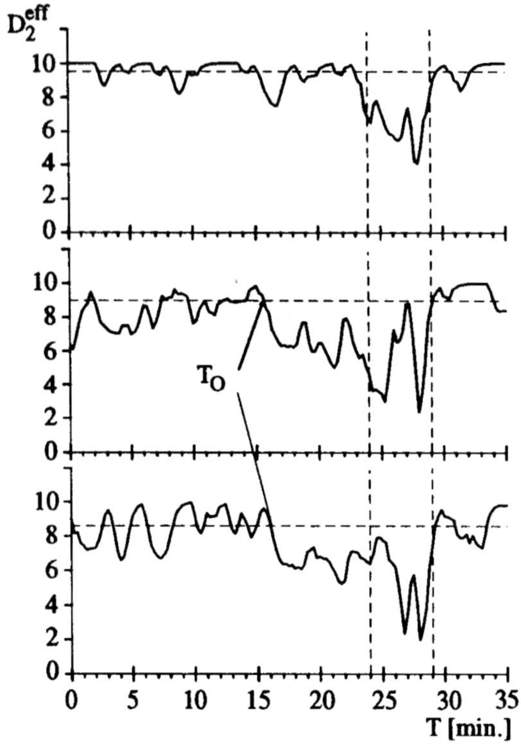

Fig. 12.2. Same as Fig. 12.1 but for state-2 recordings. Seizure onset and ending is indicated by the **vertical dashed lines**. Figure taken from [740] with permission

nor as long lasting as those occurring during Type 2 interictal epochs.) On the basis of these results, the authors conclude that there is evidence that this intrinsically nonlinear measure may be useful in predicting the onset of seizures, minutes before the seizure occurs.

Although the findings reported in this paper are suggestive, there are a number of issues that make the arguments less than compelling. First, there is little systematic discussion of statistical significance of the results. This is due, in part, to the relatively low number of cases reported. It is also due to fact that the indicators used to predict a seizure are not precisely defined, and there is apparently no out-of-sample testing. Rather, there is a general tendency for D^* to fall below, and stay below a baseline value prior to the seizure. Precise threshold values that would be taken to be indicative of an impending seizure are not determined out of sample. This is not a fatal criticism, insofar as this papers findings are intended to be suggestive. However, it is indicative of the preliminary and tentative nature of the work, so far in this area. We will have more to say about this below.

Martinerie et al. [742] have also reported results suggesting that nonlinear quantities may show significant changes minutes before a seizure. In this work, the authors again study intracranial electrodes from patients with TLE. They study a quantity similar to that looked at by Lehnertz and Elger, namely, a value of the correlation integral, $C(r)$, evaluated at a specific, moderate, value of r, r_o, the choice being determined in a reasonable way by the data. The work of Martinerie and co-workers differs from that of Lehnertz and Elger in the data they use to reconstruct the phase space in which $C(r_o) \equiv C$ is calculated. In particular, Martinerie and co-workers use four time lags from each of four different electrodes, all in the neighborhood of the seizure focus to reconstruct a 16-dimensional phase space. Their C is then calculated in this reconstruction space and used to estimate a correlation density,[2] D. In Fig. 12.3 we reproduce a figure from their paper in which they show one seizure from each of two patients with TLE. It is clear that the correlation density, D, begins to decrease well before visual onset of the seizure. Autocorrelation *functions* (which are strictly linear measures), by contrast, do not show a persistent change minutes before the seizure. The small triangles in the figure indicate the first moment at which, the authors claim, the seizure is predicted. The criterion for this earliest prediction comes from their application of the philosophy described in Manuca and Savit [743] in which they imagine that the neurodynamics can be divided into fast and slow components. We will describe this philosophy in more detail below. For the moment, it is sufficient to say that Martinerie and co-workers consider a plot of D_j vs. D_{j+1}, where j labels the time window over which D is computed. They find that, a long time before a seizure, points in this plot are clustered in a well-defined region of the (D_j, D_{j+1}) plane, and abruptly migrate out of that region several minutes before seizure onset. By defining a specific boundary for this region, they are able to define a criterion of first prediction of the seizure onset. This is the point indicated by the triangles in Fig. 12.3.

Using these methods, the authors are able to anticipate seizures from 2 to 6 min prior to seizure onset in 17 out of 19 cases. Unfortunately, there are a number of statistical and logical issues that are not addressed in this paper. First, there is no explicit discussion about the results obtained during interictal epochs. In particular, it would be helpful to know if these methods produced any false positives. Although the authors do show significance with respect to linear surrogates, a comparison with interictal epochs would be more germane.

In any case, both the work by Lehnertz and Elger [740,741] and Martinerie et al. [742] suggest that nonlinear quantities related to the correlation integral may be useful in anticipating seizures in TLE. In addition, both groups are fairly optimistic about the robustness and applicability of their methods.

[2] Note that these authors use D to refer to a correlation density. Other authors tpically use D to refer to a dimension.

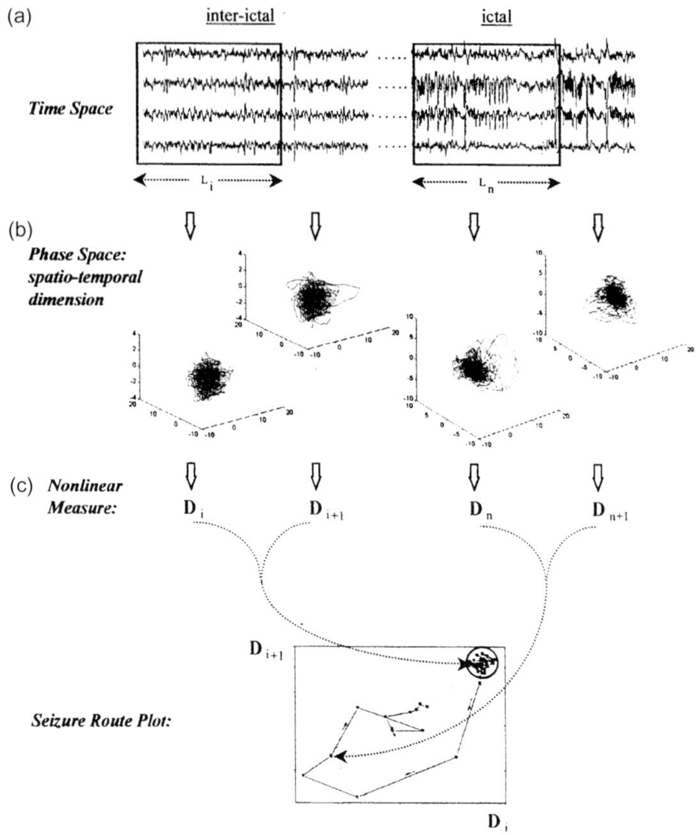

Fig. 12.3. (**a**) Autocorrelation and (**b**) correlation density as function of time prior to a seizure in two different patients. The *vertical line* indicates the time of seizure onset. The **small triangles** in (**b**) indicate the earliest time at which the authors claim to be able to anticipate the impending seizure. The small graph above (**b**) indicates significance of the correlation density, D, with respect to linear surrogates for one of the curves in (**b**). Figure from [742] with permission

Unfortunately, the situation may not be quite that simple. A paper by Pijn et al. [744] studies three seizures using data from intracranial electrodes in one patient with TLE. They compute a number of different quantities, which suggest several conclusions about the nonlinear structure of TLE. But their conclusion that is most germane to our discussion is that they do not see consistent evidence for a decrease in complexity (as measured, for example, by a lowered value of D_2 (formally, the correlation dimension), or some related quantity) prior to seizure onset. They do observe lower values of D_2 during a seizure, particularly from electrodes near the region of seizure onset. However, they state "we have not been able to ascribe any value to the D2-plots in predicting the onslaught [*sic*] of a complex partial seizure".

There could, of course, be many reasons why these authors fail to see a systematic lowering of D_2 prior to a seizure. First, precise parameter choices and methods of analysis are not the same among different groups, and it is possible (barely) that different parameter settings for their analysis would yield the systematic results seen by Lehnertz and Elger [740] and Martinerie et al. [742]. Second, maybe Pijn et al. [744] were just unlucky. They only analyzed three seizures from one patient. Martinerie et al. [742], for example, did not find precursors to seizure onset in 2 of the 19 cases they studied. Nevertheless, this study, as limited as it is statistically, should be taken seriously as raising at least some questions about the robustness of the seizure precursors optimistically suggested by Lehnertz and Elger [740] and by Martinerie et al. [742].

Indeed, a recent paper by Widman et al. [745] suggests that there may indeed be different mechanisms of seizure onset, some of which show lowered values, pre-ictally, of measures such as D_2, while others do not. In this work four different epilepsy models in hippocampal slices from guinea pigs were studied. It was found that in two of the models (seizure-like activity induced by xanthine and penicillin) intracellular recordings manifested reduced signal complexity, as measured by lowered values of D_2^{eff} prior to paroxysmal depolarization shifts (the analogue of seizure onset). In the other two models (seizure-like activity induced by low magnesium and veratridine), however, such a precursor signal was absent. The authors attribute these differences to the (presumably) different dynamical paths that the neuron takes from normal to epileptic-like activity in the different models. One can certainly question the relevance of this study to seizure prediction in human epilepsy. Aside from the perennial problem of extrapolating from in-vitro guinea pig hippocampal slices to in vivo human brains, these in-vitro studies analyze recordings from single cells, while even the intracranial recordings analyzed from human patients with TLE represent summed activity from very many neurons. Nevertheless, the suggestion that TLE in humans may be strongly heterogeneous in its dynamics, and that only some of those dynamics admit of lowered pre-ictal complexity that can be observed with simple measures such as D_2 is certainly reasonable and deserving of very serious consideration.

12.4 Role of Non-stationarity and Non-autonomy

From these results, it appears that even the relatively well-defined disease of temporal lobe epilepsy is so heterogeneous that nonlinear measures taken from fairly noise-free intracranial electrodes cannot robustly predict seizure onset. This might lead one to be pessimistic about the prospects for seizure prediction for TLE, and completely depressed about the prospects for seizure prediction in general. It may well be true that differences in the dynamical route to the seizure state make it difficult to generate a single, robust, nonlinear precursor. But the change from the non-seizure state to the seizure state is nonetheless a change. And even though the dynamics may be heterogeneous, we may be able to use the fact that the dynamics changes as a basis for seizure prediction. In particular, if there are pre-ictal dynamical changes,[3] then we may be able to exploit that fact by seeking measures that are sensitive to structural changes in the dynamics and largely independent of the nature of those changes. For a statistician this suggests looking for measures of non-stationarity. For a physicist (or mathematician), it suggests looking for measures of non-autonomy.

One may argue that the methods discussed above are, in fact, measures of non-stationarity, since the authors look for changes in some nonlinear quantity over time. However, there are many examples of systems in which the value of some quantity persists even though the dynamics changes, and conversely, examples of autonomous dynamical systems in which some quantities change over time (for example, systems with intermittency). Consequently, looking at the time variation of a single quantity is no guarantee that one will be sensitive to dynamical changes. From the point of view of nonlinear dynamics, the fundamental nature of any given dynamics is associated with the evolution of trajectories on an attractor.[4] The more sensitive a measure is to changes in the attractor and the attendant changes in the trajectories associated with the attractor, the more likely it will be that changes in that measure will reflect changes in the underlying dynamics. The approach in the papers discussed in the last subsection is to evaluate a quantity in a sequence of windows and look for variations of that quantity from one window to the next. The measures used therein are averages of a quantity over an attractor. Although these measures are sensitive to nonlinearity in the dynamics, they are, nonetheless, simple averages over the attractor, and so changes in the attractor may not be simply and apparently reflected in changes in these averages. In the next subsection, we will describe an approach that relies on the evaluation of a nonlinear quantity that incorporates a more direct comparison between two attractors reconstructed from two different windows. Such quantities result in more direct and sensitive measures of dynamical change between the two windows.

In order to avoid confusion later, I will state here that our work suggests that, defined in a certain way, there is a persistent type of non-autonomy in EEG recordings both from interictal and pre-ictal epochs in patients with TLE which does not appear to be present in healthy subjects. There are also

[3] Of course, it is possible that there simply is no pre-ictal change in epileptogenic brain; that the transition to the seizure state occurs, in general, within a few seconds of seizure onset. But this possibility is just too depressing to entertain.

[4] Here and in the rest of this section we use attractor in an informal sense that includes those objects produced in a reconstruction space from data in a given window. Such an object may not, strictly speaking, be an attractor in a precise dynamical sense.

studies suggesting that the nature of that non-autonomy itself changes with time in the patient with epilepsy, and, in particular, undergoes qualitative, observable, changes minutes to tens of minutes prior to a seizure. I will also comment on recent published work by another group [746] that supports this picture.

12.5 Non-autonomy and the Metaphase Space

In an attempt to understand and exploit the non-autonomy in epileptic EEG, we have devised an approach based on cross-correlation integrals [743]. In words, we suggest a measure that is sensitive to the extent to which two temporal windows of a time series are dynamically similar. The philosophy behind this approach is embodied in the idea that there may be more than one time scale in a system, and those time scales may be approximately separable. Thus, without too much loss of generality, we imagine that our system can be described by a dynamical equation of the form

$$\frac{\partial u}{\partial t} = G(u; v) , \tag{12.3}$$

where u is a set of internal variables, and v is an external driving force. As will become clear below, this expression should be considered an effective description of the system, and the identification of u and v will depend on the values of some parameters chosen for our analysis.

We consider, then, a time series that is divided into windows labeled W_i, each of duration, T. We suppose that there are some variables, u, that vary significantly over time T. Other variables, v, we suppose are essentially constant over a time, T. These are the variables that we identify with the driving force in (12.3). We will seek a method to determine whether the variables, v, vary from window to window. We will do this by comparing the reconstructed dynamics among windows and measuring the extent to which those reconstructed dynamics differ from window to window. Variation of v from window to window is another way of saying that the underlying dynamics is different in different windows. Insofar as we expect that there may be subtle changes in the dynamics preceding a seizure, a method such as this, that looks for dynamical changes, but makes no assumptions about the nature of those changes, may be quite useful.

To implement this philosophy, we consider the cross-correlation integral defined as

$$C_m(X,Y) = P(\| x^{(m)}(i) - y^{(m)}(j) \| < \epsilon) , \tag{12.4}$$

where $P(.)$ denotes the probability of the argument, $x^{(m)}(i) = (x_i, \cdots, x_{i-d+1})$ is an m-dimensional vector reconstructed from data from window X, and similarly $y^{(m)}(j)$ is an m-dimensional vector reconstructed from data from window Y. In this expression, we explicitly compare reconstructed data from different windows. Thus, the value of $C_{\mathrm{m}}(X, Y)$ carries information about how similar the reconstructions of different windows are. Unlike average quantities calculated window by window, $C_{\mathrm{m}}(X, Y)$ contains comparative point-by-point information (in the reconstruction space) about the structure of the attractors reconstructed from two different windows. We therefore anticipate that quantities based on $C_{\mathrm{m}}(X, Y)$ may more closely reflect the similarity or dissimilarity of the attractors, and thus may be more faithfully sensitive to the extent to which the underlying dynamics is the same or different in the two windows. Clearly, $C_{\mathrm{m}}(X, X)$ and $C_{\mathrm{m}}(Y, Y)$ are the ordinary correlation integrals for windows X and Y, respectively. By comparing values of $C_{\mathrm{m}}(X, Y)$ with $C_{\mathrm{m}}(X, X)$ and $C_{\mathrm{m}}(Y, Y)$, we can arrive at a measure of how similar the reconstruction of window X is to window Y, which can be taken to be a measure of how similar the underlying dynamics is in the two windows. There are clearly many ways in which such a measure can be constructed. We have found that one simple function of $C_{\mathrm{m}}(X, Y)$ seems to do well at discriminating between different dynamics, while at the same time not requiring inordinate amounts of data. In particular, we consider

$$d^m(X, Y) = \sqrt{\max(|R_m(X, X) - R_m(X, Y)|, |R_m(Y, Y) - R_m(X, Y)|)}\,, \tag{12.5}$$

where

$$R_m = \frac{C_{m+1} C_{m-1}}{C_m^2}\,. \tag{12.6}$$

Clearly, in the language of (12.3), the identification of u and v, the fast and slow variables, respectively (or the internal variables and the external driving force), will depend on our choice of analysis parameters such as T, m, and ϵ. Thus, in general, one must consider a range of different analysis parameters in searching for seizure precursors.

Equation (12.5) can be used by itself as a measure of the similarity or difference between the dynamics (under a certain reconstruction) in two different windows. However, we can go somewhat further if we take $d^{\mathrm{m}}(X, Y)$ seriously as a distance measure. If $d^{\mathrm{m}}(X, Y)$ were a distance measure, then we could use it to construct, by triangulation, a geometric representation associated with the degree to which different windows in the time series had similar dynamics. It turns out that, strictly speaking, $d^{\mathrm{m}}(X, Y)$ cannot be regarded as a distance measure, but in the cases we have studied, we have found that the deviations of $d^{\mathrm{m}}(X, Y)$ from a true distance measure are small enough to still allow us to give meaning to a geometric representation of their values.

To construct a geometry from the values of $d^{\mathrm{m}}(X, Y)$, therefore, we imagine that $d^{\mathrm{m}}(X, Y)$ is a distance between two points X and Y in a space of a certain number of dimensions that we call a metaphase space (MPS). The points in this space are associated with the windows of the time series, and their distances from each other in the space represent the extent to which

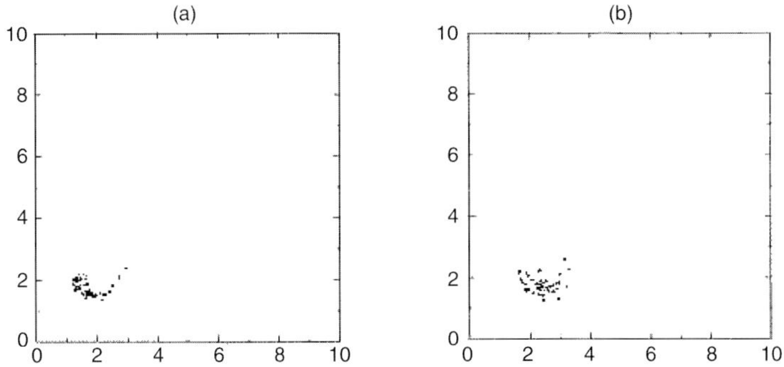

Fig. 12.6a,b. MPS plots from 30 min of Stage I and II non-REM sleep from two different healthy subjects

Although these results are preliminary, they are very tantalizing. In particular, they suggest that there are systematic dynamical changes in the nature of the non-autonomy that are observable both in intracranial and in at least some scalp recordings that can be observed minutes to tens of minutes prior to a seizure. These changes appear to be similar across seizures within a patient, but differ from patient to patient. This is all nicely consistent with the view suggested earlier. Namely, there may be observable dynamical precursors to a seizure, but the precise nature of the dynamics is likely to be different in different patients. Therefore, specific nonlinear measures that are sensitive to certain characteristics of the dynamical change may be useful for seizure prediction in some patients but not in others. The MPS plots which rely on cross-correlation integrals are more sensitive to changes in the detailed structure of the reconstructed attractors than are comparisons of quantities calculated independently in different windows. Moreover, the MPS approach makes no strong a priori assumption about the nature of the changes that one should look for, and so might be expected to be more broadly sensitive to changes in the dynamics from epoch to epoch.

Some recently reported work by Hively et al. [746] also supports this view. They use a couple of related nonlinear measures, namely, χ^2 and the L_1 norm between the invariant measures of two attractors, one constructed from a reference window, and one from a target window. They look for changes in these measures over time as an indication of a change in the nature of the underlying dynamics associated with the time series. They apply these measures to scalp EEG recordings from patients with epilepsy, and report significant changes in both χ^2 and L_1 about 10–30 min prior to a seizure in nine data sets. They also state that they observe no such changes in χ^2 and L_1 in scalp recordings from healthy subjects, although no further details about controls are given.

These observations are consistent with those of our group. To see this clearly, we have taken a number of EEG epochs and calculated both the L_1

norm, which is part of the basis of the work of Hively et al., and the cross-correlation integral, which forms the basis of our nonautonomy measures. We have found that there is a strong monotonic relationship between this L_1 norm and the cross correlation integral, showing that these two measures, while not identical, are largely sensitive to the same information.

12.7 Conclusion

Much of the work on seizure prediction using ideas from nonlinear dynamics is still, frankly, preliminary. We certainly do not have anything like a detailed or even qualitatively complete understanding of seizure genesis, nor have we yet established a robust, verifiable and consistent method of seizure prediction. (Some suggestions for what is needed and for directions of research are presented below.) However, there are some general themes that seem to be emerging from work so far, and that seem in broad outline to be consistent with some of the received wisdom about seizure genesis.

First, our work suggests that, with certain reconstruction parameters, there is an underlying non-autonomy observable in both intracranial and at least some scalp EEG recordings from patients with TLE. There is, on the contrary, no similar observed non-autonomy in the scalp recordings of healthy subjects. Furthermore, there appear to be changes in the nature of the non-autonomy as a seizure is approached, but the detailed nature of that change appears to differ among patients with epilepsy. This is all consistent with the commonly held view that a focal seizure begins with a pathological synchrony of neurons in an ictal onset zone. The development of the seizure progresses with recruitment of neurons near the seizure focus, leading to clinical and obvious electrographic manifestations of the seizure. For some set of reconstruction parameters, this neuronal recruitment can clearly be considered to indicate a change in system dynamics. However, the precise nature of that change may differ among patients. One would, though, expect more homogeneity in dynamical changes across seizures within a patient.

As for practical methods of seizure prediction, the work of several groups [740–742,744,745] taken together seems to suggest that a simple, single non-linear dynamical indicator may not be a universal predictor of seizure onset. On the other hand, preliminary work by our group [743,747] as well as the recent work of Hively et al. [746] suggests that measures that are more generally sensitive to aspects of the non-autonomy of the system, rather than specific dynamical changes may be more robust at anticipating seizures. Of particular note for practical purposes is the fact that these changes appear to be observable in at least some scalp recordings.

12.8 Prospective

As tantalizing as these results are, there are still many outstanding issues, practical, statistical and theoretical that must be addressed before one can be said to have either a practical method of seizure prediction, or a good (even empirical) understanding of the underlying dynamics.

As regards practical methods of seizure prediction, there are several important problems with the existing literature. First, there is not nearly enough data to draw firm conclusions about the efficacy of any method. In particular, one needs data from many more patients, as well as more extensive data from each patient. Second, there are clear lacunae in some of the reported methodologies. Not enough attention is paid in some publications to controls or to meaningful nulls. For example, a number of researchers use either linear or IID surrogates as bases for comparison. But it is not clear that such comparisons are meaningful. More meaningful, for seizure prediction, would be comparisons of analyses of interictal with pre-ictal data, as well as comparisons with EEG from healthy subjects. There is also little discussion in the literature of false positives. How often do the proposed methods of seizure prediction predict a seizure when one is not present? In addition, because of the variety of approaches to the problem of seizure prediction, it is difficult to make precise comparisons among the claims of different authors. All of these methodological and statistical issues, which must be addressed in any data analytic study, are even more problematic when dealing with nonlinear phenomena, which we certainly are.

Researchers must pay closer attention to the statistical and logical requirements of their investigations. In addition, the community would be greatly helped by the establishment of a large, maintained and well-documented database for seizure prediction studies. Such a database would help in at least three ways. First, data from patients with epilepsy is difficult and time consuming to collect, and this is one of the reasons that the statistics in the published literature is so limited. An accessible, maintained set of data would certainly help address this problem. Second, work on seizure prediction is also hampered by the heterogeneity of available data, which complicates the problem of directly comparing results from different research groups. Most data used in this research is collected primarily for clinical purposes. The wide variation in clinical protocols and methodologies from one clinic to the next leads to a strong heterogeneity in the data sets. In addition, a number of different devices for collecting data are in common use, each with its own acquisition, compression and filtering idiosyncrasies that further complicates the problem. A maintained and documented database would allow different researchers access to the same data, allowing better communication and synergy among research groups. Third, the database would facilitate direct comparisons among the work of different research groups. Currently, the variety of approaches to the problem of seizure prediction, in combination with the heterogeneous nature of the data sets, makes it difficult to precisely compare or validate the claims of different researchers. There is no uniformly established methodology and there are no established benchmarks for efficacy of prediction. A central database would make it possible to establish reasonable guidelines for efficacy as well as benchmarks and standards for direct comparison of different methods.

Despite these statistical and methodological problems, there are clear indications that it may be possible to develop useful and robust tools for seizure prediction. Two possible strategies suggest themselves. The first is to continue to study predictors that may be sensitive to dynamical change in the state of the brain prior to a seizure, as our group and the group of Hively, et al. are doing. A second strategy is to seek to develop a set of nonlinear predictors, some of which may be useful in some patients and others of which may be useful in other patients. As a practical matter, this may be at least as useful as a single predictor based on dynamical change, since it may be possible to tailor the predictor to the patient.

The work on seizure prediction may also have important theoretical implications. Little is understood about the way in which seizures are generated and spread, although the qualitative idea of neuronal recruitment in cases of focal epilepsy does seem to be widely accepted. The onset of seizures is clearly a nonlinear phenomenon, so, developing the ability to predict seizures using the methods of nonlinear dynamics should go hand-in-hand with a deeper understanding of the underlying neural mechanisms responsible for the seizure. As we discussed earlier, a parsimonious description of a nonlinear system may depend on the parameter ranges and scales over which one chooses to discuss the system, as well as the precise questions one is asking. This further raises the question of whether a theoretical understanding, not just an empirical description of a complex system also depends on the scales one is studying and the questions one is asking. One might well expect this to be particularly true when dealing with a complex system with emergent properties. But another epistemological complication is possible: It may be the case that, even with a given scale and set of questions, more than one theoretical explanation is correct. There are examples of such situations even for the relatively simple systems studied in the physical sciences.[7] So, for example, there may be a sense in which the picture of bistability described may be correct [747], and the picture of unstable periodic orbits may also be correct [748]. This sentiment has also been expressed by Milton [110,749]. Thus, when considering the theoretical implications of attempts to predict seizures, we must keep an open mind about where our work will lead, bearing

[7] For example, in statistical mechanics or in field theory there is a complementarity between different descriptions of a physical system. A field theory can be described either in terms of a statistical ensemble of a sum over states, or in terms of stochastic differential equations. The two "theories" are mathematically equivalent, but are, in a sense, different ontologically.

in mind the fact that there really is no well-developed epistemology for the study of complex systems in general, and neurological systems in particular.

I believe we are on the threshold of making substantial progress in seizure prediction, and probably in seizure control. To speed progress, I would urge our community to cooperate more fully in sharing data and data analytic technologies, and to work at setting clear and reproducible standards that will allow straightforward comparisons among different analyses and prediction methods. I also think it is likely that the work on nonlinear seizure prediction will contribute to a deep understanding of the dynamics of epilepsy and of neurological systems in general, although I have very little idea what the nature and scope of that understanding ultimately will be.

12.9 Addendum

Several papers have appeared since this chapter was written and are of interest. Hively et al. [750] have extended their work in [746] to compare success in anticipating seizures from scalp EEG recordings using their methods to that achieved using traditional nonlinear indicators such as measures of entropy or correlation dimension. They report that the methods devised in [746] are more consistently successful in anticipating seizures than more traditional nonlinear measures. Le Van Quyen et al. [751] have also extended their earlier work [742]. Still relying on notions of non-stationarity and the philosophy described in [743], they have refined the indicator developed in [742] and have shown some success at anticipating seizures from intracranial electrodes in a cohort of 13 people with TLE. Both of these recent papers are strongly encouraging, although there is still very considerable variability in the time for seizure anticipation. Finally, our group at the University of Michigan is continuing to study seizure anticipation, focusing primarily on scalp recordings (D. Li, W. Zhou, I. Drury and R. Savit, in preparation). We have also had some success in anticipating seizures, but have also found that our success rate is rather variable. This may not be too surprising, since our methods are not entirely dissimilar from those of Hively et al. [746,750] and Le Van Quyen et al. [751,742], although each group's methods do focus on somwehat different aspects of the data. In my opinion, there is an emerging convergence among a number of different groups on seizure prediction, using a variety of different, but related methods. This represents significant progress on this very difficult problem, and holds out the real hope for development of a reliable strategy for ambulatory seizure detection.

13 Comparison of Methods for Seizure Detection *

K.K. Jerger, T.I. Netoff, J.T. Francis, T. Sauer, L. Pecora, S.L. Weinstein, and S.J. Schiff

Whether epileptic seizures can be predicted by quantitative analysis methods applied to EEG has been a focus of much recent interest [740,751,752]. This resurgence of interest has been motivated by several factors, including the proliferation of powerful new methods for analyzing nonlinear system dynamics, as well as interest in developing epilepsy-control devices. Nevertheless, attempts to detect seizures automatically from EEG are not new, and older linear analysis methods showed promise nearly twenty years ago [753,754].

Rather than referring to a declaration in advance of the time and location of a seizure, the term "prediction" has frequently been used to refer to the process of identifying a state from the EEG that precedes a clinical seizure that is known to have occurred (see also Chap. 12). Using this meaning, the period of prediction refers to the time between identification of a pre-seizure state and either the onset of the clinical seizure or the time at which a well-trained clinician can pick up evidence of changes by visual inspection of EEG. From the point of view of designing a control device, the distinction between clinical and neurologist-determined onset may be meaningless – the device simply needs to detect the seizure dynamics early enough to permit effective intervention. From a clinical perspective, what is meaningful is whether the fact that a seizure is about to occur can be reliably determined when the future is unknown. Successful completion of this goal, which entails "prediction" in a slightly but crucially different sense from that described above, can only be shown by validating the method on out-of-sample data, where the location or even the presence of seizures is unknown to the tester, and false positive/negative detection rates can be established to determine sensitivity and specificity.

To facilitate meaningful comparisons between EEG analysis techniques commonly reported in the seizure-prediction literature, we performed a set of analyses on a common data set of twelve intracranially recorded seizures from four children undergoing presurgical evaluation for intractable epilepsy, chosen to represent a range of spatial extent, location, and type of underlying pathology. A battery of both linear and nonlinear methods (power spectral density, cross-correlation, principal component analysis, phase synchronization, wavelet packet analysis, correlation dimension, and mutual nonlinear prediction) was applied to the same set of twelve seizures to characterize

* Research was supported by grants from the National Institutes of Health and the Whitaker Foundation.

the earliest dynamical changes leading up to the clear onset of an epileptic seizure.

Clinical EEG interpretation and time series analysis of seizure EEG have traditionally been conducted separately, by specialists with very different training and without awareness of each other's techniques and conclusions. This division has contributed to a fragmented knowledge base and has limited the practical application of results to relieving patient suffering. Seeking to avoid such fragmentation, close consultation with a clinician was maintained throughout these investigations, and the final results were compared to judgements made by a neurologist board-certified in electroencephalography (S.W.), still the "gold standard" in seizure identification.

Inspection of the raw ictal EEG suggests that an increase in coherent neuronal behavior occurs during seizures (see also Chaps. 5 and 6). Each of the seven methods presented has previously been applied with variable success to characterizing this coherent behavior, although the theoretical bases of their approaches differ significantly. In our opinion, the extreme complexities involved in developing an understanding of seizure dynamics warrant addressing this problem from a variety of viewpoints.

13.1 Methods for Seizure Prediction

13.1.1 Power Spectrum

Power-spectral analysis has been applied to EEG more frequently than any of the other techniques applied here and provides the basis for all analyses in the frequency domain. Almost two decades ago, spectral analysis was used to distinguish between epochs not associated with seizures and those preceding the spike-wave bursts of absence seizures with up to eighty per cent accuracy [755]. They pointed out, however, that each subject's preburst EEG seemed to be characterized by a unique pattern of changes, and that no common prodromal pattern was found that could be applied uniformly across patients. But in 1994, when Inouye et al. [756] applied a non-stationary power spectral analysis (which computed the instantaneous power spectrum every 0.1 s for 10 s preceding a spike and wave complex), their results reflected a relative increase in arrhythmic slow activity leading up to spike and wave complexes in all ten of their patients. Increased spectral power at high frequencies (40–150 Hz) has been noted at the start of seizures beginning with a well-defined pattern of low-amplitude signal, referred to as the electrodecremental event [386].

The mechanisms of the electrodecremental EEG pattern remain obscure, despite having been recognized for decades [757]. One central question is whether this period reflects an overall decrease in activity. Fisher et al. [386] suggested that the electrodecremental period does not indicate a lack of signal, but rather a shift of the spectral energy from lower to higher frequencies. In 1995, Alarcon et al. [758] correlated characteristics extracted from the power spectrum of patients with partial epilepsy to surgical outcome. They noted that one of the most common early ictal manifestations was the generalized electrodecremental event, present in 12 of their 15 patients. They suggested, based on their surgical outcome, that these events may not be part of the ictal process itself. Instead, they might reflect generalized cerebral changes which enhance the likelihood of seizure formation in susceptible tissue–consistent with the "two-hit" hypothesis of seizure initiation and propagation. Application of spectral methods to interictal EEG has revealed increased power in lower frequencies (0.25–8 Hz) relative to higher frequencies (8.25–30 Hz) when EEG from epileptic subjects was compared to EEG from normal subjects and headache patients with normal EEG [759]. Spectral analysis of bilateral interictal recordings taken over several days from the mesiobasotemporal lobes showed a striking asymmetry in the variability of the power spectrum which persisted for hours, with what they termed the relatively "rigid" side showing 80% coincidence with lateralization of the seizure based on PET [760].

13.1.2 Cross-correlation

To our knowledge, autocorrelation was first applied to EEG by Norbert Weiner and his colleagues in 1968 [761]. More recently, autocorrelation was used to assess the likelihood of future neuronal bursts [762]. They reported decreased likelihood of bursting near the site of seizure onset from sites located within the hippocampus or entorhinal cortex. Shortly after Rogowski et al. [754] predicted seizures by several seconds using an autoregressive model, cross-correlations between interictal spikes from homologous brain structures were used to demonstrate changes in the EEG up to tens of minutes prior to the clinical onset of some epileptic seizures [753]. Over the past twenty years, the cross-correlation technique has been applied to EEG in countless studies as one of the more conventional analysis tools against which newer tools are compared. A few of these applications include using cross-correlation for determining the location of epileptogenic foci [763], investigating interdependence of EEG signals [764], estimating time delays between channels [765], and characterizing dynamical properties of sleep EEG [766].

13.1.3 Principal Components Analysis

Principal components analysis (PCA) is a linear method that has been used in EEG research to combine information across channels and reduce the dimensionality of the original multi-channel EEG to a smaller set of theoretically meaningful component variables. Accomplishing this involves constructing a linear composite of the original variables by selecting a set of weights that maximize the variance of the original data. In this case, the number of variables equals the number of electrodes or amplifier channels. In order to explain 100% of the variance expressed by the original data, the number of principal components extracted from the correlation matrix will be equal to the number of channels, and no reduction in dimensionality will occur. Thus, one must choose an acceptable percentage of the overall variance that one seeks to preserve by calculating its principal components. In 1987, when Maier et al. [767] used PCA for source localization of human visual evoked potentials, they chose a level of 95%, assuming a noise level of 5% "after prolonged averaging". In the same year, Freeman and van Dijk [768] reported that using only the first principal component was sufficient for their comparison between spatial patterns in the visual cortex and the olfactory bulb of a rhesus monkey. Nine years later, Barrie et al. [769] compared results obtained from applying PCA, a modified FFT method, and calculation of root mean square amplitudes to extraction of the broad-spectrum waveform common to all channels of an 8×8 electrode array placed on the cortical surface of a rabbit, and found the three methods to yield equivalent spatial patterns. In this same study, they reported the first principal component to account for 90–99% of their data's variance [770]. Jobert et al. [771] found that after performing PCA on the results of their spectral analysis, the first two principal components retained 89.0–99.4% of the initial variance for their sixteen subjects, a sufficient amount for their automatic analysis of sleep EEG. A sophisticated algorithm for choosing the number of principal components to extract can be found in [772].

13.1.4 Wavelets

Originating from the field of seismology [773], in the last decade and a half wavelet transforms have been applied to a number of problems including image compression [774,775], turbulence [776], and speech processing [777]. After Gotman's original work using decomposition of the EEG into half-waves for automatic seizure detection [778], there followed a number of applications of wavelet transforms to EEG analysis and seizure detection [779–786]. In 1998, Osorio et al. [787] reported their method based on Daubechies' DAUB4 wavelet [788] allowed prediction of clinical onset in 92% of 125 seizures of mesial temporal origin by a mean of 15.5 s.

13.1.5 Phase Correlation

Methods of measuring phase synchrony include those based on spectral coherence [789,414] that incorporates both amplitude and phase information, detection of maximal values after filtering [790] and wavelet filtering [791]. In their 1996 investigation of phase synchronization of chaotic oscillators, Rosenblum et al. [792] pointed out that "the notion of synchronization itself lacks a unique interpretation" but settled on Blekhman's general description of synchronization as "an appearance of some relations between functionals of two processes due to interaction" [793]. They then showed that for weakly coupled nonlinear equations, a condition exists in which the phases are locked, but the amplitudes vary chaotically and are practically uncorrelated. Tass et al. [794] developed a technique based on this work that they applied to noisy non-stationary bivariate data from magneto-encephalograms and muscle activity in Parkinson's disease. Stratonovich [795] described synchronization of noisy systems as the"appearance of peaks in the distribution of the cyclic relative phase" that point to preferred phase difference values. In order to characterize the strength of synchronization, Tass et al. [794] proposed two indices, one based on Shannon entropy and one based on conditional probability that builds on Stratonovich's idea, aiming to quantify the degree of deviation of the relative phase distribution from a uniform phase distribution.

13.1.6 Correlation Dimension

With the advent of nonlinear time series analysis tools that could be applied to experimental data, particularly the correlation integral [796], much interest arose in investigating nonlinear dynamics of electroencephalogram activity. There has been much discussion aimed at establishing the appropriate and optimal application of these methods. It has been suggested that estimating fractal dimension, a characteristic associated with chaotic systems, might provide additional insight to define a preseizure state or the seizures themselves [797]. It has also been pointed out that obtaining evidence of chaotic activity may not be trivial. Osborne and Provencale [798] showed by obtaining a finite correlation dimension for colored noise that the sole observation of a finite fractal dimension from the analysis of a time series is not sufficient to infer the presence of chaos in the system dynamics. Along the same lines, Theiler's [799] analysis of an EEG time series previously reported to be chaotic emphasized that interpreting a calculated dimension as the number of degrees of freedom of a system may be misleading. In this study, Theiler found that the estimated correlation dimension and Lyapunov exponent were essentially the same for the original data and the surrogate data sets, created by shuffling the phases of the original data set and thereby ensuring that no dynamical correlation was present from one spike-and-wave pattern to another. With these caveats in mind, Lehnertz and Elger [800], in their study published in the same year as Theiler's analysis, were careful to point out that they were only interested in relative dimension changes over time, and were not considering the dimension estimates to represent absolute degrees of freedom of the system. In 1998, they reported a marked drop in estimated dimension up to several minutes prior to seizures which persisted until seizure offset [740].

Another consideration regarding correlation dimension is whether time series from individual electrodes should be treated separately, reconstructing by time lags alone, or with other channels, forming multi-channel reconstructions based on time lag and spatial position. The majority of applications have used separate channels, despite the conclusion of Lachaux et al. [801], based on simulated EEG data, that single channels did not do as well as the multi-channel method in quantifying spatially extended dynamics. Lerner articulated another central issue of EEG analysis that transcended estimation of dimension when he stated, "The fundamental problem [leading to poor reproducibility of results] lies in the fact that the time series associated with the EEG are not stationary over periods of sufficient length to permit reliable estimation of the quantitites of interest. Indeed, the most interesting feature of the EEG is its non-stationary character" [802].

13.1.7 Mutual Prediction

Nonlinear systems may synchronize in complex ways that require methods designed specifically for their detection [803,804]. In order to characterize nonlinear dynamical interdependence between two neuronal systems, Schiff et al. [805] derived a method based on mutual nonlinear prediction that they applied to spinal cord motoneurons. This method defines the nonlinear predictability of each system based on knowledge about the other system, and it provides information on the directionality of the coupling. This is done by using time-delay reconstructions of two simultaneously sampled time series. Similar states found in a small neighborhood in one reconstruction are checked to determine whether they correspond to similar states in the second reconstruction and to what degree. This method of mutual prediction has been applied to intracranial EEG of patients with medial temporal lobe epilepsy by Le Van Quyen et al. [806], and their results indicated a marked difference between the degree of linear interaction measured by the cross-correlation coefficient (which was low) and the degree of nonlinear interaction measured by mutual prediction (which was high) at seizure onset.

13.2 Comparison of Seizure Detection Methods

These seven methods (power spectral density, cross-correlation, principal components analysis, wavelet packet analysis, phase analysis, correlation dimension, and mutual predicition) were applied as described elsewhere [807] to EEG recordings from subdural and depth electrodes from four children with medically intractable epilepsy during presurgical evaluation (Fig. 13.1). Three of the children had lesional epilepsy arising from neocortical or hippocampal structures as defined by MRI scan and confirmed by pathology at the time of resection. One showed glioses by pathology only (normal MRI). Following surgical removal of the electrographic seizure focus, each patient either remained seizure-free or experienced markedly reduced seizure frequency and severity. Raw data were band-pass filtered between 0.5–30 Hz and referenced externally. The data were segmented into ten-second half-overlapping blocks (2000 datapoints per window).

There was considerable heterogeneity in the EEG at seizure onset between these patients (data not shown). The state of all patients before and during

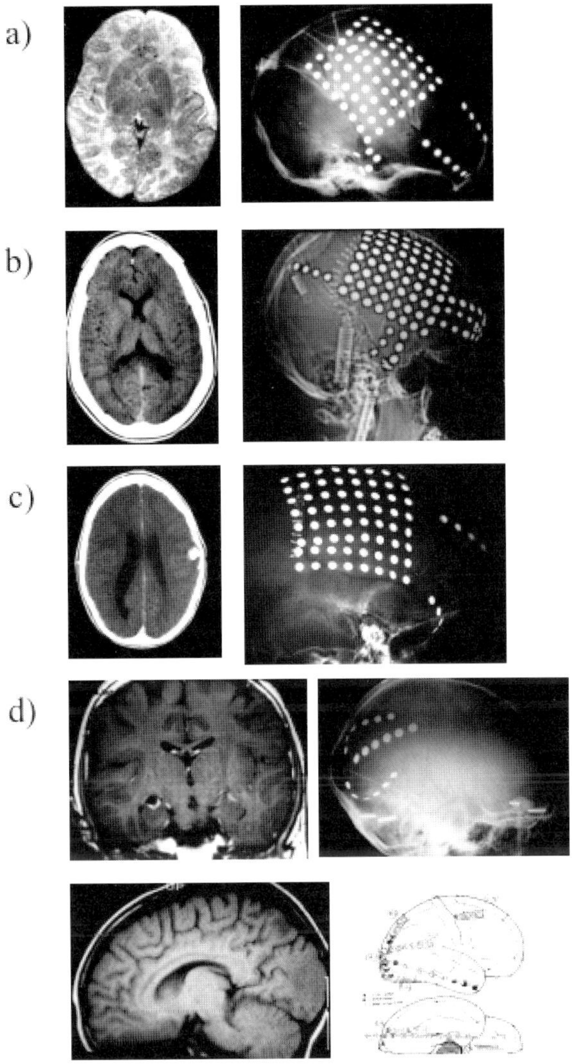

Fig. 13.1. MRIs of brain and lateral skull X-rays showing electrode placement. (**a**) Patient A, MRI demonstrating left temporal lobe cortical dysplasia extending into the parieto-occipital region, and electrodes are shown by X-ray to overlie this region; (**b**) Patient B, although MRI was normal, scalp and subdural ictal mapping revealed seizure generation from the left anterior-inferior frontal lobe and X-ray shows grid placement over this area; (**c**) Patient C, MRI shows a 2×2 cm densely calcified lesion in the surface of the left inferior parietal lobe, eroding the inner table of the skull; (**d**) Patient D, MRI shows dysgenetic right occipital lobe and a small cyst in the right choroid fissure near the anterior atrophic hippocampus. Because of this apparent dual pathology, a single long depth electrode was placed to record simultaneously from both areas, running from the occipital lobe to the anterior hippocampus

13.3 Discussion

These seven methods were successful in detecting changes leading to a seizure up to two and a half minutes before the first visual evidence of electrographic seizure onset noted by the neurologist. It should be emphasized that the time most frequently reported in the literature as that of seizure onset is the time rhythmic activity is evident in the EEG. Here we have used instead the time the first epileptiform changes were noted by the neurologist that always preceded rhythmic activity by an average of twenty-five seconds. Thus, the average lead time given by phase correlation, before clinical onset, is greater than two minutes.

Interestingly, the threshold-crossing times given by the linear methods often occurred earlier than or at the same time as some of the nonlinear methods. We found this surprising since many behaviors of neurons are known to follow nonlinear dynamics, from all-or-none firing to communication through synaptic transmission. Thus, one might expect that there is always underlying nonlinear behavior in EEG that is not fully characterized by linear tools – activity that should be reflected in our nonlinear measures. It is likely that the EEG signals are so complex that nonlinear reconstruction methods may not accurately capture the dynamics. Nonlinear tools are often designed to reveal structure from low-dimensional nonlinear systems. Faced with a truly complex system (with many degrees of freedom), they may fare worse than a linear analysis, as seen for correlation dimension and mutual prediction in these results. In addition, the method of obtaining the initial measurement may exclude relevant information. Using higher than standard sample frequencies and electrode placement density as well as utilizing distant recording sites to improve the resolution and widen the spatial range of sampling might lead to reconstructions that better reflect the true system dynamics.

Linear methods also have weaknesses. For example, a limitation of PCA is that it is restricted to defining directions that are orthogonal to each other. In their analysis of fMRI, McKeown et al. [808] found that their method based on Bell and Sejnowski's [809] independent component analysis (ICA) algorithm, which allows for non-orthogonal directions, as well as a related fourth-order decomposition technique [810] were superior to PCA in determining the spatial and temporal extent of task-related activation. A method for performing a nonlinear form of principal component analysis has recently been proposed [811] that involved the use of integral operator kernel functions; an open question remains as to how to choose the ideal kernel for a particular application.

The substantial differences between patients appeared to play a greater role in seizure predictability than the method selected. This result may be due in part to our choice not to "individualize" parameters. Had we chosen settings most appropriate for a particular patient, we may have found particular methods to be patient-specific. In any case, it is likely that results could be improved by optimizing parameters for a particular patient, and by "training" the algorithm on known seizures before presenting it with test data.

In conclusion, all seven methods were successful in indicating seizure onset before the neurologist for all but a few seizures, most of them one to three minutes in advance of electrographic onset. Since clinical seizure onset occurred an average of 25 s after electrographic onset, the lead time before clinical onset given by these methods was even longer and may prove to be sufficient for incorporation into future control devices. Nevertheless, these procedures should be applied to recordings spanning many minutes or hours to see how they fare when blind to whether or not a seizure will occur. Only with such validation can conclusions regarding their usefulness for true seizure prediction be reached. Finally, although no significant differences were found between linear and nonlinear methods, our analysis of phase performed slightly better than the other methods, which may reflect its sensitivity in detecting weakly coupled nonlinear systems.

14 Direct Deep-Brain Stimulation: First Steps Towards the Feedback Control of Seizures

S.A. Chkhenkeli

In many cases, patients with medically intractable epilepsy may not be candidates for respective epilepsy surgery. For example, the epileptic focus may lie within eloquent cortex or there may be multiple distinct epileptic foci. One therapeutic option remains available to these patients: brain stimulation. There is convincing experimental and clinical evidence to indicate that powerful seizure inhibitory mechanisms exist in the brain [757,114,812,184], in particular, diffusely projecting systems from the brainstem. Structures that have been identified as having an inhibitory influence on neocortical and deep interictal and ictal epileptic activity include the caudate nucleus [175,813–827], the cerebellar cortex and deep cerebellar nuclei [174,230,298,302,824,828–845], the caudal reticular formation [184,846,847], the centromedian nucleus of the thalamus [177,848–850], the raphe nuclei [851–854], the locus coeruleus [855–863], the substantia nigra [180,230,864], and the nucleus of the solitary tract via the vagal nerve [178].

The basic idea of brain stimulation is to try to enhance these inhibitory mechanisms by the use of electrical stimulation introduced through chronically implanted electrodes. An obvious advantage of this approach over surgical resection is that the apparatus can be removed if it has no beneficial effect.

Extensive reviews have already been published that examine the effects of stimulation of a variety of brain structures of patients with medically intractable seizures (see, for example, [865]): cerebellar cortex and deep nuclei [174], centromedian nucleus of the thalamus [177], vagal nerve stimulation (via the nucleus of the solitary tract) [178]. Here I discuss our experience with stimulation of the caudate nucleus and the dentate nucleus. In addition I briefly review our experience with the feedback control of the activity of the epileptic focus by an implanted stimulator.

14.1 Background

14.1.1 Stimulation of the Caudate Nucleus

A consensus concerning the role of the caudate nucleus for regulating epileptic activity does not exist. Some view the caudate nucleus as a universal inhibitor of cortical functions [813,814,819,825,826,866], in particular of epileptiform discharges generated in the thalamus, hippocampus and in the mesencephalic reticular formation [815,818,867–869]. In humans, stimulation of the caudate nucleus leads to speech arrest, motor inhibition, and the development of a sleep-like state [826,870,871] (the so-called "arrest response" [827]). Others feel that the caudate is a source of both excitatory and inhibitory influences [822,872–875]. Some believe that the caudate nucleus may participate in the generation of "petit mal" seizures [876,877], but do not exclude the possibility that the caudate may participate in the inhibition of an already developed seizure [876]. In contrast to these observations obtained from the study of animal models of epilepsy, our work on humans with medically intractable epilepsy clearly demonstrates that stimulation of the head of the caudate nucleus is an effective way to suppress epileptic activity [176,191,230,298,823,824,878–881].

The inhibitory influences of the caudate nucleus on epileptic activity is thought to be related to a decrease of afferent excitatory influences to the cortex. Stimulation of the head of the caudate nucleus can elicit long-lasting waves of hyperpolarization in cortical neurons [882,883]. These waves have been considered to reflect inhibitory postsynaptic potentials (IPSPs) mediated by the non-specific thalamocortical system. Buchwald and co-workers [813,814] concluded that the non-specific thalamic nuclei together with the caudate nucleus and neocortex formed a functional entity referred to as the "caudate loop" [866]. The demonstration that the required anatomical connections exist supports [884–888].

Morphological evidence supports the hypothesis that different parts of the head of the caudate nucleus may have different functions. The ventral parts of the caudate are phylogenetically connected to the olfactory brain, whereas the dorsal parts are connected to the cortex [889,890]. Monosynaptic connections exist between the hippocampus, amygdala and the ventral part of the head of the caudate nucleus [891,892]. It has been proposed that a functional chain: olfactory brain → caudate nucleus → olfactory brain; exists with the function of suppressing the spread of epileptic activity [820]. Morphological data from mammals [893] and humans [894] have demonstrated the existence of four distinct caudate cytoarchitectonic fields in the caudate suggesting that an even richer specialization of the different parts of the caudate.

Stimulation of different parts of the head of the caudate nucleus in animal models of epilepsy has produced conflicting results. In some studies stimulation of the ventral part of the head of the caudate suppresses epileptic activity [895], in others it is stimulation of the dorsal part that is effective [896]. Our studies on humans with medically intractable epilepsy clearly demonstrate that stimulation of the ventral part of the head of the caudate suppresses epileptic activity, whereas stimulation of the dorsal portion promotes epileptic activity [175,191,230,298,823,878–881,897–899].

Figure 14.1 compares the effects of low- and high-frequency stimulation of the head of the caudate nucleus on interictal epileptic activity measured using chronically implanted electrodes in human patients with medically intractable epilepsy. High-frequency stimulation (50 Hz, 4-6 V, 0.2–2 ms, 2–5 s) elicits the

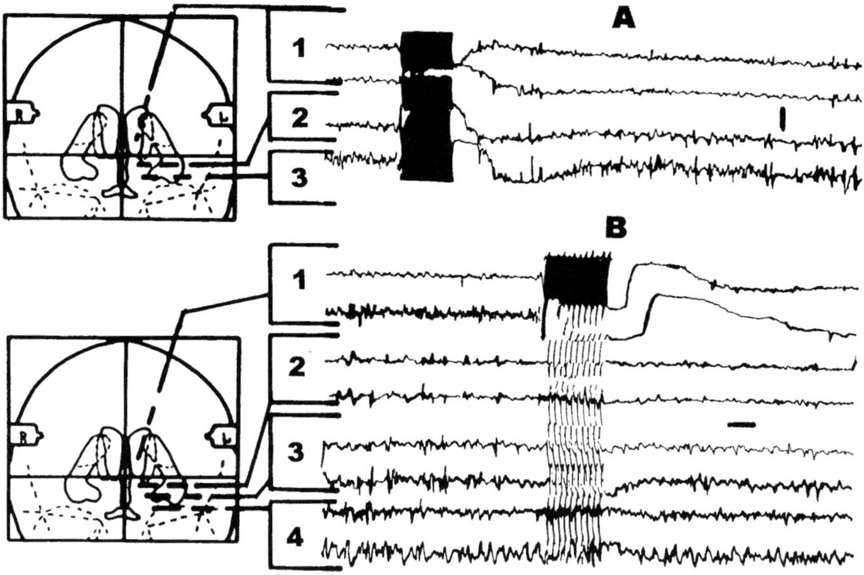

Fig. 14.1. The influence of (**A**) high-frequency and (**B**) low-frequency electrostimulation of the head of the caudate nucleus on interictal activity of the mesial basal temporal lobe structures. Channel 1- SEEG of the left caudate nucleus; Channel 2- SEEG of the amygdala; Channel 3- SEEG of hippocampus; Channel 4- SEEG of fasciculus uncinatus

appearance of isolated spikes and sharp waves as well as enhancement of pre-existing epileptic activity in the caudate nucleus and the ipsilateral hippocampus and amygdala (Fig. 14.1A). In contrast, low frequency stimulation (4–6 Hz, 2–6 V, 0.2–2 ms, 2–5 s) leads to a decrease in the frequency of the sharp transients and attenuation of the background activity in the ipsilateral mesial basal temporal lobe structures (Fig. 14.1B).

Low-frequency stimulation of the head of the caudate nucleus is also capable of suppressing epileptic discharges evoked by electrical stimulation. Figure 14.2 demonstrates that stimulation of the caudate nucleus completely suppresses focal epileptic discharges arising from the amygdala or hippocampus. Moreover, stimulation of the caudate nucleus is also able to suppress evoked epileptic activity that had spread to the contralateral hippocampus and amygdala (Fig. 14.2). Finally, low frequency stimulation of the caudate nucleus is effective in suppressing epileptic activity that arises from the contralateral mesial basal temporal lobe structures (data not shown).

These studies confirm the suggestion that the inhibitory effect of the caudate loop depends exclusively on low-frequency stimulation [813,814]. The inhibitory effects of low-frequency stimulation are more reliably elicited from its ventral part.

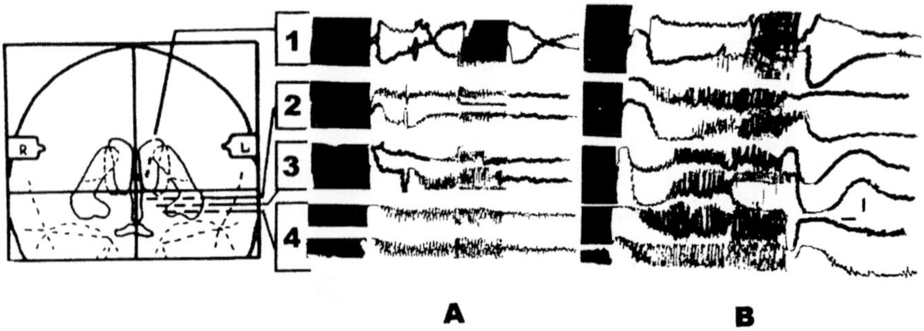

Fig. 14.2. Suppression of spreading evoked (**A**) amygdalar and (**B**) hippocampal afterdischarges by low frequency (6 Hz, 6 V, 0.5 ms, 4 s) stimulation of the head of the caudate nucleus. 1- caudate nucleus; 2- SEEG of amygdala; 3- SEEG of hippocampus; 4- SEEG of fasciculus uncinatus; 5- Artifacts of epileptogenic g; 6- Artifacts of caudate stimulation

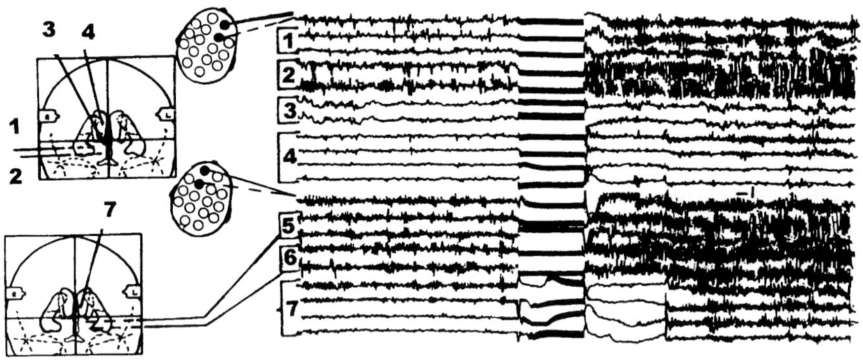

Fig. 14.3. Development of a generalized epileptic seizure following high-frequency (50 Hz, 2 V, 2 ms, 10 s) stimulation of the head of the caudate nucleus. 1,2,3,4- respectively, right amygdala, hippocampus, Forel's H-field, head of caudate nucleus; 5,6,7- respectively, left amygdala, hippocampus and caudate nucleus

Contrasting effects of low- and high-frequency stimulation of the caudate nucleus on epileptic activity have also been observed in animal models of epilepsy. However, in rabbits [872] and cats [900,820] it is high-frequency stimulation of the caudate nucleus that inhibits epileptic activity. Figure 14.3 shows the effects of prolonged high-frequency stimulation of the caudate nucleus in a patient with medically intractable epilepsy. Not only is there an enhancement in the interictal epileptic activity but also the development of a generalized epileptic seizure. In contrast, low-frequency stimulation of the same electrode stimulated in Fig. 14.3, completely suppresses the evoked amygdalo-hippocampal discharge (Fig. 14.4).

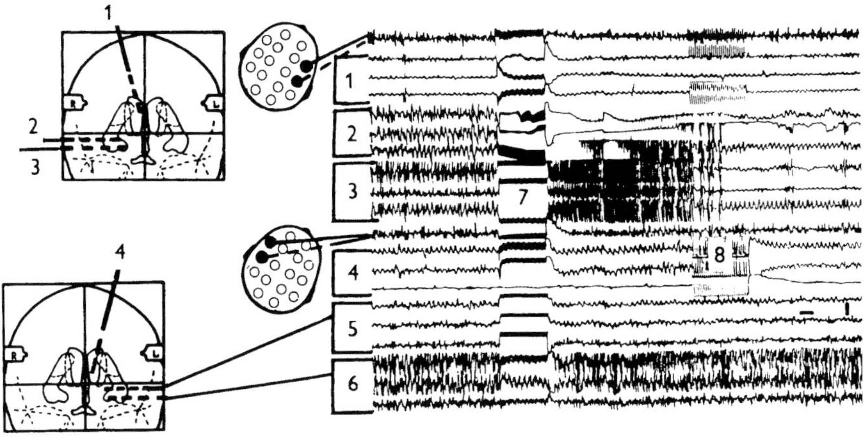

Fig. 14.4. Suppression of focal amygdalo-hippocampal discharge by low-frequency stimulation of the contralateral head of the caudate nucleus. Channels 1,2,3- SEEG of, respectively, the right caudate nucleus, amygdala, and hippocampus; Channels 4,5,6- SEEG of, respectively, the left caudate nucleus, amygdala, and hippocampus; Channel 7-Artifact from stimulation of the right amygdala to evoke the afterdischarges; Channel 8- Artifact from stimulation of left caudate nucleus used to suppress the seizures

14.1.2 Stimulation of the Dentate Nucleus

Our group had been studying the effects of paleocerebellar stimulation on the supra-tentorial EEG background activity. These studies were done during surgery with varying kinds of sub-tentorial occlusive processes, mostly tumors. Contrary to previous studies that showed a suppressive effect of cerebellar cortical stimulation on supratentorial electrical activity [822,838,901–907], we did not observe visually detectable inhibition of the EEG. Indeed, in some cases cerebellar stimulation resulted in the development of epileptiform activity in the sensorimotor area (Fig. 14.5). Cerebellar induced sensorimotor seizures have been described previously [908–910] and confirmed in a variety of animal models of epilepsy [911–913]. Consequently we focused our attention on the effects of stimulation of the deep cerebellar nuclei, in particular the dentate nucleus.

A number of investigators have shown that stimulation of the cerebellar deep nuclei can be more effective for inhibiting supra-tentorial epileptic activity than direct stimulation of cerebellar cortex. Direct bilateral fastigial efferents to the hippocampus (principally CA3), basolateral amygdala and temporal lobe neocortex have been shown to exist using histological and evoked response methods [914]. The fastigiobulbar and dentatothalamic influences on hippocampal cobalt epileptic focus in relation to the stages of epileptic activity have been studied [901]. During interictal events stimulation of the fastigial and dentate nucleus results in either a prolonged desynchro-

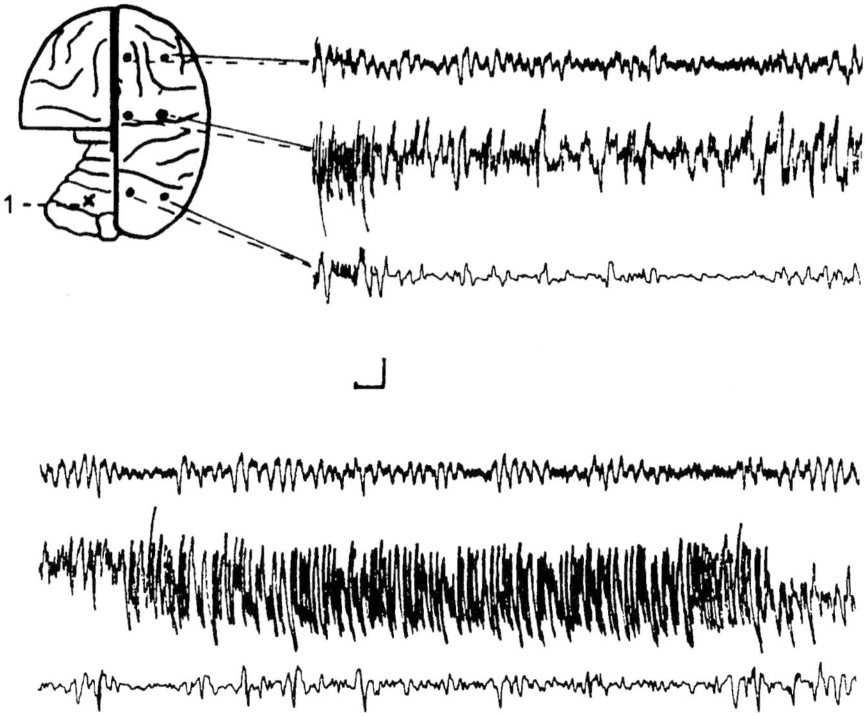

Fig. 14.5. Development of epileptiform discharge in the human sensorimotor area subsequent to electrical stimulation (50 Hz, 2 ms, 4 V, 5 s) of the contralateral cerebellar cortex. 1 - point of cerebellar cortical stimulation

nization or an interruption of epileptiform activity. During the ictal discharge fastigiobulbar stimulation often stopped the seizure in the clonic phase. Dentatothalamic stimulation during the seizure appeared to either terminate the seizure in the clonic phase or to prolong the seizure. It was concluded that the fastigiobulbar pathways have a new inhibitory influence on seizure activity and that the dentatothalamic pathways exert an excitatory influence on the ventral hippocampus. No inhibitory effects on supratentorial spike activity have been observed [915,916].

Both vermal cortical and fastigial nucleus high frequency (100 Hz) are effective in reducing the number and duration of paroxysmal events; the effects of stimulation of the dentate nucleus were variable [917,918]. It was suggested that effective seizure inhibition appeared to be associated with increased fastigial output and resultant reticular excitation. These studies suggest that the rostral vermis and fastigial nucleus play the main role in inhibiting hippocampal and amygdalar activity.

Other studies have emphasized the anti-epileptic role of the dentate nucleus [919]. The effect of unilateral dentate nucleus lesion on amygdaloid kindling in rats demonstrates the discrepancy between the behavioral and electrographic activity in the amygdala [845]. The behavioral development of kindling was facilitated by the dentate lesion, whereas there was only a slight increase in the duration of after-discharge and the spontaneous epileptiform activity of the contralateral amygdala. Recently it has been demonstrated that the dentate and fastigial nucleus, but not the interpositus nucleus, may retard the process of epileptogenesis and inhibit clonic behavior [837].

Direct pathways pass through the fastigial and dentate nucleus to mediate cerebellar influences on supra-tentorial structures [918]. The effects of transaction of dentato- and interposito-thalamic pathways and fastigial pathways (transected at the level of the superior cerebellar peduncle) have been studied [920]. Contralateral lesions significantly slowed amygdalar kindling while ipsilateral lesions decreased kindled seizure duration. Thus the cerebellum exerts a tonic effect over circuitry participating in the kindling process through the cerebellar peduncle. It should be noted that it is extremely difficult to assess the real participation of each of these pathways because of the practical impossibility to perform the complete transection of all cerebello-cerebral efferents.

In view of the above observations we elected to study the effects of chronic stimulation of the dentate nucleus for the treatment of patients with intractable epilepsy in particular very frequent complex partial and generalized tonic-clonic epileptic seizures. Figure 14.6 shows that high-frequency (100 Hz) stimulation of the dentate nucleus leads to the complete cessation of supra-tentorial deep epileptiform discharges. Similar to the caudate nucleus, the inhibitory effects of stimulation of the caudate nucleus are exerted

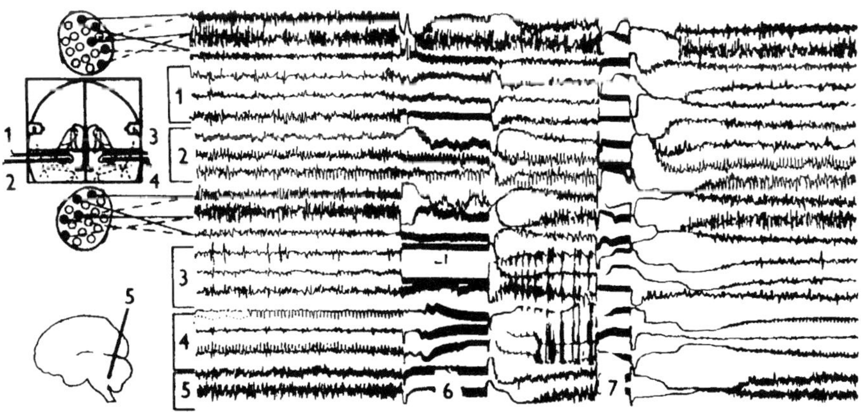

Fig. 14.6. Cessation of the amygdalo-hippocampal epileptiform discharge by high-frequency (50 Hz, 2 ms, 2 V, 3 s) stimulation of the dentate nucleus. 1,2 – the right amygdala and hippocampus; 3,4,5 – the left amygdala, hippocampus, and dentate nucleus, respectively; 6 – artifact of the epileptogenic stimulation of the left hippocampus; 7 – artifact of the inhibitory stimulation of the dentate nucleus

over both hemispheres. Beside the suppression of epileptic activity, the most common electrographic pattern of the supra-tentorial response to dentate nucleus stimulation was generalized desynchronization of the EEG.

14.2 Clinical Experience

Over the last 25 years our group has demonstrated the beneficial effects of low frequency stimulation of the caudate nucleus for the inhibition of interictal epileptic activity and focal and generalized epileptic discharges [175,191,230,298,823,878–881,897–899]. Details of the surgical procedures used to implant the stimulating electrodes are published elsewhere [230,897,921–923]. All of the procedures were performed in the Republican Center of Functional Neurosurgery and Epilepsy Surgery of the Institute of Clinical and Experimental Neurology in Tbilisi. The studies were approved by the Committee of Science and Technology of the Council of Ministry.

Initially, the implantable neurostimulators we used were designed by our group and manufactured first in Tbilisi by the "Mion" Corporation. After 1978 we used neurostimulators manufactured by the "Telsa" concern (former Czechoslovak Republic). These stimulators (LSV 340) were used with the following parameters: frequency of stimulus: 1–200 Hz; amplitude: 0–5 V; duration of impulses: .02–0.4 ms; inter-electrode distances: 3–5 mm; diameter of implantable electrode: < 1.3 mm. No serious surgical complications, no undesirable neurological or mental, or vegetative symptoms, or clinical seizures were elicited (or exacerbated) by therapeutic brain stimulation. During the adjustment of the stimulating parameters for the caudate nucleus an "arrest response" was noticed (total of 95) when the current was 8–10 mA. These behavioral changes lasted 1–3 min.

A total of 85 patients underwent deep brain stimulation and of these 54 patients had brain stimulators implanted for periods from 1 to 25 months. Tables 14.1 and 14.2 summarize the clinico-electro-physiological data of patients who underwent therapeutic deep-brain stimulation. In all of these patients, acute and/or chronic SEEG, EEG-video monitoring had documented the existence of complex organized multifocal epileptic systems. In many of these patients a dominant epileptic focus could not be localized.

Table 14.1. Types of seizures in patients who underwent therapeutic brain stimulation

Clinical seizure type	Number of Patients
Generalized tonic-clonic seizures	28
Generalized tonic seizures	14
Partial motor seizures	11
Complex partial seizures	32

Table 14.2. Types of abnormalities recorded on scalp EEG

Location of EEG abnormality	Number of Patients
Frontal bilateral	21
Temporal unilateral	17
Fronto-temporal unilateral	27
Fronto-temporal bilateral	20

14.3 Feedback Control of Deep Brain Stimulation

A simple feedback control to allow the patient to regulate the deep-brain stimulation is shown in Fig. 14.7. Briefly, the patient is able to assess the level of activity in their epileptic focus by monitoring the output of chronically implanted diagnostic electrodes using a specially designed integrator of electrical activity (Fig. 14.8). This integrator monitored the power spectra of epileptic activity in the epileptic focus. Because of the different electrographic

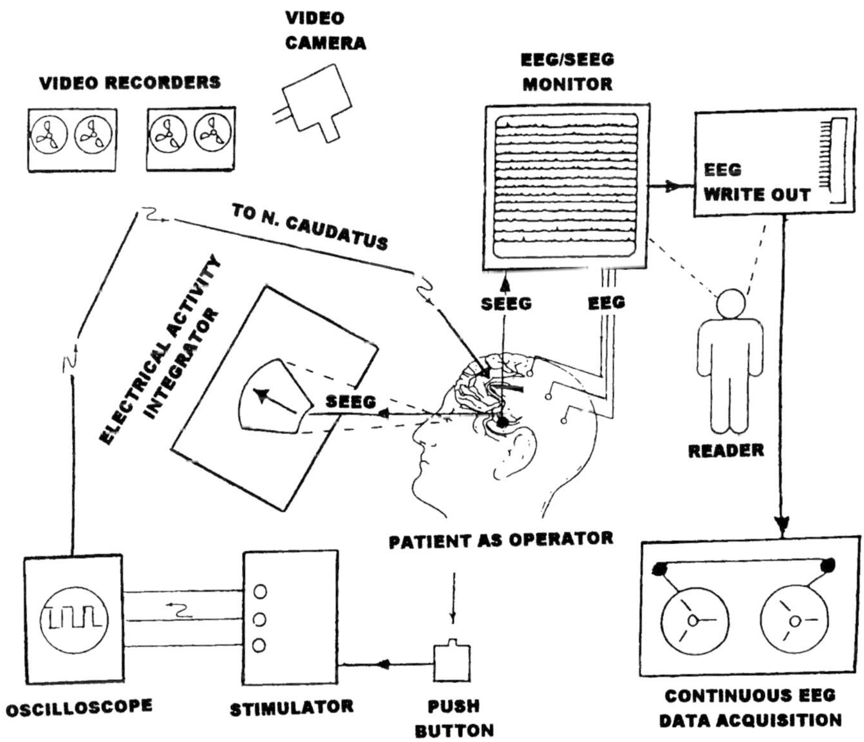

Fig. 14.7. Schematic diagram illustrating the negative feedback principle used for controlling seizures

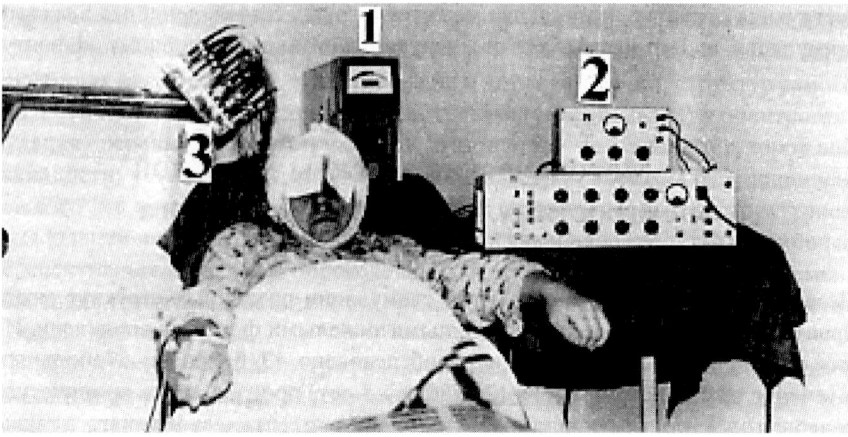

Fig. 14.8. The procedure of feedback control/stimulation. 1- the integrator of electrical activity; 2- electrostimulator "Nihon Kohden" connected with the electrodes implanted in the caudate nucleus; 3- SEEG and EEG electrodes connected through the preamplifier to the integrator. The patient is turned toward the camera to make visible the designed environment. The face has been covered with cotton for ethical purposes. In normal conditions the patient is looking at the integrator arrow and uses the push button for self-stimulation

variants of the incipient epileptic discharges it was necessary to record in each patient several spontaneous focal discharges and seizures in order to adjust the individual control parameters of the programmable integrator. When the epileptic activity becomes dangerously high (the arrow enters the red-colored zone on the integrator's display) the patient activates the deep brain stimulator. The level of activity for which stimulation was to be turned on differed for different patients. Using contemporary methods of digital analysis of brain electrical activity such as spike detection, inter-spike intervals and coherence measurement it is possible to adjust the individual threshold of epileptic discharge to the input of the stimulating device at the earliest, perhaps invisible, stage of seizure development. The implementation of this strategy was much easier for patients with clinically evident auras.

Figure 14.9 shows the placement of electrodes in ambulatory patients. The therapeutic effects of direct stimulation of the caudate nucleus resulted in these patients being seizure free for 25 months after the surgery.

14.4 Comparisons of Therapeutic Effects of Brain Stimulation

The inhibitory effects of different brain structures (head of the caudate nucleus, dentate nucleus, centromedian nucleus of the thalamus) was studied for all 85 patients. Very reliable inhibitory effects were obtained in the ma-

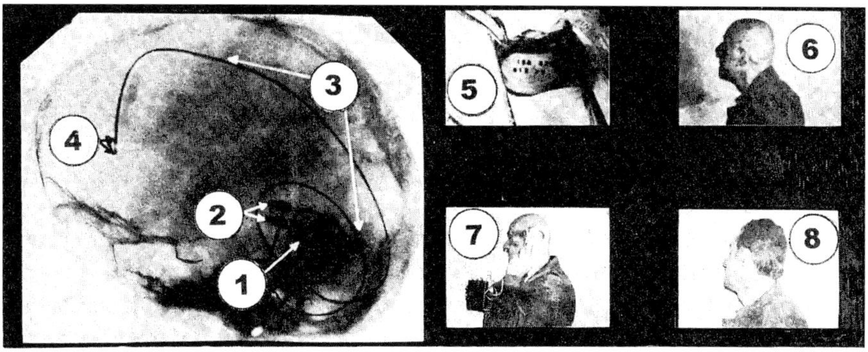

Fig. 14.9. Chronic therapeutic stimulation of the head of the caudate nucleus with subcutaneously implanted neurostimulator. (**A**) Lateral radiogram showing 1- subcutaneously implanted device; 2- electrode junction to the device; 3- subcutaneous part of electrode; 4- stimulating contacts of electrode; (**B**) the final stage (5) of implantation of device; 6,7,8- general view of the patient after surgery and during the process of self-stimulation

jority of these cases with the exception of stimulation of the central median nucleus. Of these patients 54/85 underwent chronic implantation of a brain stimulator. The remaining 31/85 patients underwent conventional treatment (lobectomy, cryodestruction of the presumed epileptic focus).

Thirty-eight patients (38/85) had brain stimulators implanted in the head of the caudate nucleus (Table 14.3). Of these patients, 35/38 experienced a greater than 50% reduction in their seizure frequency; 25/38 remain seizure free. Eleven patients (11/85) had brain stimulators implanted in the dentate nucleus (Table 14.4). Of these, 10/11 had a greater than 50% in their seizure frequency; 5/11 remained seizure free.

Five patients (5/38) had stimulators implanted in the centromedian nucleus of the thalamus. Only 2/5 of these patients experienced a reduction in their seizure frequency and no patient became seizure free. In view of the better efficacy of stimulation of either the caudate or dentate nucleus for seizure reduction we no longer implant stimulators in the centromedian nucleus of the thalamus.

Table 14.3. Results of chronic stimulation of caudate nucleus in patients with medically intractable epilepsy (1976–1996)

Surgery	Seizure free	Improved	No improvement
Stimulator alone	9	5	3
Stimulator + resection	4	1	0
Stimulator + cryodestruction	12	4	0

Table 14.4. Results of chronic stimulation of dentate nucleus in patients with medically intractable epilepsy (1976–1996)

Surgery	Seizure free	Improved	No improvement
Stimulator alone	1	2	1
Stimulator + resection	4	1	0
Stimulator + cryodestruction	12	4	0

14.5 Conclusion

These clinical investigations demonstrate the beneficial effects of low-frequency stimulation of the head of the caudate nucleus and high-frequency stimulation of the dentate nucleus for the control of seizures. Unilateral stimulation is sufficient to cause bilateral inhibition. It is likely that stimulation of these deep-brain structures activates the diffuse ascending inhibitory systems located primarily in the brainstem and cerebellum and that project diffusely and bilaterally [230,864,924] (Fig. 14.10).

It should be noted that seizures can also be suppressed by directly stimulating the epileptic focus [298,823,899,925,926]. For example, stimulation of the epileptic focus suppresses epileptic activity [508,590,927–929]. Moreover,

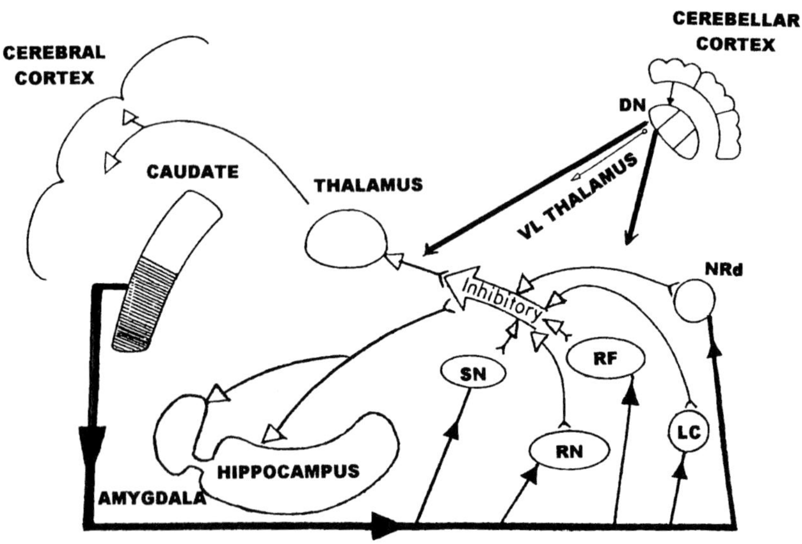

Fig. 14.10. Schematic representation of the possible mechanisms of the inhibitory influence of the caudate and dentate nuclei on epileptic activity. Abbreviations: Am – amygdala; Hipp – hippocampus; SN – substantia nigra; Thal – thalamus; RN – red nucleus. The *black solid thick lines* indicate the caudate and cerebellar influences triggering the diffuse inhibitory system (*thin black lines*)

the application of an additional stimulus at the onset of provoked cortical after-discharge abruptly stopped the after-discharge in 115 of 116 cases [930]. The mechanism for the seizure-suppressing effects of direct stimulation of the epileptic foci is thought to be related to the desynchronizing effects of the applied electrical stimulus on the neuronal pool [931].

The aim of studies of the mechanisms of the inhibition of interictal epileptic activity and seizures is to develop a reliable method to increase the activity of these inhibitory mechanisms in order to suppress seizures in patients with medically intractable epilepsy. As for the excitatory kindling, the "kindling" of the inhibitory system should elevate and stabilize the level of its activity. We believe that the local and systemic neuronal memory mechanisms based on biochemical and biophysical processes accompanying the suppression of epileptic activity at the inhibitory stimulation are able to provide and support the long-term positive therapeutic effect. These processes developed on both the neuronal membrane level and synaptic level should serve as a backbone for the "education" of the inhibitory system and normalize the balance of excitatory/inhibitory relationships in the epileptic brain. The stabilization of dynamic epileptic system on the new steady-state level of activity is the pathophysiological mechanism of beneficial either medical, resectional or stimulating treatment.

The therapeutic brain stimulation at epilepsy needs outstanding efforts to turn it from accidental studies to a reliable clinical method. It is necessary to perform the active controlled experimental and clinical studies to evaluate the indications for treatment of patients with different clinical forms and stages of disease, efficacy of brain stimulation and possible complications and contra-indications. The special issue should be the study of optimal combination of therapeutic brain stimulation with medication and/or surgical resectional procedures. The new opportunities could be offered for the treatment of multifocal epilepsy, and pediatric epilepsy with cortical dysplasia. The very preliminary experimental and clinical data could serve as a framework for the design of new clinical and experimental research to prove the load capacity of brain stimulation for the successful treatment of intractable epilepsy.

Why do stimulators seem to work well for some patients and not others? Are there differences between patients in the background activity of brainstem inhibitory mechanisms and hence different in the extent to which these mechanisms can be augmented by stimulation? Can stimulation protocols be made more effective by incorporating some of the principles from nonlinear dynamics discussed in this text Certainly the answers to these questions will require carefully controlled experimental and clinical studies.

15 Seizure Control using Feedback and Electric Fields *

B.J. Gluckman and S.J. Schiff

15.1 Introduction

If we consider the brain to be a dynamical system, and epilepsy one of its possible behaviors, then seizure prevention may be understood in terms of a complex control problem in which we perturb the system to prevent one behavior while minimizing the impact on other activity. To choose an analogy, consider a car and driver as a dynamical system and the act of driving off the road as a particular behavior we would like to prevent, while minimizing impact on normal driving tasks such as getting to destinations, choosing proper speeds, and avoiding other cars on the road. Numerous approaches can be taken, but two questions must be answered: first, how to interact with the system, and second, when to interact with it?. The operating set point of the system can be statically changed by modulating the performance of the car – slowing it down, for example. If additional information about the particular behavior of driving off the road is considered, specifically that the driver falls asleep, then more intricate control strategies can be chosen specifically to prevent just the precursor dynamic (falling asleep). One could imagine administering large doses of coffee or continuously ringing an alarm to prevent drowsiness, but in both cases, the driver is likely to filter out and thereby accommodate to the perturbation, making it ineffectual and otherwise annoying. On the other hand, if one were able to detect the driver becoming sleepy – by tracking eye motion, for example – then one could sound an alarm to interrupt the process. This is the essence of feedback control. Some measurement of the system's dynamic is made, and perturbations are administered to suitably modify system behavior in response to those measurements. Built in to the control law – the rules determining the perturbations from the measurements – can be limits on amount and time course of the applied perturbations, and algorithms for minimizing impact on normal behavior.

Design of a feedback controller for epilepsy requires two basic elements: a method for measuring and characterizing ongoing brain activity, and a method to simultaneously interact with or modulate brain activity on useful time scales. We satisfy these requirements by using a combination of extracellular electro-physiological measurements with a spatially broad low-frequency

* Research was supported by grants from the National Institutes of Health and the Whitaker Foundation.

applied electric field. When the electric field is applied aligned with a population of pyramidal neurons, it acts to make the whole population either more or less excitable. The amount of excitation (or suppression) is roughly in proportion to the amplitude and sign of the electric field. The time scale over which this modulation can be observed runs from milliseconds to hundreds of seconds ($\sim .001 < f < 100\,\text{Hz}$). The qualities of graded, reversible modulation available over a broad range of time scales makes the electric field a highly versatile control parameter for modulating brain activity. In addition, it is straightforward to modify standard measurement techniques in order to minimize stimulus artifacts in the measurements.

Using these tools along with a relatively simple feedback algorithm, we have succeeded in controlling seizure-like activity in hippocampal slice preparations. In this chapter, we will discuss the physics underlying the electric field modulation of neuronal activity, how the instrumentation is done, our recent results, and prospects for future seizure control devices.

15.2 Electric Field Modulation of Neuronal Activity

15.2.1 History and Physics

Since the experiments of Rushton in 1927 [932], it has been recognized that electrical fields can influence the threshold of excitability of neurons [666,933–935]. The physics of these effects has been well worked out [665,936,937]. An electric field oriented parallel to the somatic-dendritic axis of a neuron whose spike initiation zone is asymmetrically placed with respect to its centroid will modulating the neuron's firing. The details of this interaction, illustrated in Fig. 15.1, are as follows: An electric field parallel to the dendrite-soma axis of neurons will separate charged ions along this axis. This is just like electrophoresis – the gel is the cytoplasm of the neuron, and the cell membrane acts like a container holding the ions. Because the ions are trapped by the cell membrane, ions accumulate, thereby shifting the transmembrane potential in proportion to the applied field. If the cell body of the neuron is not symmetrically placed in relation to its dendrites, the spike initiation zone near the soma will hyperpolarize or depolarize in response to a steady electric field directed along the dendritic-somatic axis. Although this only works for asymmetric neurons, most of the large outflow neurons (principal cells) in neuronal layers in the brain are very asymmetric. The effect on these neurons is to change the threshold for action potential initiation.

The electric field magnitude required to modify the activity of a neuron is much smaller than the magnitude required to provoke neuronal firing from rest [666]. The pyramidal cells of the neocortex and the hippocampus individually have favorable geometry for electrical field manipulation. Furthermore, within a local network, the pyramidal neurons are oriented with their dendritic-somatic axis parallel to each other. Therefore, an electric field can

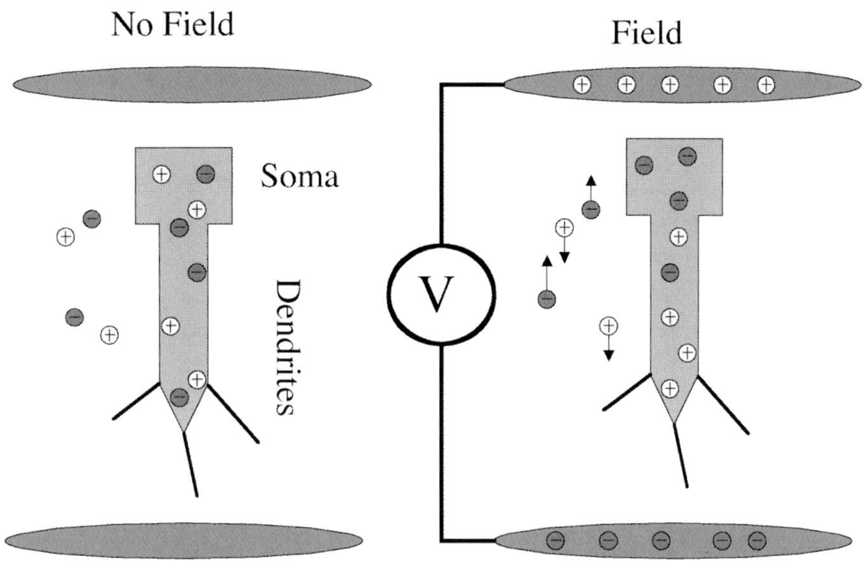

Fig. 15.1. Schematic representation of the interaction between an electric field and a neuron. Positive and negative ions sit both inside and outside the neuron. With no field applied (*left*), the distribution of ions between the soma and long dendrites is somewhat uniform. When a field is applied (*right*), a force is felt by the ions both inside and outside the neuron. The ones outside are free, so there is no net separation of charge. On the inside, the ions are trapped by the membrane; therefore charge builds up at the ends. This translates into a net change in transmembrane potential at the soma, where action potentials are initiated

be oriented parallel to this common axis and thereby modulate the activity of the whole population of pyramidal neurons.

15.2.2 Modulation with DC

Although the potential technical advantages of using electric fields for neural prosthetics were well recognized by the 1980s [937], until recently there seem to have been no attempts to suppress seizures using electric fields. In 1996 we demonstrated that DC electric fields applied external to brain tissue suppress spontaneous epileptiform activity [590]. Examples of such suppression of activity in in-vitro hippocampal slices are shown in Fig. 15.2. Details concerning slice preparation methods can be found in the reference. We have interacted with both large interictal-like bursts from the CA3 region (Fig. 15.2A) and with more seizure-like burst from the CA1 (Fig. 15.2B), for both transversely and longitudinally (not shown) cut slices. In each case, the application of a small ($\sim 10\,\mathrm{mV/mm}$) field suppressed the activity within the layer of interest. The orientation of a positive (suppressive) field with respect to the

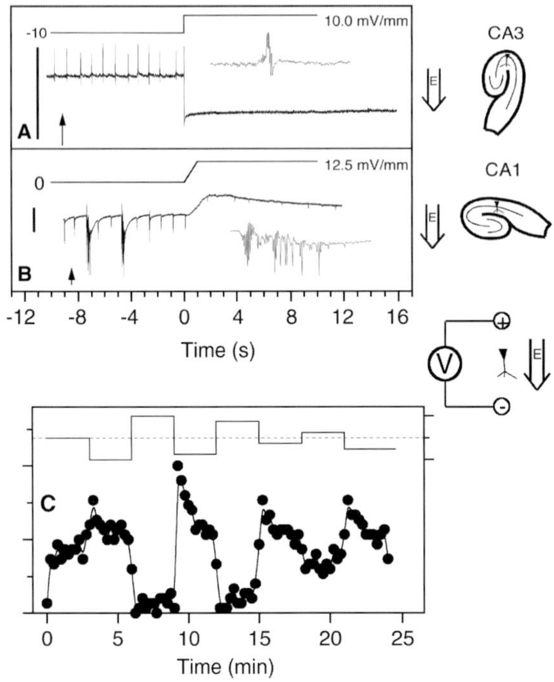

Fig. 15.2. Examples of suppression of spontaneous neuronal population activity (*lower traces each panel*) as a function of DC field (*upper traces*). Results are shown in transverse slices for (**A**) CA3 and (**B**) CA1, with time base indicated along abscissa and individual 1 mV vertical calibration bars to left of each trace. Insets show activity at expanded time base (0.5 s long, arbitrary vertical scale). Origins of insets are indicated by the *solid arrows* (↑). Slice anatomy and positive field direction (*open arrows* (⇓)) are illustrated, along with the orientation of soma and apical dendrites of the pyramidal cells monitored. In all cases of suppression, the electric field was directed along the axis defined from soma to apical dendrite, as shown in the schematic at the *bottom right*. This orientation and polarity of field would hyperpolarize the soma in each example. (**C**) Burst rate (*connected dots* •) from same transverse CA3 as (**A**) under the influence of external electric field (*solid line*), where overbar indicates the time corresponding to (a). Burst rate is determined from the number of bursts occurring within nonoverlapping 15 s time-windows. At baseline, without applied field, the pyramidal cells synchronously discharge at about 1 Hz. Negative field amplitudes accelerate while positive amplitudes suppress the burst rate. (Adapted from [590])

hippocampal layer measured is illustrated to the right for each of these examples, and the orientation of a positive (suppressive) field with respect to the dendritic-somatic axis of the pyramidal cells within the targeted network is illustrated.

Although in Figs. 15.2A,B we demonstrate that an applied field can be used to apparently suppress seizure-like activity, the response of the network to the applied field is actually both graded and reversible. This is illustrated in Fig. 15.2C, in which we plot the average population burst rate as a function of time while we changed the amplitude of the DC field. Higher-amplitude positive fields produced lower event rates, while negative fields produced higher event rates. In addition, we observe that the suppressive or excitatory response of the network to a DC field is transitory, i.e. it wears off over time. We attribute this to a combination of both electrode and tissue polarization, making the field felt by the neurons transitory, as well as neuronal adaptation.

15.2.3 Computational Confirmation of Modulation

In order to illustrate the physics of these experiments, we built a computational model of a network of the main excitatory neurons from the hippocampal CA3 region. The model is based on Pinsky and Rinzel's [473] two-chamber lumped parameter reduced version of Traub et al.'s [938] 19-compartment cable model of CA3 pyramidal neurons, and is detailed in [939]. In the model, each neuron is envisioned to have two chambers, one characterizing the soma, where action potentials are initiated, and one for the dendrites, as illustrated in Fig. 15.3A. The two chambers are electrically coupled by a finite resistance,R^i_{sd}, and are coupled to the extracellular space both capacitively and by parallel voltage-dependent variable resistors, each in series with a voltage source to model active ionic channels and concentration gradients. In addition, each chamber has a passive leakage resistance (L), and can have an additional current injected into it (I_s and I_d). A main advantage of the reduced model is its computational efficiency, as these simulations are run on desktop PCs.

As originally conceived, the reduced pyramidal cell model had no spatial extent. In order to couple in the electric field, we allow the extracellular potential outside the dendritic chamber to differ from that outside the somatic chamber, separating them with a finite extracellular resistance, R_{ds}. A network of these individual neurons is assembled into a linear resistive array as illustrated in Fig. 15.3B. In addition to the resistive coupling, the neurons are synaptically coupled to their four nearest-neighbors on each side (not shown). An electric field is now applied to this network by making the potential of the upper isopotential, V_{top}, different from the potential of the lower isopotential, which is held at zero.

The population of neurons in the array tends to fire synchronously in bursts. If a time-varying electric field is applied to this array, the burst rate is significantly modulated, as shown in Fig. 15.4. In the raster plots of Fig. 15.4A, we have plotted a bar (|) at the spike onset times for each neuron. As the field (Fig. 15.4B), in units mV/dendrite-soma length, is varied, the network burst rate is modulated (Fig. 15.4C), as in the actual experiments (Fig. 15.2C).

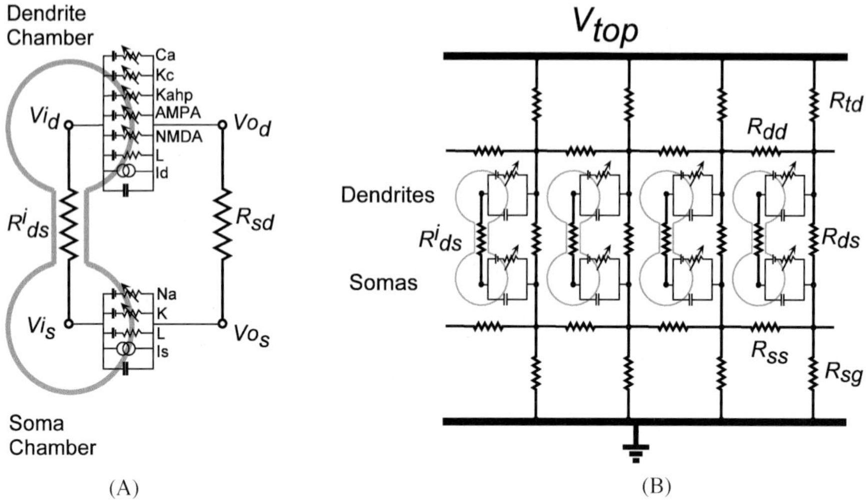

Fig. 15.3. Schematic of two-chamber model for CA3 pyramidal neurons incorporating electric field effects. (**A**) A detailed circuit diagram for a single neuron, including 2 active conductances in the soma chamber and 5 in the dendrite chamber, plus capacitance and passive leakage conductances in each. Symbols as in the text. (**B**) Circuit diagram of a linear array of neurons. An electric field is applied by imposing a potential difference between the upper and lower isopotentials. (Adapted from [939])

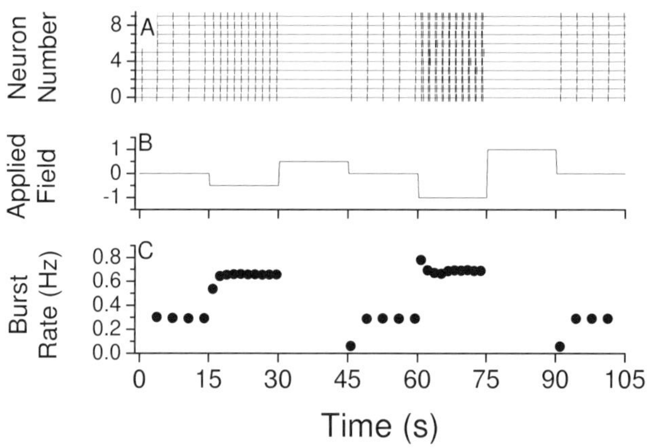

Fig. 15.4. Response of a network of 10 neurons under the influence of a DC field. (**A**) Onset times of spikes (|) as a function of time and neuron number. (**B**) Applied field, in units of (mV/dendritic-somatic distance). (**C**) Instantaneous burst rate, calculated as the inverse interburst interval, as a function of time. As observed in the experiment (Fig. 15.2C), as the field is increased, the burst rate increases, and as it is decreased, the rate decreases. (Adapted from [939])

15.3 Key Instrumentation and Experimental Methods

15.3.1 Measurement Technique

One of the technical difficulties in implementing electric field stimulation of neuronal networks is doing so without disrupting the ability to monitor ongoing neural activity. Typically, neuronal activity is characterized through electrical potentials measured either extracellularly or intracellularly. To be explicit, such measurements are performed by placing an electrode in electrical contact with the extracellular or intracellular space, and measuring potential changes through that electrode *with respect to some other place.* In many cases, just one electrode is explicitly used. Such single-ended recordings are actually differentially measured with respect to some "ground" position. The challenge when applying an electric field on the neuronal network is that, in general, such a field translates also into a potential difference between the differential measurement positions. In physics terms, the potential difference between the electrodes equals the line integral of the field $V = \int E \cdot \mathrm{d}x$.

We can solve this problem by differentially recording between two electrodes explicitly aligned along the same isopotential of applied electric field. In this case, the potential difference due to the applied field is zero. Such isopotential planes are perpendicular to the electric field lines. For an electrically homogeneous medium, the geometry of the field lines is set only by the geometry of the electrode plates imposing the field and is independent of the amplitude of the field imposed. Therefore, two points that are isopotential for one amplitude applied field will be isopotential for all applied fields. The homogeneous medium assumption applies fairly well to brain for a broad range of frequencies. We are thereby able to record activity recording while simultaneously applying relatively large (50–100 mV/mm), time-dependent (0–30 Hz) electric fields with minimal stimulus artifact.

15.3.2 Application of an Electric Field

Another technical difficulty encountered in using electric fields to modulate neuronal activity is to properly address how to actually apply an electric field across the tissue containing the network. First one needs to design an electrode configuration such that the electric field lines are parallel to dendritic-somatic axis of the pyramidal cells within the network targeted. This is not difficult in the hippocampus, since this polarization axis is typically perpendicular to the cell body layer and the local cortical surface. Therefore fields resulting from plates parallel to the cortical surface and sandwiching the layer will modulate the confined network. For the work described here we used two large parallel plates, as illustrated in Fig. 15.5, which result in a fairly uniform electric field across the slice. We then aligned the hippocampal slices with the cell body layer perpendicular to the field. A second technical issue comes from the fact that the tissue is naturally conductive. Therefore, an electric field

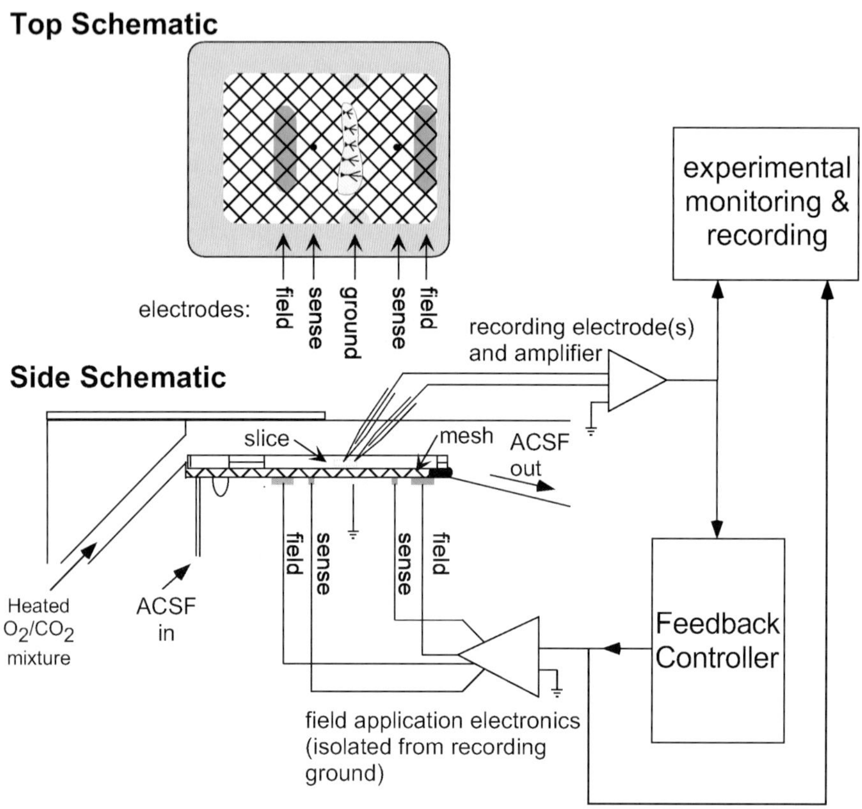

Fig. 15.5. Schematic of the hippocampal slice and electrodes in the perfusion chamber (modified Haas style) as viewed from the side (*middle*) and above (*top*), with the wiring for the field feedback amplifier indicated and block diagram of controller. The brain slices rest on a nylon mesh just below the upper surface of the perfusate of artificial cerebrospinal fluid. The atmosphere above the perfusate is warmed to the bath temperature of 35°C and saturated with 95% O_2 – 5% CO_2. An electric field is imposed on the slice by a set of Ag-AgCl electrodes embedded in the floor of the chamber. The current applied between these parallel-plate field electrodes is feedback-controlled so that the potential difference between the sensing electrodes is proportional to a program voltage. An additional pair of electrodes is used as recording ground. (Adapted from [940])

will be accompanied by an electric current (motion of ions in the fluid). Then, in order to maintain an electric field, the electrodes must be able to impart charge to the fluid both without developing a large junction potential, which would hydrolyze water and create large pH shifts, as well as potentially create other biologically harmful byproducts. In the experiments described here, we used non-polarizing Ag-AgCl electrodes, satisfying the first condition but provide long-term toxic byproducts. For in-vivo work, newly developed high electrical capacity biocompatible electrodes will have to be utilized, such as activated iridium oxide (AIROF). Even the best such materials have limited current-passing capability for safe use. Our back-of-the-envelope calculations estimate the limits of DC field application with AIROF electrodes at approximately 5 mV/mm for 10 s [941]. Therefore, neural prosthetics for electric field seizure control will have to incorporate such limitations within their control laws.

15.4 Feedback Control of In-vitro Seizure Activity

In order to address both the desirability of a dynamical interaction for seizure control and the limitations observed with DC electric field suppression, we have instituted a feedback-control algorithm, and applied it to seizure-like activity in hippocampal slices perfused at an elevated potassium concentration [940]. Our objective in this work was to demonstrate that, when applied in a feedback fashion, control of seizure-like network behavior could be achieved for extended periods of time.

15.4.1 Feedback Algorithm

For feedback purposes we characterized the neuronal activity associated with seizures as the root-mean-square (rms) of the recorded activity measured within a frequency band of 100–500 Hz, averaged over a time that varied from 0.1–1.5 s. The applied field was proportional to the positive difference between this rms activity and a threshold value. The threshold was set by an average ($\sim$ 30–3000 s) of the measured rms power. The frequency content of the applied field was restricted to less than 10 Hz. For practical purposes, a maximal (saturation) field amplitude was enforced, though when control worked best, the applied field rarely reached this maximum. In some applications, the output field was half-wave rectified (i.e. when the rms was below threshold, no field was applied). Both the gain and the threshold were set empirically. In general, optimal control was found with a moderate gain which could be estimated by (50 mV/mm)/(peak recorded power of a seizure).

Field strengths are presented in units of mV/mm, with positive field correspondingly aligned with the primary dendritic-somatic axis to produce a suppressive effect, as illustrated in Fig. 15.2. Gains are presented in arbitrary units, with positive gain corresponding to negative feedback mode.

15.4.2 Field Characteristics

Critical to performing these experiments was our ability to record neuronal activity independent of the applied time-varying electric field stimulus with minimal field stimulation artifact in the recording. We achieve this with the use of DC differential recordings from paired electrodes aligned to be nearly on the same isopotential of the applied field as described earlier. Since such alignment was always imperfect, some stimulus artifact remained. Therefore, we further restricted our applied field to have frequency content in a band distinct from that of the signal in which we were interested. The applied field was restricted to have frequency content below 10 Hz, while our primary characterization of the neuronal activity for the seizure suppression algorithm was based on the frequency band 100–350 Hz.

15.4.3 Overview of Control Phenomena

There is a characteristic low-frequency negative potential shift of the tissue associated with these seizure-like events in-vitro [942] that is quite similar to the slow low-frequency potential shifts observed during in-vitro seizures [943]. Typical seizure-like events in these slices exhibited durations of order 5–25 s and inter-event intervals of order 40 s, and low-frequency (0.01–1 Hz) potential shifts of order 0.2–5 mV. Recording to recording variations in the morphology and amplitude of DC deflection can be attributed to the details of the measurement electrode location with respect to both the origin of the seizure and to the position of the reference electrode.

15.4.4 Seizure Suppression

In Figs. 15.6A and 15.6B we show examples that illustrate how an electric field can be used to adaptively suppress seizure-like activity within the CA1. Suppression is achieved by using negative feedback. In both cases the high-frequency activity, towards which the suppression algorithm is directed, is significantly attenuated. The DC shift was completely eliminated (Fig. 15.6A) during suppression for some slices, while it was partially retained (Fig. 15.6B) for others. During control, some non-zero level of network activity is still observed from the field potentials (third inset in each). We have documented successful suppression in 20 of 30 seizing slices with which we applied adaptive control.

Control can often be maintained for prolonged periods of time. To date, the longest we have maintained control is 16 min in a slice otherwise exhibiting seizures approximately every 40 s. Since the amplitude, duration and interval between of the events slowly change over the course of an hour (see Fig. 15.7), 16 min. is near the limit for reliable suppression testing in this system.

15.4.5 Seizure Enhancement

Positive feedback, set by changing the sign on the gain that reverses the applied field polarity, can be used to either enhance seizures or even create seizures where none were observed beforehand. In Fig. 15.6C, we show an example of the characteristic population burst-firing events seen in high

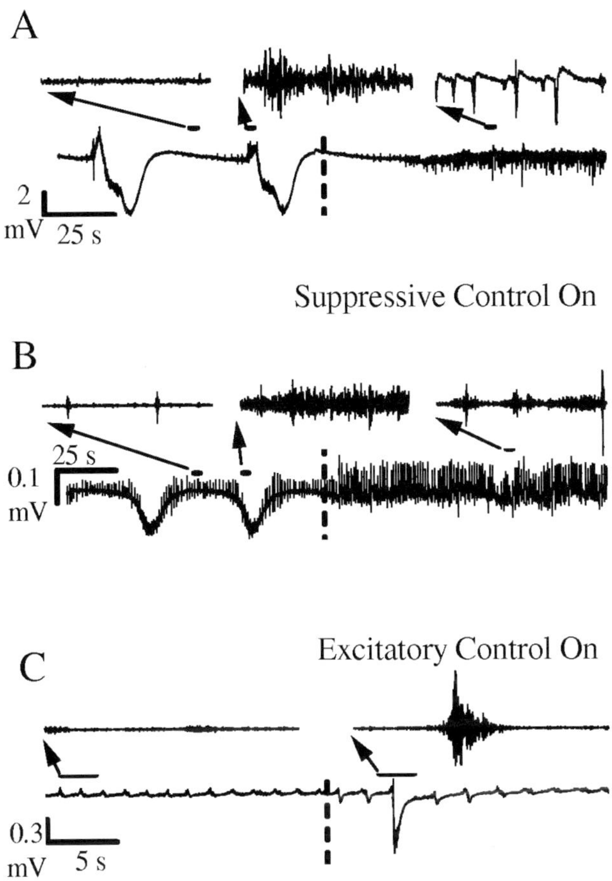

Fig. 15.6. Adaptive control of seizure activity using applied electric fields. In each panel, the main trace is the raw extracellular potential recording. Insets are tracings of activity, filtered to illustrate the high-frequency activity, shown at expanded scales. In each case, a *dashed line* is used to demarcate when control is turned on. **A**,**B**: Examples of seizure suppression from separate experiments using electric fields applied as a negative feedback parameter. Electrographic seizures are observed as an increase in high-frequency activity atop large low-frequency deflections [942]. In **B**, seizures occur interspersed among frequent short network bursts [944]. **C**: Example of seizure induction achieved using positive feedback. (Adapted from [939])

$[K^+]$ hippocampal slices [944] in the uncontrolled state. With positive feedback control, the adaptively applied field now enhances the brief network bursts into large seizure-like events with the substantial low-frequency potential shifts characteristic of seizures. We have documented seizure generation in all 4 non-seizing slices with which we applied positive feedback control.

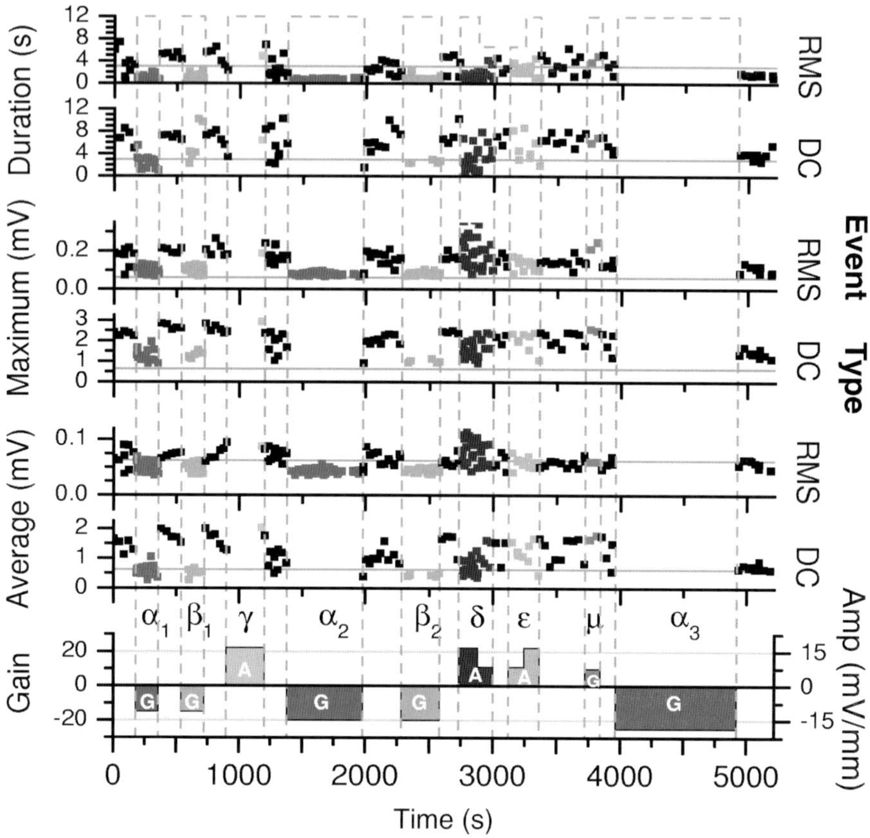

Fig. 15.7. Event detection results for a single 90-min. recording, with different electric field stimuli applied. The lower trace indicates feedback gain (G, left axis) or amplitude (A, right axis) of the applied stimulus. Greek letters and color indicate type of stimulus: baseline (*black, no letter*); full-wave feedback control (*red*, α); half-wave rectified feedback control (*green*, β); constant-amplitude suppressive field (*light blue*, γ); low-frequency noise (*dark blue*, δ); suppressive half-wave rectified low-frequency noise (*orange*, ϵ); positive feedback control (*magenta*, μ). Two types of event detection were used to identify synchronous neuronal activity from the recorded field potentials. "RMS events" were detected from variations in the rms power in the frequency band 100–350 Hz. "DC events" were detected by threshold detection after low-pass filtering the recordings at 10 Hz. The character of both types of events, as quantified by their average and maximal amplitudes as well as their duration, was visibly changed from baseline when control was applied. No events of either type were observed during the final and longest (16 min.) application (α_3) of full-wave control. (Adapted from [939]) (The colour plate of this figure appears on page 365)

15.5 Comparison of Parameters: a Single Experiment

Detailed event extraction results for a 90-min. recording from a single experiment are shown in Fig. 15.7. In this experiment, we compared the application of negative feedback both with and without half-wave rectification of the applied field at various gains, application of a constant-amplitude suppressive field and random waveform fields, as well as positive feedback control. From this experiment, we extracted events both from the rms power in the frequency band $100 < f < 350\,\text{Hz}$, which we term "RMS events," and events from the low frequency ($f < 10\,\text{Hz}$) potential shifts, which we term "DC events." For details of event extraction methods, see [940].

The type of stimulus applied is indicated in the lower trace, where the height of the blocks indicate either the gain (G, left axis) used in the proportional feedback routine, or the amplitude (A, right axis) of the waveform applied. Both the colors and the Greek letters indicate the type of stimulus applied, as indicated in the figure caption. Baseline recordings of 1–4 min were made between stimuli. In the upper plots are shown the duration, maximum and average deflections (DC or RMS power) of all events extracted either from the RMS power ("RMS events", upper trace for each pair) or low-frequency deflections ("DC events") as a function of time. Values for all extracted events are plotted. For the maximum and average deflections, the gray horizontal lines correspond to the trigger threshold for defining an event. As expected, the maximum deflections are always greater than or equal to the trigger threshold. In contrast, the average deflection need not be larger than the trigger threshold. Therefore, the trigger threshold provides a logical dividing line between large and small events in the average deflection plots. In the duration plots, a horizontal line at 3 s is plotted as a rough threshold for distinguishing seizure-like episodes from smaller burst-like events.

15.5.1 Feedback Suppression

Negative (i.e. suppressive) feedback, indicated by a negative gain, was applied with both full-wave (α) and half-wave (β) rectification. Even at the smallest gain used (α_1, β_1), all six types of event characteristics are distinct from the baseline activity (black) for both detection schemes. At the intermediate gain used, no DC events were observed during the non-rectified control (α_2), while only short, low-power RMS events were observed. For half-wave rectified control at comparable gain (β_2), short, small events were observed from both the DC and the RMS event extraction. At the highest gain used for non-rectified control (α_3, starting at time 3960 s), no DC or RMS events were detected throughout the 16 min of control application.

Examples of activity for this experiment with and without control are shown in Fig. 15.8. The upper pair of traces (A) correspond to the measured field potential (lower) and applied field (upper) starting 2min prior to the last application of non-rectified control (α_3). The baseline activity, without

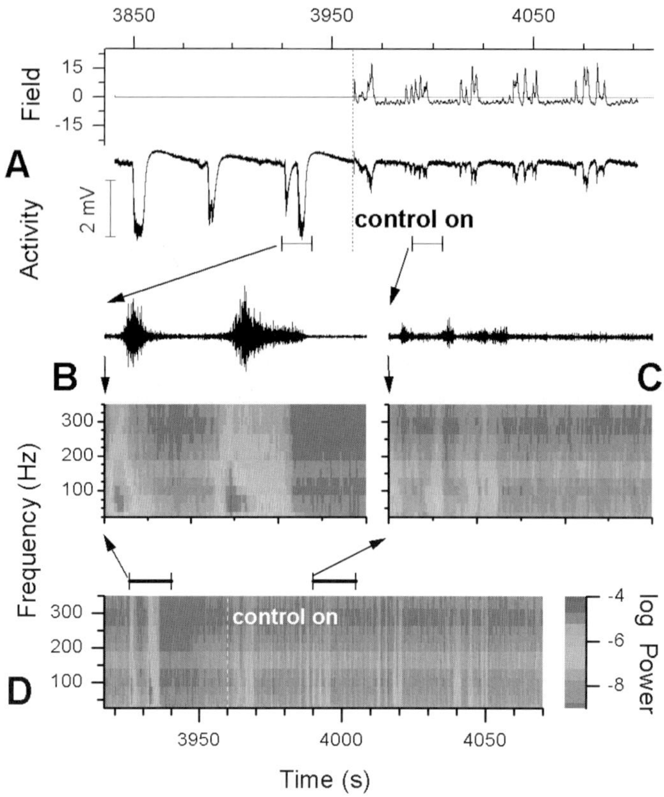

Fig. 15.8. Traces and spectrograms of activity with and without control for same experiment as Fig. 15.4. (**A**) Activity (*lower trace*) and applied field (*upper trace*) from the final application of full-wave control (α_3) from Fig. 15.4 and the baseline preceding it. (**B**,**C**) A 15-s long trace and spectrogram of a seizure-like event (**B**) and of activity during control (**C**) from (**A**). The upper traces in (**B**) and (**C**) are the activity, high-pass filtered at 100Hz. The spectrograms (**B**, **C**, **D**) are calculated in overlapping vertical frequency bins 50 Hz tall from 25–350 Hz, and in overlapping horizontal time windows 0.05 s wide. (**D**) Spectrogram for longer period illustrating contrast between baseline and controlled activity. (Adapted from [939]) (The colour plate of this figure appears on page 366)

control, is characterized by large seizure-like events that start with a burst of high-frequency activity, which are accompanied by large low-frequency potential shifts. Details of one of these events are shown in the trace of Fig. 15.8B at an expanded scale (15 s), high-pass filtered at 100 Hz, along with a spectrogram of the activity covering frequencies from 25–350 Hz. The power associated with these seizures can be observed in the spectrogram to start at high frequencies (near 120 Hz) and progress toward lower frequencies, a characteristic known as a 'spectral chirp'. Similar spectral chirps have been observed to be spectral signature of human seizures [945]. The neuronal activity following the seizure-like events in our experiments, as measured by the rms power, is depressed across all frequencies.

Expanded views for recorded neuronal activity during control are shown in Fig. 15.8C with the same scales as Fig. 15.8B. Although the rms power fluctuates during control (Fig. 15.8C), it never approaches the level observed in baseline (Fig. 15.8B). Note that the color scale is logarithmic. This behavior continues throughout the 16-min of this control application (Fig. 15.7, α_3), where the fluctuation are never large enough to trigger the rms event detection. A spectrogram corresponding to a longer period (150 s) crossing from baseline to control is shown in Fig. 15.8D. Throughout the control period, the rms power activity lacks both the characteristic highs and lows observed during non-controlled activity. We note that this power reduction/stabilization occurs across all frequencies displayed (25–350 Hz), whereas the applied field was constrained to have frequency content only below $\sim$ 10 Hz. The rms amplitude of the applied field averaged over the full control period was $\sim$ 4.8 mV/mm, and typically much smaller than the allowed maximum of 17.5 mV/mm.

15.5.2 Suppression with Constant Field

A relatively large suppressive constant (DC) field (16.7 mV/mm) was applied starting at time 900s (Fig. 15.7, γ). As was observed in earlier work [590], this had the effect of suppressing the large seizure-like events observed with no field. However, the effect had limited duration, as a large seizure-like event was observed 276 s after initiation of the field, as shown in Fig. 15.9A. This is in contrast to the 600 s period of control initiated at time $t = 1400$ s, during which no large events were observed (Fig. 15.7, α_2).

15.5.3 Stimulation with Low-Frequency Noise

One hypothesis might be that any low-frequency field might elicit a similar suppressive effect on the neuronal activity. We have tested various non-adaptive periodic and random signals. Although such signals do tend to modulate neuronal activity, we have observed little effective suppressive effect on seizures. Examples of random signals were used in the experiment of Fig. 15.7. Application δ corresponds to a full-wave (suppressive and enhancing) random field, while ϵ corresponds to a half-wave rectified (only suppressive) random field. Each was restricted to have frequency content below 1 Hz. Examples of activity from each of these applications are shown in Fig. 15.9B,C. The full-wave random field (Fig. 15.9B) did have the overall effect of breaking up the seizures in time and decreasing their duration as measured by the RMS event extraction (top of Fig. 15.7). However, the maximum amplitude of those events as measured in the rms was typically larger than baseline,

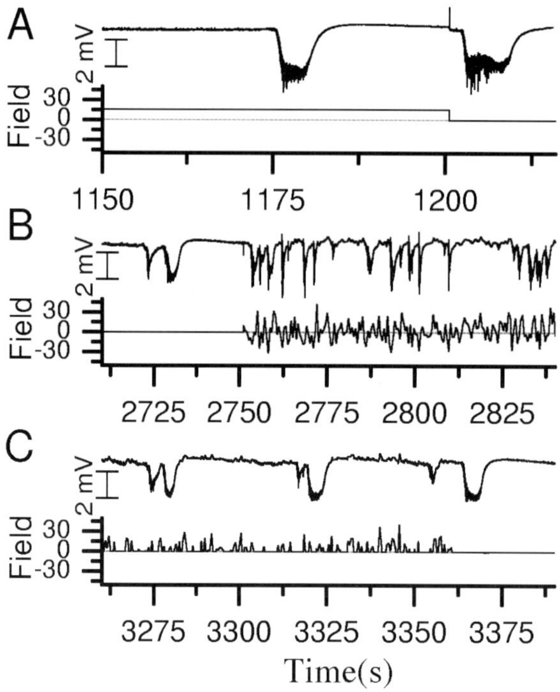

Fig. 15.9. Examples of activity during non-feedback electric field stimulus for the same recording as Fig. 15.4. For each set, the *upper trace* is of the recorded activity, while the applied field is shown in the lower trace. (**A**) Application of constant-amplitude (DC) suppressive field (4γ). (**B**) Application of full-wave low-frequency noise field (4δ). (**C**) Application of half-wave rectified low frequency noise field (4ϵ). In each case, large neuronal events are observed, though the full-wave noise field did have the effect of breaking up the seizure-like events into shorter durations. (Adapted from [939])

and comparable findings were reflected in the low-frequency deflections (DC events). The half-wave rectified field (Fig. 15.9C) had little effect at either amplitude used.

15.5.4 Positive Feedback Control

We applied a positive feedback for a short duration during this experiment. During this time, two events were observed, both of which were relatively large as measured from the average and maximum deflection for both rms and DC detection methods (Fig. 15.7, μ), as compared to the baseline events nearby in time.

15.5.5 Release Phenomena

The character of the activity during control varied from experiment to experiment. It depended both on variations in the network activity as well as our choice of parameters for the controller. In some cases, (Fig. 15.6A), during control, the network-controller system would be in a cyclic state. The network would begin to become more excitepd and then the controller would apply a field, causing the neural activity to become quiet. The field would then decrease, and the cycle would repeat. In these cases, large seizure-like events were observed nearly immediately when the controller was turned off. An example of such a seizure following release is illustrated in Fig. 15.10A, for the same control run as Fig. 15.6A. The upper trace is the recorded field potentials, while the lower trace is the applied field. In other cases, the amount of intervention by the controller cycled on a longer time scale (of order a minute), often reaching a point at which no field would be applied for a few seconds. In those cases, the activity when control was released depended on the phase of this cycle. If the controller was actively suppressing when shut off, then a seizure would progress (Fig. 15.10B). Otherwise, one would appear later, but within a few seconds of release.

In the majority of these experiments only half-wave rectified control was used. This has the effect of only suppressing activity when it is above the threshold. If we use the full proportional feedback control signal (full-wave control), the effect is not only to suppress when the activity level is too high, but also to excite when the activity level is too low. In the two longer experiments (2 slices from 2 rats) in which we compared full-wave to half-wave rectified control with similar parameters, upon release the network was consistently quiet for a period comparable to roughly half the baseline inter-event interval. An example of full-wave release is shown in Fig. 15.10C for comparison with half-wave release of Fig. 15.10B in the same network. During this experiment, designed to contrast the network responses to these different control algorithms, we alternated solely between rectified and non-rectified control (with baseline in-between) at constant gain. The intervals between turning off control and the next event were 0.1–6 s for rectified control (3 applications) and 14–17 s (4 applications) for full signal control. Application of a Student's t test estimates these distributions to be different with greater than 95% significance. Similar results were observed for the experiment of Fig. 15.7.

15.6 Results Summary

Clear suppression of the seizure-like activity compared to the baseline activity was achieved using feedback control through electric field stimulation in 20 of 30 seizing slices (4 whole transverse slices, 21 cut transverse slices, and 5 CA1 longitudinal slices; prepared from 21 rats). Half-wave rectified control was applied in all and full-wave control was applied in 5 of the successful

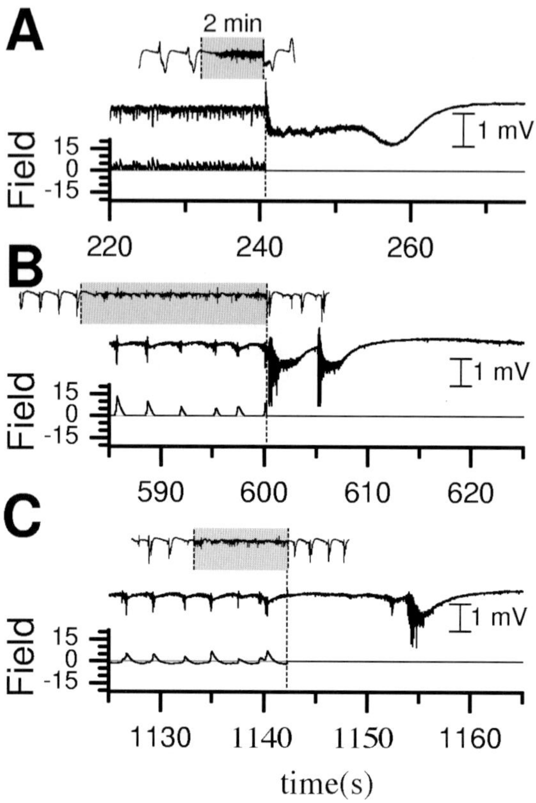

Fig. 15.10. Examples of network activity when control is released. In each panel, the inset is the activity for the full control period, indicated in *gray*, plus the baseline periods before and after. The trace in (**A**) corresponds to the same experiment as Fig. 15.3A, with half-wave rectified control. The network oscillates between excitation similar to seizure onset and being suppressed by the controller. When control is released, this activity proceeds immediately into a full seizure-like event. **B**,**C** traces from another experiment in which half-wave rectified control (**B**) was compared to non-rectified control (**C**). For half-wave rectification, seizures were observed very soon (0–3 s) after control was released, as compared to 12–18 s for non-rectified control. The time base for the insets is the same as indicated in (**A**). The inset vertical scale is half that of the main traces. (Adapted from [939])

suppression applications. We analyzed 5 experiments in detail. In each of those experiments, the rms power and power fluctuation in the frequency band 100–350 Hz during control was significantly lower than during baseline recordings. In each, there were clear differences in the character (duration, average and maximum power) of the events as extracted from the rms power, and 4 out of 5 revealed clear differences from events extracted from the DC deflections. In 6 experiments (6 slices from 6 rats), we maintained control for periods of at least 5 min without breakthrough seizures before parameters were changed. In addition, we generated seizures in non-seizing slices by applying positive feedback in 4 experiments (4 slices from 4 rats).

Our results are not due to the stimulus artifact of the applied field masking neuronal activity. First, the stimulus was restricted to frequencies well below 10 Hz. It was therefore easily distinguishable from the higher-frequency signatures used to characterize the neuronal signal for two of the three measures presented, which rely on analysis of field potential signals above 100 Hz. In addition, observation of seizures immediately upon release of control, as illustrated in Fig. 15.10A, belie the effect being artifact. The most likely explanation for these observations is that during control the network was oscillating between initiating a seizure-event and then suppressed by the controller. When the controller was turned off, the seizure progressed.

15.7 Discussion

We have demonstrated feedback electric field control of seizure-like events in neuronal networks. These population events were characterized by an increase in high-frequency activity at their onset, followed by a significant low-frequency negative shift as measured from the extracellular field potential. The control law used applied suppressive field when the rms amplitude of the measured high-frequency activity increased. This had the effect of suppressing neuronal activity at the onset of the seizures. We then used event detection both from the rms power in the frequency band 100–350 Hz and from DC deflections, along with spectral analysis, to distinguish activity during control periods from uncontrolled periods. Controlled activity revealed shorter, smaller events (if any) (Fig. 15.7), and both lower average power and power variance (see [940] for details) than the uncontrolled activity.

Previous estimates place the sensitivity of single neurons at ~ 5 mV/mm [666]. In previous work [590], we demonstrated suppression of network seizure-like activity in similar high potassium hippocampal preparation as used here with DC fields in the range of 5–10 mV/mm. Recent work by Ghai et al. [946], demonstrated suppression of seizure-like activity in a low-calcium hippocampal preparation with DC fields in the range of 1–5 mV/mm. The instantaneous amplitude used in our feedback experiments to modulate the network dynamics ranged from a few mV/mm to as high as a few tens of mV/mm. In the long full-wave control example shown in Fig. 15.7 (α_3), the rms field amplitude was ~ 4.8 mV/mm during the 16-min application, just at the lower limit of previous sensitivity estimates. Ongoing experiments in our laboratory appear to place the detection limits for single neuron and network modulation by electric fields near the 100µ V/mm range (unpublished data). We note that the physics of neuronal polarization under an electric field dictates no real threshold for interaction other than the expected loss of effect at the thermal Johnson-noise limit [947].

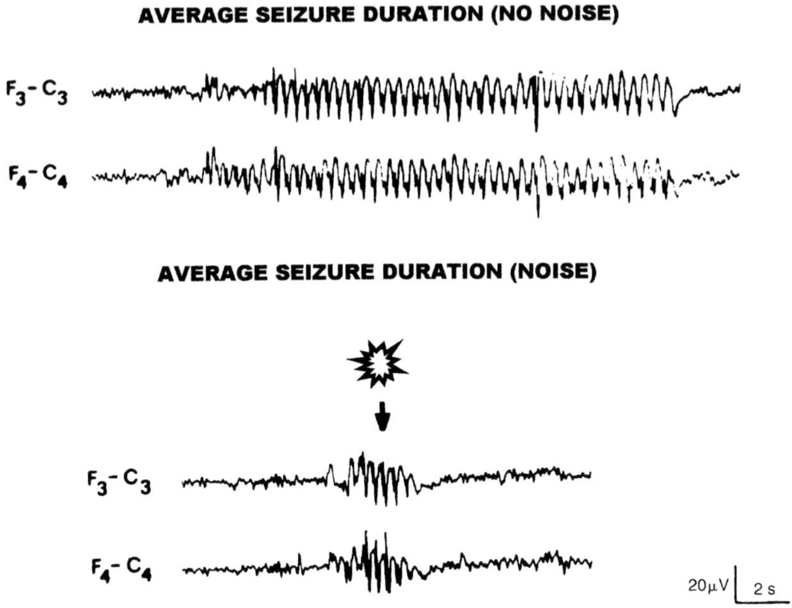

Fig. 16.1. Demonstration that an auditory stimulus (indicated by the *starburst*) can abort a seizure in an adolescent with primary generalized seizures [110]. In the absence of the auditory stimulus the seizures lasted on average 16 s

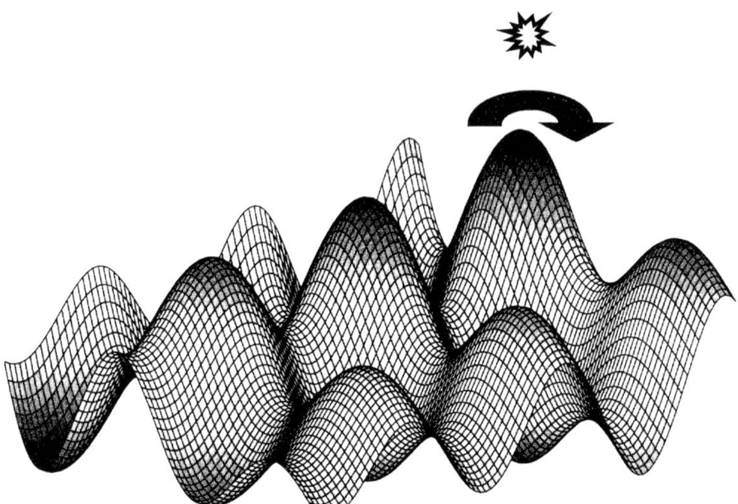

Fig. 16.2. Schematic representation of the energy landscape for a hypothetical multistable dynamical system. The auditory stimulus (*starburst*) results in a switch from one basin of attraction to another

ate mathematical models take the form of delay differential equations (DDE) [108,110,971–973]; an example is the first-order DDE

$$\dot{V}(t) + \alpha V(t) = f(V(t-\tau)), \tag{16.1}$$

where $V(t), V(t-\tau)$ are, respectively, the values of the state variable (such as membrane potential) at times $t, t-\tau$. The time delay is τ, α is a rate constant, and f describes the feedback. In order to obtain the solution of (16.1) it is necessary to specify an initial function, ϕ, on the interval $[-\tau, 0]$. Even an arbitrarily small delay can have dramatic effects on the behavior of feedback-control mechanisms modeled by (16.1) [974,975]. Here we review multistability that arises in recurrent neural loops with delay.

16.1 Delayed Recurrent Loops

Even at its most macroscopic level, the organization of the nervous system has a recurrently looped structure [976–978]. Recurrent loops relevant for epilepsy include the thalamocortical loops important for the genesis of generalized seizures [979] and the multiple loops in the hippocampus important in the genesis of temporal lobe seizures [980].

A delayed recurrent loop that has been surprisingly amenable to analysis is shown in Fig. 16.3 630–631 [68,90,971,972,981–985]. A periodically spiking neuron E makes an excitatory synapse onto the inhibitory interneuron I, which in turn makes an inhibitory synapse back onto E. The inhibitory influence of I on E depends on the activity of E at a time τ in the past. This time delay represents the sum of the conduction times along the axon and dendrites, the time required for quantal release, processing times in the interneurons, and the rise time of the inhibitory potential. Recurrent inhibitory loops arise in the olfactory cortex and in the hippocampus.

The recurrent inhibitory loop shown in Fig. 16.3 can be readily constructed in the laboratory from a single periodically spiking *Aplysia* motoneuron as follows [90]: The membrane potential of the motoneuron is monitored

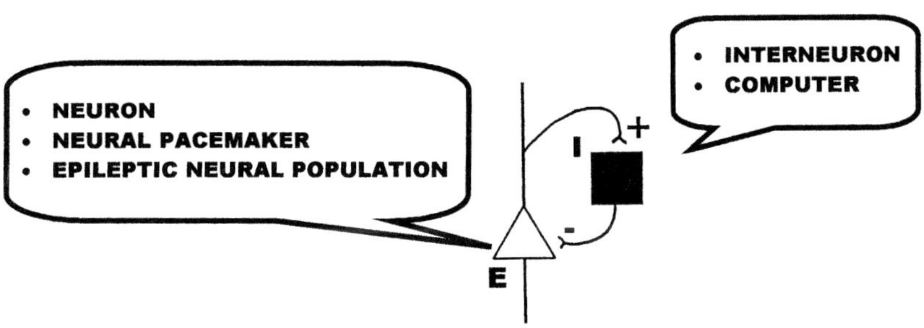

Fig. 16.3. Schematic representation of a recurrent inhibitory loop

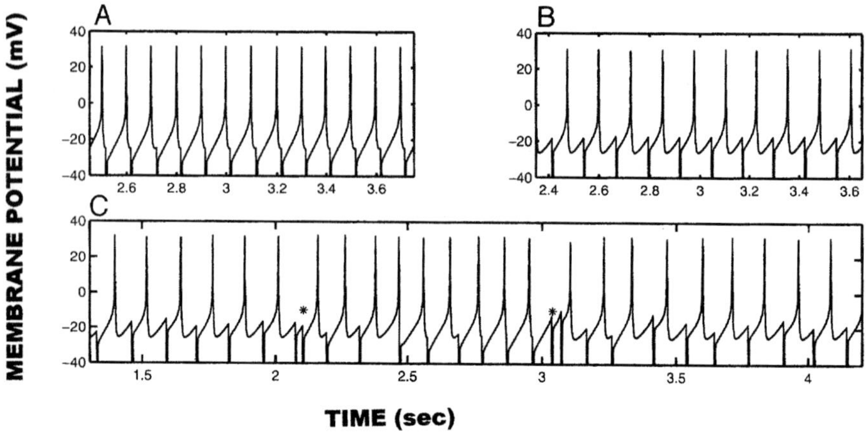

Fig. 16.4. Patterns that arise in membrane potential as a function of time for an *Aplysia* motoneuron during delayed feedback stimulation. (**A**) and (**B**) show the two attractors; (**C**) shows that it is possible to switch from one attractor to the other: see text for discussion

through the recording electrode of an AxoClamp. The analog signal proportional to membrane potential is passed to a computer via an A/D board. Each time that the computer detects a spike, it triggers the Axoclamp, after a delay time τ, to inject a single square hyperpolarizing current pulse into the motoneuron. Different initial pulse sequences can be introduced so that the co-existing qualitatively different spiking patterns can be identified. The advantage of this paradigm of recurrent inhibition is that it is possible to directly compare predictions to experimental observation.

Figure 16.4 demonstrates that this delayed recurrent neural loop exhibits multistability. A necessary condition for the occurence of multistability is that τ must be greater than the intrinsic firing period of neuron E [90,982,983,986]. Two qualitatively different spiking patterns occur when τ is 3.29 times the neuron spiking period (Fig. 16.4A and 16.4B). These spiking patterns differ in the position of the ICPs and the spiking frequency. Introduction of a perturbation ($*$ in Fig. 16.4C) causes a change from the pattern in B to that in A (Fig. 16.4C). After the new pattern is established, a second electrical perturbation causes the system to go back to the original spiking pattern.

The dynamics of the delayed recurrent neural loop can be completely understood in terms of one parameter, the time delay (τ), and a function that describes the phase-resetting properties of the neuron [90,983,987]. Since the phase-resetting properties of, for example, an epileptic population [93], can also be expressed in terms of a phase resetting curve our results are applicable to a wide range of phenomena. The feedback loop is pulse-coupled. The pulse coupling is a direct consequence of the fact that when neurons are physically separated, interactions between them are in the form of discrete synaptic potentials. The critical parameter for determining the dynamics of a pulse coupled neural loop is spike timing [988,989]. Consequently all of the physiological processes involved in the transmission of information around the loop can be incorporated into a single number, the time delay.

The effect of a stimulus on spike timing of a periodically firing neuron is summarized by the phase-resetting curve (PRC). The PRC is a summary of the phase shift, $\Delta(\phi)$, which results from perturbations that arrive at different phases, ϕ, on the oscillator's cycle, i.e.

$$\Delta(\phi) = \phi' - \phi \,,$$

where ϕ' is the new phase that results from the perturbation. The phase-resetting properties of neurons have been extensively studied both experimentally [90–93] and theoretically [439,990,485,486]. It is important to note that the phase-resetting curve for any oscillator can be measured experimentally. Thus predictions from models based on the PRC can readily be evaluated experimentally.

The phase-resetting curve that best describes the dynamics of the pulse-coupled delayed recurrent neural loop is the fixed phase-resetting curve [90]. The fixed phase-resetting curve is obtained by repeatedly injecting ICPs at a fixed phase and measuring the reset, Δ, once a stable pattern is attained. The fixed phase-resetting curve for an *Aplysia* motoneuron is shown in Fig. 16.5.

To understand the dynamics of a delayed recurrent loop it is necessary to keep track of the times that the excitatory spikes and ICPs occur. This book-keeping procedure is summarized by the map

$$\Psi_n = \frac{\tau}{T} + \sum_{i=1}^{k} \Delta(\phi_{n-i}) = F(\Psi_{n-1}, \Psi_{n-2}, \cdots, \Psi_{n-i}) \,, \tag{16.2}$$

where Ψ reflects the number of full and partial cycles through which the cell has progressed between the spike and the ICP, $\phi_i - \Psi_i$ mod 1, T is the intrinsic period of the excitatory neuron, and k is a parameter that depends

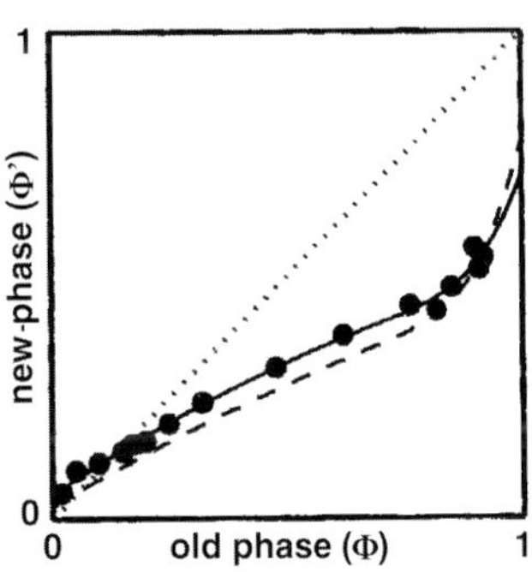

Fig. 16.5. Phase-resetting curve determined for a *Aplysia* motoneuron. This PRC was prepared by delivering ICPs at phase 0.76: see [90] for details

on when the last few ICPs arrived [90]. The basis for the multistability can be readily understood by determining the fixed points, Ψ^*, of (16.2). The fixed points, i.e. when all $\Psi_n = \Psi^*$, are the solutions of the equation

$$f(\Psi) = g(\Psi)\,, \tag{16.3}$$

where

$$f(\Psi) = \Psi - \frac{\tau}{T}$$
$$g(\Psi) = k\Delta(\Psi)\,.$$

Figure 16.6 shows the graphical solution of (16.3). Provided that τ is greater than T, multiple intersections are possible. It can be shown that the condition for the stability of these solutions is that the slope of $\Delta(\phi)$ at the fixed point, S^*, satisfies $-1 < S^* < k^{-1}$ ([987]). Moreover, numerical simulations of (16.2) indicate that when $\tau > 5T$ there are ≥ 2 attractors for all τ and as τ increases the number of attractors increases [90].

Figure 16.7 compares the predictions of (16.2) for the PRC shown in Fig. 16.5 to experimental observation. For this choice of the PRC all solutions represent periodic firing patterns in which one ICP arrives per inter-spike interval; the spiking patterns differ only by their frequency and the phase, ϕ, of the arrival of the ICP. From τ between 2.0–2.3 and τ 2.95–3.7, two spiking patterns are predicted to co-exist (solid lines). The experimental observations (•) are in quantitative agreement.

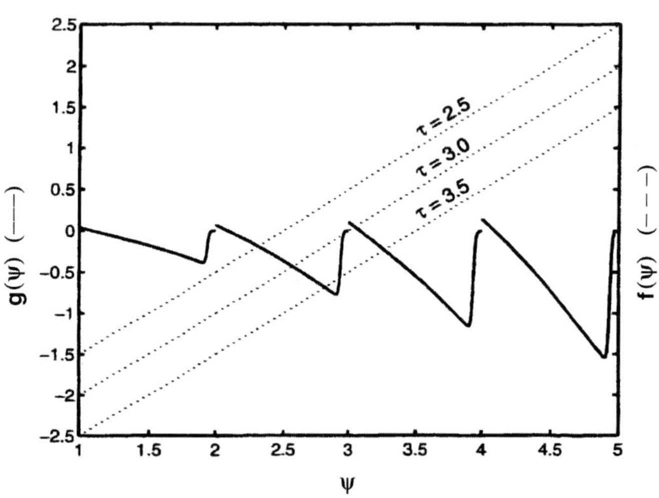

Fig. 16.6. The critical points of (16.3) are the intersections of $f(\Psi)$ (*dotted line*) with $g(\Psi)$ (*solid line*). The piecewise nature of $g(\Psi)$, comes from the dependence of k on Ψ [90]

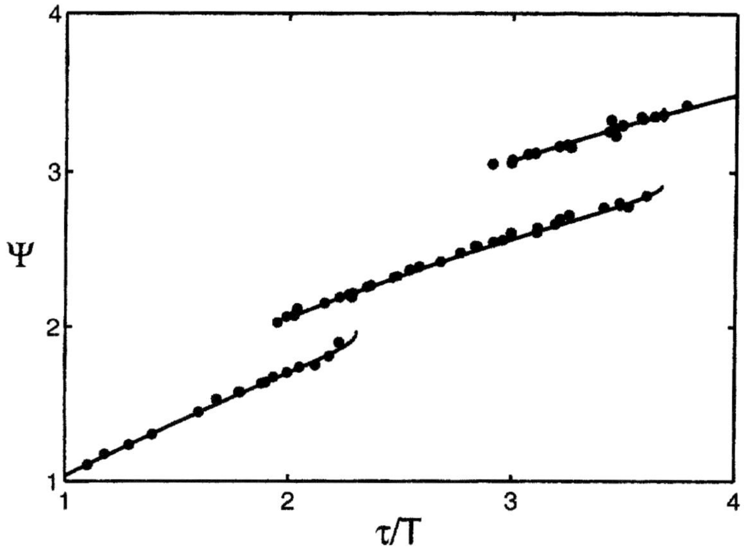

Fig. 16.7. Steady-state Ψs for the arrival of ICPs (•) as a function of the time delay for a recurrently clamped *Aplysia* motoneuron. The *solid line* gives the prediction of (16.2) using the PRC given in Fig. 16.5

16.2 Basins of Attraction

For a delayed recurrent loop the initial functions are intervals of length τ that give the times that the motoneuron spikes and the ICPs occur. For the first τ that the feedback is turned on, there are no ICPs (the earliest time that an ICP can occur is τ). Thus, during this time it is possible to use the computer to insert a given pulse pattern of length τ to initialize the delayed loop. Basins of attraction for the stable attractors exist in a space of all possible intervals of length τ.

Figure 16.8 shows a "cross section" though this infinite dimensional space that consists of only those intervals for which two ICPs occur per interval of length τ. In this case there are only two attractors and hence two basins of attraction. The separatrices that delineate the basins of attraction of the stable fixed points are determined by picking initial points in the vicinity of an unstable fixed point or unstable cycle and iterating (16.2) backward in time. The contour lines in Fig. 16.8 are the basin boundaries determined by iterating the (16.2) using the PRC shown in Fig. 16.5. The markers are the attractors measured experimentally for the same choice of the initial function. In 59/63 different initializations, the experimentally observed attractor corresponds precisely to that predicted from the calculated basins of attraction [987].

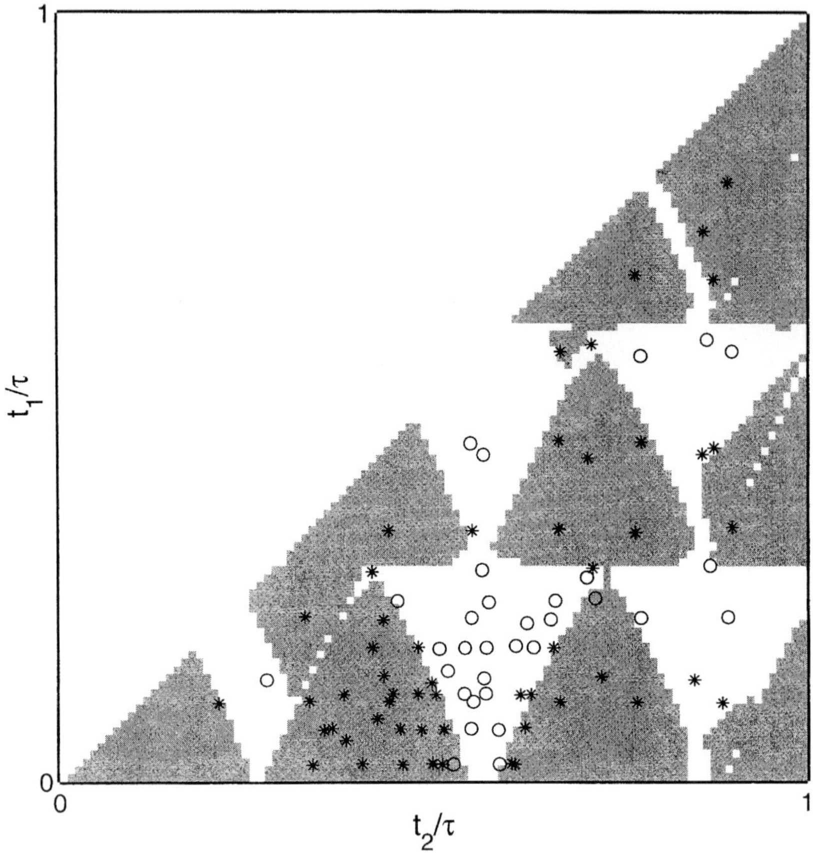

Fig. 16.8. Cross section of the basins of attraction for the two spiking patterns that co-exist in a recurrently clamped *Apylsia* motoneuron when $\tau/T = 3.1$. The function space is symmetric about the 45° line and only the lower half is shown. The area colored '*white*' corresponds to the spiking pattern shown in Fig. 16.4A and the area colored '*gray*' to the spiking pattern shown in Fig. 16.4B. The *open circles* ($\circ$) and the asterisks ($*$) designate, respectively, a choice of the initial function which is predicted by (16.2) to give rise to the spiking patterns in Fig. 16.4A and B. Agreement between theory and prediction occurs when an $\circ$ occurs in the *white regions* or an $*$ occurs in the *gray regions*. See text for further discussion

16.3 Noise-induced Switching

The nervous system is a very noisy environment [991,992]. It is possible that noise can cause a switching between basins of attraction [983,993]. The observed distribution of switching times for the perception of visually ambiguous figures provided one of the earliest demonstrations of this phenomena in the human brain [994].

Recently, switching between attractors has been detected at the onset of seizures recorded from intracranial electrodes implanted in patients with temporal lobe seizures [747]. In these studies, time series analysis suggested that the time variations of the EEG signals could be characterized by changes in a single variable. The observations were most consistent with a model of bistability in which mesoscopic collections of neurons flip between two collective states. In this context the time variation in the underlying neuronal dynamics corresponds to the time variations in the switching probabilities between two states. Since this time variation in the dynamics occurs in a similar way at all electrodes in the brain (even those far removed from the seizure focus), this view leads to the speculation that there are excitation waves that are constantly passing across the brains of these epilepsy patients, even interictally.

16.4 Statistical Periodicity

What are the expected dynamics of a complex, multistable, noisy dynamical system with retarded variables?

Recently it has been conjectured that such systems exhibit statistical periodicity [370]. To illustrate what the term 'statistical periodicity' means we refer to the one-dimensional map

$$x_{t+1} = S(x_t) = \begin{cases} ax_t & 0 \le x_t \le \frac{1}{2} \\ a(1-x_t) & \frac{1}{2} < x_t \le 1 \, , \end{cases} \tag{16.4}$$

where a is a constant, $1 < a \le 2$, and x_t, x_{t+1} are, respectively, the values of x at times t and $t+1$. This map is called the tent map because of its graphical appearance [995] and exhibits statistical periodicity when $a = \sqrt{2}$ [996,997]. The dynamics generated by the tent map are also seen in a variety of theoretical contexts [997–1001] including neuro physiologically motivated models of spatially extended excitable systems [68,1002]. The advantage of the tent map is the ease with which its properties can be studied numerically and the results verified analytically.

There are three ways that we could plot the data generated by (16.4). First, we could display the values of x as a function of time. This type of plot is shown in Fig. 16.9a and corresponds, for example, to the same type of information recorded in a neural spike train. The dynamics are obviously complex. A second way that we could plot the data is in the form of a histogram. Here the analogy is to an inter-spike interval histogram. Fig. 16.9b shows the "inter-spike interval histograms" prepared from the corresponding time series of Fig. 16.9a. This histogram approximates the stationary density of the distribution of x [998,1003].

The final way to investigate the dynamics of (16.4) is as follows: identify a large number (say N_o) of different initial values of x at $t = 0$, ${}^{o}x_1, {}^{o}x_2, \cdots, {}^{o}x_{N_o}$. We can construct a histogram from these initial values that we designate,

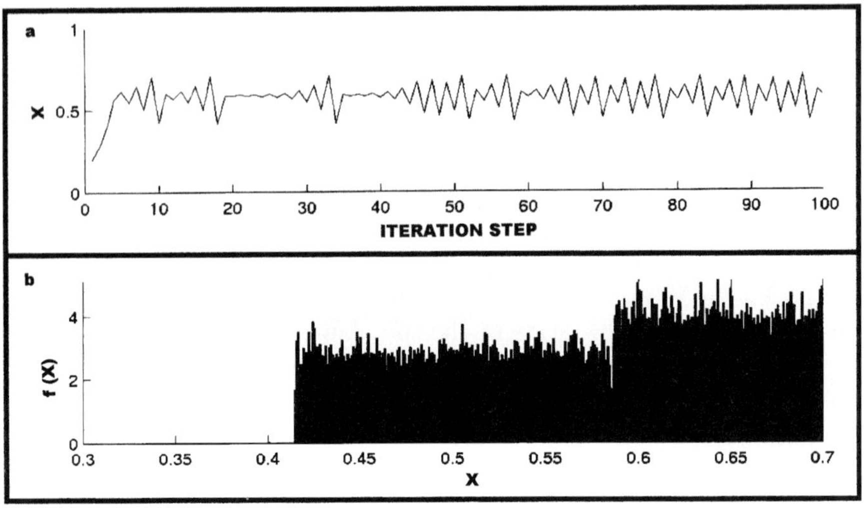

Fig. 16.9. Comparison of different representations of the dynamics generated by (16.4) for $a = \sqrt{2}$. (**a**) Time series representation; (**b**) histogram, or density, distribution determined from the time series in (**a**)

$f_o(x)$. At the next time step we obtain for each initial value of x a new value as ${}^1x_1 = S({}^ox_1), {}^1x_2 = S({}^ox_2), \cdots, {}^1x_{N_o} = S({}^1x_{N_o})$ and with these new values of x we can construct a new histogram, $f_1(x)$. We can continue in this way and follow how the histogram f changes as a function of time [1004]. For certain choices of a the histograms eventually settles down to one that does not change as a function of time. However, when we have statistical periodicity (i.e. $a = \sqrt{2}$) the limiting state is one in which the histogram itself changes with period 2 (Fig. 16.10). Indeed by changing the value of a cycling of densities can be observed with periods of 2^n, $n = 1, 2, 3, \cdots$ [1005]. This cycling of densities is the phenomenon of *statistical periodicity.*

The spatio-temporal dynamics of a spatially extended dynamical system that exhibits statistical periodicity is shown in Fig. 16.11. This figure shows the dynamics of a 2-dimensional coupled network of tent maps in which the dynamics of each element in the lattice are given by

$$\begin{aligned} x_{t+1}^{i,j} &= (1-\epsilon)S(x_t^{i,j}) \\ &+ \frac{\epsilon}{4}\left[S(x_t^{i-1,j}) + S(x_t^{i,j-1}) + S(x_t^{i+1,j}) + S(x_t^{i,j+1})\right] , \end{aligned} \tag{16.5}$$

where S is given by the right-hand side of (16.4), ϵ is the coupling coefficient, and there are periodic boundary conditions. Figure 16.11a compares the collapsed density for this 2-D coupled map lattice to the spatio-temporal distribution of activity (Fig. 16.11b) when $a = \sqrt{2}$. The collapsed density reflects the period 2 cycling of statistical periodicity. However, it is clear on examining the spatio-temporal distribution of activity that there is more

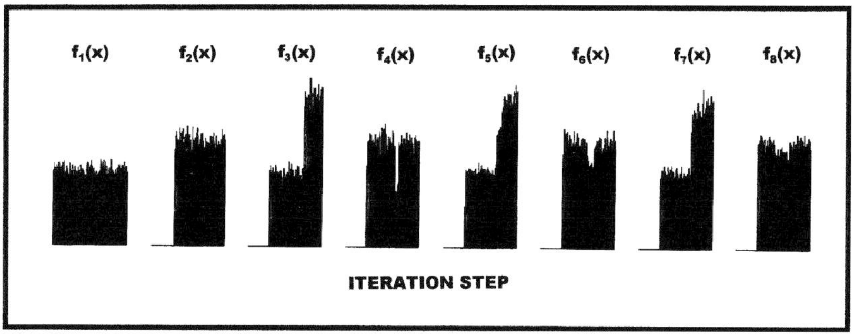

Fig. 16.10. The period 2 cycling of densities characteristic of statistical periodicity. Histograms were constructed by iterating 40000 tent maps, each with an initial condition chosen randomly from a uniform distribution in the interval from 0 to 1. At each iteration step the output of the 40000 tent maps were combined to construct the histogram shown. The histograms have been normalized so that the total area is equal to one. Parameter: $a = \sqrt{2}$. Note how quickly the histogram approaches a steady state

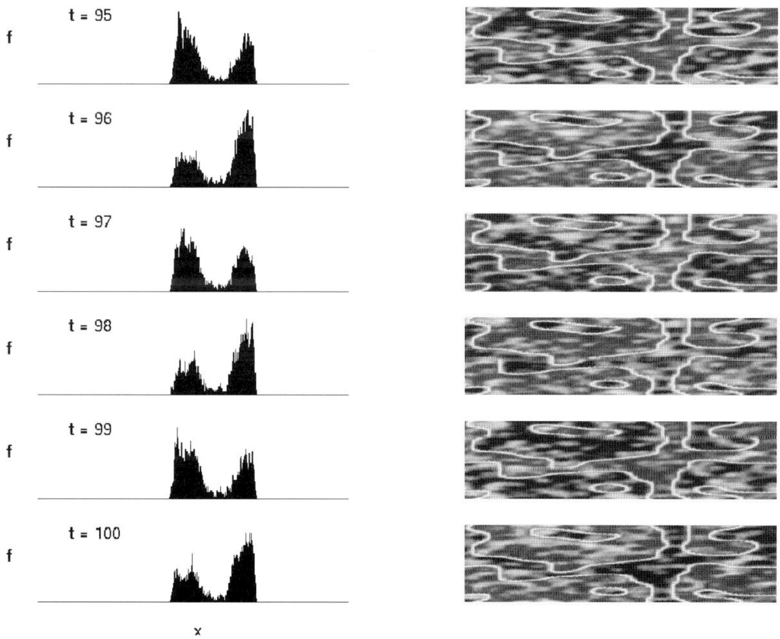

Fig. 16.11. Comparison of the (**a**) collapsed density and (**b**) the spatio-temporal distribution of activity for a 60×60 coupled map lattice of tent maps. The lattice was iterated for a total of 100 time steps and the densities and spatio-temporal patterns are shown for the last 6 time steps. In (**a**) the *solid horizontal line* indicates the interval from 0 to 1. The initial conditions were chosen from a uniform distribution over the interval from 0 to 1. For all densities the area has been normalized to one. Parameters: $a = \sqrt{2}$, $\epsilon = 0.45$ (The colour plate of this figure appears on page 367)

going on. In particular, spatio-temporal patterns are apparent. These patterns arise because of phase transitions [1002]. A necessary, but not sufficient condition, for the appearance of these patterns is that the ensemble exhibit statistical periodicity [1002,1006].

Statistical periodicity is an ensemble property, i.e. its existence can not be easily inferred by measurements of the dynamics of a single element in the ensemble. The cycling of the statistical properties of a dynamical system that exhibits statistical periodicity is reminescent of a multistable dynamical system subjected to noisy fluctuations: each distribution reflects statistical fluctuations in one of the basins of attraction, the cycling reflecting switching between the attractors. Strong connections have been demonstrated between the occurence of statistical periodicity and multistability in high-dimensional coupled map lattices [999,1002] and in lattice representations of differential delay equations [999].

The possible relevance of statistical periodicity to the dynamics measured by the electroencephalogram (EEG) has been discussed previously [370]. The activity measured by a single EEG electrode corresponds to the average activity of the collapsed density of a large underlying neural population. If this population exhibits statistical periodicity, then the EEG electrode will record an oscillation as is observed (Chap. 5). The period of these oscillations will depend on the time delay in the underlying feedback and feedforward control loops. Time delays range from 1–2 ms to $\geq$ 100–200 ms (Chap. 2) and thus a range of frequencies can appear in the EEG; changes in frequency perhaps arising as different recurrent loops dominate the population dynamics. The spatio-temporal patterns of EEG coherence (Chap. 6) resemble the spatio-temporal patterns generated by statistical periodic neural populations (Fig. 16.11).

16.5 Conclusion

It has been found that the bifurcation diagrams for delayed recurrent loop containing a large variety of different oscillators is remarkably similar to that shown in Fig. 16.7. The PRCs incorporated into a delay recurrent loop that have been examined are the Poincare oscillator, the Morris–Lecar oscillator, the *Aplysia* R15 bursting oscillator, and that measured for rat cortical neurons [90]. The insensitivity of the bifurcation curve for a delayed recurrent loop to the phase resetting curve of its component neurons suggests that it might be possible to develop a generic method for controlling multistable delayed recurrent neural loops.

Multistable dynamical systems can, in principle, be controlled and manipulated with small perturbations [1007,1008]. When the basins of attraction are known (e.g. Fig. 16.8) a multistable system could be controlled by techniques that combine state space methods with adaptive control [1009,1010]. A simple strategy would be that each time an unwanted switch between attractors occurs, the feedback is switched off for a time τ during which the initial condition for the desired attractor is re-inserted. Alternately, appropriate current pulses could be inserted into the loop at specific times to switch back to the desired attractor.

Experimentally, the possibility of resetting the properties of populations has been demonstrated [493,93]. An unsolved problem is determining what fraction of a synchronized population must be perturbed in order to reset the whole population. Presumably, the answer is more than one neuron and, hopefully, less than all of the neurons in the population. These simple considerations point to the advantages of stimulating sub-cortical structures to control seizures. Since many sub-cortical structures project diffusely over wide regions of cortex this would seem to be the most effective manner in which resetting stimuli could be introduced simultaneously to a significant proportion of an epileptic neural population.

17 Unstable Periodic Orbits (UPOs) and Chaos Control in Neural Systems*

M.L. Spano, W.L. Ditto, K. Dolan, and F. Moss

Chaos lurks on the border between regularity and randomness. While it masquerades as randomness, chaos is in reality a deterministic phenomenon. When we see irregularity, our thoughts hasten to blame randomness and disorder. Why should this be so? How can anything as marvellously ordered as the human body be dependent on processes that are completely random? Why is it that when the ubiquitous irregularity of biological systems is studied, instant conclusions are drawn about randomness? Recently we have begun to realize that irregularity is much richer than mere randomness can encompass. Thus we are brought to chaos.

Sustained irregularity has always upset our notions of how the world should behave. Yet it seems to be the canonical behavior of biological systems. One informal definition of chaos is sustained irregular behavior. Although descriptive, this definition is too vague to define the rich behavior of chaotic systems. A more precise defining feature of chaotic systems is their sensitivity to initial conditions (e.g. [1011]). It is this definition that we will utilize for the characterization and control of chaotic systems.

Fleeting glimpses of order within disorder are often observed. We have all seen short stretches of almost periodic behavior in otherwise irregular systems. A tantalizing example lies in the stock market, where many hope to reap windfall fortunes from analyzing short-term order and predicting the volatile market. But short-term order is a profound, even defining, feature of chaotic systems.

To be explicit, chaotic systems exhibit:

1. Sensitivity to initial conditions, where the behavior of the system can change dramatically in response to small perturbations of the system's parameters and/or initial values. This makes long-term (but not short-term!) prediction impossible.
2. Complex geometric structure(s) in the system's phase space (fractal objects whose composition includes an infinite variety of unstable periodic behaviors) to which the system's behavior is attracted.

The combination of sensitivity to initial conditions with complex geometry in system's state space can produce, to the casual observer, an appearance

* The authors gratefully acknowledge support from the Office of Naval Research, Physical Sciences Division. MLS also acknowledges assistance from the NSWC ILIR Program.

of randomness. However, upon closer inspection one glimpses a remarkable order. The flesh of chaotic systems adheres to a skeleton of infinite unstable periodic behaviors that seemingly come and go with no apparent pattern. That is, no apparent pattern until one looks with the tools of nonlinear dynamics. Armed with an understanding of unstable periodic motions, chaotic systems take on a more understandable, and consequently controllable, character. It is this order that allows the successful short-term prediction of a system's behavior and the subsequent control, which would be impossible in a totally random system. In the following, we will apply chaos theory to specific biological systems.

In the first part of this chapter we review the current state-of-the-art techniques for detecting unstable periodic orbits (UPOs) in short, noisy sets of nonlinear data, such as recorded from epileptic brain tissues. The second part of this chapter shows how the detected UPOs can be used to implement control of chaos strategies. In each case we will illustrate the approach with two examples: ventricular fibrillation in a rabbit heart preparation and epileptic seizures in the rat hippocampal slice.

17.1 Unstable Periodic Orbits

In both driven and autonomous chaotic systems, the motion (or orbit) of the system in phase space can at times be almost periodic. In driven systems, where the drive period is denoted as T, the motion will occasionally approach various periodic motions with period $T, 2T, 3T, \ldots$. In autonomous systems, it is also possible for multiple periodicities to appear, despite the lack of a reference drive period. But in either case, the approach to one of these periodicities is soon followed by a retreat from it and a subsequent approach to another periodic motion. This continues indefinitely as the system transiently visits all the orbits available to it. Thus in a chaotic system there are numerous UPOs [1012].

The definition of chaos, i.e. sensitivity to perturbations, implies that a chaotic attractor (the entirety of the path in phase space traversed by the system) is dense in such UPOs. Thus every point in a chaotic attractor lies arbitrarily close to a UPO of some period. Another important property of the system is that it is ergodic; that is, the system explores the entire allowed region of phase space. Thus time averages can be replaced with simpler phase space averages.

A UPO can be regarded as a local signature for chaos. Since chaotic systems consist of an infinite set of UPOs, the trajectory of the system will occasionally come close to a relatively low period orbit, causing the system to behave visibly in an almost periodic manner for a short time. Such encounters with UPOs of low period can be detected, sometimes even in the presence of very strong noise [1013]. By detecting and characterizing many of these UPOs representing different unstable periodic motions of the system, one obtains strong evidence that the dynamical system may indeed be chaotic. Moreover, due to the presence of noise and short data sets, the analysis of stable and unstable orbits in a system is often the only practical way to extract dynamical information. This can be particularly true in extremely noisy chaotic systems, where determinism in data taken from the system may be very difficult to find using more traditional nonlinear techniques. Unstable periodic orbits, and in some cases high-period stable orbits, can even influence the dynamics of stable periodic systems, if there is enough noise present.

17.2 Locating UPOs

Unstable periodic orbits have become of great experimental importance in the last decade. The original experimental use for UPOs was in the control of chaos [1014,1015]. Once a UPO has been detected, and its properties identified, it is possible to stabilize it using small perturbations, thereby forcing the system into a stable periodic behavior [1014,1016–1018]. In this section we review the techniques for detecting the presence of UPOs in experimental data, and then we illustrate their application to a cardiac and a hippocampal slice preparation.

17.2.1 Direct Observation of UPOs

In many discrete time dynamical systems, referred to herein as maps, it is quite easy to locate UPOs. However, measurements of physical systems are usually continuous. That is, trajectories of these systems trace out loops in phase space. A useful tool would enable one to convert the continuous phase space trajectory into a map for location of UPOs. Such a tool is available; it is call a Poincaré section. In essence it allows one to take a slice of the phase space of the system, thereby reducing the dimensionality of the phase space by 1. In a chaotic system, points on the section do not move continuously, but rather jump from one location to another in a seemingly random fashion. However, in a low-dimensional system, when such a point is near a UPO, one notes that it always approaches the UPO from the same direction (the stable direction or eigenvector) and departs its vicinity along another consistent direction (the unstable direction or eigenvector).[1] This structure in phase space, called a saddle point, is what Ott et al. [1015] exploited to implement their chaos control scheme [1015].

Calculating a system's Poincaré section can be done by actually plotting the intersections of the phase-space trajectory with the sectioning plane. But in driven experimental systems, it is often easier to simply strobe the continuous data at the drive frequency (an excellent example of this approach

[1] Editor's note: This is sometimes referred to as "marching in, marching" phenomena.

is [490]). This can often be done in hardware, thereby obviating a computationally intensive calculation and simultaneously sectioning the data much more precisely than can usually be done in software.

A different procedure for obtaining a map from a physical system is called delay coordinate embedding [1019–1021]. This relies on the fact that, instead of measuring all of the system's n phase-space coordinates at one time, the geometry of the attractor can often be satisfactorily reconstructed by measuring one system variable at n times. This embedding can then be used to look for UPOs in the same fashion as a Poincaré section.

Another useful technique is to construct an embedding of data that consists of intervals between events. This was first used in the water-drop experiment [1022] where the interval between two successive drops was measured to characterize the system. This method was extended to biological systems by plotting the intervals between heartbeats in a rabbit heart preparation versus the previous interval. That such an interval plot is indeed a proper embedding has recently been shown by Sauer [1023].

In low-dimensional chaotic physical systems with very low noise, it is often easy to pick out UPOs by eye. Such a technique was used in the original experiments on the control of chaos [1014], where the UPO, along with its associated stable and unstable directions, could clearly be observed. In addition it was possible to calculate the UPO's eigenvalues by taking simple ratios of the positions of successive points in the neighborhood of the UPO. Figure 17.1 illustrates this technique. As should be obvious, this method is of most use in an almost noise-free system of reasonably low dimension.

It is also possible to observe visually the pattern of motion associated with UPOs in higher-dimensional chaotic systems. If the data from such a system are projected onto a 2-D plot, one may see the typical UPO pattern. This is because one can model the stable and unstable directions with a linearization around the UPO. In a projection onto a 2-D plot, these straight lines project as straight lines, thereby preserving the essentials of the motion near a UPO. However, while such a plot may reveal the presence of UPOs, it can only do so if there is only one unstable direction and one of the stable directions is substantially stronger than the others. In addition, a projection direction along the unstable direction would prevent observation of the UPO in question. While one can observe UPOs in this fashion, one should be warned that the eigenvalues and eigenvectors measured in this fashion have more to do with the projection than with the details of the physical system itself. Thus this technique is unsuitable for a detailed analysis of higher-dimensional systems.

17.2.2 Topological Recurrence Method

A simple statistical method for detecting unstable periodic orbits is the topological recurrence (TR) method. This method addresses many of the problems associated with noisy systems by searching for the occurrences of a simple

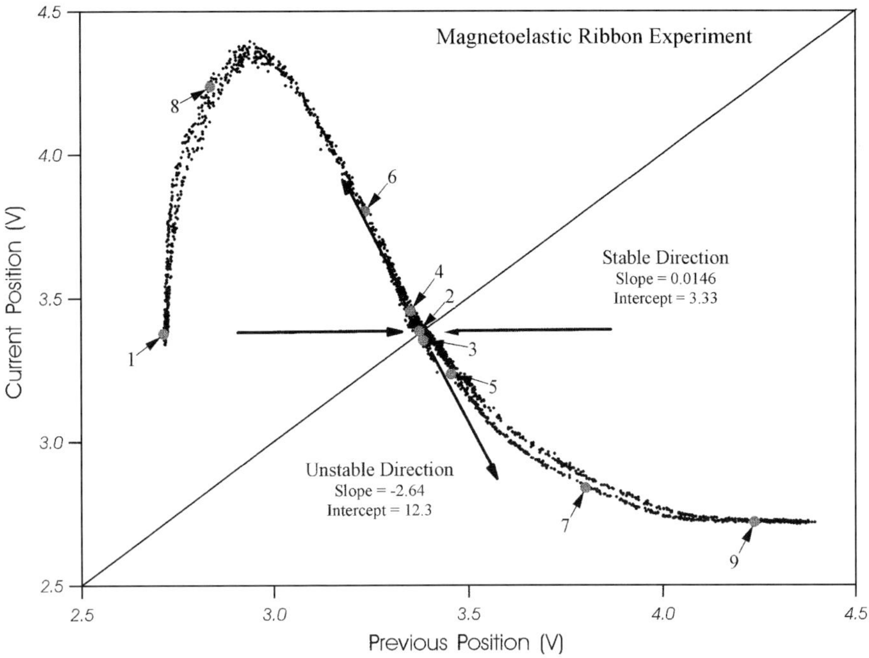

Fig. 17.1. Low-dimensional attractor consisting of 2000 points from the magnetoelastic ribbon experiment. A sequence starting at point 1280 is highlighted with *red dots*. The fixed point is at the intersection of the attractor with the diagonal. The saddle-like geometry around the fixed point is clearly visible to the eye. This pattern is repeated at the same location with the same directions over a hundred times in this clean physical data set (The colour plate of this figure appears on page 368)

pattern that is indicative of an encounter with a UPO. This method operates on time series data such as the time intervals between neuron firings or on nearly any other type of embedding. The attractor can be reconstructed from this time series as an $n+1$ dimensional return map, that is, T_i versus T_{i-1} versus $\ldots T_{i-n}$, where n is the dimensionality of the system.

In this type of embedding, encounters with a UPO will appear as a sequence of points converging towards the UPO and then diverging away from it along, respectively, the stable and unstable manifolds. This is difficult to observe since the dimension of the system is usually not known and often cannot be determined with any reliability. However, if we look only at the first return map T_i vs T_{i-1}, then the stable and unstable manifolds will project as straight lines. The overall structure of the attractor may be lost, particularly in the presence of noise, but in the vicinity of a fixed point, this converging and diverging behavior will still be present [1013,1024] as mentioned above and as shown in Fig. 17.2. Note that in such a plot the 45° line represents consecutive points that are nearly equal, so any period-1 fixed-points in the

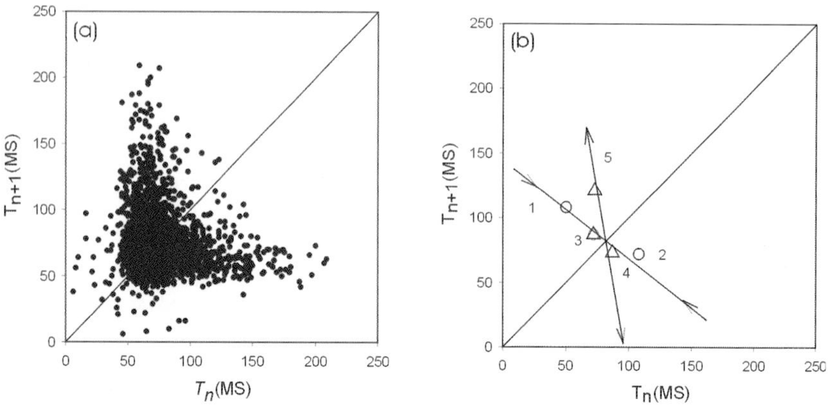

Fig. 17.2. (**a**) First return map of interspike time intervals from the rat facial cold receptor [1024]. Points near the 45° line represent consecutive time intervals that are nearly equal. (**b**) A typical encounter found in the preceding data. The converging points are shown with *circles*, and the diverging points with *triangles*. The two *arrows* denote the stable and unstable directions

system will lie on this line. Only period-1 UPOs (period-1 fixed points) will be discussed here, but this method can be modified to locate higher period orbits as well. To locate encounters with fixed points in the data, we search for the following pattern:

> Three points on the return map whose orthogonal distances to the 45° line grow consecutively smaller, followed by three points whose orthogonal distances to the 45° line grow consecutively larger. The point of nearest approach is shared, so that five points on the return map are used.

The number of times that this pattern is seen in a file is then counted. As with any statistical test, this pattern can occur even in pure noise with some probability. It is therefore necessary to determine if the number of encounters found is significantly larger than what we would expect to find just due to chance. This is usually done by comparing the number of encounters to those in an appropriate surrogate data file (see Sect. 15.3.4). Some number of surrogate data files is analyzed (typically 100), and the number of encounters found in each surrogate is tabulated. The mean and standard deviation are calculated, and a statistical significance can be determined with the following formula:

$$K = \frac{N - \bar{N}_{\mathrm{S}}}{\sigma_{\mathrm{S}}}, \tag{17.1}$$

where K is the statistical significance in units of standard deviation, N is the number of encounters found in the original data file, $\overline{N}_{\mathrm{S}}$ is the mean number of encounters found in the surrogates, and σ_S is the standard deviation of that mean. Assuming Gaussian statistics, a value of $K > 3$ indicates with a probability of 99% that the encounters in the file are not due to random chance. The assumption of Gaussian statistics is valid as long as the number of surrogates is fairly large (more then 20), and the number of encounters typically found in the surrogates is not close to zero.

One strength of this type of analysis is that it is computationally very fast. Even a relatively slow computer can analyze several hundred points of data in a fraction of a second, making the TR method ideal for real-time analysis. This is especially important for control techniques, since the location and properties of the fixed point can be determined very quickly, even in a non-stationary environment where the properties of a fixed point under control are changing with time. Alternatively, in post-experiment analysis, where speed is not as important, additional criteria can be added to sharpen the statistic. For example, we could require that the converging and diverging points do so exponentially, or put restrictions on how close the intersection of the calculated stable and unstable manifolds must be to the 45° line.

Another strength of the TR method is that it can operate on very small data files. Not only is this useful for control, since it allows you to quickly respond to changes in the system, but it is also useful for analyzing data that is sampled as some system parameter changes, e.g. when generating a bifurcation diagram. Small samples for which the parameter is essentially constant can be individually analyzed and changes in the K statistic can be used to locate bifurcations or other changes in the dynamics of the system.

To demonstrate the effectiveness of this analysis method, we apply it to a simple deterministic system, the Hénon map:

$$\begin{aligned} x_{n+1} &= 1 - ax_n^2 + y_n \\ y_{n+1} &= bx_n + \xi_n \,, \end{aligned} \tag{17.2}$$

where ξ_n is a uniform random deviate from $-\epsilon$ to $+\epsilon$. A uniform deviate is used because adding too large a random number could cause the system to diverge. Because the noise is added internally at each iteration (i.e. additive noise), it cannot simply be filtered out. This is a very common problem in physical nonlinear noisy systems. The parameters $a = 1.4$, and $b = 0.3$ were used, resulting in chaotic dynamics. Noise was added in varying quantities from $\epsilon = 0.00$ to $\epsilon = 0.010$ in steps of 0.002. Data sets of 5000 points were generated, since this is typical of the length of many of our actual biological data sets. Only the variable x_n was recorded, since it contains all of the deterministic dynamical information of the system. The resulting time series is thus already suitable for analysis by the TR method. For this analysis 100 surrogates were used. The surrogates were generated by randomizing the order in which the numbers making up the original file occur, thus maintaining the distribution of the original data, but destroying all other information contained in the file. The resulting statistics for each file are shown in Table 17.1.

Table 17.1. UPO analysis of the Hénon map using the topological recurrence algorithm

Noise ϵ	Number of Encounters N	Surrogate Mean $\bar{N}_S$	Standard Deviation σ_S	Statistic K
0.000	502	256.68	11.06	22.18
0.002	494	255.94	11.40	20.88
0.004	481	251.31	12.20	18.83
0.006	489	254.27	11.03	21.21
0.008	479	255.57	10.55	21.17
0.010	486	253.62	9.97	23.32

The TR method is able to detect the presence of the unstable fixed point with much more than 99% confidence even in the presence of very large dynamical noise. The robustness of this method with respect to noise comes from the fact that it searches for a pattern that is very rare in non-dynamical data. The presence of noise may cause some of the actual encounters with UPOs to be missed by the algorithm, but data from systems with real UPOs will typically have many more encounters in them than what we would expect to see by chance. Thus only when the noise is so high that the majority of the real encounters are not detected, should this algorithm fail.

17.2.3 Periodic Orbit Transform Method

Ott et al. [1015] recognized that, although a chaotic attractor is highly nonlinear, the local region around a UPO can be modeled as a linear map. They used this to formulate their theory for controlling chaos. So et al. [748,1025] used this fact to devise a periodic orbit transform (POT) technique applicable to time series data that concentrates the data in this local (linear) region onto the fixed point. The transformed data is then used to construct a histogram that will exhibit peaks at the locations of the fixed points. Data outside the linear region does not map to the fixed point, but rather is scattered randomly across the histogram.

Following the exposition of So et al. [748], we will initially describe this technique in a 1-D map. Start by considering a small region around a fixed point. The linear approximation to the map is $X_{n+1} = AX_n + B$, where A is the slope of the map and B is the intercept. The fixed point is $X^* = B/(1-A)$. Substituting for B, we obtain $X^* = (X_{n+1} - AX_n)/(1 - A)$. This estimate transforms onto the fixed point. The n-th estimate for the fixed point, $\hat{x_n}$, is

$$\hat{x}_n = [x_{n+1} - s_n\,(k)\,x_n]\,/\,[1 - s_n\,(k)]\ , \qquad (17.3)$$

where $s_n(k)$ is an approximation of the slope, i.e.

$$s_n(k) = (x_{n+2} - x_{n+1}) / (x_{n+1} - x_n) + k (x_{n+1} - x_n) \,. \tag{17.4}$$

Once we have transformed the data, it only remains to construct a histogram of the transformed data. This histogram is, however, subject to spurious singularities due to the differences in the denominators in (17.3) and the first term of (17.4). Thus the second term in (17.4) was introduced by So et al. to avoid these divergences. The value $k = \kappa R$, is introduced, where R is a uniformly distributed random number in the range [-1,1], and κ is the magnitude of the randomization. The transformation is applied to a given data point a large number of times, typically between 100 and 500, each time with a different random number R. This is repeated for each data point in the set. When the histogram is plotted, points that are not near a fixed point are scattered by this randomization unless $X_{n+1} \approx X_n$, where the second term in (17.4) vanishes. In this fashion spurious singularities are removed, but peaks in the histogram due to fixed points are retained.

The constant κ must be chosen carefully. It is highly data dependent and must be chosen large enough to remove the unwanted singularities, while being kept small enough not to destroy the peaks marking a fixed point. This is necessary because, although the second term in (17.4) does vanish at the fixed point, it erodes the size of the linear region around the fixed point, and thus the quantity of data that is transformed onto the fixed point.

The POT method can be generalized to evaluate higher dimensional data sets [748,1025]. In this case the transform is given by:

$$\hat{\boldsymbol{z}}_n = (\boldsymbol{1} - \boldsymbol{S}_n)^{-1} (\boldsymbol{z}_{n+1} - \boldsymbol{S}_n \boldsymbol{z}_n) \,, \tag{17.5}$$

where $\boldsymbol{z}_n = (x_n, x_{n-1}, \ldots, x_{n-d+1})^\dagger$ is a delay coordinate vector constructed with a suitable choice of dimension d and $\boldsymbol{z}_n^\dagger$ denotes the transpose of $\boldsymbol{z}_n$. The vector $\boldsymbol{S}_n$ is defined as:

$$\boldsymbol{S}_n = \begin{pmatrix} a_n^1 & a_n^2 & a_n^3 & \cdots & a_n^{(d-1)} & a_n^d \\ & & \boldsymbol{1} & & & \boldsymbol{0} \end{pmatrix} + \kappa \, \boldsymbol{R} \, ||\boldsymbol{z}_{n+1} - \boldsymbol{z}_n|| \,, \tag{17.6}$$

where for computational simplicity the norm was arbitrarily chosen to be the L_1 norm, $||\boldsymbol{z}_n|| = \sum_{j=1}^{d} \left| z^j \right|$, and where

$$\begin{pmatrix} a_n^1 \\ \vdots \\ a_n^d \end{pmatrix} = \begin{pmatrix} (\boldsymbol{z}_n - \boldsymbol{z}_{n-1})^\dagger \\ \vdots \\ (\boldsymbol{z}_{n-(d-1)} - \boldsymbol{z}_{n-d})^\dagger \end{pmatrix}^{-1} \begin{pmatrix} z_{n+1}^1 - z_n^1 \\ \vdots \\ z_{n-(d-1)}^1 - z_{n-(d-1)}^1 \end{pmatrix} . \tag{17.7}$$

Here κ is again the magnitude of the randomization, but $\boldsymbol{R}$ is a $d \times d$ random matrix with each matrix element chosen randomly as for the scalar R.

It should be noted that So et al. [1025] later revised the POT method slightly. In their longer paper, they chose the second term of (17.6) to be $\kappa \boldsymbol{R} \cdot$

$(z_{n+1} - z_n)$. Although this choice is analytically more tractable, it increases the time needed for computation significantly.

As with the TR method, the question also arises as to the significance of the UPOs found by the POT method. Once again we resort to surrogates to address this issue. For this analysis 100 surrogates were used. As done previously, the surrogates were generated by randomizing the order in which the numbers making up the original file occur, thus maintaining the distribution of the original data, but destroying all other information contained in the file.

In order to facilitate the comparison of the transformed surrogate data with the transform of the original data, we calculate a statistic T_i at each bin i of the histogram:

$$T_i = \frac{(\text{Transformed data})_i - (\text{Avg. transformed surrogates})_i}{(\text{Std. dev. transformed surrogates})_i} . \tag{17.8}$$

A value of T_i above 3 indicates the location of a UPO with a statistical confidence greater than 99%. So et al. [748,1025] calculate a similar statistic, but choose to divide by the maximum value of the standard deviation over the entire histogram.

Figure 17.3 shows the POT method applied to the same Hénon data presented in Table 17.1. The results agree perfectly with the analytic calculation of the UPO when $a = 1.4$, $b = 0.3$, and zero noise, giving a UPO position of 0.631.

The POT method is elegant in its application and has several advantages over the other methods discussed so far. Unlike the TR method, which uses only the stable and unstable directions, the POT method takes all the geometry around the fixed point (eigenvectors and eigenvalues) into account. In addition, since the POT method performs the transform on each data point a large number of times (typically 100 to 500), there is an effective "amplification" of the peaks in the histogram. This amplification makes the POT method very useful on short data sets such as those typically found in biological systems. However, because of this, the POT method is also computationally intensive and thus is unsuitable for use in real-time applications such as chaos control or anti-control.

17.2.4 Surrogate Analysis

In the previous sections we discussed the need to test against surrogate data files in order to establish the statistical significance of our results. Whichever method is chosen for locating UPOs, the question of how to generate the surrogates must eventually be faced. There are many different methods for generating surrogate data, and it is a non-trivial task to determine which surrogate is appropriate for these types of analysis. In any statistical test, the type of surrogate that should be used depends on what you are trying

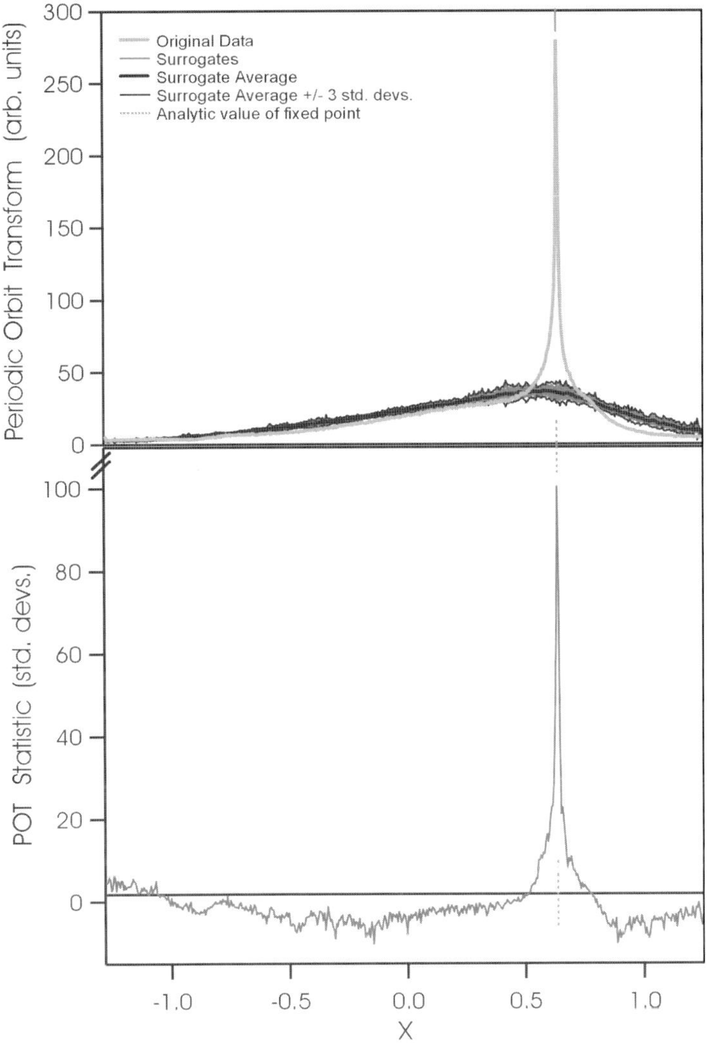

Fig. 17.3. Periodic orbit transform for the same Hénon data presented on the first line of Table 17.1. The transform of the original data (*upper pane – light gray*) is well above the band of the transformed surrogate data (*upper pane – dark gray*). The *lower pane* shows the POT statistic (*gray*) calculated from the values in the *upper pane*. The result agrees well with the analytic value (*dotted line*) calculated from the Hénon equations

to prove. If we are trying to detect the presence of UPOs in a data set, then what we want to prove is that what we have found actually are UPOs. In order to prove such a strong statement, one would have to create surrogate data that does not contain any periodic orbits, but which is in every other way identical to the original data set. Unfortunately this is for all practical purposes impossible. Both of the algorithms described above are based on the fact that if a UPO is present in the system, then at various locations in the data file that UPO will be visited. The most we can hope for from a surrogate is that it destroys any and all "real" encounters with fixed points and also preserves any statistical properties of the data that could contribute to false positive results from the analysis.

Here we will discuss the two types of surrogates used most often in this kind of analysis, shuffled surrogates and amplitude-adjusted Fourier transformed (AAFT) surrogates. Each of these techniques has its own strengths and weaknesses.

The simplest of these methods is the shuffled surrogate. Here the order in which the data appears in the file is randomized, but the distribution (shape, mean, width, etc.) of the data is unchanged. Maintaining the distribution of the original data file is very important, since it is clear from the descriptions of both the TR algorithm and the POT algorithm that the distribution can have a significant influence on the number of false UPO encounters we would expect to see due to chance [1026]. For this reason, shuffling the data is preferable to simply generating a sequence of random numbers. A simple shuffling algorithm that accomplishes this is as follows:

> Given a time series $(x_1, x{+}_2, \ldots, x_{L-1}, x_L)$ where L is the length of the data file, for each x_i select a random integer j from 1 to L. The elements x_i and x_i are then switched. The result of this is that each element x_i from the original file will appear as element X_j in the surrogate file with each j a random number uniformly distributed from 1 to L. To ensure complete randomness, this procedure may be repeated a number of times, although this is not necessary if a good random number generator is used.

Although preserving the distribution is a step in the right direction, shuffled surrogates are still not perfect. The distribution of the surrogate will match that of the original file, but in every other respect the surrogate is simply white noise. Therefore, technically the only null hypothesis that can be rejected is that the original data is white noise. A statistically significant result from either of the two tests described previously will certainly suggest the presence of either stable or unstable periodic orbits, but it does not prove it.

There are many other statistical properties of the original data that could, in principle, affect the results of our test. One property that has been known to have an impact on nonlinear analysis tests is the power spectrum. To address this possibility, we apply a type of surrogate originally developed for correlation dimension analysis [1027]. These surrogates, which we will refer to as AAFT (amplitude adjusted Fourier transformed) surrogates, preserve both the power spectrum and the distribution of the original data set. The following algorithm will generate such surrogates:

1. Generate L points of Gaussian white noise.
2. Reorder the white noise sequence so that its ranks match those of the original data. This is done by sorting the white noise (determining which element is the smallest, next smallest, etc...) using the rank order of the original data as the sorting key.
3. Phase randomize the ranked data from step 2 by taking the Fourier transform, and assigning random phases (uniformly distributed from 0 to 2π) to each frequency component.
4. Re-rank the original data so that its ranks agree with those of the surrogate created in step 3. This assures that the distributions of the surrogate and original data are identical, but has very little effect on the power spectrum.

Using these surrogates we can reject the null hypotheses that the data is a linear stochastic process passed through a static nonlinear transformation [1027]; i.e. what would commonly be referred to as colored noise. Given this more advanced surrogate, one might wonder, "Why bother with shuffled surrogates at all?" The answer is that the AAFT surrogates are much more computationally intensive. For post-experiment analysis, the time required to generate AAFT surrogates is usually not a problem, but for real-time analysis this computation time can become a serious issue. The TR method is ideal for real-time analysis since it can operate on very small data files, and can do so in a fraction of a second on a moderately fast computer. This type of analysis is very important in the control of chaos, where the dynamical parameters of the system may be changing with time and the control must compensate for these changes. If surrogate analysis is to be used in this type of situation, shuffled surrogates are currently the only ones that can be generated quickly enough to be useful.

We have stated that a good surrogate should destroy any real UPOs in the original file, while preserving any statistical properties that could influence the results of the algorithm being used. This raises the question of whether the power spectrum of the data does in fact have such an influence, specifically on the TR algorithm. To test this property, random data files were generated with non-white power spectra using a two-dimensional Ornstein–Uhlenbeck process [1028] as follows:

$$\begin{aligned}\frac{\mathrm{d}y}{\mathrm{d}t} &= -\frac{1}{\tau}(y-x)\\ \frac{\mathrm{d}x}{\mathrm{d}t} &= -\frac{1}{\tau}x+\frac{1}{\tau}\sqrt{2D}\xi(t)\,,\end{aligned} \tag{17.9}$$

where $\xi(t)$ is Gaussian white noise of unit variance, D is the noise intensity, and τ is the correlation time of the resulting process. The data is then embedded by measuring the time between positive zero crossings of the variable y, thus making it suitable for analysis with the TR method. Various correlation times were used, and both shuffled surrogates and AAFT surrogates

Table 17.2. Two-dimensional Ornstein–Uhlenbeck process analyzed with the TR method, using 100 surrogates

τ (ms)	25	50	75	100
SS surrgoates	0.39	0.85	0.57	1.05
AFT surrogates	0.22	0.56	0.69	0.82

Table 17.3. Two-dimensional Ornstein–Uhlenbeck process with inserted encounters

τ (ms)	25	50	75	100
SS surrogates	7.71	5.39	6.59	7.42
AAFT surrogates	6.35	7.56	6.52	9.26

were tested to determine if the power spectrum had any effect on the results (see Table 17.2). Both surrogates produce statistically indistinguishable results and are not "fooled" by the colored noise into thinking that UPOs are present.

To determine if the surrogates are also capable of properly detecting UPOs when they are present, we use the same data files, but this time with encounters added to the file. This is done by taking the following pattern of numbers $\{50, 108, 73, 87, 72, 121\}$ which meets the criterion stipulated above for an event and inserting it into the data file many times. This sequence is inserted between every encounter found in the data file, overwriting whatever data was there originally, in order to preserve the file length. The event described is scaled in order to match the distribution of the file; that is, each number in the sequence is multiplied by the same constant, which is chosen to make the sequence match the original data distribution as closely as possible. The sequence still remains an encounter because rescaling of this type does not affect order statistics. These new data sets were then analyzed with both the shuffled and AAFT surrogates, and the results appear in Table 17.3. In this case both surrogates are able to detect the UPOs in our new files. This test may seem somewhat artificial, but it should be noted that many real biological systems are very difficult to distinguish from the Ornstein–Uhlenbeck process described here, indicating that the TR method can be very useful for identifying dynamics in these types of systems.

Some biological systems that are also difficult to distinguish from noise differ in that they possess a very strong peak in their power spectrum. Since the power spectrum in these types of systems is qualitatively very different from the Ornstein–Ohlenbeck process described above, we will perform the same tests using another random process known as harmonic noise [1029,1030]. This data is produced by finding the positive zero crossings of the phase

Table 17.4. Harmonic noise analyzed with the TR method using 100 Surrogates

ω (rad/s)	251	126	84	63
Shuffled surrogates	0.10	−0.75	0.68	−0.62
AAFT surrogates	0.46	−0.50	1.21	−0.55

Table 17.5. Harmonic noise with inserted encounters

ω (rad/s)	251	126	84	63
Shuffled surrogates	6.23	3.51	6.15	4.50
AAFT surrogates	5.87	4.34	7.16	5.20

space trajectory of a damped linear harmonic oscillator driven by Gaussian white noise:

$$\frac{\mathrm{d}^2 x}{\mathrm{d}t^2} + \Gamma \frac{\mathrm{d}x}{\mathrm{d}t} + \omega_0^2 x = \sqrt{2D\Gamma}\xi(t) , \tag{17.10}$$

where $\xi(t)$ is Gaussian white noise of unit variance, D is the noise intensity, Γ is the damping, and ω_0 is the natural frequency. Four different values of ω_0 were used, and the data were analyzed just as in the previous example. The results are shown in Tables 17.4 and 17.5. Once again the two surrogates both produce essentially the same results, again suggesting that the power spectrum of a data set does not have a significant influence on the results of the TR algorithm. Of course it is impossible to test every possible type of colored noise, but these results nevertheless demonstrate that both the AAFT and shuffled surrogates seem to work equally well.

17.3 UPOs: Applications to Biology

Unstable periodic orbits (UPOs) have been detected in a variety of biological preparations ranging from the crayfish's caudal photoreceptor [1026] to the beating heart [1031].

17.3.1 Canine Ventricular Fibrillation

Since the heart was traditionally the first dynamical disease studied with the tools of nonlinear dynamics, we shall use it here as a primer for the techniques of determining whether a system is chaotic and, if so, of controlling that chaos.

Witkowski et al. [1031] published a study examining a dog heart preparation undergoing ventricular fibrillation (VF). VF is a frenzied and irregular heart rhythm that renders the heart incapable of sustaining life and is always fatal within minutes unless externally terminated by passing a large electrical

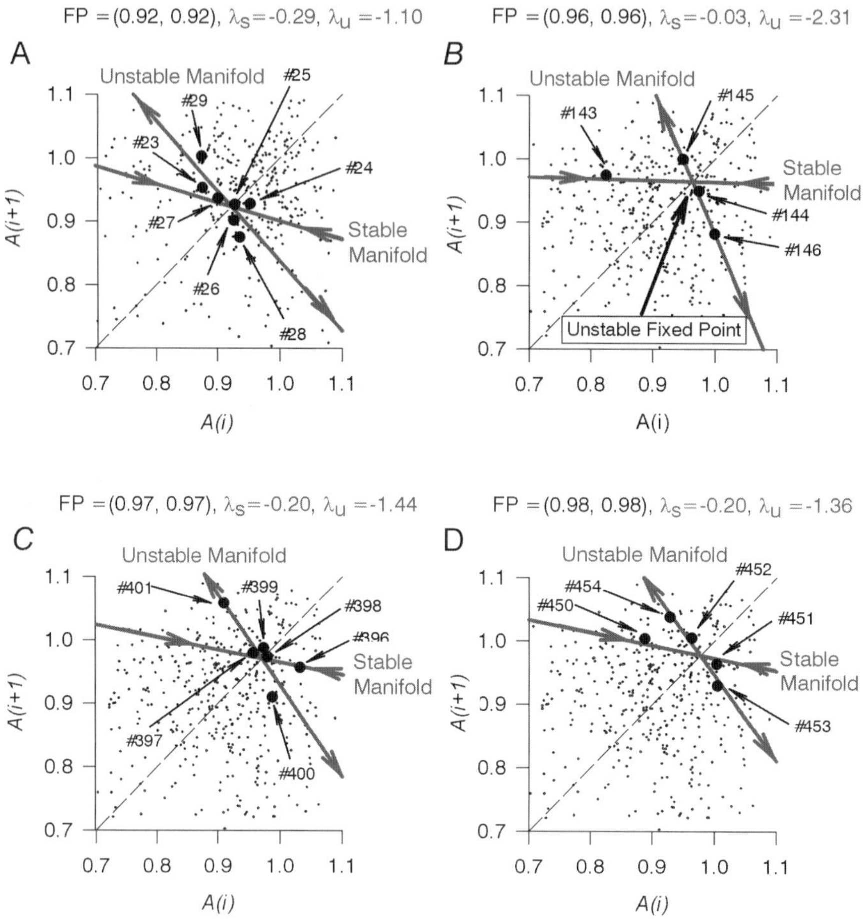

Fig. 17.4A–D. Four examples of return plots for the time series generated by the ratio of negative to positive area A(i) under the transmembrane current for a typical 1 min interval of in vivo ventricular fibrillation illustrating the local structure of the chaotic attractor. Coordinates for the calculated unstable fixed point (FP), stable eigenvalue, and unstable eigenvalue for this visitation of the unstable fixed point are provided above the plot (The colour plate of this figure appears on page 369)

current through the heart. Witkowski and Penkoske [1032] were examining the time series formed from the intervals between beats as well as the time series formed by successive measurements of the ratio of the negative to positive areas under the transmembrane current curve ($A(i)$) for each beat. They sought signs of deterministic chaos as a preliminary to trying to control the fibrillation. Figure 17.4 shows typical return plots of $A(i)$ for their data. The saddle-like structure is clear.

17.3.2 Rat Brain Hippocampus

In a paper on the control of chaos in neural tissue [1030], the authors reported controlling chaos in the electrical population spiking of the in-vitro rat brain hippocampus. As a preliminary to control, they sought UPOs of period one in the signal. The intervals between successive population spikes were plotted on a delay diagram, as shown in Figs. 17.5 and 17.6. Sequences of points approaching the diagonal from a single direction and departing along another definite direction were identified by direct observation. The sequences were clearly seen, despite the fact that these data appear to be of dimensionality greater than 2 and are relatively noisy. Nevertheless, as shown in Fig. 17.5, unstable saddle points could be observed, and, as shown in Fig. 17.6, these sequences were repeated over and over again. A later paper [1025] examined

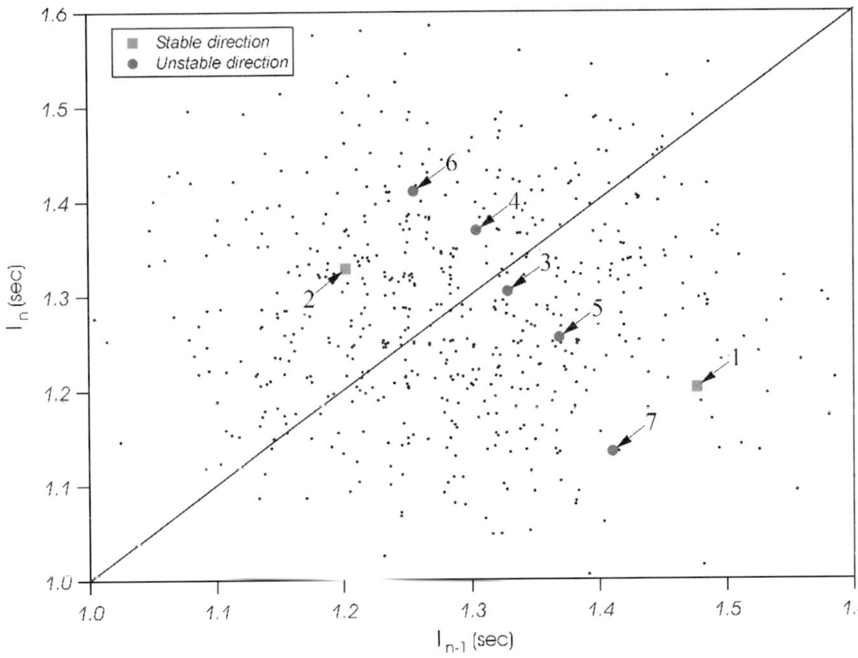

Fig. 17.5. Return plots of inter-burst intervals. Seven sequential points are color-coded. Note that as the trajectory crosses the line of identity along the direction from 1 to 2, the next points take a peculiar sequence that starts close to the line of identity at point 3, and then alternates on either side of the line of identity and progressively diverges from it along a nearly straight line for points 3–7. The points colored in *square* (1 and 2) plus point 3 (*cirlce*) define a stable direction or manifold, while the points with *circles* (3–7) define an unstable manifold. The intersection of these manifolds with the line of identity defines the unstable fixed point (The colour plate of this figure appears on page 370)

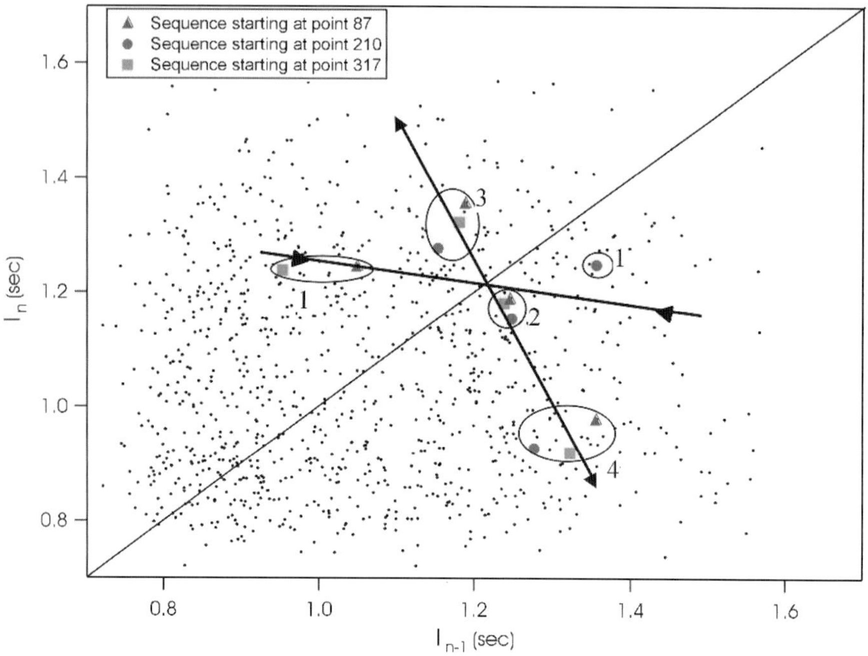

Fig. 17.6. Return plot showing multiple color-coded trajectories that did not follow each other in time. The starting point for each sequence, numbered 1 in each, began at spike numbers 87 (*triangle*), 210 (*circle*), and 317 (*square*), out of a total series of 320 spikes. The stable manifold is shown with *arrows* pointing towards the unstable fixed point, and the unstable manifold has arrows drawn in the direction away from the unstable fixed point. Note that each sequence starts at a point roughly the same distance from the unstable fixed point in a region close to the stable manifold. Each trajectory follows the same path, and after closely approaching the unstable fixed point for the points labeled 2, there evolves for each trajectory a sequence of exponentially diverging jumps (on alternating sides of the line of identity along the unstable manifold). Although the trajectories are clustered along the unstable manifold for four spikes, the fifth spikes are widely scattered (not shown). This is the manifestation of the sensitivity to initial conditions typical of chaotic systems. These plots show clear evidence for non-random structure in these trajectories, and their patterns bear the hallmarks of deterministic chaos

data from the same preparation and confirmed these earlier results using the POT method.

17.4 The Control of Chaos

The presence of the infinite number of unstable periodic motions is the key to control of chaotic systems [1014,1033,1034]. This form of stabilization is analogous to a baseball on a saddle. Control is achieved through movement of the saddle's position or adjustment of the ball's motion to keep the baseball constantly rolling back toward the unstable equilibrium point in the center of the saddle. This theoretical technique was originally pioneered by Ott et al. [1015]. Their technique (and variations thereof) has been successfully applied to control chaos in the vibrations of a magnetoelastic ribbon [1014], electrical circuits [1016], the output of a solid-state laser [1018], chemical oscillations [1017], drug-induced arrhythmias of in-vitro rabbit ventricle [1035], seizure-related population spiking of hippocampal slices [1036], and many other systems. These applications all share in common a proportional feedback control around unstable periodic state space trajectories.

17.4.1 Control of Chaos in Cardiac Tissue

It is possible to control a chaotic cardiac arrhythmia using chaos control. Our cardiac preparation [1035] consisted of an isolated perfused portion of the interventricular septum from a rabbit heart as shown in Fig. 17.7. Cardiac arrhythmias were induced by the addition of ouabain and epinephrine to the perfusing mechanism. The mechanism for the cardiac arrhythmias induced in this way is probably a combination of triggered activity and non-triggered automaticity caused by progressive intracellular Ca^{2+} overload from Na^{+} pump inhibition and increased Ca^{2+} current. Typically the ouabain/epinephrine combination induced spontaneous beating, initially at a constant interbeat interval, and then progressing to period-2 and higher-order periodicity before developing a highly irregular aperiodic pattern of spontaneous activity. The duration of the aperiodic phase was variable, lasting up to several minutes before spontaneous electrical activity irreversibly ceased.

The irregular, aperiodic arrhythmias induced by ouabain/epinephrine in this preparation showed a number of features symptomatic of chaos. Most importantly, in progressing from spontaneous beating at a fixed interbeat interval to highly aperiodic behavior, the arrhythmia passed through a series of transient stages that involved higher-order periodicities. These features are illustrated in Fig. 17.8 in which the n-th interbeat interval (I_n) has been plotted against the previous interval (I_{n-1}) at various stages during ouabain/epinephrine-induced arrhythmias. As before, this Poincaré return map allows us to view the dynamics of the system as a sequence of pairs of points (I_n, I_{n-1}), thus converting the continuous dynamics of our system to a map. On such a map, chaotic data (Fig. 17.8E) can easily be distinguished from periodic data (Fig. 17.8A–D). Additionally, the sequence of the data points on such a plot reveals the stable and unstable directions, knowledge of which is required to implement control (Fig. 17.9).

In this rabbit preparation, each control attempt consisted of a learning phase and a control phase. During the learning phase, an UPO was identified and characterized [1035]. During the subsequent control phase, the computer waited until a close approach to the UPO was detected $|I_n, I_{n-1}| < \epsilon$, where ϵ defines the control region and is a fraction of the total attractor size. The

Fig. 17.7. Photograph of an isolated well-perfused portion of the interventricular septum from a rabbit heart, arterially perfused through the septal branch of the left coronary artery with a physiologically oxygenated Kreb's solution at 37°C. The heart was stimulated by passing a 3 ms constant-voltage pulse, typically 10–30 V, at twice threshold between platinum electrodes embedded in the preparation. Electrical activity was monitored by recording monophasic action potentials with Ag-AgCl wires on the surface of the heart. Monophasic action potentials were digitized at 2 kHz and processed in real time by a computer to detect the activation time of each beat from the maximum of the first derivative of the voltage signal. Arrhythmias were induced by adding 2–5 μM ouabain with or without 2–10 μM epinephrine to the arterial perfusate (The colour plate of this figure appears on page 371)

algorithm then initiated a control stimulus that moved the next point on the Poincaré plot (as predicted by the local dynamics of the UPO) onto the stable manifold (i.e. the contracting direction) of the UPO, thereby allowing the natural dynamics of the system to subsequently draw the system state onto the UPO itself. Stimuli were administered on subsequent points to keep

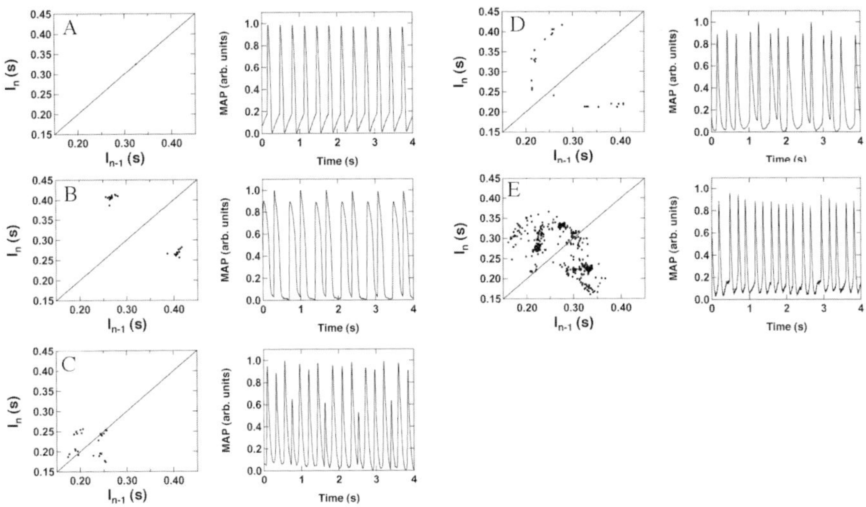

Fig. 17.8. Poincaré maps of interbeat intervals and time series of monophasic action potentials at various stages during arrhythmias induced by ouabain/epinephrine in typical rabbit septa. Typically the arrhythmia was initially characterized by spontaneous periodic beating at a constant interbeat interval (**A**), proceeded through higher-order periodicities such as (**B**) period-2, (**C**) period-4, and (**D**) period-5, and ending in an (**E**) completely aperiodic pattern. Note that, in the Poincaré map of the final stage, the points form an extended structure that is neither point-like nor a set of points (i.e., is not periodic) and is not space filling (i.e. is not random). This is indicative of chaos [1035]

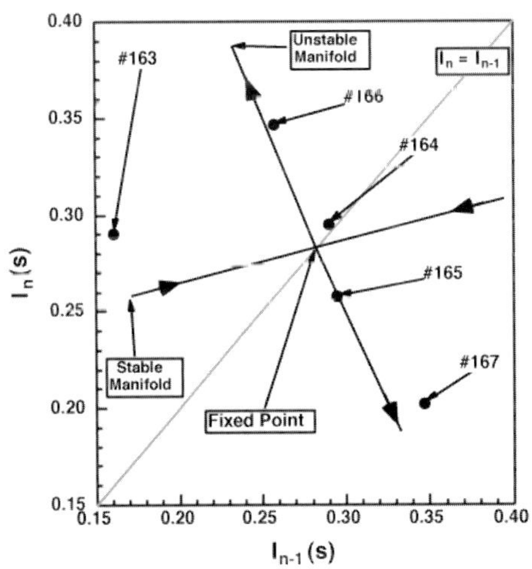

Fig. 17.9. The Poincaré map of the aperiodic phase of a ouabain/epinephrine induced arrhythmia in a typical rabbit heart preparation illustrating the structure of the chaotic attractor [1035]

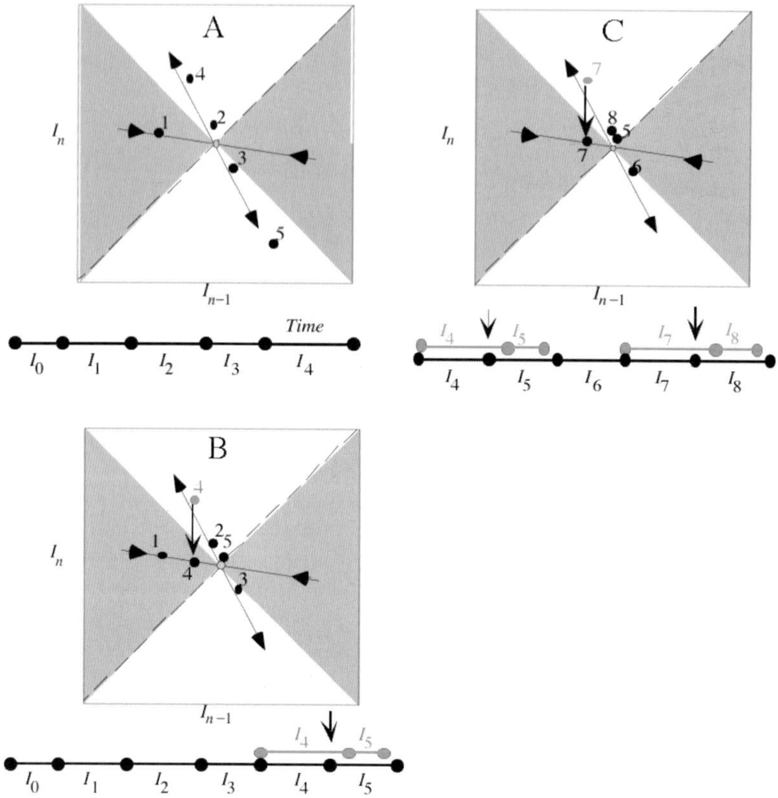

Fig. 17.10. Schematic representation of the chaos control technique $\boldsymbol{A} \rightarrow \boldsymbol{B} \rightarrow \boldsymbol{C}$ in excitable tissue. *Downward arrows* denote control interventions pulling the interval onto the stable manifold as shown

the system on the stable manifold. Thus the system state was continually contracted towards the UPO. This is shown schematically in Fig. 17.10.

If the current point on the Poincaré plot strayed outside the control region, stimuli were then discontinued until the system state point re-entered the control region. It should be noted here that on-demand pacing is simply the special case of this algorithm where the stable manifold is horizontal. An example of applying this algorithm to the rabbit heart is shown in Fig. 17.11.

With this approach, chaos control was complicated by the fact that the interventions were, of necessity, unidirectional. By delivering an electrical stimulus before the next spontaneous beat, the interbeat interval could be shortened, but it could not directly be lengthened. This is because a stimulus, which elicits a beat from the heart, shortens the interbeat interval between the previous spontaneous beat and the beat elicitepd by the stimulus. The effects of this limitation are apparent in Fig. 17.11, where the best control we could achieve was period-3.

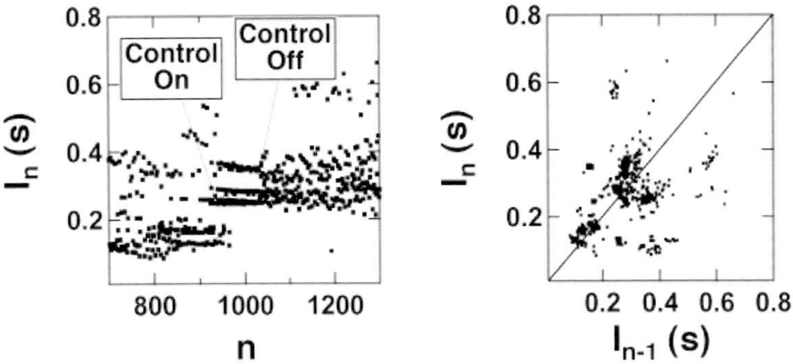

Fig. 17.11. (**A**) Interbeat interval In versus beat number n during the chaotic phase of the ouabain/epinephrine-induced arrhythmia in a typical control run. The region of chaos control is indicated. (**B**) The corresponding Poincaré map for the time series on the left. The *light points* represent interbeat intervals during the uncontrolled arrhythmia while the *dark points* represent interbeat intervals during the chaos control. Note the transient period between the initiation of the chaos control and the establishment of a period-3 pattern that is apparent on both plots [1035]

Several observations should be made about the pattern of the stimuli delivered by the chaos control program. First, these stimuli did not simply overdrive the heart. Stimuli were delivered sporadically, not on every beat and never more than once in every three beats on average. In contrast, periodic pacing, in which stimuli were delivered at a fixed rate, was never effective at restoring a periodic rhythm and often made the original aperiodicity more marked. For non-chaos control, irregular pacing was similarly ineffective at converting chaotic to periodic behavior.

Recently we have extended our control of chaos techniques to the control of atrial fibrillation in human hearts. This work is detailed in another paper [1037] and will not be discussed in detail here, save to note that the identification of an unstable fixed point [1035,1036] and its subsequent characterization [1013] were much improved by the intervening advances in technique. The results of the UPO analysis are shown in Fig. 17.12.

17.4.2 Control of Chaos in Brain Tissue

A similar strategy can be used to control chaotic behavior in brain tissue [1036]. Several types of in-vitro brain slice preparations, usually after exposure to convulsant drugs that reduce neuronal inhibition, exhibit population burst-firing activity similar to the interictal spike. One of these preparations is the high potassium K^+ model, where slices from the hippocampus of the rat brain (a frequent site of epileptogenesis in the human) are exposed to artificial cerebrospinal fluid containing high K^+ that causes spontaneous bursts of synchronized neuronal activity that originate in a region known as the third part of the cornu ammonis or CA3 as shown in Fig. 17.13.

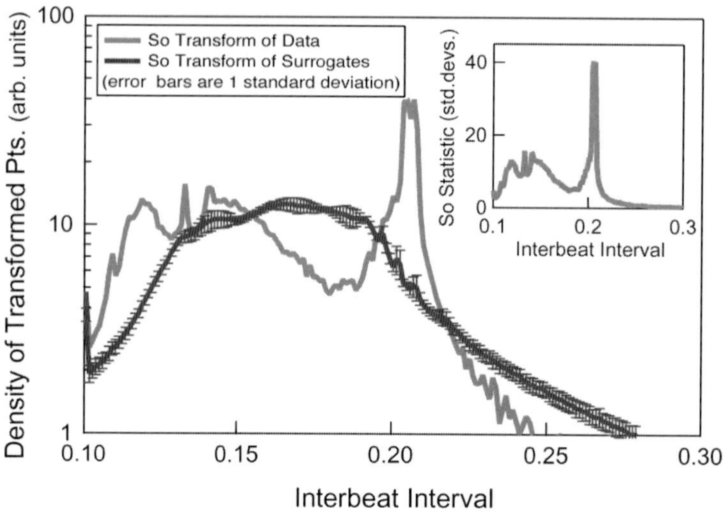

Fig. 17.12. The So transform of the human AF data (*light gray*). The *solid dark gray line* indicates the average of 100 surrogates, with error bars denoting the standard deviation at each point. (Inset) The So statistic, which is the height of the So transform minus the average of the surrogates and then normalized by the standard deviation of the surrogates for each value of the interbeat interval [1038]. Note that this is a slightly different definition from the one in [748]

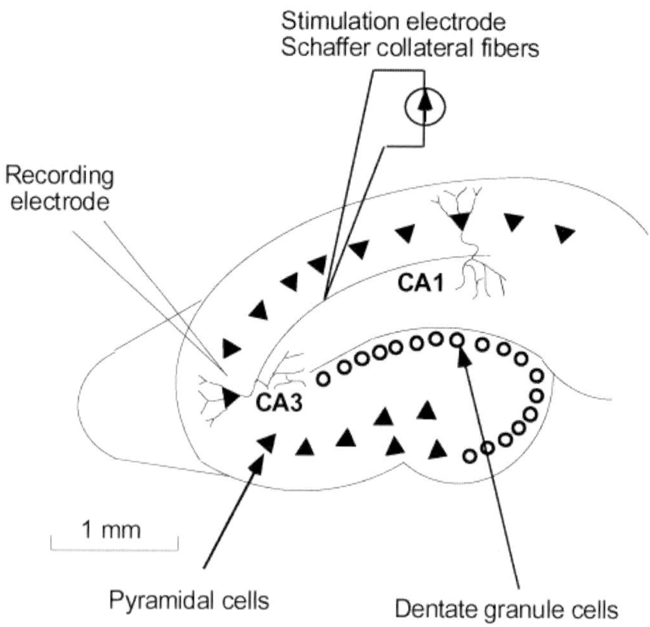

Fig. 17.13. Schematic diagram of the transverse hippocampal slice and arrangement of recording electrodes [1036]

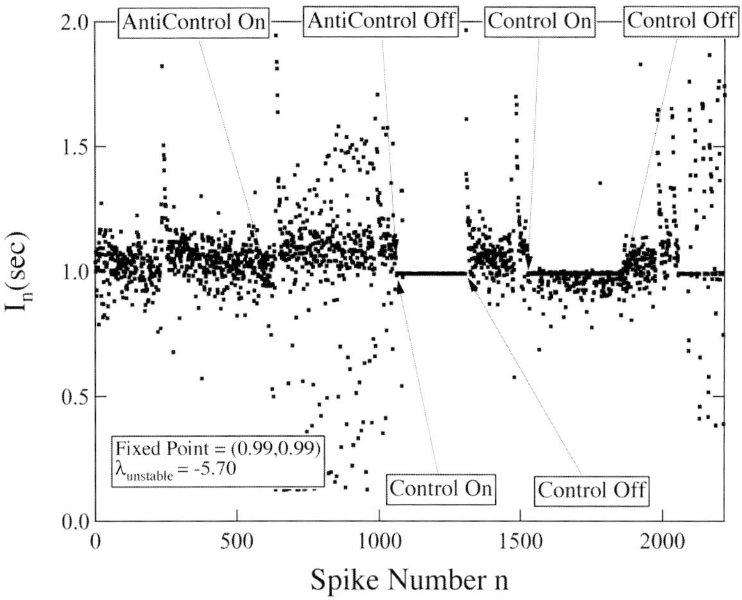

Fig. 17.14. Demonstration of chaos anti-control and chaos control in a hippocampal slice of a rat brain exposed to artificial cerebrospinal fluid containing high [K^+] and undergoing spontaneous chaotic population burst firing or spiking [1036]

If one observes the timing of these bursts, clear evidence for UPOs is seen in the return map. As reported we were able to regularize the timing of such bursts through intervention with stimuli delivered by micropipette with timing as dictated by the chaos control algorithm to put the system onto the stable direction. As shown in Fig. 17.14, not only were we able to regularize the intervals between spikes, but we were also able through a chaos "anti-control" strategy to make the intervals more chaotic. It is the latter that might serve a useful purpose in breaking up epileptic activity through the prevention or eradication of pathological order in the timing of the spikes. This original anti-control or chaos maintenance strategy [1036] has been further elaborated and expanded to high dimensions in the past few years [1039–1041].

17.5 Discussion

The analysis of periodic orbits can be used for more than just the detection and control of dynamical behavior in noisy systems. Unstable periodic orbits can also be used for the anti-control, or maintenance, of chaos. This may be particularly relevant to epilepsy since many investigators regard the normal EEG as chaotic (). Sometimes a chaotic attractor will have two or more basins of attraction, which merge during a dynamical crisis. Often this crisis is mediated by a UPO. Or a chaotic attractor may have two basins separated by a small region of phase space. Noise can then drive transitions between two different types of behavior (the two basins) even in stationary conditions (see also Chap. 14). By analyzing the unstable periodic orbits in the region of either transition, small perturbations can be used to lock the system's dynamics into one basin of attraction [1042]. As an example, it has been speculated that the heart has two different basins – one that encompasses the normal heart rhythms and one that yields fibrillation. If correct, this underlines the importance of understanding these basins and how they behave.

Another use for UPO analysis is "chaotic spectroscopy". Although all chaotic systems have an infinite number of UPOs, the weights given to each UPO (i.e. the time the system spends near each UPO) are different for each system. Thus one can characterize the chaos by identifying not only the predominant UPOs, but also by noting the weight given to each UPO.

One hope is that these results might shed some light on the means by which the brain communicates both externally and internally. As a side benefit, we have been pushed to develop new techniques for analyzing chaotic data. The methods described here for finding UPOs in chaotic data opens the door towards the development of novel approaches for the diagnosis and prediction of dynamical disease [94]. We have found that, while the techniques of nonlinear dynamics can be applied with success to many different biological systems, it is also true that these same biological systems can teach us much about the nonlinear dynamics of complex systems. Thus there is a synergy between our efforts to apply the techniques of nonlinear dynamics to biology and the development of an understanding of nonlinear dynamical systems.

18 Prospects for Building a Therapeutic Cortical Stimulator*

J.R. Hetling

Ventricular fibrillation is an undesirable bioelectrical event in excitable tissue. Implantable cardiac pacemakers have extended the lives of countless patients. Such resounding clinical successes suggest the possibility of treating other adverse bioelectrical phenomena with implantable devices. In this chapter we consider the technical challenges to developing an implantable device to treat epilepsy via cortical stimulation. Following a brief review of recent literature supporting this approach, we will form a general description of such devices. Each stage in our proposed device will be introduced with a general discussion of design considerations, and, where possible, related to examples of existing or emerging implantable therapeutic stimulators.

By this late in the text, it should be evident that we are not absolutely certain what such a device should do with regard to monitoring and modulating the physiology of the patient. However, regardless of the final design there will be general functional components common to any such device. Without making specific recommendations for design details, we will, then, focus on the general issues relavant to the fabrication of each functional stage in an implantable epilepsy neuroprosthesis.

18.1 Cortical Stimulation as a Treatment Approach

Much of this book makes arguments for the possibility of treating epilepsy with artificial cortical stimulation. Here, we briefly review a number of studies that directly support the use of artificial stimulation to suppress or arrest seizures associated with epilepsy. While not an exhaustive review, we will summarize below some of the current literature in this area.

In temporal lobe epilepsies, a large percentage of focal seizures begin in the hippocampus. Accordingly, a great deal of research has been done in-vitro using hippocampal slices. In cases involving both evoked seizures [666,933,1043], and spontaneous seizures [590,1044,1045], application of external electric currents successfully suppressed the epileptiform activity. The rationale, in brief, is that the external electric field couples with the hippocampal neurons, but to different degrees related in part to each neuron's physical distance to the

* I thank S. Cranstoun of the Neuroengineering Group at the University of Pennsylvania for helpful discussions

stimulating electrodes. This serves to differentially stimulate the neurons involved in the coordinated activity, thereby dis-coordinating the seizure activity [484]. This then serves the opposite function of a heart defibrillator, which strives to uniformly depolarize all the heart tissue in order to promote coordinated activity.

Moving one step closer to the clinical situation, there has also been a good deal of work done in vivo in animals, using the kindling model of epilepsy. Here, seizures are induced by a series of low-amplitude, short-duration electrical stimuli to the brain. If these stimuli are applied long enough, spontaneous seizures continuing beyond the cessation of stimulation can result [1046]. These studies make use of a second stimulation site, somewhat removed from the site of kindling, but within the region demonstrating seizure activity. This second "therapeutic" stimulation has been shown to interfere with the kindled seizure [1047,1048]. Stimulating regions in the brain outside of the region generating the seizure, such as the mammillary nuclei or the anterior thalamic nuclei, have also been attempted, but with less success. In the case of the mammillary nuclei, 100 Hz stimulation resulted in protection against chemically induced seizures, and disruption of ongoing seizures in rats [1049]. High-frequency stimulation (100 Hz) of the anterior thalamic nuclei also produced partial protection against chemically induced seizures; evidence suggests this stimulation site may not disrupt or prevent seizures, but it may prevent spread of seizures.

Vagus nerve stimulation has been shown to be of benefit [519,1050], and has been approved for use in epilepsy by the FDA. A pilot study of anterior thalamic nucleus (ATN) stimulation for epilepsy has been launched by R. Fisher and others. They have implanted 3 patients in the last year and anecdotal evidence indicates that it prevents seizure spread but does not reduce the frequency of seizures. Stimulation to the thalamus at low frequencies can induce cortical changes, which resemble spike-wave discharges. Stimulation at higher frequencies can lead to desynchronization [1051].

One small trial of five patients with ATN stimulation was reported as an abstract [1052]; three patients were said to benefit from the treatment. One controlled study of centromedian thalamic stimulation has been reported [177]. One patient had a significant reduction in seizures during the stimulation period and the reduction persisted after the stimulator was turned off. No kindling was seen in these patients.

Two other recent approaches to brain stimulation to abort seizures have been reported. One involves evoking an after-discharge in an epileptogenic zone by stimulation and then using high-frequency local stimulation to abort the after-discharge [1053]. The second consists of direct stimulation of the epileptogenic hippocampus in patients with depth electrodes to suppress seizures [1054]. These studies support the possibility of disrupting epileptic seizures with surface, shallow or deep-brain stimulation, and so our discussion will consider each possibility.

18.2 Devices for Therapeutic Stimulation

A number of implantable devices that modulate neural or neuromuscular systems are already commercially available, and it will be illustrative to first briefly describe each. Five prominent examples are noted below; in later sections in this chapter, specific references will be made to these existing commercial devices concerning the details of design. Although not referred to here for brevity, a robust literature exists describing the evolution of each of these devices.

Cardiac Pacemaker. Implantable pacemakers consist of a battery-powered pacing circuit housed in a hermetically sealed titanium case which is located near the heart, generally beneath the collar bone. An insulated wire extends from a pin connector in the case and is threaded through a vein, terminating in one or more electrodes embedded in the myocardium. Features unique to the pacemaker leads are the use of a coiled conductor that can withstand the repeated strain caused by cardiac wall movement, and a corticosteroid-eluting electrode tip, which releases the steroid slowly over years to reduce inflammation and fibrous incapsulation of the electrode tip. Stimulating electrodes are also used to sense cardiac potentials; thus, artificial pacing is provided only when needed. Some pacemakers monitor other physiological variables (e.g. respiration rate, pH, blood O_2 saturation, temperature) in order to adjust the pacing rate to accommodate increased exercise. Modern pacemakers incorporate two-way radio frequency (RF) telemetry systems that allow clinicians to monitor device function and reprogram the pacing circuitry. Representative manufacturer: Medtronic.

Vagus Nerve Stimulator. Steady stimulation of the vagus nerve can be effective in reducing the frequency of epileptic seizures. The unit consists of a battery-powered pulse-generating circuit contained in a titanium case, an insulated lead several centimeters long, terminating in two helical electrodes that are carefully positioned around the vagus nerve. The case is positioned in the shoulder area. Noninvasive reprogramming by the clinician is achieved using an RF link, and patients can start or stop pulse delivery by placing a magnetic "key" on the skin over the device. Representative manufacturer: Cyberonics, Inc.

Deep-Brain Stimulator. Stimulation of the ventral intermediate nucleus of the thalamus has been used for tremor-control therapy in Parkinsons patients. These devices use a quadripolar lead inserted deep in the brain with the aid of a tungsten stylus. The pulse-generating circuitry is housed in a titanium case located in the upper chest cavity. High-frequency stimulation is constant; stimulus parameters can be adjusted noninvasively through an RF link. Representative manufacturer: Medtronic.

Cochlear Implant. Cochlear implants are the most successful neuroprosthetic device to date. The representative system consists of a microphone located near the ear which is connected to processing circuitry housed either in a small case located in a pocket, or which hangs inconspicuously behind the ear. The sound information is processed, encoded, and transmitted to the internal component using RF telemetry. The internal receiver decodes the RF transmission and accordingly activates approximatley 16 electrodes located in the cochlea. Electrodes are positioned along a flexible strip that follows the coiled shape of the cochlea upon insertion. Representative manufacturer: Clarion.

Implantable Drug Pump. These devices were developed for long-term local drug delivery for treatment of chronic pain, chemotherapy, diabetes, etc. A reservoir and propellant chamber are housed in a titanium case and located near the site of delivery. The propellant chamber is filled with a gas that exerts a near constant pressure on the reservoir, forcing the drug through a catheter – there is no battery needed. The tip of the catheter is located at the desired delivery target. Factors affecting the rate of drug delivery include atmospheric pressure, viscosity of the solution being delivered, the pressure difference between the reservoir and the catheter tip, and the temperature. The reservoir can be refilled through a percutaneous needle. Representative manufacturer: Medtronic.

What additional technology is required to build an implantable device to treat epilepsy via cortical stimulation? To answer that question, consider what we would expect such a device to do. Unlike the vagal nerve stimulator, our device should be reactive, responding to an imminent seizure with a therapeutic stimulus that would avert the seizure. For brevity, we will refer below to any representative of this theoretical class of implantable devices as a Seizure-Aversion-Implant, or SAI.[1]

Such a device would presumeably contain three stages, described generally as follows:

- An input stage would continually monitor some indicator(s) of brain function (most likely bioelectric, but possibly chemical or optical);
- A processing stage (the brain of the device) would compare the input signal to a set of parameter values, and make a decision regarding the necessity for active intervention;
- An output stage that would provide a modulatory stimulus to the brain when called upon to do so by the processing stage.

This could be an open-loop system, in which the output is predetermined (once the decision has been made to provide an output), or closed-loop, in

[1] Editor's note: In other parts of the book this device is referred to as a "brain defibrillator". See Chap. 19 for discussion of how to name this device.

which the output is, at some frequent interval, adjusted based on new information from the input stage (feedback). We would also like to include the capability for noninvasive adjustments by the clinician, and a patient override, which would allow the user to start or stop the stimulus on command.

The general functions of our SAI are currently performed by FDA-approved, commercially available devices. What are the additional requirements, if any, for the production of a SAI? Disregarding, for the moment, the details of fabrication, the effectiveness of an SAI would be determined by the nature of the disease and the limits of our knowledge of the disease. That is, is there a signal to measure, do we know what that signal is; is there a beneficial stimulus, and do we know what that stimulus is? These four questions have been approached by the other contributors to this text. Without attempting to identify what the signals and stimuli are, this chapter will assume that they have been identified, and discuss the issues related to design and fabrication. From this perspective, the following features of an SAI may distinguish it from existing devices:

- Recording and stimulation sites may be different from each other, and different from those targeted by existing devices.
- Recording and stimulation modalities may be different (e.g. bioelectric recording, chemical stimulation).
- Possible requirement for substantial computational power to carry out seizure identification algorithms.

What follows is a discussion of the design issues pertaining to each stage in our proposed SAI. In addition to giving details of existing examples where possible, we will briefly describe the fundamental concepts underlying the design of each stage

With the sweeping assumptions made above, sufficient progress has been made in microfabrication, biomaterials and neuro-biology such that construction of an SAI would become largely a matter of details (with full acknowledgment that the devil often resides in the details!). We must forewarn the reader that often the state of the art in fabrication of microdevices is not available to the scientific community. Physical devices and methods of producing the same are more intuitively seen as having commercial value, and therefore more often patented or considered proprietary knowledge than other types of scientific results. This may not be the obstacle it first appears to be, however, as neuroprosthetics are by nature collaborative efforts, and an individual investigator is likely to outsource the actual fabrication work. It then becomes necessary to know not how to do things, but only what can be done. A summary of the technological possibilities is then also an objective of this chapter. We will close with some speculation on the potential impact of interfacial molecular bioengineering, an emerging research focus that holds promise for enhancing the functional interface between an artificial device and living cells at the molecular level.

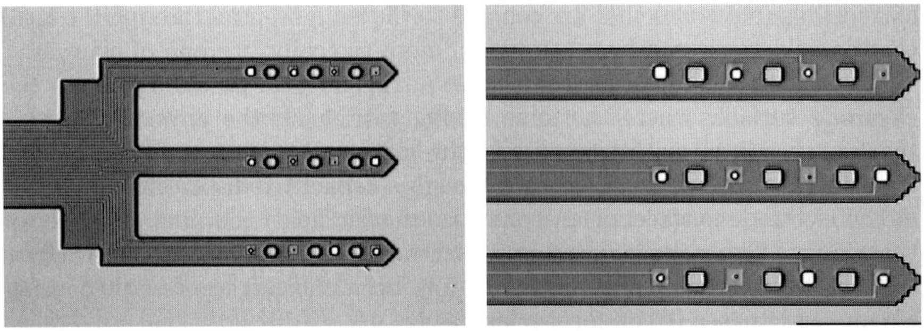

Fig. 18.1. Flexible microelectrode array developed by Pellinen et al. [1060]. *Left* and *right* panels show just one possible configuration, a trident with 12 active electrodes ranging in size from 10 × 10 μ microns to 40 × 40 μ. Three 50 × 60 μ holes (seeding wells) are located on each tine to accommodate tissue ingrowth in chronic applications. Each tine is 170 μ wide, substrate and insulator are polyimide. Each electrode is addressable via a breakout out of view to the left; a commercially available connector is used to make the connection to external instruments (The colour plate of this figure appears on page 372)

One approach being developed to reduce the relative movement between electrodes and soft tissue is to make the electrodes and their supporting substrates of the same stiffness as the surrounding tissue, so that as the brain flexes, so would the electrodes, and relative movement would be minimized. Pellinen et al. [1060] have reported development of a class of implantable microelectrodes designed for chronic recording in brain (see Fig. 18.1). The device consists of metal film microelectrodes and conductors patterned on a very thin substrate of polyimide, with the conductors being insulated with a top coating of polyimide. The polyimide can be made precisely to any thickness. Polyimide has been demostrated to be very biocompatable, durable in vivo, and amenable to surface modification techniques (described in the last section). Three-dimensional arrays of active elctrodes have been demonstrated using this fabrication technique, and it holds great promise for the field of implantable microelectrodes for both recording and stimulating.

The geometry of the recording electrode array will depend on the source of the signal of interest, i.e. where the source is located in the brain. The optimum placement of a recording electrode is adjacent to the neurons creating the signal. The microelectrode arrays described above would be optimal if high spatial resolution was needed to record the needed information to detect seizure onset, and would be suitable for surface or extremely superficial recording. If high spatial resolution is not required, or sampling over a larger area is required, a smaller number of macro (millimeter scale) electrodes may be sufficient, and would be easier for the surgeon to work with. Deep-brain recording would require a geometry compatible with surgical placement, and would probably be reminiscent of the deep-brain stimulating electrodes used for tremor control (a small number of cylindrical electrodes located sequentially along a shaft that penetrates the brain with the aid of a metal stylus). All of the commercially available electric stimulator devices described in Sect. 18.3 that make use of recorded bioelectric signals recorded those signals through the stimulating electrodes. In general, an electrode suitable for stimulating will generally be adequate for recording, but the reverse is not true.

18.3.2 Chemical Signals: Ion-selective Electrodes

Ion-selective electrodes, as their name implies, are sensitive to only certain charge-carrying species in the solution being measured. The most familiar ion-selective electrodes, like the intracellular electrodes described above, allow for the passage of ions across the electrode surface. However, unlike the open lumen of the glass capillary electrode, ion-selective electrodes use a semi-permeable membrane to isolate the solution being measured from the metal electrode at which the ion-electron transduction event occurs. The selectivity of such electrodes varies considerably, and is primarily a function of the specificity of the semi-permeable membrane. This is the principle upon which the ubiquitous benchtop pH meter operates (H^+ ion-selective), and similar devices exist to measure Na^+, K^+, Li^+, Ca^{++}, Cl^-, F^-, NH_4^+ and CO_2. Micro-scale ion-selective electrodes have recently been used in self-referencing probe systems to measure local concentrations of oxygen near cultured cells (commercially available from Diamond General Development Corp., Ann Arbor, MI). It seems reasonable that this technology could be incorporated into a long-term implantable device.

A second type of ion-selective electrode has been developed to measure concentrations of nitric oxide, and is based on a carbon fiber electrode modified with O-phenylenediamine (OPD) and Nafion [1061]. It is not immediately evident what advantage either of these electrode types might afford, but they should be kept in mind.

18.3.3 Alternate Signals: Optical

Real-time cortical activity has been measured using a variety of signal types, ranging from noninvasive functional magnetic resonance imaging (fMRI), to measures of inherent changes in optical properties of the cortical matter requiring a window to be opened in the skull. While MRI machines are currently approaching desk-top dimensions, it is as yet not feasible to consider this an implantable technology. However, optical technology is becoming ever smaller, to the point where entire optical benches are now being fabricated on a single silicon chip measured in millimeters. Thus, we will briefly consider this technology here as an emerging possibility, although with uncertain advantages.

As optical sources and detectors have evolved over the past two decades, it has become increasingly apparent that physiological substances and processes have inherent optical signatures. The inherent signature often consists of some change in the scattering property of the cell or tissue, as opposed to a luminescence. These inherent signatures are distinct from optical techniques requiring the use of dyes, such as the potentiometric fluorescent probes commonly used to monitor membrane potential or calcium concentrations (e.g. Fluo-4, Molecular Probes, Inc.). Because most current fluorescent probes (dyes) are phototoxic and subject to photobleaching, they are not ideally suited to long-term in vivo use. Thus, we will consider only inherent light-scattering optical methods here.

Penetration depth of long-wavelength light exceeds that currently available with silicon microelectrode arrays (2–3 mm). At approximately 1300 nm, there exists a window between the strong absorption bands of haemoglobin and water where light penetrates several millimeters in muscle tissue. Inherent light-scattering or birefringent properties of neural tissues have been reported, e.g. cortical [189,1055] and retinal [1062]. These techniques have so far been used to monitor activity at the tissue and cellular level. Optical monitors of cortex probably reflect activity-related recruitement of microvasculature. As an epileptic seizure is composed of the coordinated activity of large groups of neurons, it appears that optical monitoring may be well suited to this purpose. We are not aware of current efforts to develop an implantable optical monitor. However, sensor probes for optical coherence tomography, an infrared imaging technique, require only optical fibers and a prism, and have been fabricated small enough to be inserted into blood vessels. While the light sources and associated instrumentation currently required to record and analyze light scattering signals are substantial, this still seems to be a technology worthy of further investigation.

18.4 Processing Stage

The processing stage of our hypothetical SAI would contain the brains of the device, and the potential need for extensive computational requirements is one of the features that most distinguishes this device from existing vagal nerve or deep brain stimulators. Once the desired computational tasks of the SAI have been defined, the production of a device to perform the task becomes a matter of skilled bio-engineers and electrical engineers sitting down to the task. We cannot predict whether this device can be made small enough, with power requirements low enough, for implantation. As a point of reference, modern pacemakers use complementary metal oxide semiconductor (CMOS) circuit technology, requiring up to about 500 bytes of memory, which is housed (along with a battery and telemetry circuitry) in a case approximately 60 mm in dameter, and 10 mm thick.

The extent of the computational power required will depend on the methods of seizure detection and suppression chosen, and so here we will discuss in general, the minimal requirements of any SAI, namely to predict seizure onset, and then react with some stimulus. This requires that the processor at least coordinate the following functions:

- Sample the physiological parameter(s) being monitored,
- Condition the sampled signal,
- Analyze the sampled signal to determine the likelihood of imminent seizure,
- Make a decision as to whether to provide a stimulus, and
- Control the delivery of a fixed stimulus when warranted.

Additional computational tasks could include modulation of the delivered stimulus based on further input (in a closed-loop system), updating of parameter values that predict seizure onset (in an adaptive or learning system), accommodating a patient manual-override for initiation or cessation of the stimulus, and encoding/decoding radio-frequency (RF) signals for bidirectional telemetry for clinician reprogramming of the device.

All of the above operations could potentially be carried out on any desktop PC. With this in mind, the possibility of using telemetry, and thereby decreasing the amount of computational power that must be placed within the body, makes a first-generation device using this strategy much nearer at hand. As an example, several retinal neuroprostheses being developed require a substantial amount of processing of the image signal before it can be encoded into a stimulus pattern at the electrodes. These devices are being designed to include both external and internal components. Outside the body is a camera for sampling the visual scene, a signal processor that attempts to mimic the signal processing of the outer retina, and an RF encoder and transmitter. Inside the eye is an RF receiver/decoder, power converter and stimulating electrode array placed in contact with the retina. In this example, the bulk of the signal processing takes place in the external processor worn on a belt [1063,1064].

RF telemetry between the in vivo and ex vivo environments is commonplace in existing implantable stimulators. In the last ten years, bidirectional telemetry between cardiac pacemakers and external programmers has become standard, allowing the clinician to obtain information about implant function such as lead impedance, battery condition, and pulse parameters. The clinician can then in turn reprogram the implant to meet the changing requirements of the patient.

18.5 Output Stage

One of the assumptions upon which our theoretical SAI device is based is that we are aware of a cortical stimulation that will be effective at averting an imminent seizure. The two most likely stimulus classes are electrical and chemical. Once again, we need to emphasize that the specific parameters of such a stimulus are not yet well characterized, and so our discussion is restricted to an overview of the available technology relevant to each of these general classes. As with the input stage interface, the stimulating components must be biocompatible, durable in an in vivo environment, be small, and have reasonable power-consumption requirements. These requirements have all been met previously with devices in routine clinical use. Below we briefly describe the fundamental challenges, and list examples of successful devices that have met these challenges.

18.5.1 Electrical Stimulation: Mechanisms

Artificial stimulation of cortical neurons has been demonstrated using electrodes placed on the cortical surface, and by inserting fine electrodes into the cortex. Regardless of electrode position, the same principles determine the effectiveness of the stimulation pulse: the rate of change of the applied current, with the requirements that a minimum charge density and total amount of charge are reached. The pulse width, W, is related to the threshold current, I_{th}, by the relation

$$I_{\mathrm{th}} = \frac{I_{\mathrm{rh}}}{1 - \exp\left(-\mathrm{W}/\tau\right)} \tag{18.1}$$

and to the threshold charge, Q_{th}, by

$$Q_{\mathrm{th}} = \frac{I_{\mathrm{rh}} W}{1 - \exp\left(-\mathrm{W}/\tau\right)}\,, \tag{18.2}$$

where I_{rh} is the rheobase current, i.e. the smallest current required to cause excitation, and τ is a time constant describing the membrane and, for extracellular stimulation, the space constant of the extracellular resistance. These relations are well known and can be derived either experimentally [1065] or theoretically from the cable equations [1066].

Smaller electrodes, and intimate contact between electrodes and neurons, result in higher effective current density, and therefore lower power requirements. Increased surface area, accomplished by roughening the electrode surface at the molecular level, can dramatically increase charge-transfer capacity; this technique has been used on commercially available multielectrode arrays designed for both recording and stimulation of cell culture or tissue slice preparations (Multichannel Systems, Reutlingen, Germany, [1067–1069]). The continuing refinement of microfabricated electrodes for chronic recording and stimulation has opened up the question of how to construct an optimal functional electrical interface at the micro or nano level. Electrode parameters such as array configuration, electrode shape, molecular surface roughness, material, and surface derivitization must be further examined in the context of the intended application. Several research groups are combining computational modeling of microelectrode arrays and neurons, in-vitro measurements from neurons cultured on microelectrode arrays, and in vivo measurements from cells adjacent to implanted microelectrode arrays in order to answer these questions (e.g. [1070,1071]). While much of this effort is focused on the optimization of neuroprosthetic interfaces in the eye, the multi-faceted approach can be applied to any area of artificial neural stimulation. As interfacial molecular bioengineering redefines electrode-neuron interfaces (Sect. 18.9), we need to continue to search for new optimal designs.

The most effective stimulation site(s) will determine the configuration of the implanted electrodes. General quantities to be considered are number of stimulation sites, depth relative to the brain surface, and suitability for routine surgical placement. If the optimum stimulation sites are also suitable for recording bioelectric indicators of imminent seizure, then one set of electrodes could be used for recording and stimulating. Below we will consider the various electrode materials, with the understanding that they can each be incorporated into electrodes optimized for surface, shallow or deep stimulation, similar to the recording electrodes described above.

18.5.2 Electrical Stimulation: Metal Electrodes

Electrical stimulation would be delivered via one or more metal electrodes. An extensive literature describing the use of different metals to form stimulating electrodes was motivated by early cardiac pacemaker development (e.g. [1072,1073]). The suitability and effectiveness of a metal electrode for stimulating neural tissue depend largely on the electrochemistry of the electrode surface. Polarizable electrodes such as platinum and titanium do not easily produce charge carriers at the electrode-electrolyte interface, and so nearly all of the current passing from the electrode to the electrolyte is displacement or capacitive current. These inert metals also tend to be very durable in the body, and reasonably biocompatible. Gold, while slow to oxidize, has proven a poor choice for long-term in vivo applications as it is subject to dissolution. The displacement current of polarizable electrodes has been increased through techniques that greatly increase the surface area of the elctrode by roughening it on a molecular level.

Non-polarizable electrodes rely on faradaic mechanisms at the electrode-tissue interface. They operate near the equilibrium potential of the electrode, and above the threshold for reversible oxidation-reduction reactions to occur. Care must be taken not to exceed the threshold for non-reversible mechanisms (far from the equilibrium potential of the electrode), which may result in electrode corrosion and oxygen or hydrogen evolution. General considerations with any metal stimulating electrode in the body include the possibility of electrotoxicity; that is, are toxic ions released as a result of the oxidation and reduction reactions occurring as current is passed between the electrode and the surrounding environment. This is minimized by using short, biphasic

these materials are currently used in prototype retinal prostheses, and are the major structural materials in the flexible cortical electrodes described in Sect. 18.4.1.

The electronics of an implantable device must be protected from the infiltration of bodily fluids, and the industry standard packaging is a hermetically sealed casing (one from which no helium gas can leak). Techniques for hermetically sealing implantable devices are well established. All of the implantable devices described at the beginning of this chapter use titanium as the case material, which is both biocompatible and durable in vivo. An important feature of these implantable devices is a labeling scheme that can be interpreted non-invasively. This becomes important as it allows the clinician (who may be performing emergency care) to identify the type of device in the patient. Commonly used options are distinctive case shapes or radiopaque symbols or ID codes.

If electrodes are to be used for monitoring the bioelectricity of the cortex, and for providing a therapeutic stimulus, the materials of their construction must be carefully considered. Implanted electrodes consist of four major components – the active area where current passes between the electrode and the surrounding tissue, the conductor connecting the control circuit to the active area, the insulation around the conductor, and the connector where the conductor makes contact with the circuitry through the hermetically sealed case. The materials of the active area have been described above. Insulation of the leads for vagal nerve stimulators has been accomplished with silicone rubber. The conductors within the lead of deep-brain stimulators have been insulated with a fluoropolymer, and encased in a polyurethane tube. For penetrating microelectrode arrays, the structural material is silicon, which is quite biocompatable. The flexible planar electrode arrays being developed by Kipke at Arizona State University [1060] are metal films on a polyimide substrate. Polyimide is amenable to microfabrication procedures, and its use is sure to expand in chronic in vivo applications. Conductor materials are commonly gold, platinum or platinum/iridium. International standards have been established for connector pins used in cardiac pacers to facilitate lead replacement; similar connectors are used in the other devices described above.

18.8 Closing and Speculation

The essential component parts of an implantable cortical stimulator, namely a set of electrodes connected to control circuitry housed in a suitable casing, have been previously developed for other clinical applications. Cardiac pacemakers, vagal nerve stimulators, and cochlear implants have all reached a significant level of success. Implantable drug delivery devices for steady delivery exist, and implantable technology to allow controlled bolus drug release is being developed. Development of an implantable device that would counteract a developing epileptic seizure seems to be a natural extension of existing, FDA-approved devices, within the constraints described above.

The relatively new field of interfacial molecular bioengineering (IMB) is defining the forefront of a new approach to integrating artificial constructs into biological environments. Advances in microfabrication, cellular and molecular biology, biochemistry and biomaterials are supporting the emergence of a new generation of hybrid implants and replacement tissues and organs. Using a variety of techniques, we are beginning to design interfaces between artificial components and living systems on cellular and molecular scales.

While the field of IMB is still defining itself, a general theme of these efforts can be described as forming an artificial substrate or matrix, the surface chemistry of which is modified by coating with selected proteins, and which is then seeded with cells, in an effort to form a seamless interface between the artificial and the biological. The artificial component may serve simply as a structural guide, as when endothelial cells are cultured in a polymer mesh tube to form replacement blood vessels, or as an electronic device such as our hypothetical seizure aversion implant. In this case, the electrical stimulation of the implant may be directed at a population of "interface cells" that have been optimized to respond to a field potential with a release of neurotransmitter from neurites that have grown out into the surrounding cortical tissue. The benefit of this approach is that implants are not just tolerated by the body, but actually have improved functionality, becoming integrated into the natural physiology of the patient. Bioinformatics and genetic engineering are opening the door to biological implants for chronic treatment of diseases such as diabetes, where pancreatic islet cells can be engineered to over-express for insulin, and then be placed in the body in an artificial immunoprotective capsule. If growth factors or neurotransmitters effective at reducing epileptic seizures are some day identified, such a capsule (containing the appropriately designed cells) placed beneath the skull might be a possibility.

Even restraining ourselves to the currently available technology, it seems highly unusual that prototype cortical stimulators are not more common in the research community, not to mention start-up biotech companies. Returning to our original assumptions, the two intellectual issues to be resolved are identification of the appropriate physiological indicators of imminent epileptic seizure, and defining the appropriate stimulus to avoid the seizure. Once these have been accomplished, the development of a seizure-aversion-implant seems to be not far down the road.

19 Brain Defibrillators: Synopsis, Problems and Future Directions

J. Milton and P. Jung

The goal of this text was to challenge experts in epilepsy, bioengineering and dynamical systems theory to discuss the possibility of building a brain defibrillator to treat epilepsy. Of course nobody can claim that such a device exists, and for that matter is even possible. However, even the most skeptical reader must admit that surprising progress has been made on the three most important components of such a device: 1) seizure prediction; 2) treatment strategies; and 3) delivery of treatment strategies. Moreover, initial attempts at the feedback control of seizures in epileptic tissue (Chaps. 15 and 17) and the human brain (Chap. 14) have been surprisingly successful.

In this chapter we discuss some clinical and research problems which are as yet unanswered. We leave the readers to determine whether these topics lead to interesting and fruitful research or to a bottomless abyss.

19.1 Brain Defibrillator or Fibrillator?

By this stage, the careful reader will likely have begun to raise doubts about the choice of the term "brain defibrillator" as the name for a device which performs the feedback control of seizures. We initially picked this term by analogy with the use of programmable, implantable cardiac defibrillators used to treat ventricular fibrillation. Ventricular fibrillation represents a state with an irregularly irregular electrical rhythm coupled with disorderly desynchronized contractions of cardiac muscle. An electrical stimulus is delivered by the cardiac defibrillator to restore the orderly electrical rhythm and synchronized contractions of the heart muscle cells. On the other hand, during a seizure there is increased synchronization of massive numbers of neurons. This is why the EEG becomes dominated by high amplitude rhythmic changes in potential. Thus the goal of a brain defibrillator seems to be exactly the opposite of a cardiac defibrillator, i.e. it must deliver an appropriate stimulus to prevent this massive synchrony from occurring.

We could, of course, devote this chapter to the task of trying to pick a more appropriate and catchy name for a seizure-aborting device. For example, we could call it a "brain fibrillator". However, we can quickly find fault with this term: the initial part of the seizure involves a desynchonization of neurons; synchronization occurs slightly later. Thus the choice of the term "defibrillator" versus "fibrillator" versus "some other term" (e.g. Seizure Aversion Implant (SAI) introduced in Chap. 18) might depend on which stage of the seizure the aborting stimulus was to be delivered and which aborting mechanism was accessed.

Our philosophy to stick with the term "brain defibrillator" is based on two practical considerations. First, this term immediately conjures up in the mind of the clinically oriented reader and the patient with epilepsy what the device does. Second, since we cannot claim that we know how this device will work, it is likely better to initially use a term which all agree points to the wrong mechanism, rather than introduce a term which seems to be right at the moment, but ultimately leads investigators in the wrong direction!

19.2 What to Treat: Seizures or Spikes?

It is important not to equate the term seizures with epilepsy. A seizure represents a paroxysmal and transient alteration in the collective dynamics of a large neural network. On the other hand, temporal lobe epilepsy is a slowly evolving disease of the nervous system in which the tendency to have recurring seizures is the most dramatic manifestation (Chap. 1). Whereas a seizure can occur as an isolated event in a normal brain, temporal lobe epilepsy is a progressive neural network disorder which involves changes in both the local connectivity of neurons as well as alterations in the interplay between neurons located in the cortex and subcortex (Chaps. 1–3). In view of this distinction we can ask the question as to whether the goal of a brain defibrillator should be to abort seizures or to prevent the development of medically intractable epilepsy (MIE). In terms of a familiar metaphor, should we treat the weather or the climate?

The occurrence of seizures often constitutes a medical emergency. However, it is important to realize that most seizures have little long-lived significance for the sufferer. Unless, of course, the seizure occurs in the wrong place and at the wrong time, e.g. while driving, walking on a tightrope, giving an important business presentation. Thus we could consider epilepsy to be a "social disease" since the when and where of seizure occurrence may be more important than the seizure itself. It is the social consequences of seizure recurrence which drive current research into developing algorithms to predict seizure occurrence (Chaps. 12 and 13). The hope is that, given enough warning, the patients with epilepsy might be able to remove themselves from danger before the seizure occurs.

Much in the same spirit, the "brain defibrillator" attempts to abort a seizure when it occurs. However, despite all of the drama that surrounds the occurrence of a seizure, it is important not to lose sight of the big picture. Anyone who has followed patients with epilepsy for many years will be aware that temporal lobe epilepsy is much more than its seizures. It is clearly a disease that slowly and progressively eats away at memory, and results in an inevitable decline in socioeconomic status. This epileptic syndrome takes years to develop and likely continues to change through the lifetime of the sufferer (Chap. 1). These features of epilepsy attract little attention because they occur imperceptibly over time-scales much longer than those required for a seizure, or for that matter the tenure of a hospital administrator.

As the epileptic syndrome evolves, it becomes progressively less responsive to anticonvulsant medications, and eventually becomes completely resistant (Chap. 1). A fundamental unsolved question is whether the development of MIE is a function of the seizure frequency, the total number of seizures, or some other, less apparent, factor. Does better seizure control alter the subsequent development of MIE? It could equally be argued that the goal of a 'brain defibrillator' should be to prevent the occurrence of epileptic spikes in order to prevent MIE. In fact if preventing MIE is the major goal then it is not even clear whether the stimuli should be directed solely at neurons. Neuroglia are increasingly recognized as playing important roles in epilepsy (Chap. 4).

19.3 Seizure Prediction

One of the mantras of this textbook is that seizure prediction is necessary for seizure control. This is because a finite time is required to activate a brain defibrillator and deliver the seizure-aborting stimulus. As we have learned, linear time series methods appear adequate if the required lead time is < 5 min (Chap. 13); longer-range prediction appears to require the use of methods of nonlinear time series analysis (Chap. 12). A closely related issue is the identification of the critical time when a seizure is most vulnerable to being aborted. The use of sensory and electrical stimuli suggests that seizures can be aborted when the stimuli are applied just after a clinical seizure begins (Chaps. 15–17). An unsettled question is whether the application of controlling stimuli prior to clinical seizure onset is as effective.

A number of methodological problems have plagued research into seizure prediction (Chap. 12). The most important of these are the very small numbers of EEGs that have been analyzed, the fact that the same EEGs have not been analyzed by different groups, and the extent to which changes in the state of arousal of the subject affect the utility of the algorithms. A confounding problem is that EEG changes similar to those seen in patients with epilepsy occur, for example, in patients with migraines and those taking certain medications, e.g. antidepressants.

All of the studies on seizure prediction utilize measurements of the EEG. It is important to keep in mind that the EEG may not be that well suited to the task of seizure-prediction in the context of the development of a brain defibrillator (Chap. 5). The inclusion of the filters and differential amplifiers necessary to measure the EEG would likely quickly overwhelm the resources of an implantable brain defibrillator (Chap. 18). Moreover, the EEG is not very sensitive to seizures which arise in many areas of the brain, including the amygdala and the depths of the cortical sulci (Chap. 5). However, the EEG is not the only technology that can be used to monitor brain activity in order to predict seizure occurrence. Other possibilities include measurements of neural activity with extracellular electrodes and the use of optical methods. These latter methods are fundamentally different from the EEG since they directly measure neural spiking. Can the results from current seizure prediction algorithms based on EEG be directly extrapolated for use with these other technologies? Finally, it may be important to recognize that specially trained dogs seem to do quite well in predicting seizure occurrence without direct access to the EEG or neural spiking!

19.4 Sudden Changes in Neural Dynamics

The hallmark of epilepsy is the paroxysmal nature of seizure occurrence. In mathematical models, there are few mechanisms that can produce sudden changes in qualitative dynamics: 1) changes in control parameters (e.g. bifurcations); 2) switching between attractors; and 3) changes in the relative frequency of neural oscillators. The paroxysmal nature of seizure occurrence provides an important insight into the nature of possible treatment strategies.

Sudden changes in qualitative dynamics occur in mathematical models when the value of an important control parameter is shifted into a critical range. For example, as a threshold is crossed a stable steady state becomes unstable and is replaced by a stable limit cycle. Thus we could use the brain defibrillator to change parameters back into ranges associated with healthy brain dynamics. This is the idea behind the concept of a dynamic disease [94,106–111]. Unfortunately, the identity of the critical control parameters for epileptic seizures is not known. However, it is of interest to note that many of the putative control parameters, e.g. extracellular K^+, pH, local fluctuations in neural transmitter, are regulated by the activity of neuroglial cells (Chap. 4). This raises the interesting possibility of developing treatment strategies based on altering the activity of these cells.

19.5 The Case for Brief Stimuli

Since the critical control parameters for seizure generation are not known, investigators have shifted their attention to the possibility of developing control strategies that exploit the properties of the dynamics that arise after the seizure develops (i.e. after the bifurcation has occurred). Two examples of this approach have been discussed: 1) the control of multistability (Chap. 16); and 2) the control of chaos (Chap. 17).

The observation that brief sensory or electric stimuli can abort a seizure suggests the possibility of multistability (Chap. 16). Theoretical studies suggest that in models of populations of neurons the onset of self-maintained traveling wave phenomena is heralded by a subcritical Hopf bifurcation, i.e. there is a range of parameters for which a stable steady state coexists with a stable limit cycle. When time delays are present, this bifurcation can lead to multiple limit cycles. Which attractor is observed depends on the initial conditions. The basins of attraction for each of the stable attractors are separated by an unstable separatrix. Brief stimuli change the initial conditions and hence switches between the basins of attraction become possible. For example, analysis of EEGs has shown that spontaneous switching between attractors has been observed in the early stages of the development of temporal lobe seizures. If seizure development is associated with some, but not all, of the basins of attraction then a carefully honed and timed stimulus could be used to switch neural dynamics from a bad basin of attraction (i.e. one associated with a seizure) to a good one (i.e. a basin of attraction associated with normal brain dynamics).

A second situation in brief stimuli that can be used to affect control arises when chaotic dynamics arise (Chap. 17). When chaos arises an infinite number of unstable periodic orbits are present. Suppose that we choose the initial conditions so that we were on the Period 2 unstable orbit. We could follow this trajectory and every time that the dynamics begin to deviate from Period 2 we could apply a small perturbation to drive it back onto the Period 2. This is the basis of control of chaos. Of course, if we wish to disrupt a seizure we might want to design our stimuli so that the system never settled down on a periodic orbit for a long time (anticontrol of chaos). This strategy would prevent massive synchrony of neurons from occurring.

Although from a theoretical point of view multistability and chaos are quite different, from a practical point of view it would be very difficult to distinguish a noisy dynamical system possessing a large number of stable attractors (multistability) from a chaotic system possessing an infinite number of unstable periodic orbits. The distinction ultimately rests on whether plausible physiological mechanisms can be identified. In the case of chaos, evidence in support comes from time series analysis of the EEG and other neural signals; however, a neurophysiologically plausible mechanism which generates chaos has not been convincingly identified. The evidence for multistability is stronger (Chap. 16); however, it has not been shown that noise-induced switching between attractors can account for all of the time series phenomena reported from analysis of the EEG. Of course it is possible that one of the coexistent attractors may itself be chaotic.

Both of these approaches rely on the use of brief perturbations to effect control. Thus these methods are very attractive for application in a brain defibrillator. A disadvantage of these methods is that the dynamics must be very precisely monitored and the stimuli very carefully designed. It is not yet clear how such techniques could be applied at a population level in which considerable heterogeneity is present and for which information about population dynamics is only partially known. This problem may be partially offset by the development of methods to identify unstable periodic orbits (UPOs) (Chap. 17). Although the application of these methods to chaotic dynamical systems has been stressed, they likely apply, with little modification, to the control of noisy multistable dynamical systems as well.

19.6 Changing Neural Synchrony

EEG measurements suggest that the initial stage of seizure development involves desynchronization of neural activity (Chap. 5). Subsequently, synchronization of massive numbers of neurons occurs. Measurements of EEG coherence indicates that this synchronization does not involve the neural population uniformly (Chap. 6). Instead the cortical surface becomes tessellated with regions having differing extents of synchrony. The basis for this tessellation is not known; however, it may be related to either the histological regionality of the cortical surface (i.e. Brodman's areas) and/or to the somatotopic organization of projections of the underlying subcortical structures. Clearly one way to disrupt seizure development would be to interfere with the desynchrony-synchrony transition.

The degree of synchronization between neural oscillators depends on their relative frequencies and the strength of coupling between them (Chaps. 7–8). Very small changes in the relative frequencies can result in dramatic changes in synchrony. This observation opens the door for rate control of neural synchrony (Chap. 8). For example, periodic, or in principle aperiodic, electrical stimuli could be used to desynchronize the neural population. It is quite possible that subcortical nuclei, such as the caudate and dentate nuclei, exert their seizure-aborting effects in a similar manner (Chaps. 3 and 14). This possibility has also been discussed in the context of the use of deep brain stimulation to treat some of the symptoms of Parkinson's disease.

19.7 Controlling Waves of Neural Activity

The propagation of seizures as monitored by EEG does not seem to behave in the same way as expected for wave propagation in excitable media (Chaps. 9–11). In an excitable media, the leading edge of the traveling wave is trailed by a region of refractoriness. In contrast, during the propagation of a seizure the trailing edge has the same rhythmicity as the leading edge (see Figs. 5.7, 5.8, 6.2 and 6.5). There does not even seem to be a substantial time lag (i.e. a time lag > 1 ms) between the rhythmic activity measured by electrodes at the "front" and those nearest the "source". Moreover, when a seizure stops the EEG rhythmicity typically disappears over the involved cortical areas simultaneously. The EEG changes imply that during a seizure there is a propagation of synchrony over the cortical surface (Chaps. 5 and 6). These observations support the possibility that synchrony in cortical neurons is driven by a common periodic source, perhaps a subcortical nucleus (Chap. 3). If this is the case then perhaps the search for the wave phenomena characteristic of excitable media needs to be extended to an examination of the dynamics of neurons in the subcortical nuclei during seizures.

On the other hand, recent optical imaging studies of epileptic brain and brain slices have pointed to wavelike properties of epileptic propagation at the level of "small" neuronal populations (Chap. 10), i.e. via Pathway 1 in the terminology of Chap. 2. Does this observation imply that the propagation of activity waves and neural synchrony are not tightly coupled phenomena? Does the existence of different propagation pathways for seizures (Chaps. 2–3) imply that different dynamics occur simultaneously during the seizure on different temporal and spatial scales? Do different methods for measuring seizure propagation differ in their ability to monitor the activity in these various pathways? It is important that the reader keep in mind that the changes in the EEG recorded during a seizure may represent the activity of only a few percent of the underlying neurons (still a very large population!). This begs the question: what are the other neurons doing during the seizure? Analogies to the observation that cellular activity and synchrony are different arise in the context of the phase locking of oscillators (Chap. 8) and in the case of electromechanical dissociation, a nearly always fatal cardiac arrhythmia.

One way in which different dynamical phenomena can coexist in excitable media arises when time delays are present. Time delays arise in the brain because central myelinated and unmyelinated axons have differing conduction velocities (Chaps. 2 and 16). Little work has appeared on the characteristics of wave phenomena in excitable media with retarded variables. It could be argued that this is a major oversight since propagation delays must be a component of all real spatially extended dynamical systems (U. an der Heiden, personal communication). Recent observations suggest that the presence of time delays increase the spatial complexity of excitable media and can result in irregular wave propagation, including multistability and chaos.

19.8 Electrical Versus Chemical Stimuli

In principle there are two types of stimuli which can be used to abort a seizure: 1) chemical, and 2) electromagnetic.

The advantage of a chemical stimulus, e.g. a bolus of anticonvulsant medication, is that most physicians are familiar with the dogma "see a seizure, inject a benzodiazepine". Thus the idea of injecting a small amount of anticonvulsant medication directly into an epileptic focus or its blood supply to abort a seizure is an attractive one. However, there are a number of potential disadvantages. First, the activation and drug delivery time of a drug pump requires that sufficient time be available. This puts a greater strain on seizure-prediction algorithms than for electrical stimuli (see below). Second, once delivered the dispersal of the seizure aborting drug is primarily determined by diffusive processes which themselves are slow. The fact that diffusion plays a role in the delivery of the aborting stimulus adds yet another

their occurrence (Chap. 13). Neither of the two attempts the feedback control of seizures; self-adjusted stimulation of the caudate and dentate nucleus (Chap. 14) and the "brain defibrillator" use of electric fields (Chap. 15) involve the use of nonlinear control methods. Finally even some of the control of chaos methodology discussed in Chap. 17 involves the application of linearization methods.

These observations do not imply that nonlinear phenomena need not be considered in the design of a brain defibrillator. Linear approaches provide a benchmark against which advances offered by nonlinear approaches can be measured. The final construction of a workable brain defibrillator will likely very much depend on the details (Chap. 18). Obviously a brain defibrillator can only be so big, and will have restricted computational abilities and power supply. Trade-offs will need to be made between the costs of monitoring the signal versus the costs of aborting the seizure. It is in this framework that the relative benefits of linear versus nonlinear control approaches must eventually be judged.

19.11 Problems

19.11.1 Methodological Woes

The major output of the superficially located neurons in the cortical gray is vertical and hence away from the cortical surface. A variety of studies emphasize that many features of epilepsy reflect an interplay between cortical neurons and those located in subcortical structures including the brainstem. How is this interplay to be monitored? Presently, the only technique which is acceptable for patient care and which has sufficient spatio-temporal resolution is the EEG. However the EEG is totally incapable of directly monitoring the dynamic interplay between cortex and subcortex unless supplemented with the use of subcortical electrodes. It is difficult to imagine how the true nature of the epileptic syndrome can be understood without the development of methods capable of monitoring activity at all levels of the nervous system simultaneously. It is equally difficult to imagine how such a device could be developed.

On the other hand it is quite possible that we do not need to know all of this detail in order to accomplish our goal of developing a brain defibrillator. The reports of successful self-regulated deep brain stimulation for seizure control (Chap. 14) and the recent feedback control of seizures by manipulating electric fields (Chap. 15) are encouraging.

19.11.2 Measurement Guided Mathematical Models

Mathematical models of neural populations typically assume that the activity of single neurons, e.g. neural spiking, is the variable which is measured. However, the only methods for the study of human epilepsy which are available, safe for patient care, and possessing sufficient spatio-temporal resolution, are those based on EEG. However, the EEG is not a measure of neural spiking, it is a measure of neural synchrony. To make matters worse it is a statistical measurement of the properties of a large population of neurons. This observation points to the need for better mathematical models of neural populations which focus on predictions which can be measured using the EEG.

19.11.3 Stage of Patient's Epileptic Syndrome

It is sad that in the United States the primary factor in selecting patients as candidates for epilepsy surgery is based on their socioeconomic status. Can the patient afford the surgery and presurgical workup? Is the patient aware that epilepsy surgery is an option? Presently, patients undergoing a temporal lobectomy are simply classified as left versus right, neocortical versus hippocampal onset. Consequently there is essentially no information available to determine why surgery seemed to be successful for one patient and not for another.

We have emphasized that the epileptic syndrome continually evolves over the lifetime of a patient with epilepsy. It follows from this observation that in order to rationally compare the effectiveness of a therapy it is obviously necessary to know at which stage the illness was when treated. This is equivalent to asking whether we are dealing with a discrete epileptic focus or a spatially extended epileptic system. A crude method for estimating the stage might be to use the change in seizure frequency (estimated, for example, from seizure diaries) following introduction of a new anticonvulsant medication. The time it takes for the seizures to become refractory to this medication would be expected to be inversely related to the extent to which an epileptic system has developed. An objective measurement might be to use depth electrodes to evaluate the participation of subcortical nuclei in the development of the seizure.

19.11.4 Animal Models

The development of a brain defibrillator would be greatly facilitated by the availability of a suitable animal model for medically intractable epilepsy. Here we do not debate the relative strengths and weaknesses of the various animal models of epilepsy that have been developed. We draw attention to the fact that the same lack of characterization of the epileptic syndrome as discussed above is the characteristic of these studies.

19.12 Conclusion

This textbook began innocently at a small symposium on wave phenomena in excitable media which was held in conjunction with the March meeting of the American Physical Society in 1997. Following presentations by Drs. H. Levine and A.S. Mikhailov, one of us (JM) presented data on the propagation of human epileptic seizures measured using subdural grids of electrodes placed directly on the surface of the brain. It was argued that many of the properties of seizure propagation seemed to differ in many ways from those expected from studies of other excitable media. In the audience was an editor from Springer-Verlag (Claus Ascheron). He approached us at the end of the symposium and suggested the possibility of putting together the symposium in the form of a textbook. The goal of the textbook would be to introduce physical scientists to problems in epilepsy and epileptologists to ideas from physics. Being naive and not anticipating the four year's effort that this project would take, we hastily agreed.

The next 1–2 years were quite disappointing since only two of the original presenters of the APS symposium (Drs. Levine and Mikhailov) actually submitted their chapters. At our lowest point of despair we were suddenly struck by the concept of the "brain defibrillator". Certainly great progress has been made in both epilepsy and in the control of complex dynamical systems. Thus rather than restrict the book to presenters at the original APS symposium we decided that we would recruit experts whose speciality was related in some way to the construction of this hypothetical device. We proposed this idea to Springer-Verlag and they agreed. In retrospect it is likely that since they had yet to receive something from us, they had nothing to lose!

Over the next two years the book steadily grew from its initial conception of 8 chapters to its present 19 chapters. We thank all of the contributors for putting up with all of our pleading and in many cases putting a great deal of effort into an idea that was only briefly discussed to them over the telephone or by e-mail. Moroever, we thank those authors who submitted chapters nearly three years ago for their patience in waiting for the addition of the more recent chapters. It is quite possible that the process of adding new chapters, and hence new perspectives, might have continued indefinitely had it not been for our long-suffering wives who had become sick and tired of listening to us talk about the book. Enough already!

Depending on one's point of view, a workable brain defibrillator is either just around the corner or a silly pipe dream. However, dogs can predict seizures and mothers can treat abort absence seizures in their children by simply shaking and talking to them at the right time. Surely a brain defibrillator must be possible. The only unanswered question is what form the device will take.

Color Plates

Chapter 4

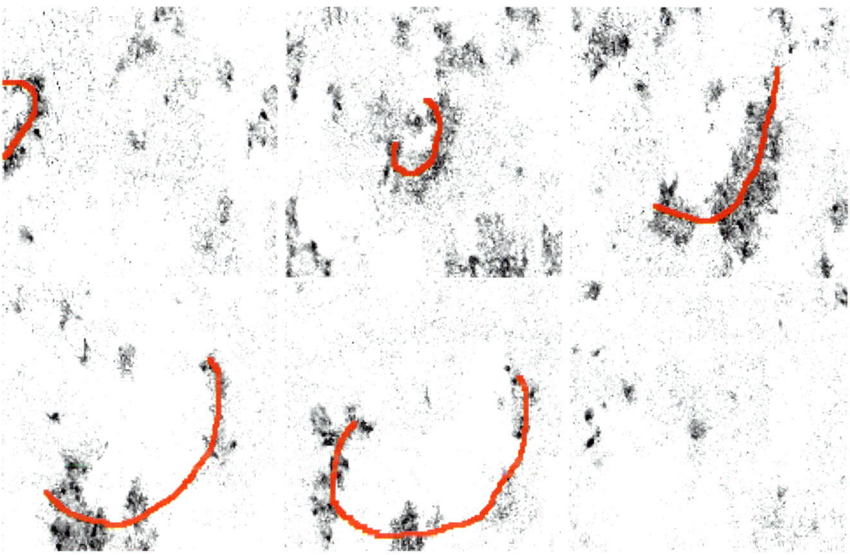

Fig. 4.2. Three positive-gradient images (corresponding to the images in Fig. 4.1 with an image delay of $n = 10$ images (17 s) are shown. The gray-scale has been inverted: *darker pixels* correspond to higher calcium activity. The waves are strongly fractured; the *red lines* indicate how the fragments belong to one large wave that fades out in the last frame

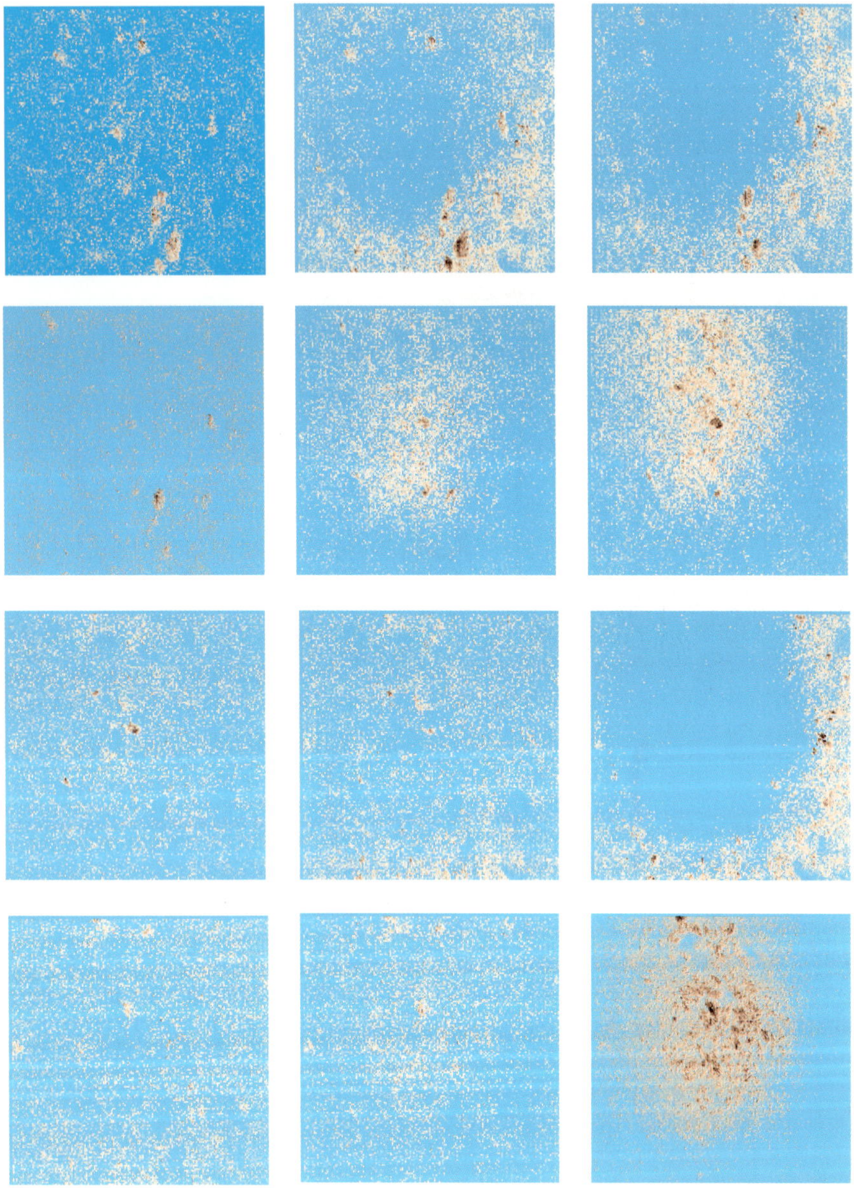

Fig. 4.6. A sequence of subtracted snapshots of calcium activity in a culture of astrocytes from TS-tissue is shown from *left top* to *right bottom*. The time interval between two snapshots is 1.7 s. Higher calcium activity is coded in larger gray-scale. One can observe two dramatic events in the form of a large-scale organization of calcium activity in form of circular waves

Chapter 6

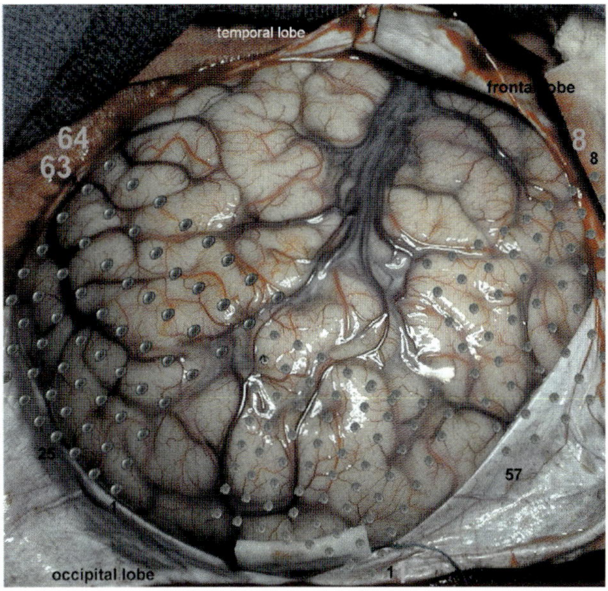

Fig. 6.1. A photograph of the exposed surface of the brain revealing the placement of the 8×8 grid relative to the abnormal gyral morphology. Electrode 64 is over the left inferior temporal lobe; prominent blood vessels indicate the left Sylvian fissure

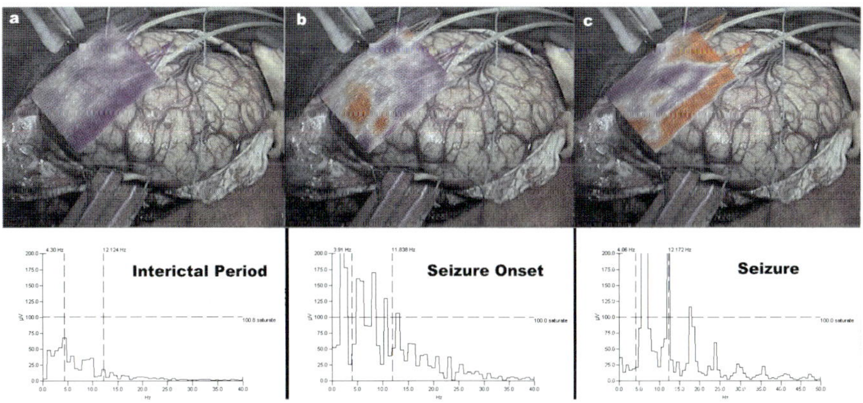

Fig. 6.3. Changes in ECoG power during the course of a seizure. The seizure (*red*) begins in the posterior aspect of the grid (**a**) and spreads along its superior edge and covers the anterior edge of the grid. The successive epochs are labelled (**a**–**c**) and correspond, respectively, to the ECoG segments shown in Fig. 6.2

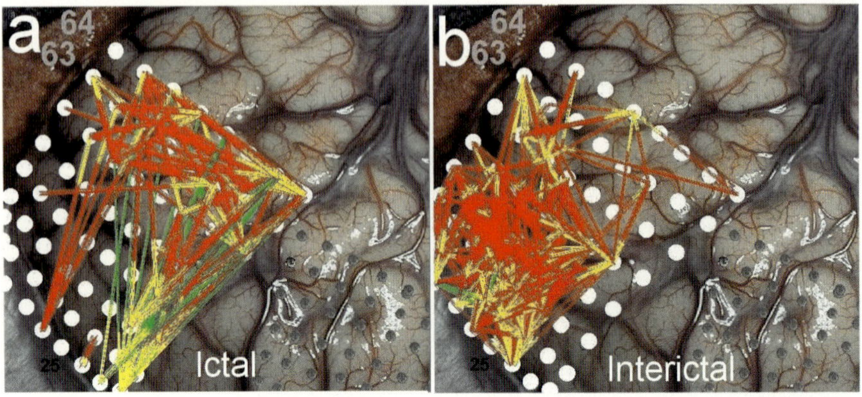

Fig. 6.4. Coherence patterns recorded during the ictal (**a**) and interictal (**b**) periods were largely complimentary, suggesting that during the interictal period the areas that would seize were disorganized and became synchronous during the seizure

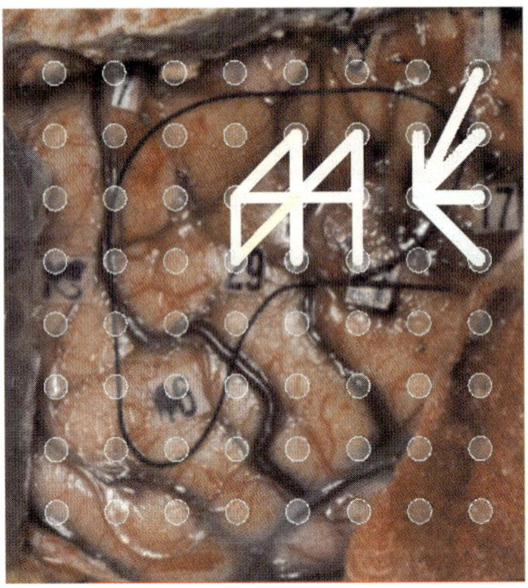

Fig. 6.8. A portion of the interictal coherence patterns calculated in the delta band recorded from an epileptic frontal lobe. The area to be resected is marked with a black string. Within that area, there was a region between electrodes 18 and 19 where the coherence was dramatically reduced. This region had a striking reduction in staining, suggesting a reduction in the number of viable neurons

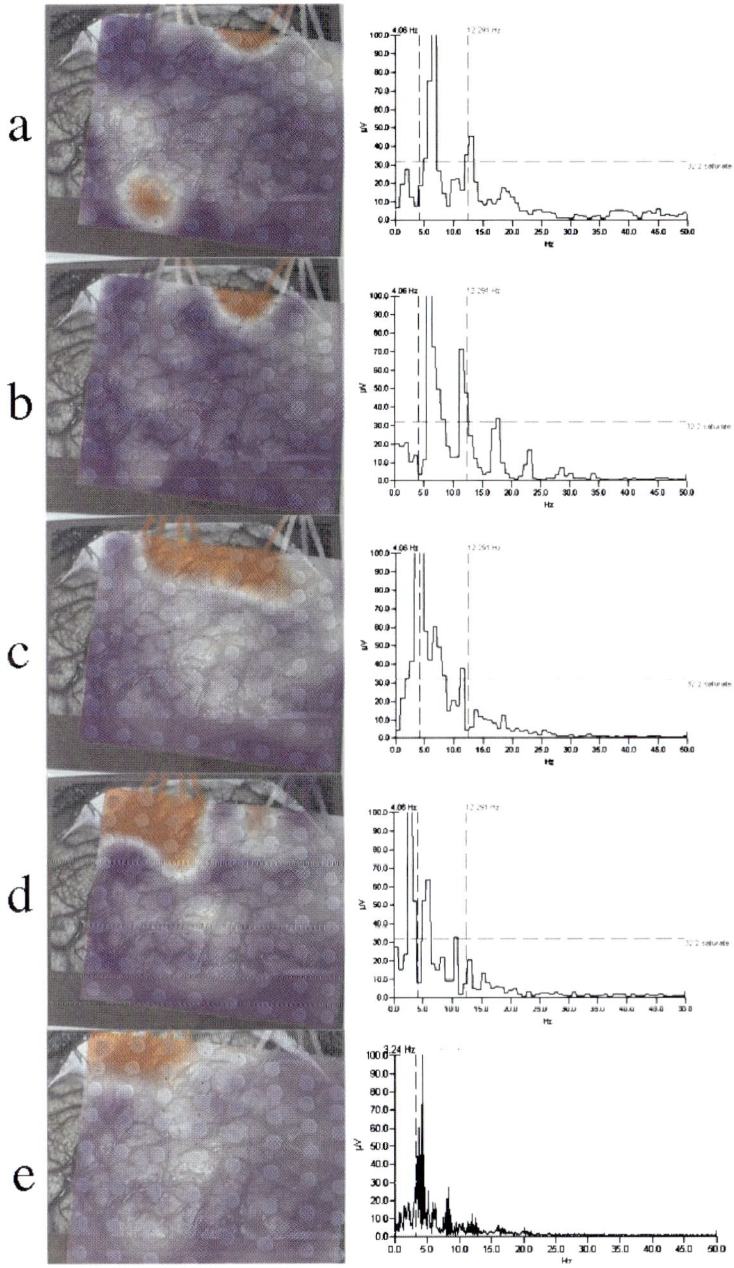

Fig. 6.6. The distribution of increased power (*red*) in the theta and alpha bands as a spontaneous seizure progresses. The seizure moves along the anterior edge of the grid (**a**–**c**) and plays out in the initial area while continuing to progress in the subsequently involved area (**d**–**e**). The successive epochs are labelled (**a**–**e**) and correspond, respectively, to the ECoG segments shown in Fig. 6.5

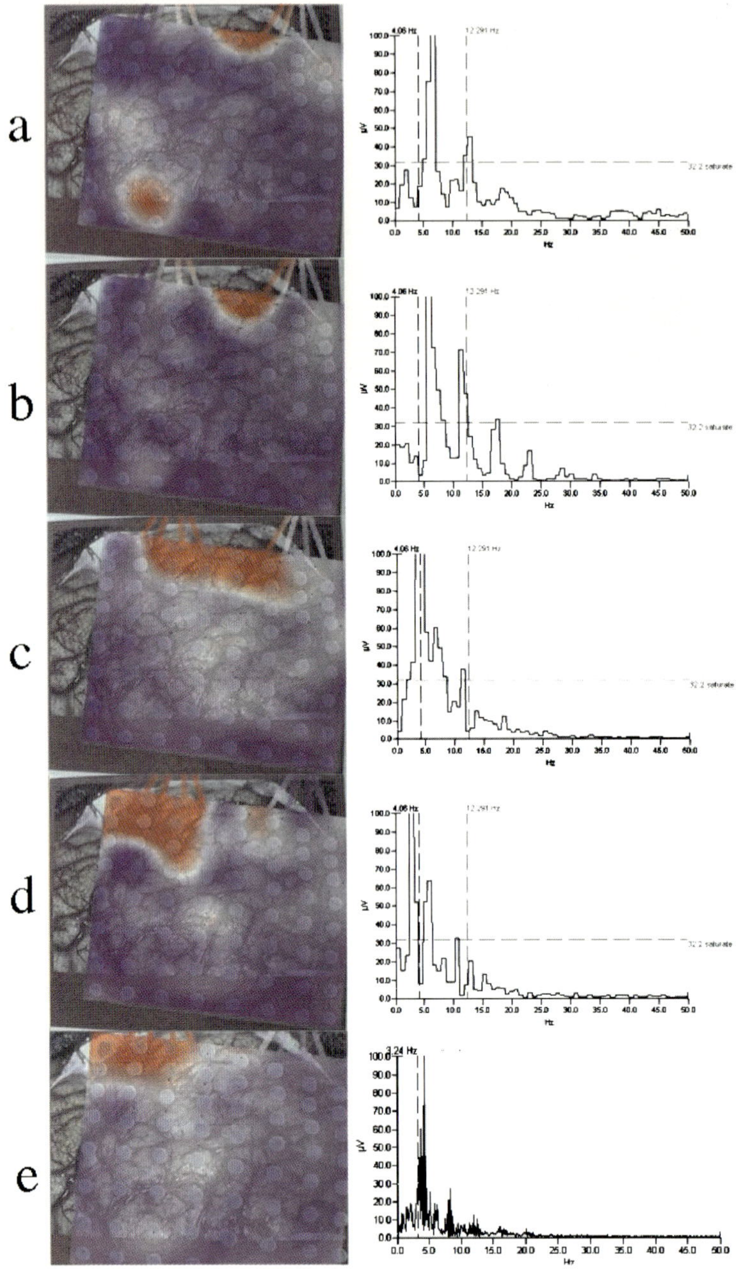

Fig. 6.7. Coherence patterns during the interictal (**a**) and ictal (**b**) periods. Because this seizure did not behave as a unitary phenomenon it does not reveal increased coherence during the seizure. However, during the interictal period before the seizure began, there is increased coherence in the region that would subsequently seize

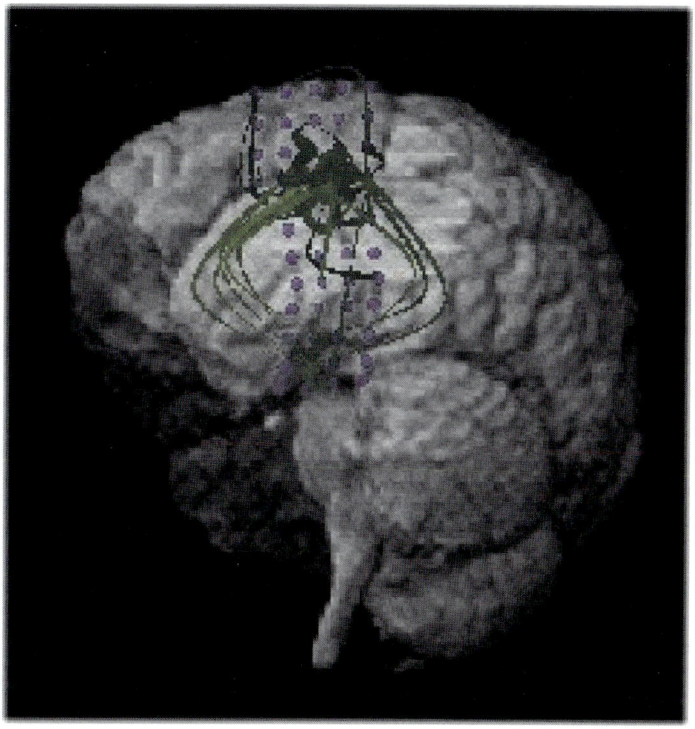

Fig. 6.10. Coherence patterns suggesting that an area near a mesial temporal lobe lesion was driving a distant area on the superior temporal gyrus that was related to expressive speech

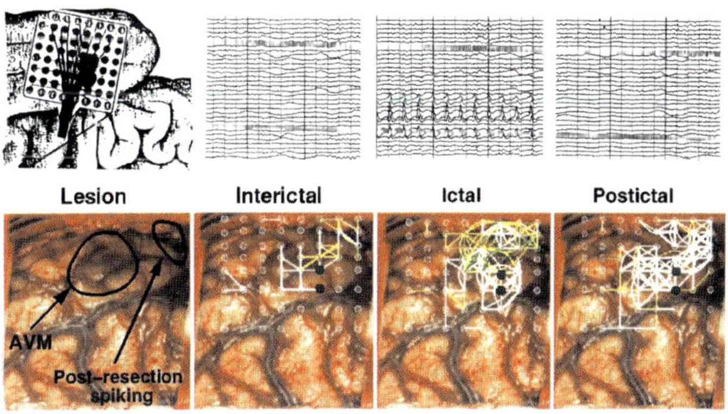

Fig. 6.11. Raw ECoG (*top*) and coherence patterns (*bottom*) observed before and during an after-discharge. The discharges in the conventional recordings were localized in the area of the AVM, but the coherence revealed functional connections between the site of the stimulation (*green dots*) and an area that began spiking only after the resection was completed. The interictal background coherence pattern heralded the appearance of the post-resection focus

Chapter 8

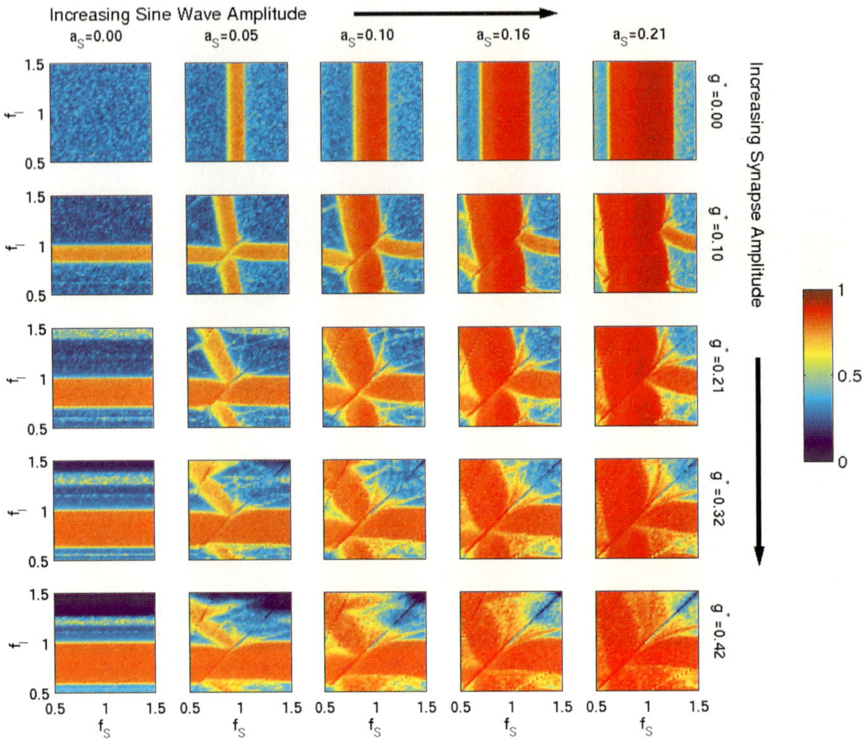

Fig. 8.9. Investigation of the synaptic control of synchrony in the Morris–Lecar model. A periodic inhibitory synaptic input is used to control synchrony in a population of sinusoidally driven neurons. The two experiments in Fig. 8.8 correspond to two points in the parameter space investigated here. Colorbar to the right indicates the population synchrony. *Parameters*: synaptic frequency f_I, conductance g, sine frequency f_S and amplitude a_S

Chapter 11

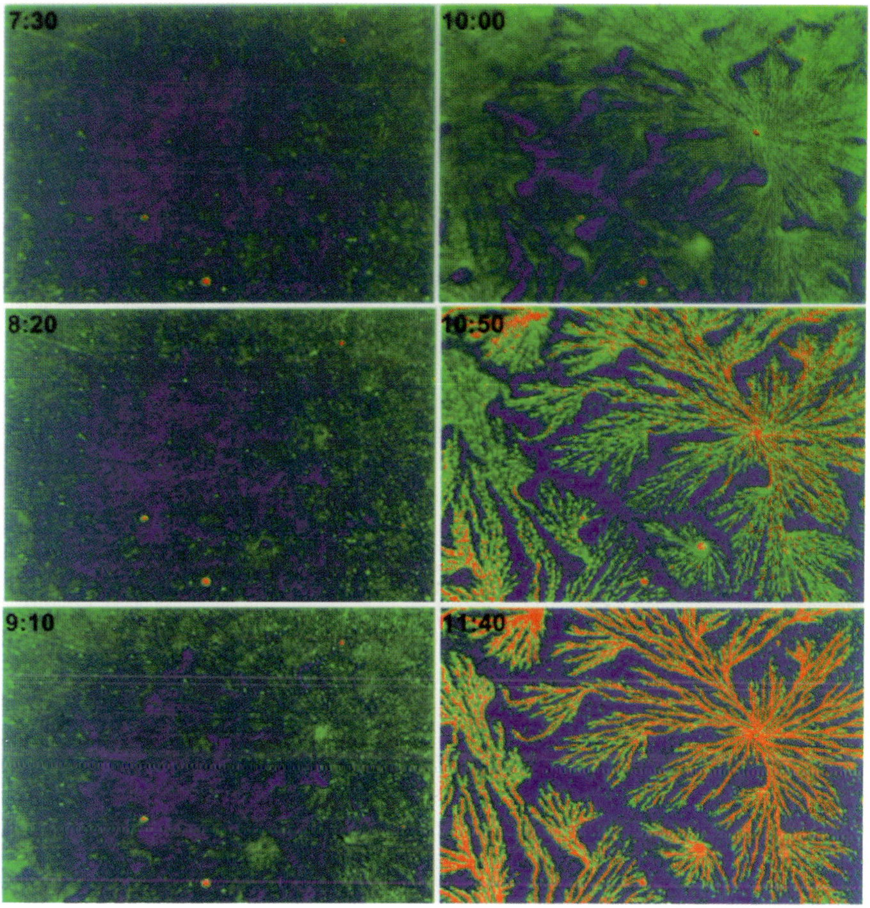

Fig. 11.7. Series of cell density images showing the formation of streams starting from an initially homogeneous density (courtesy of R. Kessin)

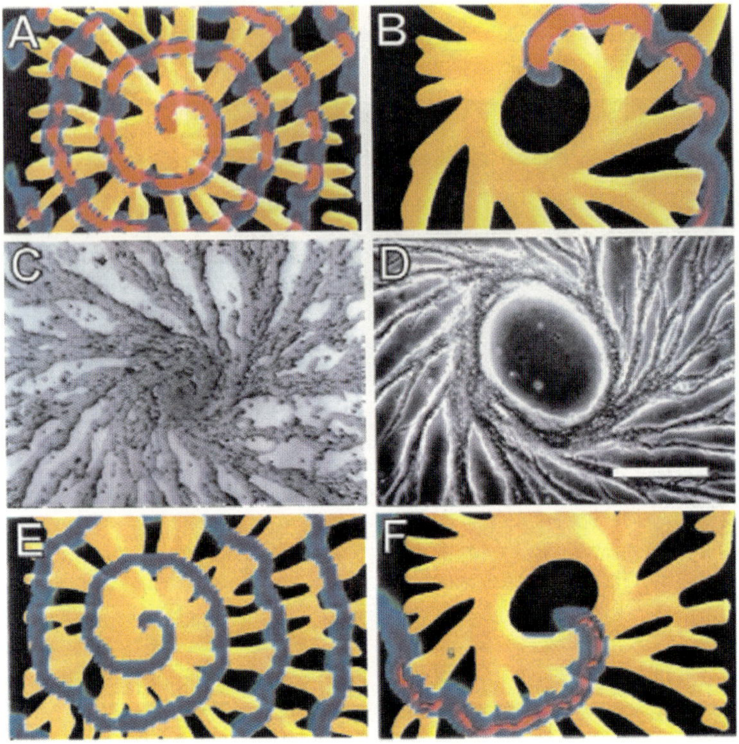

Fig. 11.8. Close-up view of the aggregation center in experiments (**C,D**) and a model (**A,B**; also **E,F**) due to Vasiev et al. [707,708]. The hole has been enhanced in going from **A** to **B** and **C** to **D** by lowering the excitability of the system, accomplished experimentally by adding caffeine. (Courtesy of C. Weijer)

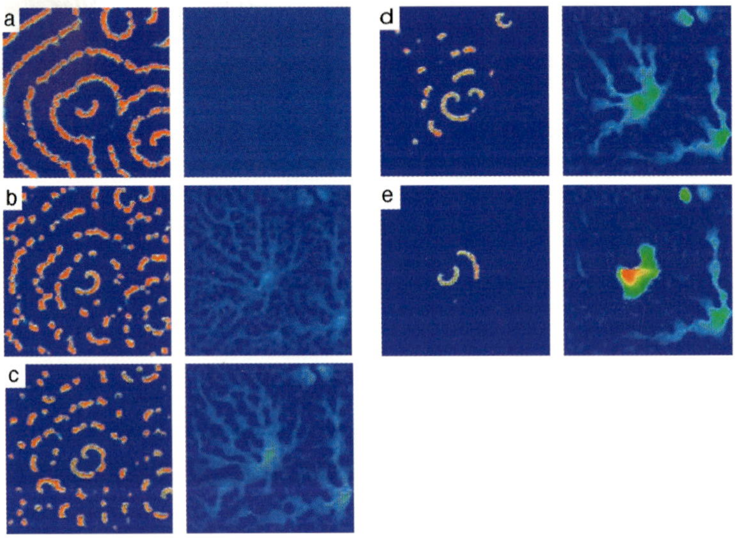

Fig. 11.9. Stream formation in the simple automata model – taken from [699]

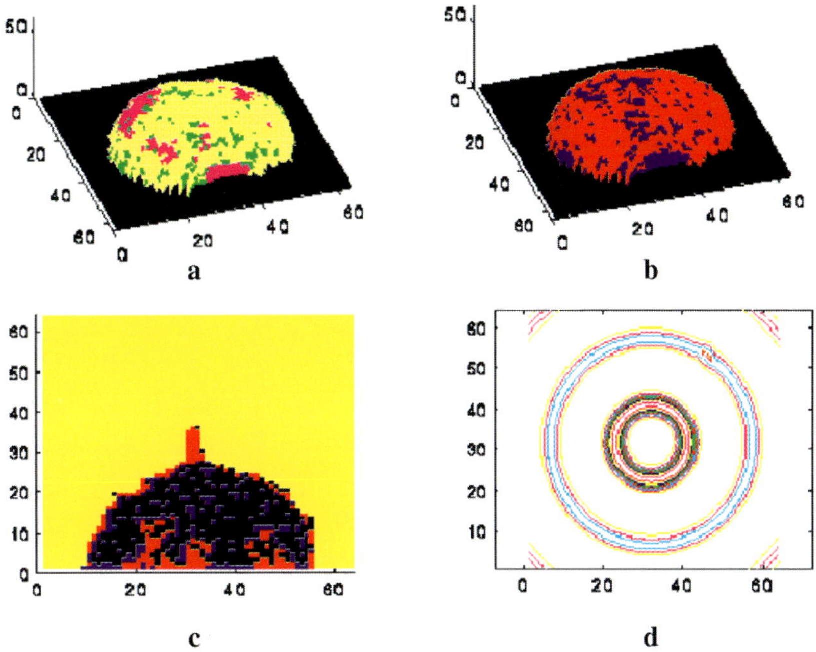

Fig. 11.13. Snapshot of a simulation with the Potts model framework, showing sorting and tip-formation due to a combination of differential adhesion and chemotaxis. Part (**a**) shows the three dimensional surface plot of the mound, color codes indicate the different states of cells: *green* for quiescent, *purple* for active and *yellow* for refractory. (**b**) shows the cell type distribution on the surface of the mound: *blue* represent pre-spore cells and *red*, pre-stalk cells. (**c**) is a vertical cross section through the mound; we can see *red* (pre-stalk) cells forming the protruding tip. (**d**) shows the chemical concentration (a view from the top) at the surface of the substrate

Chapter 13

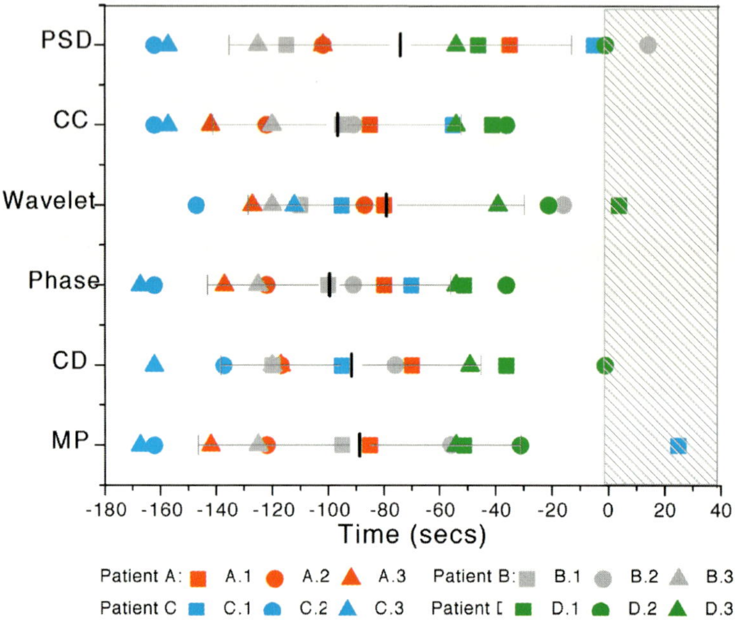

Fig. 13.3. Difference between threshold crossing time and neurologist's judgement of electrographic seizure onset. Results for all twelve seizures are shown for each method, with one color corresponding to each patient, and one shape corresponding to the seizure number. One standard deviation is indicated by the *horizontal bar*, with the *vertical bar* at its center indicating the mean result over all seizures for that method. The *vertical gray line* at 0 s corresponds to the time given by the neurologist as electrographic seizure onset. This time preceded clinical seizure onset by an average of 25 s

Chapter 15

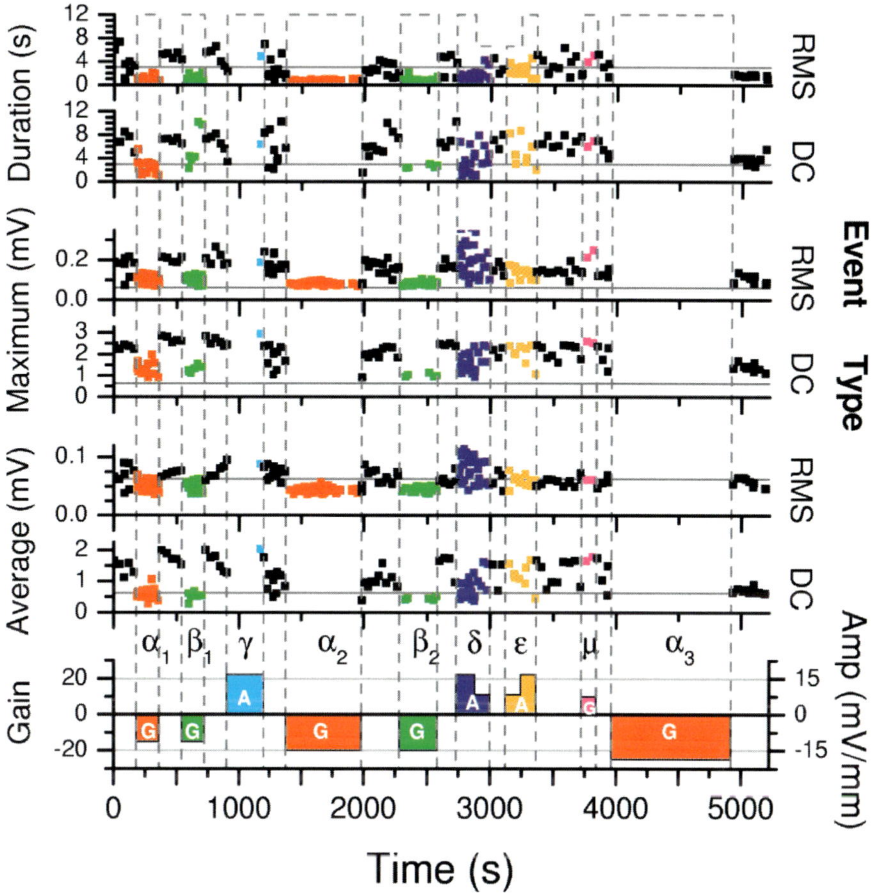

Fig. 15.7. Event detection results for a single 90-min. recording, with different electric field stimuli applied. The lower trace indicates feedback gain (G, left axis) or amplitude (A, right axis) of the applied stimulus. Greek letters and color indicate type of stimulus: baseline (*black, no letter*); full-wave feedback control (*red*, α); half-wave rectified feedback control (*green*, β); constant-amplitude suppressive field (*light blue*, γ); low-frequency noise (*dark blue*, δ); suppressive half-wave rectified low-frequency noise (*orange*, ϵ); positive feedback control (*magenta*, μ). Two types of event detection were used to identify synchronous neuronal activity from the recorded field potentials. "RMS events" were detected from variations in the rms power in the frequency band 100–350 Hz. "DC events" were detected by threshold detection after low-pass filtering the recordings at 10 Hz. The character of both types of events, as quantified by their average and maximal amplitudes as well as their duration, was visibly changed from baseline when control was applied. No events of either type were observed during the final and longest (16 min.) application (α_3) of full-wave control. (Adapted from [939])

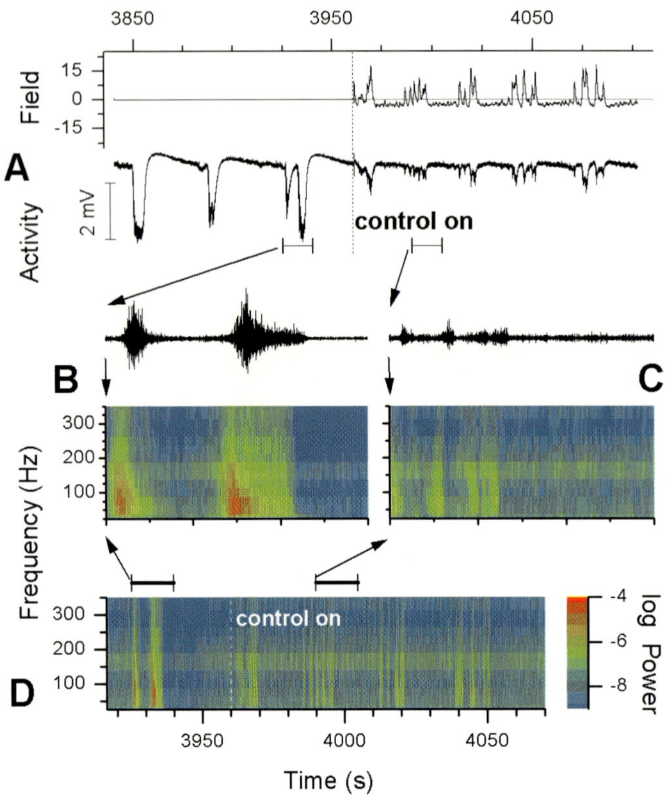

Fig. 15.8. Traces and spectrograms of activity with and without control for same experiment as Fig. 15.4. (**A**) Activity (*lower trace*) and applied field (*upper trace*) from the final application of full-wave control (α_3) from Fig. 15.4 and the baseline preceding it. (**B**,**C**) A 15-s long trace and spectrogram of a seizure-like event (**B**) and of activity during control (**C**) from (**A**). The upper traces in (**B**) and (**C**) are the activity, high-pass filtered at 100Hz. The spectrograms (**B**, **C**, **D**) are calculated in overlapping vertical frequency bins 50 Hz tall from 25–350 Hz, and in overlapping horizontal time windows 0.05 s wide. (**D**) Spectrogram for longer period illustrating contrast between baseline and controlled activity. (Adapted from [939])

Chapter 16

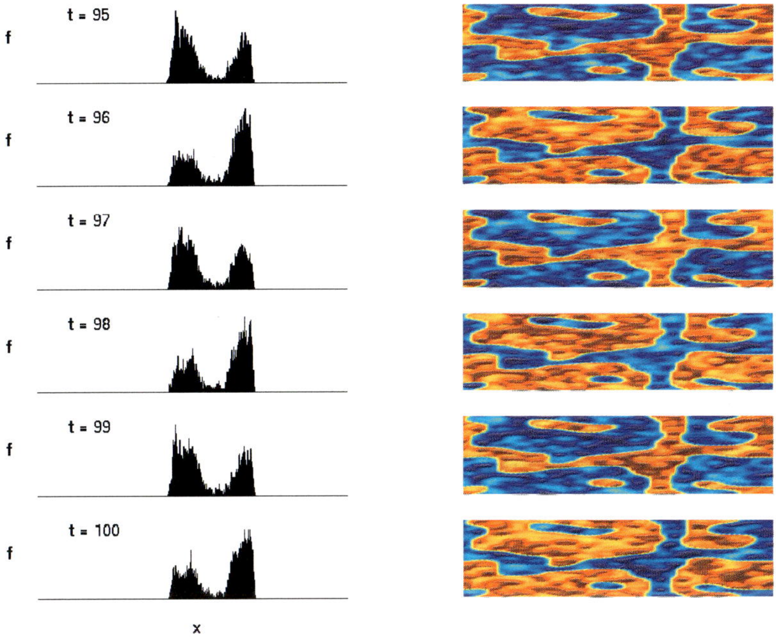

Fig. 16.11. Comparison of the (**a**) collapsed density and (**b**) the spatio-temporal distribution of activity for a 60×60 coupled map lattice of tent maps. The lattice was iterated for a total of 100 time steps and the densities and spatio-temporal patterns are shown for the last 6 time steps. In (**a**) the *solid horizontal line* indicates the interval from 0 to 1. The initial conditions were chosen from a uniform distribution over the interval from 0 to 1. For all densities the area has been normalized to one. Parameters: $a = \sqrt{2}$, $\epsilon = 0.45$

Chapter 17

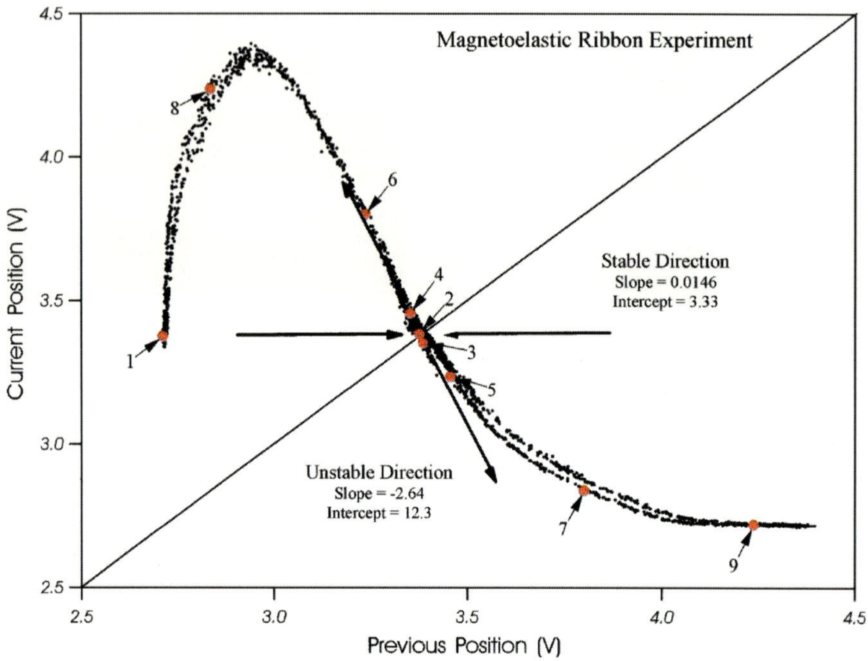

Fig. 17.1. Low-dimensional attractor consisting of 2000 points from the magnetoelastic ribbon experiment. A sequence starting at point 1280 is highlighted with *red dots*. The fixed point is at the intersection of the attractor with the diagonal. The saddle-like geometry around the fixed point is clearly visible to the eye. This pattern is repeated at the same location with the same directions over a hundred times in this clean physical data set

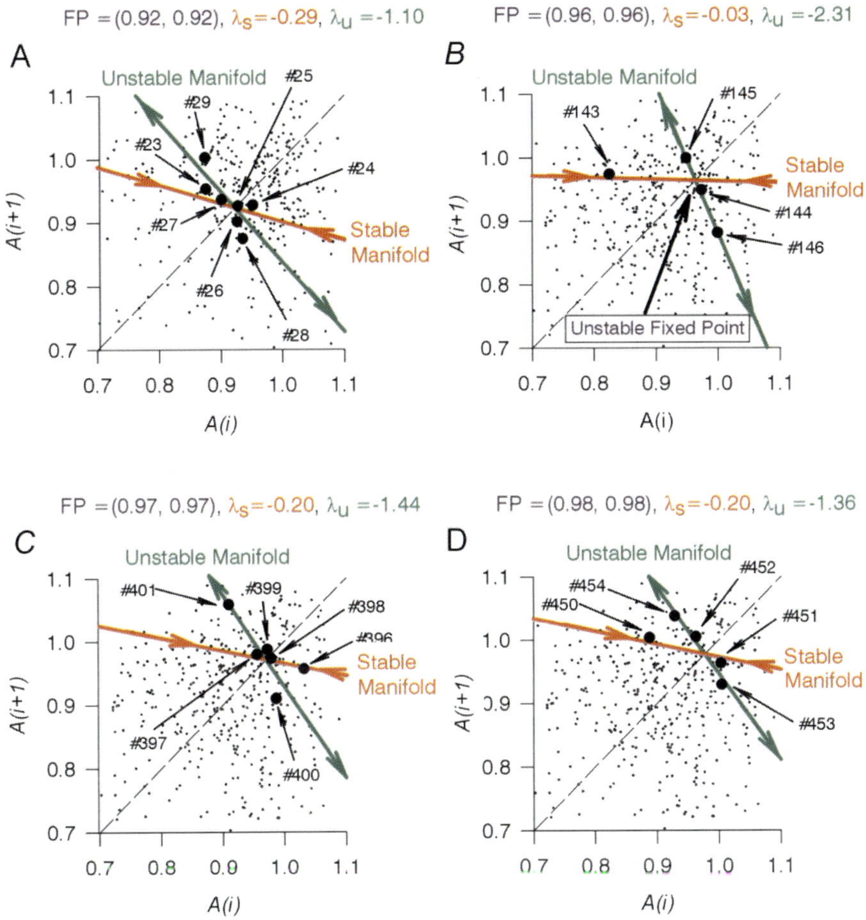

Fig. 17.4A–D. Four examples of return plots for the time series generated by the ratio of negative to positive area A(i) under the transmembrane current for a typical 1 min interval of in vivo ventricular fibrillation illustrating the local structure of the chaotic attractor. Coordinates for the calculated unstable fixed point (FP), stable eigenvalue, and unstable eigenvalue for this visitation of the unstable fixed point are provided above the plot (The colour plate of this figure appears on page 369)

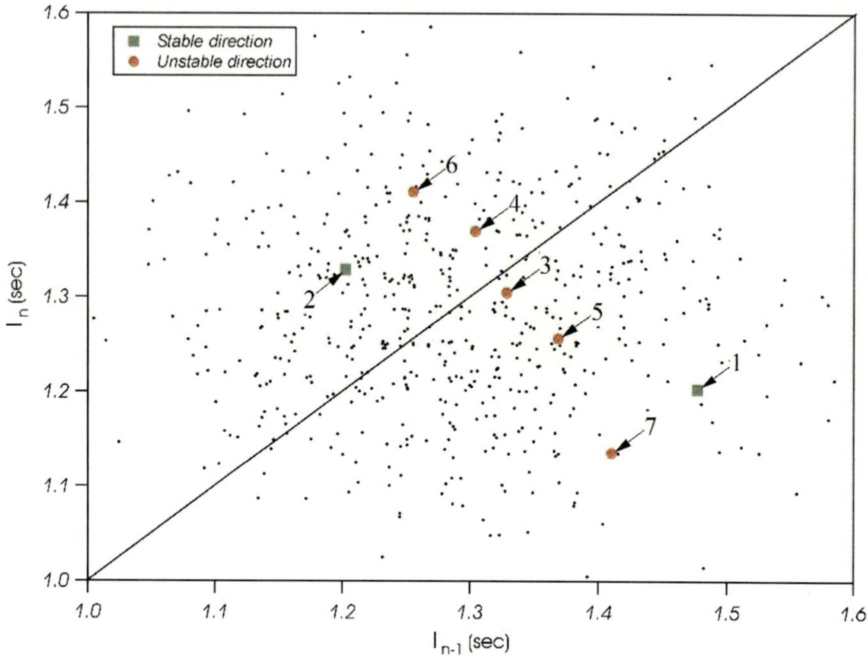

Fig. 17.5. Return plots of inter-burst intervals. Seven sequential points are color-coded. Note that as the trajectory crosses the line of identity along the direction from 1 to 2, the next points take a peculiar sequence that starts close to the line of identity at point 3, and then alternates on either side of the line of identity and progressively diverges from it along a nearly straight line for points 3–7. The points colored in *square* (1 and 2) plus point 3 (*cirlce*) define a stable direction or manifold, while the points with *circles* (3–7) define an unstable manifold. The intersection of these manifolds with the line of identity defines the unstable fixed point

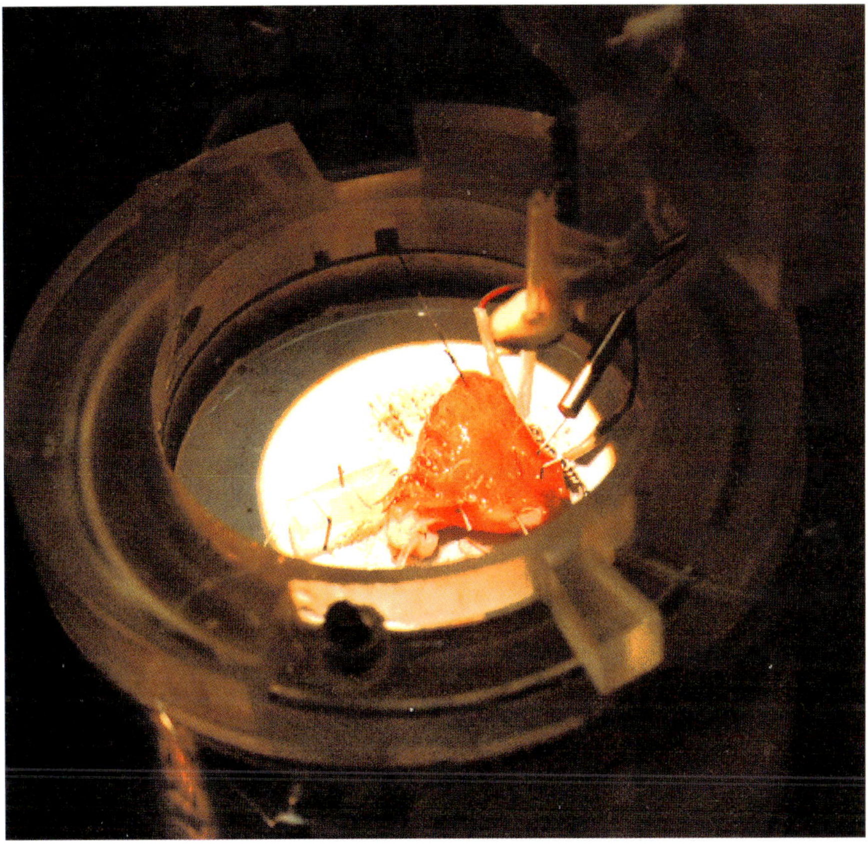

Fig. 17.7 Photograph of an isolated well-perfused portion of the interventricular septum from a rabbit heart, arterially perfused through the septal branch of the left coronary artery with a physiologically oxygenated Kreb's solution at 37°C. The heart was stimulated by passing a 3 ms constant-voltage pulse, typically 10–30 V, at twice threshold between platinum electrodes embedded in the preparation. Electrical activity was monitored by recording monophasic action potentials with Ag-AgCl wires on the surface of the heart. Monophasic action potentials were digitized at 2 kHz and processed in real time by a computer to detect the activation time of each beat from the maximum of the first derivative of the voltage signal. Arrhythmias were induced by adding 2–5 μM ouabain with or without 2–10 μM epinephrine to the arterial perfusate

Chapter 18

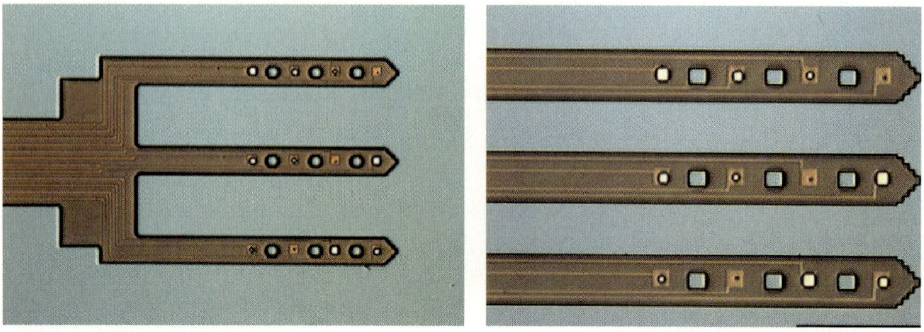

Fig. 18.1. Flexible microelectrode array developed by Pellinen et al. [1060]. *Left* and *right* panels show just one possible configuration, a trident with 12 active electrodes ranging in size from $10 \times 10\,\mu$ microns to $40 \times 40\,\mu$. Three $50 \times 60\,\mu$ holes (seeding wells) are located on each tine to accommodate tissue ingrowth in chronic applications. Each tine is $170\,\mu$ wide, substrate and insulator are polyimide. Each electrode is addressable via a breakout out of view to the left; a commercially available connector is used to make the connection to external instruments

References

1. A. H. Hayes, *Disease of the Nervous System* (Peabody Medical Institute, Boston, 1875)
2. F. Gilliam and R. Kuzniecky and E. Faught and L. Black and G. Carpenter and R. Schrodt, Epilepsia **38** (1997) 233–236
3. F. Gilliam and R. Kuzniecky and K. Meador and R. Martin and S. Sawrie and M. Viikinsalo and R. Morawetz and E. Faught, Neurology **53** (1999) 687–694
4. H. H. Jasper, Mechanisms of propagation: extracellular studies, in *Basic Mechanisms of the Epilepsies*, ed. H. H. Jasper and A. A. Ward and A. Pope, Little Brown, Boston, 1969, pp. 421–440
5. J. H. Jackson, *Selected Writings* (Hodder & Stoughton, London, 1931)
6. Classification of seizures, Epilepsia **22** (1981) 489–501
7. Classification of syndromes, Epilepsia **30** (1989) 389–399
8. P. Kotagal, Psychomotor seizures: Clinical and EEG findings, in *The Treatment of Epilepsy: Principles and Practice*, ed. E. Wyllie, Lea & Febiger, Philadelphia, 1993, pp. 378–392
9. B. R. Tharp, Epilepsia **13** (1972) 627–642
10. T. L. Babb and E. Halgren and C. Wilson and J. Engel and P. Crandall, Electroenceph. Clin. Neurophysiol. **51** (1981) 104–107
11. M. A. Mikati and G. L. Holmes, Temporal lobe epilepsy, in *The Treatment of Epilepsy: Principles and Practice*, ed. E. Wyllie, Lea & Febiger, Philadelphia, 1993, pp. 513–524
12. M. G. Harbord and J. I. Manson, Pediatr. Neurol. **3** (1987) 263–268
13. P. Kotagal and A. D. Rothmer and G. Erenberg and R. P. Cruse and E. Wyllie, Arch. Neurol. **44** (1987) 1177–1180
14. W. Wyllie and A. D. Rothmer and H. Lüders, Pediatr. Clin. North Am. **36** (1989) 343–364
15. J. G. Milton and J. Gotman and G. M. Remillard and F. Andermann, Epilepsia **28** (1987) 471–478
16. S. S. Spencer, Temporal lobectomy: Selection of candidates, in *The Treatment of Epilepsy: Principles and Practice*, ed. E. Wyllie, Lea & Febiger, Philadelphia, 1993, pp. 1062–1074
17. T. L. Babb and W. J. Brown, Pathological findings in epilepsy, in *Surgical Treatment of the Epilepsies*, ed. J. Engel, Jr., Raven Press, New York, 1987, pp. 511–540
18. C. J. Bruton, *The Neuropathology of Temporal Lobe Epilepsy* (Oxford University Press, New York, 1988)
19. A. T. Berg and S. Shinnar, J. Clin. Neurophysiol. **14** (1997) 102–110

20. T. L. Babb and J. K. Pretorius, Pathological substrates of epilepsy, in *The Treatment of Epilepsy: Principles and Practice*, ed. E. Wyllie, Lea & Febiger, Philadelphia, 1993, pp. 55–70
21. D. W. Choi, J. Neurobiol. **23** (1992) 1261–1276
22. M. Roy and R. Sapolsky, Trends Neurosci. **22** (1999) 419–422
23. J. E. Cibula and R. I. Gilmore, J. Clin. Neurophysiol. **14** (1997) 111–127
24. J. Engel, Jr. and P. A. Schawartzkroin and S. L. Moshé and D. H. Lowenstein, *Brain Plasticity and Epilepsy* (Academic Press, New York, 2001)
25. F. Morrell, Epilepsia **1** (1959/60) 538–60
26. F. Morrell, Lasting changes in synaptic organization produced by continuous neuronal bombardment, in *CIOMS conference on brain mechanisms*, ed. J. F. Delafresnaye and A. Fessard and A. Konorski, Blackwell Scientific Publications, Oxford, 1961, pp. 375–392
27. F. Morrell, Arch. Neurol. **42** (1985) 318–335
28. J. A. Wada and K. Hamada, Epilepsia **40** (1999) 669–676
29. J. A. Wada and M. Sato, Neurology **24** (1974) 565–574
30. J. A. Wada and M. Sato and M. E. Corcoran, Epilepsia **15** (1974) 465–478
31. F. Morrell and Y. Geinisman and L. de Toledo-Morell, Epilepsia **28** (1987) 617
32. G. V. Goddard, Nature **214** (1967) 1020–1021
33. S. A. Chkhenkeli and O. A. Sigua and A. I. Chubinidze and Z. V. Gobechiya and E. Sh. Bregvadze, Zurnal Voprosy Neirokhirurgii Imeni N-N–Burdenko **(6)** (1977) 23–29
34. F. Morrell and T. Rasmussen and P. Gloor and L., Electroenceph. Clin. Neurophysiol. **54** (1983) 26
35. E. F. Sperber and K. Z. Haas and P. K. Stanton and S. L. Moshe, Dev. Brain Res. **60** (1991) 88–93
36. F. U. Knudsen and A. Paerregaard and R. Andersen and J. Andersen, J. Pediatr. **36** (1985) S216
37. N. P. Rosman and J. L. Labazzo and T. Colton, Ann. Neurol. **34** (1993) 452
38. S. M. Wolf and A. Forsythe, Acta Paediatr. Scand. **78** (1989) 291–295
39. P. A. Schwartzkroin, Basic mechanisms of epileptogenesis, in *The Treatment of Epilepsy: Principles and Practice*, ed. E. Wyllie, Lea & Febiger, Philadephia, 1993, pp. 83–98
40. J. Engel, Jr., Electroenceph. Clin. Neurophysiol. **76** (1990) 296–316
41. T. L. Babb, Adv. Neurol. **79** (1999) 763–779
42. T. Sutula and J. Cavazos and G. Golarai, J. Neurosci. **12** (1992) 4173–4187
43. T. Sutula and X. X. He and J. Cavazos and G. Scott, Science **239** (1988) 1147–1150
44. T. Sutula and G. Cascino and J. Cavazos and I. Paranda, Ann. Neurol. **26** (1989) 321–330
45. T. L. Babb and W-R. Kupfer and J. K. Pretorius, Soc. Neurosci. Abst. **14** (1988) 881
46. G. Cascino and T. Sutula and J. Cavazos and I. Parada and I. Ramierz, Epilepsia **29** (1988) 684
47. F. E. Dudek and M. Spitz, J. Clin. Neurophysiol. **14** (1997) 90–101
48. F. Morrell, Physiol. Rev. **41** (1961) 443–494
49. M. E. Scheibel and A. B. Scheibel, Hippocampal pathology in temporal lobe epilepsy, in *Epilepsy: Its Phenomena in Man*, ed. M. A. B. Brazier, Academic Press, New York, 1979, pp. 315–355

50. L. E. Westrum and L. E. White, Jr. and A. A. Ward, J. Neurosurg. **21** (1964) 1022–1046
51. P. R. Huttenlocher and P. T. Heydemann, Ann. Neurology **16** (1984) 595–602
52. E. Fifkova, Cell. Mol. Neurobiol. **5** (1985) 47–63
53. P. Buser and J. Bancaud, Electroenceph. Clin. Neurophysiol. **55** (1983) 1–12
54. P. Buser and J. Bancaud and J. Talairach, Electrophysiological studies on the limbic system with multiple multilead stereotaxic electrodes in epileptic patients, in *Neurophysiology Studied in Man*, ed. G. G. Somjen, Excerpta Medica, Amsterdam, 1972, pp. 112–125
55. P. Gloor, Electroenceph. Clin. Neurophysiol. **7** (1955) 223–242
56. J. E. Krettek and J. L. Prince, J. Comp. Neurol. **172** (1977) 687–722
57. R. Lorente de Nó, Cerebral cortex: architecture, intracortical connections, motor projections, in *Physiology of the Nervous System*, ed. J. F. Fulton, Oxford University Press, Oxford, 1949, pp. 288–312
58. P. Gloor and J. T. Murphy and J. J. Dreifuss, Ann. N.Y. Academy of Sciences **157** (1969) 629–641
59. D. A. Simpson, J. Neurol. Neurosurg. Psychiat. **15** (1952) 79–92
60. L. W. Swanson and W. M. Cowan, J. Comp. Neurol. **186** (1979)
61. J. P. Aggleton and M. J. Burton and R. E. Passingham, Brain Res. **190** (1980) 347–368
62. U. an der Heiden, *Analysis of Neural Networks* (Springer-Verlag, New York, 1979)
63. R. L. Beurle, Phil. Trans. Royal Soc. Lond. **240** (1956) 55–94
64. R. L. Beurle, Functional organization in random networks, in *Principles of Self-ogranization*, ed. H. von Foerster and G. W. Zopf, Pergamon Press, New York, 1962, pp. 291–314
65. P. H. Chu and J. G. Milton and J. D. Cowan, Int. J. Bifurc. Chaos **4** (1994) 237–243
66. B. G. Farley, Comp. in Biomed. Res. **2** (1965) 265–294
67. B. G. Farley and W. A. Clark, Activity in networks of neuron-like elements, in *Information Theory*, ed. C. Cherry, Butterworths, Washington, 1961, pp. 242–251
68. J. G. Milton, *Dynamics of Small Neural Populations* (American Mathematical Society, Providence, Rhode Island, 1996)
69. J. G. Milton and T. Mundel and U. an der Heiden and S-P. Spire and J. D. Cowan, Traveling activity waves, in *The Handbook of Brain Theory and Neural Networks*, ed. M. A. Arbib, MIT Press, Cambridge, MA, 1995, pp. 994–997
70. H. R. Wilson and J. D. Cowan, Biophys. J. **12** (1972) 1–23
71. H. R. Wilson and J. D. Cowan, Kybernetik **13** (1973) 55–80
72. J. G. Milton and P. H. Chu and J. D. Cowan, Spiral waves in integrate-and-fire neural networks, in *Advances in Neural Information Processing Systems 5*, ed. S. J. Hanson and J. D. Cowan and C. L. Giles, Morgan Kaufmann, San Mateo, CA, 1993, pp. 1001–1007
73. D. A. Sholl, *The Organization of Cerebral Cortex* (Methuen, London, 1956)
74. U. Heinemann and H-D. Lux and M. J. Gutnick, J. Neurophysiol. **1977** (1977) 225–239
75. R. D. Traub and R. Miles, *Neuronal Networks of the Hippocampus* (Cambridge University Press, New York, 1991)

76. D. A. Turner and P. A. Schwartzkroin, J. Neurosci. **3** (1983) 2381–2394
77. J. O. McNamara, J. Neurosci. **14** (1994) 3413–3425
78. I. Mody, Brain Pathol. **3** (1993) 395–403
79. C. E. Ribak, Brain Res. **326** (1985) 251–260
80. R. S. Sloviter, Hippocampus **1** (1991) 41–66
81. R. S. Sloviter, Ann. Neurol. **35** (1994) 640–654
82. D. Giaretta and G. Kostopoulos and P. Gloor and M. Avoli, Neurosci. Lett. **59** (1985) 203–208
83. C. E. Ribak, Comtemporary methods in neurocytology and their application to the study of epilepsy, in *Basic Mechanisms of the Epilepsies: Molecular and Cellular Approaches (Advances in Neurology, vol 44)*, ed. A. V. Delgado-Escueta and A. A. Ward, Jr. and D. M. Woodbury and R. J. Porter, Raven Press, New York, 1986, pp. 739–764
84. T. L. Babb and W. J. Brown and J. K. Pretorius and W. R. Kupfer and P. H. Crandall, J. Neurosci. **9** (1989) 2562–2574
85. K. Z. Haas and E. F. Sperber and S. L. Moshe and P. K. Stanton, J. Neurosci. **16** (1996) 4250–4260
86. M. Isokawa-Akesson and C. Wilson and T. L. Babb, Epilepsy Res. **3** (1989) 236–247
87. J. Engel, Jr. and C. L. Wilson, Evidence for enhanced synaptic inhibition in epilepsy, in *Neurotransmitters, Seizures, and Epilepsy III*, ed. G. Nistico and P. L. Morselli and K. G. Lloyd and R. G. Fariello and J. Engel, Jr., Raven Press, New York, 1986, pp. 1–13
88. U. Ernst and K. Pawelzik and T. Giesel, Phys. Rev. Lett. **74** (1995) 1570–1573
89. J. Rinzel and D. Terman and X-J. Wang and B. Ermentrout, Science **279** (1998) 1351–1355
90. J. Foss and J. Milton, J. Neurophysiol. **84** (2000) 975–985
91. H. M. Pinsker, J. Neurophysiol. **40** (1977) 527–543
92. H. M. Pinsker, J. Neurophysiol. **40** (1977) 544–556
93. K. A. Schindler and C. A. Bernasconi and R. Stoop and P. H. Goodman and R. J. Douglas, Z. Naturforsch **25a** (1997) 509–512
94. J. Bélair and L. Glass and U. an der Heiden and J. Milton, CHAOS **5** (1995) 1–7
95. T. L. Babb and C. L. Wilson and M. Isokawa-Akesson, Electroenceph. Clin. Neurophysiol. **66** (1987) 467–482
96. E. W. Lothman, Hippocampus **4** (1994) 286–290
97. J. L. Stringer and E. W. Lothman, Exp. Neurol. **116** (1992) 198–203
98. T. Iijima and M. P. Witter and M. Ichikawa and T. Tominaga and R. Kajiwara and G. Matsumoto, Science **272** (1996) 1176–1179
99. S. Dupont and R. Semah and S. Clemenceau and C. Adam and M. Baulac and Y. Samson, Arch. Neurol. **57** (2000) 1331–1336
100. Y. Schiller and G. D. Cascino and E. L. So and W. R. Marsh, Neurology **54** (2000) 346–349
101. P. Wolf, Behavioral therapy, in *Epilepsy: A Comprehensive Textbook*, ed. J. Engel, Jr. and T. A. Pedley, Lippincott-Raven, Philadephia, 1997, pp. 1359–1364
102. W. R. Gowers, *Epilepsy and Other Chronic Convulsive Disorders* (Churchill (republished by Dover. N.Y., 1964), London, 1881)
103. R. Efron, Brain **79** (1956) 267–281

104. F. M. Forster, *Reflex Epilepsy, Behavioral Therapy and Conditioned Reflexes* (Thomas, Springfield, Il., 1977)
105. J. Milton and D. Black, CHAOS **5** (1995) 8–13
106. L. Glass and M. C. Mackey, Ann. N. Y. Acad. Sci. **620** (1979) 22–44
107. M. C. Mackey and U. an der Heiden, Funkt. Biol. Med. **1** (1982) 156–164
108. M. C. Mackey and L. Glass, Science **197** (1977) 287–289
109. M. C. Mackey and J. G. Milton, Ann. N. Y. Acad. Sci. **504** (1987) 16–32
110. J. G. Milton, Epilepsy: multistability in a dynamic disease, in *Self-organized Biological Dynamics and Nonlinear Control*, ed. J. Walleczek, Cambridge University Press, New York, 2000, pp. 374–386
111. J. G. Milton and M. C. Mackey, J. Royal Coll. Phys. Lond. **23** (1989) 236–241
112. R. Larter and B. Speelman and R. M. Worth, CHAOS **9** (1999) 795–804
113. R. Guttman and S. Lewis and J. Rinzel, J. Physiol. **305** (1980) 377–395
114. A. Kreindler, *Experimental Epilepsy* (Elsevier, New York, 1965)
115. N. A. Gorelova and J. Bureš, J. Neurobiol. **14** (1983) 353–363
116. M. Shibata and J. Bures, J. Neurophysiol. **35** (1972) 381–388
117. M. Shibata and J. Bures, J. Neurobiol. **5** (1974) 107–118
118. P. E. Kunkler and R. P. Kraig, J. Neurosci. **18** (1998) 3416–3425
119. G. Nicholson and R. P. Kraig, The behavior of extracellular ions during spreading depression, in *The Application of Ion-selective Microelectrodes*, ed. T. Zeuthen, Elsevier/North Holland Biomedical Press, New York, 1988, pp. 217–238
120. H. C. Tuckwell and R. M. Miura, Biophysics J. **23** (1978) 257–276
121. A. C. Charles and S. K. Kodali and R. F. Tyndale, Mol. Cell Neurosci. **7** (1996) 337–353
122. A. H. Cornell-Bell and S. M. Finkbeiner, Cell Calcium **12** (1991) 185–204
123. A. H. Cornell-Bell and S. M. Finkbeiner and M. S. Cooper and S. J. Smith, Science **247** (1990) 470–473
124. R. Miles and R. D. Traub and R. K. S. Wong, J. Neurophysiol. **60** (1988) 1481–1496
125. J. L. Novak and B. C. Wheeler, Brain Res. **497** (1989) 223–230
126. R. D. Chervin and P. A. Pierce and B. W. Connors, J. Neurophysiol. **60** (1988) 1695–1713
127. D. Golomb and Y. Amitai, J. Neurophysiol. **78** (1997) 1199–1211
128. M. J. Gutnick and W. J. Wadman, Soc. Neurosci. Abst. **12** (1986) 349
129. P. Anderson and H. Silfvenius and S. H. Sundberg and O. Sveen and H. Wigstrom, Brain Res. **144** (1978) 11–18
130. A. Grinvald and A. Manker and M. Segal, J. Physiol. **333** (1982) 269–291
131. W. D. Knowles and R. D. Traub and B. W. Strowbridge, Neurosci. **21** (1987) 441–455
132. P. A. Schwartzkroin and D. A. Prince, Brain Res. **147** (1978) 117–130
133. J. E. Desmedt and G. Cheron, Electroenceph. Clin. Neurophysiol. **50** (1980) 382–403
134. R. E. Foster and B. W. Connors and S. G. Waxman, Dev. Brain Res. **3** (1990) 371–386
135. B. W. Connors and Y. Amitai, Generation of epileptiform discharge by local circuits of neocortex, in *Epilepsy: Models, Mechanisms, and Concepts*, ed. P. A. Schwartzkroin, Cambridge University Press, Cambridge, 1993, pp. 388–423

136. B. W. Connors and D. J. Pinto and A. E. Telfeian, Local pathways of seizure propagation in neocortex, in *Brain Plasticity and Epilepsy*, ed. J. Engel, Jr. and P. A. Schwartzkroin and S. L. Moshé and D. H. Lowenstein, Academic Press, New York, 2001, pp. 527–546
137. Y. Chagnac-Amitai and B. W. Connors, J. Neurophysiol. **61** (1989) 747–758
138. S. G. Waxman and M. V. L. Bennett, Nature New Biology **238** (1972) 217–219
139. H. A. Swadlow, Exp. Neurol. **43** (1974) 424–444
140. H. A. Swadlow, Exp. Neurol. **43** (1974) 445–451
141. H. A. Swadlow, J. Neurophysiol. **66** (1991) 1393–1409
142. H. A. Swadlow and T. G. Weyand, J. Comp. Neurol. **203** (1981) 799–822
143. R. Miller, What is the contribution of axonal conduction delay to temporal structure in brain dynamics?, in *Oscillatory Event-related Brain Dynamics*, ed. C. Pantev, Plenum Press, New York, 1994, pp. 53–57
144. P. L. Nunez, *Neocortical Dynamics and Human EEG Rhythms* (Oxford University Press, New York, 1995)
145. J. Tomasch, Anat. Rec. **119** (1954) 119–135
146. J. J. Wright and A. Sergejew, Electroenceph. Clin. Neurophysiol. **79** (1991) 403–412
147. P. L. Nunez, *Electric Fields of the Brain: The Neurophysiology of EEG* (Oxford University Press, New York, 1981)
148. C. Baumgartner and G. Lindinger and A. Ebner and S. Aull and W. Serles and A. Olbrich and S. Lurger and T. Czech and R. Burgess and H. Lüders, Neurology **45** (1995) 118–122
149. R. G. Emerson and C. A. Turner and T. A. Pedley and T. S. Walczak and M. Fergione, Electroenceph. Clin. Neurophysiol. **94** (1995) 338–348
150. W. W. Sutherling and D. S. Barth, Ann. Neurol. **25** (1989) 373–381
151. J. Gotman, Electroenceph. Clin. Neurophysiol. **56** (1983) 501–514
152. P. Y. Ktonas and R. Mallart, Electroenceph. Clin. Neurophysiol. **78** (1991) 105–110
153. K. Zilles, Cortex, in *The Human Nervous System*, ed. G. Pixinos, Academic Press, New York, 1990, pp. 757–802
154. H. Petsche and P. Rappelsberger and R. Trappl, Electroenceph. Clin. Neurophysiol. **29** (1970) 567–578
155. H. Petsche and J. Sterc, Electroenceph. Clin. Neurophysiol. **25** (1968) 11–22
156. E. S. Goldensohn and A. M. Salazar, Advances in Neurology **44** (1986) 559–582
157. B. W. Connors, Nature **310** (1984) 685–687
158. B. W. Connors and M. J. Gutnick, Trends in Neurosci. **13** (1990) 99–104
159. F. Morrell, Multiple subpial transections and other interventions, in *Epilepsy: A Comprehensive Textbook*, ed. J. Engel, Jr. and T. A. Pedley, Lippincott-Raven, Philadelphia, 1997, pp. 1877–1890
160. F. Morrell and W. W. Whisler and T. P. Bleck, J. Neurosurg. **70** (1989) 231–239
161. J. Aicardi, *Epilepsy in Children* (Raven Press, New York, 1986)
162. J. Bancaud and J. Talairach and P. Morel and M. Bresson and A. Bonis and S. Geier and E. Hemon and P. Buser, Electroenceph. Clin. Neurophysiol. **37** (1974) 275–282
163. J. P. Lieb and J. Engel, Jr. and T. L. Babb, Epilepsia **27** (1986) 286–293

164. C. Ajmone-Marsan, Focal electrical stimulation, in *Experimental Models of Epilepsy: A Manual for the Laboratory Worker*, ed. D. P. Purpura and J. K. Penny and D. Tower and D. M. Woodbury and R. Walker, Raven Press, New York, 1972, pp. 147–172
165. B. W. Konigsmark and A. F. Abdullah and J. D. French, Electroenceph. Clin. Neurophysiol. **10** (1958) 687–696
166. D. A. Williams, Brain **76** (1953) 50–59
167. T. C. Ruch and H. D. Patton and J. W. Woodbury and A. L. Towe, *Neurophysiology, 2nd Edition* (W. B. Saunders, Philadelphia, 1965)
168. P. Gloor, J. Clin. Neurophsyiol. **2** (1985) 327–354
169. E. W. Dempsey and R. S. Morison, Electroenceph. Clin. Neurophysiol. **94** (1942) 37–48
170. H. H. Jasper and J. Droogleever-Fortuyn, Assoc. Nerve Ment. Dis. Proc. **26** (1947) 272–298
171. K. Gale, J. Clin. Neurophysiol. **9** (1992) 264–277
172. R. S. Lichtenstein and C. Marshall and A. E. Walker, Arch. Neurol. **1** (1959) 288–302
173. M. J. Gutnick and D. A. Prince, Science **176** (1972) 424–426
174. I. Cooper, *Cerebellar Stimulation in Man* (Raven Press, New York, 1978)
175. S. A. Chkhenkeli, Bull. of Georgian Academy of Sciences **90** (1978) 406–411
176. S. A. Chkhenkeli, Physiol. Hum. Anim. **4** (1978) 406–411
177. R. S. Fisher and S. Uematsu and G. L. Krauss and B. Cysyk and R. McPherson and R. Lesser and B. Gordon and P. Schwerdt and M. Rise, Epilepsia **33** (1992) 841–851
178. A. Handforth and C. M. McGeorge and S. C. Schachter and B. M. Uthman and D. K. Naritoku and E. S. Tecoma and T. R. Henry and S. D. Collins and B. V. Vaughn and R. C. Gilmartin and D. R. Labar and G. L. Morris and M. C. Salinsky and I. Osorio and R. K. Ristanovic and D. M. Labiner and J. C. Jones and J. V. Murphy and G. C. Ney and J. W. Wheeler, Neurology **51** (1998) 48–55
179. J. L. Stringer and E. W. Lothman, J. Neurophysiol. **62** (1989) 136–143
180. J. O. McNamara and M. T. Galloway and L. C. Rigsbee and C. Shin, J. Neurosci. **4** (1984) 2410–2417
181. M. G. Pierson and K. L. Smith and J. W. Swann, Brain Res. **486** (1989) 381–386
182. J. W. Miller and J. A. Ferrendelli, Brain Res. **508** (1990) 297–300
183. A. Fernandez-Guardiola and E. Manni and J. H. Wilson and R. S. Dow, Exp. Neurol. **6** (1962) 48–69
184. H. Gaustaut and T. Naquet and M. Fisher-Williams, J. Nerv. Mental Dis. **127** (1958) 21–33
185. H. Gaustaut and H. Regis and S. Dongier and A. Roger, Electroenceph. Clin. Neurophysiol. **8** (1956) 720
186. R. Jung, Electroenceph. Clin. Neurophysiol. **9** (1957) 352
187. T. Shinamoto and M. Verzeano, J. Neurophysiol. **17** (1957) 278–288
188. P. Tass, *Phase Resetting in Medicine and Biology* (Springer-Verlag, New York, 1999)
189. M. M. Haglund and G. A. Ojemann and D. W. Hochman, Science **358** (1992) 669–671
190. D. Malonek and U. Dirnagl and U. Lindauer and K. Yamada and I. Kanno and A. Grinvald, Proc. Natl. Acad. Sci. USA **94** (1997) 14826–14831

191. S. A. Chkhenkeli, *Differential Surgical Treatment of Epilepsy* (Metzniereba, Tbilisi, 1982)
192. W. K. Doyle and D. D. Spencer, Anterior temporal resections, in *Epilepsy: A Comprehensive Textbook*, ed. J. Engel, Jr. and T. A. Pedley, Lippincott-Raven, New York, 1998, pp. 1807–1817
193. M. A. Falkoner and D. Hill and A. Meyer and W. Mitchell and D. A. Pond, Lancet **i** (1955) 827–835
194. O. Foerster and W. Penfield, Brain **53** (1930) 99–119
195. M. M. J. Guiliaume and G. Mazars, Rev. Neurol. **81** (1949) 900–903
196. R. A. Krynauw, J. Neurol. Neurosurg. Psychiat. **13** (1950) 243–267
197. H. O. Luders, ed., *Epilepsy Surgery* (Raven Press, New York, 1992)
198. C. Munari and S. Francione and P. Kahane and D.Hoffman and L. Tassi and G. Lo Russo and A. L. Bahabid, Multilobar resections for the control of epilepsy, in *Operative Neurosurgical Techniques*, ed. H. H. Schmidec and W. H. Sweet, W. B. Saunders Company, Philadephia, 1995, pp. 1323–1339
199. P. Niemeyer, The transventricular amygdalohippocampectomy in temporal lobe epilepsy, in *Temporal Lobe Epilepsy*, C. C. Thomas, Springfield, Ill., 1958, pp. 461–482
200. A. Olivier, Surgery of extratemporal epilepsy, in *The Treatment of Epilepsy: Principles and Practice*, ed. E. Wyllie, Lea & Febiger, Philadephia, 1993, pp. 1092–1105
201. W. Penfield and M. Baldwin, Ann. Surg. **136** (1952) 625–634
202. W. Penfield and H. Flanigin, Arch. Neurol. Psychiatry **64** (1950) 491–500
203. W. Penfield and H. Steelman, Ann. Surg. **126** (1947) 740–742
204. P. M. Saradzhishvili and O. A. Sigua and S. A. Chkhenkeli, Klinicheskaja Meditzina. **50** (1972) 172–179
205. D. D. Spencer and G. A. Ojemann, Overview of therapeutic procedures, in *Surgical Treatment of the Epilepsies*, ed. J. Engel, Jr., Raven Press, New York, 1993, pp. 455–471
206. S. S. Spencer, Epilepsia **39** (1998) 114–123
207. A. E. Walker, *Posttraumatic Epilepsy* (C. C. Thomas, Springfield., Ill., 1949)
208. H. G. Wieser and M. G. Yasargil, Surg. Neurol. **17** (1982) 445–457
209. O. J. Andy and M. F. Gurko and L. P.Ciurintano, Confinia Neurologia **36** (1974) 106–112
210. J. Bancaud and J. Talairach and S. Bonis and C. Schaub and G. Szikla and P. Morel and M. Bodras-Ferrer, *La Stereoelectroencephalograpie dans L'epilepsie* (Masson, Paris, 1965)
211. S. A. Chkhenkeli, Pathophysiological basis and results of differentiated surgical treatment of limbic epilepsy, in *Modern Problems of Neurology and Neurosurgery*, ed. V. Gabashvili, Metzniereba, Tbilisi, 1988, pp. 117–132
212. P. L. Gildenberg and R. R. Tasker, *Textbook of Stereotactic and Functional Neurosurgery* (McGraw-Hill, Toronto, 1998)
213. S. A. Chkhenkeli and V. L. Towle and J. G. Milton and J.-P. Spire, XIII Congress of ESSFN (Freiburg, Germany) (1998) 4.21
214. R. Jelsma and C. Bertrand and S. Martines and P. Molina-Negro, J. Neurosurg. **39** (1973) 42–51
215. S. Kalyanaraman and B. Ramamurthi, Neurology (India) **18(Suppl.1)** (1970) 34–41
216. F. Mundinger, Z. Allg. Med. **50** (1974) 852–855

217. H. Narabajashi and T. Mizutani, Confinia Neurologia **32** (1970) 289– 297
218. G. A. Ojemann and A. A. Ward, Jr., Advances in Neurology **8** (1975) 241–263
219. O. A. Sigua and S. A. Chkhenkeli, Zhurnal Voprosy Neirokhirurgii Imeni N-N-Burdenko **(1)** (1976) 17–22
220. E. A. Spiegel and H. T. Wyeis and M. Marks and S. Lee, Science **106** (1947) 349–350
221. J. Talairach and M. David and P. Tournoux, *L'exploration chirurgicals stereotaxique du lobe temporal dans l'epilepsie temporale* (Masson, Paris, 1958)
222. E. H. Bertram, Epilepsia **38** (1997) 95–105
223. R. G. Bickford, Electroenceph. Clin. Neurophysiol. **8** (1956) 526–527
224. M. A. B. Brazier, Ann. N. Y. Acad. Sci. **112** (1964) 33–59
225. S. A. Chkhenkeli, Proc. XIIth Cong. World Soc. Stereo. and Functional Neurosurg. (Lyon) (1997) 135–136
226. J. Engel, Jr., New concepts of the epileptic focus, in *The Epileptic Focus*, ed. H. G. Wieser and E. G. Speckman and J. Engel, Jr., John Libbey Eurotext., London, 1987, pp. 83–94
227. J. Engel, Jr., Electroenceph. Clin. Neurophysiol. **10** (1991) 90–98
228. B. I. Ludwig and C. Ajmone-Marsan and J. Van Buren, Neurology **26** (1976) 1085–1099
229. P. M. Saradzhishvili and S. A. Chkhenkeli, Zhurnal Nevropatologii I Psykhiatrii Imeni S S Korsakova **77** (1977) 1281–1287
230. S. A. Chkhenkeli, Zhurnal Nevropatologii I Psikhiari Imeni S-S-Korsakova **82** (1982) 91–101
231. Z. Huber, Neurol. Neurochirurgia Polska **1** (1967) 379–383
232. R. C. Schneider and E. C. Crosby and S. M. Farhat, J. Neurosurg. **22** (1965) 246–263
233. G. F. Rossi, Brain Res. **95** (1975) 395–402
234. J. Talairach and J. Bancaud, Confinia Neurologia **27** (1966) 91–94
235. A. G. Zemskaja, *Focal Epilepsy in the Childhood* (Meditzina Publ., Leningrad, 1971)
236. W. S. McCulloch, Electroenceph. Clin. Neurophysiol. **1** (1949) 19–24
237. J. D. Green and R. Naquet, Acta Med. Belgica **1957** (1994) 225
238. C. Ajmone-Marsan, Electroenceph. Clin. Neurophysiol. **15** (1963) 197–208
239. R. K. Stone and E. Paramesh, Bull. Los Angeles Neurol. Soc. **40** (1975) 71–80
240. C. Ajmone Marsan and J. Stoll, Arch. Neurol. Psychiat. **66** (1951) 669–686
241. P. Bailey and F. A. Gibbs, JAMA **145** (1951) 365–378
242. P. D. MacLean and A. P. Arellano, Electroenceph. Clin. Neurophysiol. **2** (1950) 1–16
243. R. C. Collins and T. V. Caston, Neurology **29** (1979) 705–716
244. G. D. Jackson and A. Connelly and J. H. Cross and I. Gordon and D. G. Gadian, Neurology **44** (1994) 850–856
245. J. Engel, Jr. and L. Cahan, Potential relevance of kindling to human partial epilepsy, in *Kindling 3*, ed. J. A. Wada, Raven, New York, 1986, pp. 37–54
246. L. F. Quesney, Epilepsia **27** (1986) S27–S45
247. N. So and P. Gloor and F. Quesney and M. Jones-Gotman and A. Olivier and F. Andermann, Ann. Neurol. **25** (1989) 423–431
248. C. A. Pagni and F. Marossero, Electroenceph. Clin. Neurophysiol. **18** (1965) 260–271
249. F. P. Vedjiev, Physiologicheskii Zhurnal SSSR **43** (1967) 743–751

250. P. Gloor, *The Temporal Lobe and Limbic System* (Oxford University Press, New York, 1997)
251. R. Tzikalova, Exp. Med. Morphol. (Bulg.) **10** (1971) 90–94
252. L. P. Mestvirishvili, *Convulsive Activity in some Limbic Structures and their Significance in the Development of Partial and Generalized Convulsions* (Metzniereba, Tbilisi, 1969)
253. P. B. Melnichuk and B. A. Makarov, Peculiarities of the spread of epileptiform activity emerged at the stimulation of the different limbic system structures, in *Epilepsy (Clinic, Pathogenesis, and Treatment)*, Meditzina Pub., Moscow, 1972, pp. 305–308
254. R. S. G. Jones and U. Heinemann, J. Neurophysiol. **59** (1988) 1476–1496
255. W. A. Wilson and H. S. Swartwelder and W. W. Anderson and D. V. Lewis, Epilepsy Res. **2** (1988) 289–293
256. W. H. Hoffmann and L. B. Haberly, J. Neurosci. **11** (1991) 2021–2031
257. S. Piredda and K. Gale, Nature **317** (1985) 623–625
258. V. M. Fernandes de Lima and J. P. Pijn and C. Nunes Felipe and F. Lopes da Silva, Electroenceph. Clin. Neurophysiol. **76** (1990) 520–539
259. C. Ajmone-Marsan and J. M. Van Buren, Epileptiform activity in cortical and subcortical structures in the temporal lobe of man, in *Temporal Lobe Epilepsy*, ed. M. Baldwin and P. Bailey, C. C. Thomas, Springfield, Il., 1958, pp. 78–108
260. F. Angeleri and F. Ferro-Milone and S. Parigi, Electroenceph. Clin. Neurophysiol. **16** (1964) 100–129
261. M. A. Brazier, Exp. Neurology **36** (1972) 263–272
262. M. A. B. Brazier, Electrical seizure discharges within the human brain: the problem of spread, in *Epilepsy: Its Phenomena in Man*, ed. M. A. B. Brazier, Academic Press, New York, 1973, pp. 153–170
263. M. Fisher-Williams and R. A. Cooper, Electroenceph. Clin. Neurophysiol. **15** (1963) 568–587
264. C. A. Pagni, Confin. Neurol. **23** (1963) 477–499
265. C. L. Wilson and M. Isokawa-Akesson and M. L. Wang and T. L. Babb and J. Engel, Jr., Electroenceph. Clin. Neurophysiol. **61** (1985) S54–S55
266. C. L. Wilson and M. Isokawa and T. L. Babb and P. H. Crandall, Exp. Brain Res. **82** (1990) 279–292
267. J. P. Lieb and T. L. Babb, Epilepsia **27** (1986) 294–300
268. J.Gotman, Electroenceph. Clin. Neurophysiol. **67** (1987) 120–133
269. J. P. Lieb and T. L. Babb and J. Engel, Jr. and T. M. Darcey, Propagation pathways of interhemispheric seizure discharges compared in human and animal hippocampal epilepsy, in *Fundamental Mechanisms of Human Brain Function*, ed. J. Engel, Jr. and G. A. Ojemann and H. O. Luders and P. D. Williamson, Raven Press, New York, 1987, pp. 165–169
270. J. A. McCaughran and M. E. Corcoran and J. A. Wada, Epilepsia **19** (1978) 19–33
271. D. G. Amaral and R. Insausti and W. M. Cowan, J. Comp. Neurol. **224** (1984) 307–336
272. S. Demeter and D. L. Rosene and G. W. Van Hoesen, J. Comp. Neurol. **233** (1985) 30–47
273. E. Iwai and M. Yukie, J. Comp. Neurol. **261** (1987) 362–387

274. C. L. Wilson and M. Isokawa-Akesson and T. L. Babb and J. Engel, Jr. and L. D. Cahan and P. H. Crandall, A comparative view of local and interhemispheric limbic pathways in humans: an evoked potential analysis, in *Fundamental Mechanisms of Human Brain Function*, ed. J. Engel Jr. and G. A. Ojemann and H. O. Luders and P. D. Williamson, Raven Press, 1987, pp. 27–38
275. S. S. Spencer and P. D. Williamson and D. D. Spencer and R. H. Mattson, Epilepsia **28** (1987) 479–489
276. J. P. Lieb and R. M. Dasheiff and J. Engel, Jr., Epilepsia **32** (1991) 822–837
277. V. M. Okudzhava and S. A. Chipashvili, Zhurnal Nevropatologii I Psikhiatrii Imeni S-S-Korsakova **73** (1973) 1679–1684
278. B. L. Ralston and H. Longer, Electroenceph. Clin. Neurophysiol. **18** (1965) 325–333
279. L. T. Rutledge and T. T. Kennedy, J. Neurophysiol. **23** (1990) 188–195
280. L. T. Rutledge and T. T. Kennedy, Exp. Neurol. **4** (1991) 470–483
281. S. R. Cajal, Anales de la Sociedad Espanola de Historia Natural **4** (1895, erès II) 5–118
282. J. A. Cruce, Brain Res. **85** (1985) 211–219
283. R. B. Veazey and D. G. Amaral and W. M. Cowan, J. Comp. Neurol. **207** (1982) 135–156
284. B. A. Vogt, Cingulate cortex, in *Cerebral Cortex, Volume 4, Association and Auditory Cortices*, ed. A. Peters and E. G. Jones, Plenum Press, New York, 1985, pp. 89–149
285. B. A. Vogt and D. L. Rosene and D. N. Pandya, Science **204** (1979) 205–207
286. J. P. Aggleton and M. Mishkin, J. Comp. Neurol. **222** (1984) 56–68
287. W. J. H. Nauta, J. Anat. **95** (1961) 515–531
288. F. T. Russchen and D. G. Amaral and J. L. Price, J. Comp. Neurol. **256** (1987) 75–121
289. P. F. Bladin, The role of the mesial cortex in epilepsy (parasagittal), in *Limbic Epilepsy and the Dyscontrol Syndrome*, ed. M. Girgis and L. G. Kilch, Elsevier, New York, 1980, pp. 41–52
290. R. Marino, Jr. and A. Cukiert and G. Gronich, Xth Meeting of the World Society for Stereotactic and Functional Neurosurgery, Maebashi (1989) 165
291. G. Mazars, Epilepsia **11** (1970) 41–47
292. S. Locke and D. C. Kruper and P. I. Yakovlev, Arch. Neurol. **11** (1964) 1–12
293. B. J. Racine, Electroenceph. Clin. Neurophysiol. **38** (1975) 1–12
294. S. A. Chkhenkeli, Bull. of Georgian Academy of Sciences **94** (1979) 725–728
295. S. A. Chkhenkeli, The optimization of diagnosis of limbic epileptic focus in human, in *Theoretical Basis of Pathological Conditions*, ed. N. P. Bekhtereva, Nauka Pub., Leningrad, 1980, pp. 65–67
296. S. A. Chkhenkeli, Bull. of Georgian Academy of Sciences **97** (1980) 481–484
297. S. A. Chkhenkeli and M. Sramka, Proc. 9th Eur. Cong.Neurol. Surgery (Moscow) (1991) 294
298. S. A. Chkhenkeli and M. Sramka, *Epilepsy and its Surgical Treatment, Volume 1* (Slovak Academy of Sciences (VEDA), Bratislava, 1991)
299. J. P. Lieb and K. H. Hoque and C. E. Skomer C. E. and X-W. Song, Electroenceph. Clin. Neurophysiol. **67** (1987) 101–119
300. S. A. Chkhenkeli, Zurnal Voprosy Neirokhirurgii Imeni N-N–Burdenko. **(3)** (1981) 47–55

301. D. A. Prince and B. J. Wilder, Arch. Neurol. **16** (1967) 194–202
302. S. A. Chkhenkeli and M. Sramka, Proc. of the 8th Congress of ESSFN (Budapest) (1988) 42
303. N. R. Temkin and G. R. Davis, Epilepsia **25** (1984) 450–456
304. M. E. Newmark and J. K. Penry, Epilepsia **21** (1980) 281–300
305. F. Halberg, Totus Homo **10** (1978) 20–30
306. P. Passouant and J. Cadilhac, Décharges épileptiques, in *Epilepsy (Modern Problems in Pharmaco-Psychiatry), vol. 4*, ed. E. Niedermayer, S.Karger, Basel, 1970, pp. 87–104
307. I. Tomka, Z. EEG-EMG **14** (1983) 154–159
308. P. Passounat and J. Cadilhac and R. Labauge, Montpellier Med. **45** (1954) 581–597
309. G. F. Rossi and G. Colicchio and P. Pola and R. Roselli, Epilepsy Res. **Suppl. 2** (1991) 23–30
310. D. M. Fenney and F. F. Gullotta, Brain Res. **45** (1972) 254–259
311. J. A. Kusske and G. A. Ojemann GA. and A. A. Ward, Exp. Neurol. (1972) 279–290
312. C. H. Li and H. Jasper and L. Henderson, Electroenceph. Clin. Neurophysiol. **4** (1952) 513–526
313. O. Pompeiano, Sleep mechanisms, in *Basic Mechanisms of the Epilepsies*, ed. H. H. Jasper and A. A. Ward and A. Pope, Little, Brown and Co., Boston, MA, 1969, pp. 453–473
314. J. Villablanca and J. Schlag and R. Marcus, Epilepsia **11** (1970) 163–177
315. S.Y. Chiu and P. Schrager and J.M. Ritchie, Nature **311** (1984) 156–157
316. B.A. MacVicar, Science **226** (1984) 1345–1347
317. S. Bevan and S. Y. Chiu and P. T. A. Gray and J. M. Ritchie, Proc. R. Soc. London Ser. B **225** (1985) 299–313
318. B. A. Barres and W. J. Koroshetz and K. J. Swartz and LLY Chun and D. B. Corey, Neuron **4** (1990) 507–527
319. J. E. Minturn and H. Sontheimer and J. A. Black and BR Ransom and S. G. Waxman, Glia **6** (1992) 19–29
320. J. M. Ritchie and J. A. Black and S. G. Waxman and K. J. Angelides, Proc. Natl. Acad. Sci. USA **87** (1990) 9290–9294
321. J. A. Black and S. G. Waxman, Glia **1** (1988) 169–183
322. J. A. Black and B. Friedman and S. G. Waxman and L. W. Elmer and K. J. Angelides, Proc. R. Soc. London Ser. B **238** (1989) 39–51
323. H. Sontheimer and Br. Ransom and A. H. Cornell-Bell and J. A. Black and S. G. Waxman, J. Neurophysiol. **65** (1991) 3–19
324. H. Sontheimer and J. A. Black and Br. Ransom and S. G. Waxman, J. Neurophysiol. **68** (1992) 985–1000
325. H. Sontheimer and J. A. Blackman and S. G. Waxman, TINS **19** (1996) 325–331
326. E.R. O'Connor and H. Sontheimer and D.S. Spencer and N.C. de Lanerolle, Epilepsia **39** (1998) 347–354
327. S. Patt and C. Labrakakis and M. Bernstein and P. Weydt and J. Cervos-Navarro and G. Nisch and H. Kettenmann, Neuroscience **71** (1996) 601–611
328. S. Patt and H. Schmidt and C. Labrakakis and P. Weydt and M. Fritsch and J. CervosNavarro and H. Kettenmann, Acta Neuropathol. (Berl) **91** (1996) 209–214

329. C. Labrakakis and S. Patt and P.Weydt and J. Carvos-Navarro and R. Mayer and H. Kettenman, J. Neuropathol. Exp. Neurol. **56** (1997) 243–254
330. T. Amedee and E. Ellie and B. Dupouy and J. D. Vincent, J. Physiol. **441** (1991) 35–56
331. B. A. Barres and L. L. Y. Chun and D. P. Corey, J. Neurosci. **9** (1989) 3169–3175
332. S. Kirischuk and J. Scherer and T. Moller and A. Verkhratsky and H. Kettenmann, Glia **13** (1995) 1–12
333. E. Sykova, *Ionic and Volume Changes in the Microenvironment of Nerve and Receptor Cells* (Springer, Berlin, 1992)
334. A. C. Charles and J. E. Merrill and E. R. Dirksen and M. J. Sanderson, Neuron **6** (1991) 983–992
335. A.H. Cornell-Bell and S. Finkbeiner and Mark S. Cooper and S. J. Smith, Science **247** (1990) 470–473
336. S. Finkbeiner, Neuron **8** (1992) 1101–1108
337. P. Jung, Phys. Rev. Lett. **78** (1997) 1723–1726
338. P. Jung and A. H. Cornell-Bell and S. Kadar and J. Wang and K. Showalter and F. Moss, Chaos **8** (1998) 567–575
339. P. Jung and J. Wang and R. Wackerbauer, Phys. Rev. E **61** (2000) 2095–2098
340. A. H. Cornell-Bell and A. Willimson, Hyperexcitability of neurons and astrocytes in epileptic human cortex, in *Biology and Pathology of Astrocyte-Neuron Interactions*, ed. S. Federoff, Plenum Press, New York, 1993, pp. 51–65
341. S. H. Lee and S. Magge and D. D. Spencer and H. Sontheimer and A. H. Cornell-Bell, Glia **15** (1995) 195–202
342. A. Bordey and H. Sontheimer, Epilepsy Res. **32** (1998) 286–303
343. M. J. During and D. D. Spencer, Lancet **341** (1993) 1607–1610
344. R.D. Scheyer, Involvement of glutamate in human epileptic activities, in *Progress in brain research Vol. 116*, ed. OP. Ottersen and I.A. Langmoen and L. Gjerstad, Elsevier Science, New York, 1998, pp. 359–369
345. E. R. OConnor and J. H. Pizzonia and D. D. Spencer and N. C. de Lanerolle, Epilepsia **37** (1996) 51–56
346. E. R. O'Connor and H. Sontheimer and D. D. Spencer and N. C. de Lanerolle, Epilepsia **39** (1998) 347–354
347. I. E. Scheffer and S. F. Berkovich, Brain **120** (1997) 479–490
348. R. H. Wallace and D. W. Wang and R. Singh and I. E. Scheffer and A. L. George and H.A. Phillips and K. Saar and A. Reis and E. W. Johnson and G. R. Sutherland and S. F. Berkovic and J. C. Mulley, Nature Genet. **19** (1998) 366–370
349. G. M. Ronen and T. O.Rosales and M. Connolly and V. E. Anderson and M. Leppert, Neurology **43** (1993) 1355–1360
350. N.A. Singh and C. Charlierand D. Staufer and B. R. DuPont and R. J. Leach and R. Melis and G. M. Ronen and I. Bjerre and T. Quattlebaum and J. V. Murphy and M. L. McHarg and D. Gagnon and T. O. Rosales and A. Peiffer and V. E. Anderson and M. Leppert, Nat. Genet. **18** (1998) 25–29
351. C. Biervert and B. Schroeder and C. Kubisch and S. F. Berkovic and S. Propping and T. J. Jentsch and O. K. Steinlein, Science **279** (1998) 403–406
352. M.Nedergaard, Science **263** (1994) 1768–1771
353. V. Parpura and T.A. Basarsky and F. Liu and K. Jeftinija and S. Jeftinija and P.G. Haydon, Nature **359** (1994) 744–747

354. V. Parpura, Nature **369** (1994) 744–747
355. A.C. Charles, Dev. Neurosci. **16** (1994) 196 – 206
356. M.E. Harris-White and S.A. Zanotti and S.A. Frautschy and A.C. Charles, Journal of Neurophysiology **79** (1998) 1045–1052
357. D. D. Daly and T. A. Pedley, *Current Practice of Clinical Electroencephalography* (Raven Press, New York, 1990)
358. J. S. Ebersole and T.A. Pedley, *Current Practice of Clinical Electroencephalography* (Raven Press, New York, 2002)
359. E. Niedermeyer and F. Lopes da Silva, *Electroencephalography: Basic Principles, Clinical Applications, and Related Fields* (Williams & Wilkins, Baltimore, 1993)
360. R. Spehlmann, *EEG Primer* (Elsevier Biomedical Press, New York, 1981)
361. J. S. Bolton, Brain **33** (1910) 26–147
362. E. L. White, *Cortical Circuits: Synaptic Organization of the Cerebral Cortex Structure* (Birkhauser, Boston, 1989)
363. K. L. Chow and A. L. Leiman, Neurosciences Res. Prog. Bull. **8** (1970) 157–217
364. J. Szentagothai, Brain Res. **95** (1975) 475–496
365. A. L. Towe, Brain Behav. Evol. **11** (1975) 16–47
366. A. Solodkin and G. W. Van Hoesen, J. Comp. Neurol. **365** (1996) 610–627
367. V. B. Mountcastle, J. Neurophysiol. **20** (1957) 408–434
368. E. M. Harth and T. J. Csermely and B. Beek and R. D. Lindsay, J. Theor. Biol. **26** (1970) 93–120
369. R. J. Douglas and K. A. C. Martin and D. Whitteridge, Neural Comp. **1** (1989) 480–488
370. J. G. Milton and M. C. Mackey, J. Physiol. (Paris) **94** (2000) 489–503
371. M. L. Feldman, Morphology of the neocortical pyramidal neuron, in *The Cerebral Cortex. Vol. 1, Cellular Components of the Cerebral Cortex*, ed. A. Peters and E. G. Jones, Plenum Press, New York, 1984, pp. 123–200
372. C. Beaulieu and Z. Kisvarday and P. Somoygi and M. Cynader and A. Cowey, Cerebral Cortex **2** (1992) 295–309
373. M. Abeles and E. Vaadia and Bergman, Network **1** (1990) 13–25
374. D. R. Humphrey, Electroenceph. Clin. Neurophysiol. **25** (1968) 421–442
375. R. Cooper and A. L. Winter and H. J.Chow and W. G. Walter, Electroenceph. Clin. Neurophysiol. **18** (1965) 217–228
376. H. Lueders and L. A. Bustamente and L. Zablow and E. S. Goldensohn, Neurology **31** (1981) 846–851
377. E. Reichenthal and S. Hocherman, Electroenceph. Clin. Neurophysiol. **42** (1977) 248–251
378. B. R. Tharp, Electroenceph. Clin. Neurophysiol. **31** (1971) 45–55
379. S. W. Kuffler and J. G. Nicholls, Erg. Physiol. **57** (1966) 1–90
380. G. C. Somjen and M. Trachtenberg, Neuroglia as generator of extracellular current, in *Origin of Cerebral Field Potentials*, ed. E-J. Speckmann and H. Caspers, Thieme, Stuttgart, 1979, pp. 21–32
381. E. S. Goldensohn, Initiation and propagation of epileptogenic foci, in *Advances in Neurology, Vol. 11*, ed. J. K. Penry and D. J. Koffler, Raven Press, New York, 1975, pp. 141–162
382. H. Petsche and H. Pockberger and P. Rappelsberger, Cortical structure and electrogenesis, in *Dynamics of Sensory and Cognitive Processing by the Brain*, ed. E. Basar, Springer-Verlag, New York, 1988, pp. 123–129

383. P. Gloor, Am. J. EEG Technol. **17** (1977) 5–20
384. H. H. Jasper, Electroenceph. Clin. Neurophysiol. **10** (1958) 370–375
385. J. Gotman and D. J. Koffler, Electroenceph. Clin. Neurophysiol. **72** (1989) 7–15
386. R. S. Fisher and W. R. Webber and R. P. Lesser and S. Arroyo and S. Uematsu, J. Clin. Neurophysiol. **9** (1992) 441–448
387. M. Javidan and A. Katz and T. Tran and S. Pacia and D. D. Spencer and S. S. Spencer, Epilepsia **33 (suppl 3)** (1992) 59
388. S. V. Pacia and J. S. Ebersole, Epilepsia **34 (suppl 6)** (1993) 26
389. P. Gloor and C. L. Vera and L. Sperti, Electroenceph. Clin. Neurophysiol. **15** (1963) 353–378
390. J. S. Ebersole and S. V. Pacia, Epilepsia **37** (1996) 386–399
391. C. Ajmone-Marsan, Epilepsia **2** (1961) 22–38
392. B. Anziska and R. Q. Cracco, Clin. Electroenceph. **8** (1977) 206–210
393. W. T. Blume and G. B. Young and J. F. Lemieux, Electroenceph. Clin. Neurophysiol. **57** (1984) 295–302
394. L. R. Geiger and R. N. Harner, Arch. Neurol. **35** (1978) 276–286
395. D. W. Klass, Electroencephalographic manifestations of complex partial seizures and their treatment, in *Complex Partial Seizures and their Treatment*, Raven Press, New York, 1975, pp. 113–140
396. S. V. Pacia and J. S. Ebersole, Epilepsia **38** (1997) 642–654
397. R. Crabb and M. C. Mackey and A. D. Ray, Chaos **6** (1996) 477–492
398. J. S. Ebersole, J. Clin. Neurophysiol. **14** (1997) 470–483
399. J. S. Ebersole, EEG and MEG dipole source modeling, in *Epilepsy: A Comprehensive Textbook*, ed. J. Engel, Jr. and T. A. Pedley, Lippincott-Raven Publishers, Philadelphia, 1997, pp. 919–935
400. M. B. Carpenter, *Core Text of Neuroanatomy* (Williams & Wilkins, Baltimore, 1984)
401. T. W. Woodbury, Potentials in a volume conductor, in *Medical Physiology and Biophysics*, ed. T. C. Ruch and F. F. Fulton, Saunders Co., Philadelphia, 1960, pp. 83–91
402. W. H. Press and S. A. Teukolsky and S. A. Vetterling and B. P. Flannery, *Numerical Recipes in C: The Art of Scientific Computing* (Cambridge University Press, Cambridge, MA, 1992)
403. R. W. Thatcher and P. J. Krause and M. Hrybyk, Electroenceph. Clin. Neurophysiol. **64** (1986) 123–143
404. P. Rappelsberger and H. Petsche, Brain Topography **1** (1986) 46–54
405. C. M. Gray and P. Konig and A. K. Engel and W. Singer, Nature **338** (1989) 334–338
406. C. Andrew and G. Pfurtscheller, Electroenceph. Clin. Neurophysiol. **98** (1996) 144–148
407. G. Fein and J. Raz and F. F. Brown and E. L. Martin, Electroenceph. Clin. Neurophysiol. **69** (1988) 581–584
408. V. L. Towle and I. Syed and C. Berger and R. Grzesczcuk and J. Milton and R. K. Erickson and P. H. Cogen and E. Bergson and J-P. Spire, Electroenceph. Clin. Neurophysiol. **106** (1998) 30–39
409. G. Tononi and G. M. Edelman, Science **282** (1998) 1846–1851
410. F. Crick and C. Koch, Neuroscience **2** (1990) 263–275
411. P. Fries and J. H. Reynolds and A. E. Rorie and R. Desimone, Science **291** (2001) 1560–1564

412. R. B. Duckrow and S. S. Spencer, Electroenceph. Clin. Neurophysiol. **82** (1992) 415–422
413. T. H. Bullock and M. C. McClune and J. Z. Achimowicz and V. J. Iragui-Madoz and R. B. Duckrow and S. S. Spencer, Proc. Natl. Acad. Sci. USA **92** (1995) 11568–11572
414. V. Menon and W. J. Freeman and B. A. Cutillo and J. E. Desmond and M. F. Ward and S. L. Bressler and K. D. Laxer and N. Barbaro and A. S. Gevins, Electroenceph. Clin. Neurophysiol. **98** (1996) 89–102
415. V. L. Towle and R. K. Carder and L. Khorasani and D. Lindberg, J. Clin. Neurophysiol. **16** (1999) 528–547
416. B. Shen and M. Nadkarni and R. A. Zappulla, J. Clin. Neurophysiol. **110** (1999) 115–125
417. T. H. Bullock and M. C. McClune and J. Z. Achimowicz and V. J. Iragui-Madoz and R. B. Duckrow and S. S. Spencer, Electroenceph. Clin. Neurophysiol. **95** (1995) 161–177
418. H. P. Zaveri, *Generation and propagation of seizures in temporal lobe epilepsy (Ph.D. thesis)* (University of Michigan, Ann Arbor, MI, 1993)
419. M. T. Toczek and M. J. Morrell and M. W. Risinger and L. Shuer, J. Clin. Neurophysiol. **14** (1997) 499–506
420. M. Le Van Quyen and C. Adam and J. M. Lachaux and B. Renault and F. J. Varela, NeuroReport **8** (1997) 1703–1710
421. P. J. Franaszcuk and G. K. Bergey and M. J. Kaminski, Electroenceph. Clin. Neurophysiol. **91** (1994) 413–427
422. C. M. Gray, J. Comp. Neurosci. **1** (1992) 11–38
423. M. A. Wilson and J. M. Bower, J. Neurophysiol. **67** (1992) 981–995
424. U. Kim and T. Bal and D. A. McCormick, J. Neurophysiol. **74** (1995) 1301–1323
425. M. Steriade, Trends in Neurosci. **22** (1999) 337–345
426. R. D. Traub and J. G. R. Jeffreys and M. A. Whittington, *Fast Oscillations in Cortical Circuits* (MIT Press, Cambridge, MA, 1999)
427. P. S. G. Stein and D. G. Stuart and A. I. Selverston, *Neurons, networks, and motor behavior* (MIT Press, Cambridge, MA, 1997)
428. R. J. Douglas and K. A. C. Martin, Neocortex, in *The Synaptic Organization of the Brain*, ed. G. M. Shepherd, Oxford University Press, New York, 1990, pp. 389–438
429. B. W. Connors and R. C. Malenka and L. R. Silva, J. of Physiol. **412** (1988) 513–541
430. D. A. McCormick, J. Neurophys. **62** (1989) 1018–1027
431. D. A. Prince and R. A. Deisz and S. M. Thompson and Y. Amitai, Functional alterations in GABAergic inhibition during activity, in *Neurotransmitters in Epilepsy*, ed. G. Avanzini and J. Engel, Jr. and R. Fariello, Demos, New York, 1990, pp. 47–57
432. R. Miles and R. K. S. Wong, Nature **306** (1983) 371–373
433. R. K. S. Wong and R. D. Traub, J. Neurophysiol. **49** (1983) 459–471
434. X-J. Wang and J. Rinzel, Oscillatory and bursting properties of neurons, in *Handbook of Brain Theory and Neural Networks*, ed. M. Arbib, MIT Press, Cambridge, MA, 1995, pp. 686–691
435. S. K. Han and C. Kurrer and Y. Kuramoto, Phys. Rev. Lett. **75** (1995) 3190–3193

436. R. D. Traub and J. G. R. Jeffreys and R. Miles, J. Physiol. (London) **472** (1993) 267–287
437. A. Destexhe and T. Bal and D. A. McCormick and T. J. Sejnowski, J. Neurophysiol. **76** (1996) 2049–2070
438. D. Golomb and X. J. Wang and J. Rinzel, J. Neurophysiol. **75** (1996) 750–769
439. L. Glass and M. C. Mackey, *From Clocks to Chaos* (Princeton University Press, Princeton, 1988)
440. D. Hansel and G. Mato and C. Meunier, Neural Comp. **7** (1995) 307–337
441. J. Guckenheimer and P. J. Holmes, *Nonlinear Oscillations, Dynamical Systems and Bifurcations of Vector Fields* (Springer-Verlag, New York, 1983)
442. G. B. Ermentrout and N. Kopell, J. Math. Biol. **29** (1991) 195–217
443. Y. Kuramoto, *Chemical Oscillations, Waves and Turbulence* (Springer-Verlag, New York, 1984)
444. R. H. Rand and P. J. Holmes, Int. J. Nonlinear Mech. **15** (1980) 387–399
445. P. C. Bressloff and S. Coombes, Phys. Rev. Lett. **81** (1998) 2168–2171
446. P. C. Bressloff and S. Coombes, Phys. Rev. Lett. **81** (1998) 2384–2387
447. P. C. Bressloff and S. Coombes, Neural Comp. **12** (2000) 1–37
448. P. C. Bressloff and S. Coombes, Mathematical techniques for modeling biophysical neural networks, in *Biophysical Neural Networks: Foundations of Integrative Neuroscience*, ed. R. Poznanski, Mary Ann Leibert Inc., New York, 2001, p. in press
449. P. C. Bressloff, J. Math. Biol. **40** (2000) 169–188
450. D. Golomb and G. B. Ermentrout, Proc. Natl. Acad Sci. USA **96** (2000) 13480–13485
451. A. L. Hodgkin and A. F. Huxley, J. Physiol. **117** (1952) 500–544
452. L. F. Abbott and T. B. Kepler, Model neurons: from Hodgkin–Huxley to Hopfield, in *Statistical Mechanics of Neural Networks: Lecture Notes in Physics* **368**, ed. L. Garrido, Springer-Verlag, New York, 1990, pp. 5–18
453. T. B. Kepler and L. F. Abbott and E. Marder, Biol. Cybern. **66** (1992) 381–387
454. C. Morris and H. Lecar, Biophys. J. **35** (1981) 193–213
455. J. Rinzel and G. B. Ermentrout, Analysis of neural excitability and oscillations, in *Methods of Neuronal Modeling*, ed. C. Koch and I. Segev, MIT Press, Cambridge, NA, 1989, pp. 135–169
456. G. B. Ermentrout and N. Kopell, SIAM J. Math. Anal. **50** (1990) 125–146
457. G. B. Ermentrout, Neural Comp. **8** (1996) 979–1001
458. F. C. Hoppensteadt and I. Izhekevich, *Weakly Connected Neural Nets (Applied Mathematical Sciences 126)* (Springer-Verlag, New York, 1997)
459. J. P. Keener and F. C. Hoppensteadt and J. Rinzel, SIAM J. Appl. Math. **41** (1981) 503–517
460. S. Coombes and P. C. Bressloff, Phys. Rev. E **60** (1999) 2086–2096
461. A. Destexhe and Z. F. Mainen and T. J. Sejnowski, J. Comp. Neurosci. **1** (1994) 195–231
462. G. B. Ermentrout, J. Math. Biol. **23** (1985) 55–74
463. C. van Vreeswijk and L. F. Abbott and G. B. Ermentrout, J. Comp. Neurosci. **1** (1994) 313–321
464. S. Coombes and G. J. Lord, Phys. Rev. E **55** (1997) 2104–2107
465. R. E. Mirollo and S. H. Strogatz, SIAM J. Appl. Math. **50** (1990) 1645–1662
466. W. Gerstner and J. L. van Hemmen and J D Cowan, Neural Comp. **94** (1996) 1653–1676

467. L. F. Abbott and C. van Vreeswijk, Phys. Rev. E. **48** (1993) 1483–1490
468. W. Gerstner and J. L. van Hemmen, Phys. Rev. Lett. **71** (1993) 312–315
469. D. Golomb and D. Hansel and B. Shraiman and H. Sompolinsky, Phys. Rev. A. **45** (1992) 3516–3530
470. D. Golomb and J. Rinzel, Physica D **72** (1994) 259–282
471. C. van Vreeswijk, Phys. Rev. E. **54** (1996) 5522–5537
472. P. C. Bressloff and S. Coombes, Physica D **126** (1999) 99–122
473. P. F. Pinsky and J. Rinzel, Biol. Cybern. **73** (1995) 129–137
474. D. J. Amit and M. V. Tsodyks, Network **2** (1991) 259–274
475. G. B. Ermentrout, Neural Comp. **6** (1994) 679–695
476. G. B. Ermentrout, Rep. Prog. Phys. **61** (1998) 353–430
477. A. Atiya and P. Baldi, Int. J. Neural Syst. **1** (1989) 103–124
478. C. M. Marcus and R. M. Westervelt, Phys. Rev. A **39** (1989) 347–365
479. C. C. Chow, Physica D **118** (1998) 343–370
480. J. A. White and C. C. Chow and J. Ritt and C. Soto-Trevino and N. Kopell, J. Comp. Neurosci. **5** (1998) 5–16
481. G. B. Ermentrout, J. Comp. Neurosci. **5** (1998) 191–208
482. C. Fohlmeister and W. Gerstner and R. Ritz and J. L. van Hemmen, Neural Comp. **7** (1995) 1046–1055
483. M. Minsky, Theoretical concepts of synchrony, in *Basic Mechanisms of the Epilepsies*, ed. H. H. Jasper and A. A. Ward and A. Pope, Little Brown, Boston, 1969, pp. 755–767
484. D. M. Durand, Clin. Neurophysiol. **10** (1993) 281–297
485. J. D. Murray, *Mathematical Biology* (Springer-Verlag, New York, 1989)
486. A. T. Winfree, *The Geometry of Biological Time* (Springer-Verlag, New York, 1980)
487. B.H. Knight, J. Gen. Physiol. **59** (1972) 734–766
488. M. R. Guevara and L. Glass and A. Shrier, Science **214** (1981) 1350–1353
489. M. R. Guevara and A. Shrier and L. Glass, Amer. J. Physiol. **254** (1986) H1–H10
490. H. Hayashi and M. Nakao and K. Hirakawa, J. Phys. Soc. Japan **52** (1983) 344–351
491. J.M. Fellous and A.R. Houweling and R.H. Modi and R.P. Rao and P.H. Tiesinga and T.J. Sejnowski, J. Neurophysiol. **85** (2001) 1782–1787
492. H. Hayashi and S. Ishizuka, Chaos in a molluscan neuron, in *Chaos in Biological Systems*, ed. H. Degn and A. V. Holden and L. F. Olsen, Plenum, New York, 1987, pp. 157–166
493. H. Hayashi and S. Ishizuka, Brain Res. **686** (1995) 194–206
494. C. Graves and L. Glass and D. Laporta and R. Meloche and A. Grassino, Amer. J. Physiol. **250** (1986) R902–R909
495. G. A. Petrillo and L. Glass, Amer. J. Physiol. **246** (1984) R311–R320
496. B. R. Walker and A. Easton and K. Gale, Epilepsia **40** (1999) 1051–1057
497. D. M. Woodbury and J. W. Woodbury, Epilepsia **31 (Suppl 2)** (1990) S7–S19
498. L.F. Abbott and J.A. Varela and K. Sen and S.B. Nelson, Science **275** (1997) 220–224
499. J. D. Hunter and J. G. Milton, J. Neurosci. **21** (2001) 5781–5793
500. K. Sen and J.C. Jorge-Rivera and E. Marder and L.F. Abbott, J. Neurosci. **16** (1996) 6307–18

501. J.A. Varela and K. Sen and J. Gibson and J. Fost and L.F. Abbott and S.B. Nelson, J. Neurosci. **17** (1997) 7926–40
502. R.S. Zucker, Annu. Rev. Neurosci. **12** (1989) 13–31
503. R.S. Zucker, Neuron **17** (1996) 1049–55
504. W.S. Lytton and T.J. Sejnowski, J. Neurophysiol. **66** (1991) 1059–1079
505. J.P. Segundo and J.F. Vibert and M. Stiber and S. Hanneton, Neurosci. **68** (1995) 657–92
506. Destexhe, A., Neural Comp. **9** (1997) 503–514
507. J. D. Hunter, *Control of synchrony in the Aplysia buccal ganglion* (The University of Chicago (Ph.D. thesis), Chicago, 2001)
508. K. Jerger and S. J. Schiff, J. Neurophysiol. **73** (1995) 876–879
509. J.D. Hunter and J.G. Milton and P.J. Thomas and J.D. Cowan, J. Neurophysiol. **80** (1998) 1427–1438
510. R.V. Jensen, Phys. Rev. E. **58** (1998) R6907–10
511. H.L. Bryant and J. P. Segundo, J. Physiol. (London) **260** (1976) 279–314
512. Z.F. Mainen and T.J. Sejnowski, Science **268** (1995) 1503–1506
513. J.M. Goldberg and C.E. Smith and C. Fernández, J. Neurophysiol. **51** (1984) 1236–1256
514. R.B. Stein Biophys. J. **7** (1967) 37–68
515. D. Zipser and B. Kehoe and G. Littlewort and J. Fuster, J. Neurosci. **13** (1993) 3406–3420
516. F. Ratliff, ed., *Studies on Excitation and Inhibition in the Retina: A Collection of Papers from the Laboratories of H. Keffer Hartline* (Rockerfeller University Press, New York, 1974)
517. F. Ratliff and B. W. Knight and N. Graham, Proc. Natl. Acad. Sci. USA **62** (1969) 733–740
518. B. Mlinar and A. M. Maria and R. Corradetti, J. Physiol. **534** (2001) 141–158
519. E. Ben-Menachem and R. Mañon-Espaillat and R. Ristanovic and B. J. Wilder and H. Stefan and W. Mirza and W. B. Tarver and J. F. Wernicke First International Vagus Nerve Stimulation Study Group, Epilepsia **35** (1994) 616–626
520. S. J. Redman and D. G. Lampard and P. Annal, J. Neurophysiol. **31** (1968) 499–508
521. A. T. Winfree, *When Time Breaks Down* (Princeton University Press, Princeton, NJ, 1987)
522. A. M. Zhabotinky, *it Konzentratsionnye Avtokolebaniya [Concentration Self-Oscillations]* (Nauka, Moscow, 1974)
523. S. Jakubith and H. H. Rotermund and W. Engel and A. van Oerzen and G. Erlt, Phys. Rev. Lett. **65** (1990) 3013–3016
524. S. Pirkl and P. Oswald, J. Physique II **1996** (1996) 355–373
525. D. J. Yu and W. P. Lu and R. G. Harrison, Phys. Rev. Lett. **77** (1996) 5051–5054
526. A. J. Durston, Dev. Biol. **37** (1974) 225–235
527. G. Gerisch, Wilhelm Roux Archiv Entwicklungsmech. Organismen **156** (1965) 127–144
528. M. Alessie and F. Bonke and S. Shopman, Circ. Res. **33** (1973) 54–62
529. J. Lechleiter and S. Girard and E. Peralta and D. Clapman, Science **252** (1991) 123–126
530. N. Wiener and A. Rosenblueth, Arch. Inst. Cardiol. Mex. **16** (1946) 205–265

531. A. Rosenblueth, Amer. J. Physiol. **194** (1958) 491–494
532. O. Selfridge, Arch. Inst. Cardiol. Mex. **18** (1948) 177–187
533. W. E. Garrey, Am. J. Physiol. **33** (1914) 397–414
534. W. E. Garrey, Physiol. Rev. **4** (1924) 215–250
535. T. Lewis, *The Mechanism and Graphic Registration of the Heart Beat, 3rd edition* (Shaw and Sons, Ltd., London, 1925)
536. G. R. Mines, J. Physiol. **46** (1914) 349–383
537. S. Balakhovsky, Biophysics **10** (1965) 1175–1179
538. B. P. Belousov, A periodic reaction and its mechanism (1951, from his archives, english translation), in *Oscillations and Traveling Waves in Chemical Systems*, Wiley & Sons, New York, 1985, pp. 605–613
539. B. P. Belousov, Periodically acting reaction and its mechanisms, in *Sbornik Referatov po Radiatsionnoy Meditsine [Collection of Abstracts in Radiation Medicine]*, Medgiz, Moscow, 1959, p. 145
540. A. N. Zaikin and A. M. Zhabotinsky, Nature **225** (1970) 535–537
541. A. M. Zhabotinsky, Biofizika **9** (1964) 306–311
542. I. M. Gel'fand and M. L. Tsetlin, Dokl. Akad. Nauk. USSR **131** (1960) 1242–1245
543. V. I. Krinsky, Biofizika **11** (1966) 776–784
544. V. I. Krinsky, Problemy Kibernetiki [Problems of Cybernetics] (Nauka, Moscow) **20** (1968) 59–73
545. V. I. Krinsky and A. V. Kholopov, Biofizika **12** (1967) 524–528
546. A. M. Zhabotinsky and A. N. Zaikin, *Kolebatelnye Protsessy v Biologicheskikh i Khimicheskikh Sistemakh* [Oscillatory Proceesses in Biological and Chemical Systems] (Biol. Research Center, Pushchino) **2** (1971) 314–317
547. A. M. Zhabotinsky and A. N. Zaikin, J. Theor. Biol. **40** (1973) 45–61
548. A. T. Winfree, Foskning och Framsteg **6** (1971) 9–10
549. A. T. Winfree, Science **175** (1972) 634–636
550. F. B. Gul'ko and A. A. Petrov, Biofizika **17** (1972) 261–270
551. A. M. Pertsov and E. A. Ermakova and A. V. Panfilov, Physica D **14** (1984) 117–124
552. A. M. Pertsov and A. V. Panfilov and F. U. Medvedeva, Biofizika **28** (1983) 100–102
553. A. V. Panfilov and A. M. Pertsov, Biofizika **27** (1982) 866–889
554. V. A. Davydov and A. S. Mikhailov and V. S. Zykov, Kinematical theory of autowave patterns in excitable media, in *Nonlinear Waves in Active Media*, Springer-Verlag, Berlin, 1989, pp. 261–279
555. A. S. Mikhailov, *Foundations of Synergetics. I. Distributed Active System* (Springer-Verlag, Berlin, 1990)
556. A. S. Mikhailov and V. S. Zykov, Dokl. Akad. Nauk. SSSR **286** (1986) 341–344
557. V. S. Zykov, Biofizika **25** (1980) 888–892
558. V. S. Zykov, *Stimulation of Wave Processes in Excitable Media (translation from Russian)* (Manchester University Press, Manchester, 1987)
559. V. A. Davydov and V. S. Zykov and A. S. Mikhailov and P. K. Brazhnik, Izv. Vyssh. Ucheb. Zaved. **31** (1988) 574–582
560. G. H. Markstein, J. Aeronaut. Sci. **18** (1951) 199–209
561. W. K. Burton and N. Cabrera and F. C. Frank, Philos. Trans. R. Soc. London Ser. A **243** (1951) 299–358
562. A. S. Mikhailov and V. A. Davydov and V. S. Zykov, Physica D **70** (1994) 1–39

563. A. S. Mikhailov, Phys. Rev. E **49** (1994) 5875–5877
564. F. Mertens and R. Imbihl, Nature **370** (1994) 124–126
565. F. Mertens and N. Gottschalk and M. Bär and M. Eiswirth and A. S. Mikhailov and R. Imbihl, Phys. Rev. E **51** (1995) 5193–5196
566. A. S. Mikhailov and V. S. Zykov, Spiral waves in weakly excitable media, in *Chemical Waves and Patterns*, ed. R. Kapral and K. Showalter, Kluwer, Dordrecht, 1995, pp. 119–162
567. P. K. Brazhnik and V. A. Davydov and V. S. Zykov and A. S. Mikhailov, Zh. Eksp. Teor. Fiz. **93** (1987) 1725–1726
568. K. I. Agladze and V. A. Davydov and A. S. Mikhailov, JETP Lett. **45** (1987) 767–770
569. V. Peŕez-Muñuzuri and R. Aliev and B. Vasiev and V. S. Krinsky, Physica D **56** (1992) 229–234
570. O. Steinbock and J. Schütze and S. C. Miller, Phys. Rev. Lett. **68** (1992) 248–251
571. E. A. Ermakova and A. M. Pertsov, Biophysics **31** (1986) 855–861
572. E. A. Ermakova and A. M. Pertsov and E. E. Shnoll, Dokl. Akad. Nauk SSSR **301** (1988) 332–335
573. J. P. Keener and J. J. Tyson, Physica D **21** (1986) 307–324
574. V. S. Zykov, Biofizika **32** (1987) 337–340
575. A. S. Mikhailov and V. S. Zykov, Physica D **52** (1991) 379–397
576. W. Jahnke and W. Skaggs and A Winfree, J. Phys. Chem. **93** (1989) 740–749
577. V. S. Zykov, Biophysics **31** (1986) 940–944
578. T. Plesser and S. C. Müller and B. Hess, J. Phys. Chem. **94** (1990) 7501–7507
579. S. C. Müller, Experiments on excitation waves, in *Nonlinear Physics of Complex Systems*, Springer-Verlag, Berlin, 1996, pp. 133–148
580. V. I. Krinsky and A. S. Mikhailov, *Avtovolny [Self-Propagating Waves]* (Znaniye, Moscow, 1984)
581. K. I. Agladze and V. I. Krinsky and A. M. Pertsov, Nature **308** (1984) 834–835
582. J. Maselko and K. Showalter, Physica D **49** (1991) 21–32
583. M. Bär and M. Eiswirth, Phys. Rev. E **48** (1993) 1635–1637
584. M. Courtemanche and A. T. Winfree, Int. J. Bifurcation and Chaos **1** (1991) 431–444
585. A. Karma, Phys. Rev. Lett. **71** (1993) 1103–1106
586. A. V. Panfilov and P. Hogeweg, Phys. Lett. A **176** (1993) 295–299
587. A. V. Panfilov and A. Holden, Phys. Lett. A **151** (1990) 23–26
588. A. F. M. Maree and A. V. Panfilov, Phys. Rev. Lett. **78** (1997) 1819–1822
589. V. S. Zykov and A. S. Mikhailov and S. C. Müller, Phys. Rev. Lett. **81** (1998) 2811–2814
590. B. J. Gluckman and E. J. Neel and T. I. Netoff and W. L. Ditto and M. L. Spano and S. J. Schiff, J. Neurophysiol. **76** (1996) 4202–4205
591. L. F. Agnati and B. Bjelke and K. Fuxe, Amer. Sci. **80** (1992) 362–373
592. J. G. R. Jeffreys, Physiol. Rev. **75** (1995) 689–723
593. D. W. Hochman and S. C. Baraban and J. W. M. Owens and P. A. Schwartzkroin, Science **270** (1995) 99–102
594. Y. Yaari and A. Konnerth and U. Heinemann, J. Neurophysiol. **56** (1986) 424–438
595. M. A. Arbib, *Handbook of Brain Theory and Neural Networks* (MIT Bradford Press, Cambridge, MA, 1995)

596. W. J. Freeman, *Mass Action in the Nervous System* (Academic Press, New York, 1975)
597. R. Plonsey, Lect. Math. Life Sci. **21** (1989) 139–149
598. B. J. Roth, IEEE Trans. Biomed. End. **42** (1995) 1174–1184
599. S. P. Dawson and J. Keizer and J. E. Pearson, Proc. Natl. Acad. Sci. USA **96** (1999) 6060–6063
600. J. Keizer and G. D. Smith, Biophys. Chem. **72** (1998) 87–100
601. A. T. Winfree, Int. J. Bifurc. Chaos **7** (1994) 487–526
602. A. M. Pertsov and M. Wellner and J. Jalife, Phys. Rev. Lett. **78** (1997) 2656–2659
603. W. M. Kistler and R. Seitz and J. L. van Hemmen, Physica D **114** (1998) 273–295
604. F. X. Witkowski and L. J. Leon and P. A. Penkoske and W. R. Giles and M. L. Spano and M. L. Spano and W. L. Ditto and A. T. Winfree, Nature **392** (1998) 78–82
605. A. T. Winfree, Science **266** (1994) 1003–1006
606. F. Fenton and A. Karma, Chaos **8** (1998) 20–47
607. J. Jalife and R. Gray and G. E. Morley and J. M. Davidenko, Chaos **8** (1998) 79–83
608. B. J. Roth and J. M. Saypol, Int. J. Bifurc. Chaos **1** (1991) 927–928
609. J. P. Keener, J. Theor. Biol. **148** (1991) 49–82
610. M. Shibata and J. Bures, Reverberating spreading depression in pigeon forebrain, in *Biological Rhythms in Birds: Neural and Endocrine Aspects*, ed. Y. Tanabe and K. Tanaka and T. Ookawa, Springer-Verlag, Tokyo, 1980, pp. 113–120
611. C. Largo and G. C. Tombaugh and P. G. Aitken and O. Herreras and G. G. Somjen, J. Neurophysiol. **77** (97) 6–16
612. M. Lauritzen, TINS **10** (1987) 8–13
613. B. K. Rasmussen and J. Olesen, Cephalgia **12** (1992) 221–228
614. J. Olesen, Arch. Neurol. **44** (1987) 321–327
615. H. C. Tuckwell, J. Neurosci. **44** (1989) 143–148
616. M. S. Cooper, Biosystems **34** (1995) 65–85
617. A. N. van den Pol and S. M. Finkbeiner and A. H. Cornell-Bell, J. Neurosci. **12** (1992) 2648–2664
618. A. N. van den Pol and F. E. Dudek, Neurosci. **56** (1993) 793–811
619. J. Sneyd and M. Wilkins and A. Strahonja and M. J. Sanderson, Biophys. Chem. **72** (1998) 101–109
620. P. B. Guthrie and J. Knappenberger and M. Segal and M. V. L. Bennett and A. C. Charles and S. B. Kater, J. Neurosci **19** (1999) 520–528
621. A. C. Charles, Dev. Neurosci. **16** (1994) 196–206
622. A. C. Charles, Glia **24** (1998) 39–49
623. J. W. Dani and A. Chernjavsky and S. J. Smith, Neuron **8** (1998) 429–440
624. M. E. Harris-Smith and S. A. Zanotti and S. A. Frautschy and A. C. Charles, J. Neurophysiol. **79** (1998) 1045–1052
625. A. Atri and J. Amundson and D. Clapham and J. Snyed, Biophys. J. **65** (1993) 1727–1739
626. P. Lipp and E. Niggli, Biophys. J. **65** (1993) 2272–2276
627. M. Bazhenov and I. Timofeev and M. Steraide and T. J. Sejnowski, Nature Neuroscience **2** (1999) 168–174

628. G. F. A. Harding, Int. J. Psychophysiol. **16** (1994) 273–279
629. G. F. A. Harding, Nature Medicine **4** (1998) 265–267
630. W. J. Freeman, Brain Res. Rev. **11** (1986) 259–284
631. F. Crick and C. Koch, Nature **391** (1998) 245–50
632. G. M. Edelman, *The Remembered Present* (Basic Books, New York, 1989)
633. T. J. Ebner and G. Chen, Prog. Neurobiol. **46** (1995) 463–506
634. A. Grinvald and R. D. Frostig and E. Lieke, Physiol. Rev. **68** (1988) 1285–1365
635. S. V. Szapiel, Arch. Neurol. **58** (2001) 1061–1063
636. T. Tsau and L. Guan and J-Y. Wu, J. Neurophysiol. **82** (1999) 1965–1973
637. J. Dye and W. Heiligenberg, J. Comp. Physiol. A **161** (1987) 187–200
638. W. Heiligenberg, Ann. Rev. Physiol. **46** (1984) 561–583
639. T. Llinas and Y. Yarom, J. Physiol. (London) **376** (1986) 163–182
640. N. Rashevsky, Cold Spring Harbor Symp. Quant. Biol. **4** (1936) 90–97
641. W. A. H. Rushton, Proc. Roy. Soc. Lond. B. **106** (1937) 210–243
642. J. Leon and B. M. Horacek, J. Electrocardiol. **24** (1991) 1–41
643. R. Plonsey and R. C. Barr, J. Electrocardiol. **20** (1987) 219–226
644. G. R. Mines, Trans. Roy. Soc. Can. **4** (1914) 43–53
645. A. T. Winfree, Chaos **8** (1998) 1–19
646. A. S. Abakumov and A. A. Gulyayev and G. A. Gulyayev, Biofizika **15** (1970) 1074–1080
647. A. J. Durston, J. Theor. Biol. **42** (1973) 483–504
648. S. F. Lin and B. J. Roth and J. P. Wikswo, J. Cardiovasc. Electrophysiol. **10** (1999) 574–586
649. B. J. Roth, J. Cardiovas. Electrophysiol. **8** (1997) 768–778
650. B. J. Roth, J. Theor. Biol. **190** (1998) 389–393
651. B. J. Roth, IEEE Trans. Biomed. Eng. **47** (2000) 820–821
652. G. K. Moe and W. C. Rheinboldt and J. A. Abildskov, Am. Heart J. **67** (1964) 200–220
653. K. S. Cole, *Membranes, Ions, and Impulses* (University of California Press, Berkeley, 1968)
654. A. L. Muler and V. S. Markin, Biofizika **22** (1977) 699–704
655. A. I. Shcherbinov and N. I. Kukushkin and M. Y. Sakson, Biofizika **18** (1973) 519–525
656. A. T. Winfree, Rotating solutions to reaction/difusion equations, in *S.I.A.M./A.M.S. Proc. 8*, ed. D. Cohen, Amer. Math. Soc., Providence, Rhode island, 1974, pp. 13–31
657. S. I. Amari, Biol. Cybern. **27** (1977) 77–87
658. G. B. Ermentrout and J. D. Cowan, J. Math. Biol. **7** (1979) 265–280
659. B. N. Vasiev and F. Siegert and C. Weijer, Phys. Rev. Lett. **78** (1997) 2489–2492
660. B. J. Roth and S. F. Lin and J. P. Wikswo, J. Electrocardiol. **31 (Suppl. S)** (1998) 6–12
661. N. Trayanova and K. Scouibine and F. Anguel, Chaos **8** (1998) 221–223
662. W. Krassowska and J. C. Neu, Biophys. J. **66** (1994) 1768–1776
663. J. C. Neu and W. Krassowska, CRC Crit. Rev. Biomed. Eng. **21** (1993) 137–199
664. V. V. Sigaev and I. G. Bobrinskaya and V. V. Zverev and A. A. Akimon and Y. P. Stepanov and V. S. Khlgatyan, Khirurgila **(4)** (1989) 139–142
665. D. Tranchina and C. Nicholson, Biophys. J. **50** (1986) 1139–1156

666. J. G. R. Jeffreys, J. Physiol. (Lond.) **319** (1981) 143–152
667. R. Abrams, *Electroconvulsive Therapy* (Oxford University Press, New York, 1997)
668. T. Vicsek, *Fractal Growth Phenomena* (World Scientific, Singapore, 1992)
669. E. Meron, Phys. Rep. **218** (1992) 1–66
670. Q. Ooyang and Z. Noszticzius and H. L. Swinney, J. Chem. Phys. **96** (1992) 6773–6776
671. E. Ben-Jacob and O. Shochet and I. Cohen and A. Tenenbaum and A. Czirók and T. Vicsek, Fractals **3** (1995) 849–868
672. M. Matsushita and H. Fujikawa, Physica A **168** (1990) 498–506
673. E. Budrene and H. Berg, Nature **349** (1991) 630–633
674. E. Budrene and H. Berg, Nature **376** (1995) 49–53
675. H. Levine, Physica A **249** (1998) 53–63
676. W. Loomis, *The Development of Dictyostelium discoideum* (Academic Press, New York, 1982)
677. J. T. Bonner, *The Cellular Slime Molds* (Princeton University Press, Princeton, NJ, 1967)
678. P. N. Devreotes, Science **245** (1989) 1054–1058
679. P. N. Devreotes, Neuron **12** (1994) 235–241
680. G. Gerisch and B. Hess, Proc. Natl. Acad. Sci. USA **71** (1974) 2118–2122
681. K. J. Lee and E. C. Cox and R. E. Goldstein, Phys. Rev. Lett. **76** (1996) 1174–1177
682. F. Alacantra and M. Monk, J. Gen. Microbiol. **85** (1975) 321–334
683. A. A. Noegel and J. E. Luna, Experientia **51** (1995) 1135–1143
684. K. Tomchik and P. Devreotes, Science **212** (1981) 443–446
685. G. S. Skinner and H. L. Swinney, Physica D **48** (1991) 1–16
686. Y. Tang and H. Othmer, Math. Biosci. **120** (1994) 25–76
687. J. Martiel and A. Goldbeter, Biophys. J. **52** (1987) 807–828
688. D. A. Kessler and H. Levine, Phys. Rev. E **48** (1993) 4801–4804
689. H. Levine, Chaos **4** (1994) 563–568
690. O. Steinbock and H. Hashimoto and S. C. Miller, Physica D **49** (1991) 233–239
691. J. J. Tyson and J. P. Keener, Physica D **32** (1988) 327–361
692. J. J. Tyson and J. D. Murray, Development **106** (1989) 421–426
693. M. Gerhardt and H. Schuster and J. Tyson, Physica D **46** (1990) 392–415
694. L. Wu and J. Franke, Gene **91** (1990) 51–56
695. E. Palsson and K. J. Lee and R. E. Goldstein and J. Franke and R. H. Kessin and E. C. Cox, Proc. Natl. Acad. Sci. USA **94** (1997) 13719–13723
696. E. Pálsson and K. J. Lee and R. E. Goldstein and J. Franke and R. H. Kessin and E. C. Cox, Proc. Natl. Acad. Sci. USA **93** (1996) 1151–1155
697. W. Loomis, Microbiol. Rev. **60** (1996) 135–150
698. A. R. Kimmel and R. A. Firtel, Curr. Opin. Genet. Dev. **1** (1991) 383–390
699. H. Levine and I. Aranson and L. Tsimirig and T. V. Truong, Proc. Natl. Acad. Sci. USA **93** (1996) 6382–6386
700. J. Laureral and J. Halloy and A. Goldbeter, Proc. Natl. Acad. Sci. USA **94** (1997) 9153–9158
701. M. Falcke and H. Levine, Phys. Rev. Lett. **80** (1998) 3875–3878
702. W. F. Loomis, Dev. Biol. **70** (1979) 1–12
703. M. Wang and P. J. M. van Haastert and P. N. Devreotes and P. Schaap, Dev. Biol. **128** (1988) 72–77

704. I. Aranson and H. Levine and L. Tsimirig, Phys. Rev. Lett. **76** (1996) 1170–1173
705. P. C. Newell, Biology of the chemotactic response, in *Biology of the Chemotactic Response*, ed. J. M. Lackie, Cambridge University Press, Cambridge, MA, 1981, pp. 89–114
706. O. Steinbock and S. Miller, Zeit. fur Natur. C **50** (1995) 275–281
707. T. Bretschneider and B. Vasiev and C. J. Weijer, J. Theor. Biol. **189** (1997) 41–51
708. B. Vasiev and F. Siegert and C. J. Weijer, J. Theor. Biol. **184** (1997) 441–450
709. R. E. Goldstein, Phys. Rev. Lett. **77** (1995) 775–778
710. J. C. Dallon and H. G. Othmer, Phil. Trans. Roy. Soc. Lond. B **352** (1997) 391–417
711. H. Levine and W. N. Reynolds, Phys. Rev. Lett. **66** (1991) 2400–2403
712. B. N. Vasiev and P. Hogeweg and A. V. Panfilov, Phys. Rev. Lett **73** (1994) 3173–3176
713. E. F. Keller and L. Segel, J. Theor. Biol. **26** (1970) 399–415
714. E. F. Keller and L. Segel, J. Theor. Biol. **30** (1971) 235–248
715. M. P. Brenner and L. S. Levitov and E. O. Budrene, Biophys. J. **74** (1998) 1677–1693
716. L. Tsimring and H. Levine and I. Aranson and E. Ben-Jacob and I. Cohen and O. Shochet and W. N. Reynolds, Phys. Rev. Lett. **75** (1995) 1859–1862
717. C. Van Oss and A. Panfilov and P. Hogeweg and F. Siegert and C. Weijer, J. Theor. Biol. **181** (1996) 203–213
718. T. Hofer and P. K. Maini, Phys. Rev. E **56** (1997) 2074–2080
719. T. Hofer and J. A. Sherratt and P. K. Maini, Proc. Roy. Soc. B **259** (1995) 249–257
720. T. Hofer and J. A. Sherratt and P. K. Maini, Physica D **85** (1995) 425–444
721. H. Levine and L. Tsimring and D. Kessler, Physica D **106** (1997) 375–388
722. S. Bozzaro and E. Ponte, Experientia **51** (1995) 1175–1188
723. K. W. Doolittle and I. Reddy and J. G. McNally, Dev. Biol. **167** (1995) 118–129
724. N. Savil and P. Hogeweg, J. Theor. Biol. **184** (1997) 229–235
725. J. A. Glazier and F. Graner, Phys. Rev. E. **47** (1993) 2128–2154
726. B. Wang and A. Kuspa, Science **277** (1997) 251–254
727. G. Shaulsky and A. Kuspa and W. F. Loomis, Genes and Development **9** (1995) 1111–1122
728. M. S. Steinberg, Science **141** (1963) 401–408
729. J. C. M. Mombach and J. A. Glazier and R. C. Raphael and M. Zajac, Phys. Rev. Lett. **75** (1995) 2244–2247
730. J. Rietdorf and F. Siegert and C. J. Weijer, Dev. Biol. **177** (1996) 427–438
731. F. Siegert and C. J. Weijer, Curr. Biol. **5** (1995) 937–943
732. Y. Jiang and J. Glazier and H. Levine, Biophys. J. **75** (1998) 2615–2625
733. E. Ben-Jacob and O. Shochet and A. Tenenbaum and I. Cohen and A. Czirók, Nature **368** (1994) 46–49
734. F. Siegert and C. J. Weijer, Proc. Natl. Acad. Sci. USA **89** (1992) 6433–6437
735. T. Umeda, Bull. Math. Biol. **51** (1989) 485–500
736. M. Dvorkin, Microbiol. Rev. **60** (1996) 70–102
737. H. Moser and B. Wever and H. Wieser and P. Meier, Physica D **130** (1999) 291–305

738. Y. Salant and I. Gath and O. Henriksen, Med. Biol. Eng. Comp. **36** (1998) 549–556
739. D. Popivanov and A. Mineva, Math. Biosci. **157** (1999) 303–320
740. K. Lehnertz and C. Elger, Phys. Rev. Lett. **80** (1998) 5019–5022
741. C. Elger and K. Lehnertz, Eur. J. Neurosci. **10** (1998) 786–789
742. J. Martinerie and C. Adam and M. Le Van Quyen and M. Baulac and C. Clemanceau and B. Renault and F. Varela, Nature Med. **4** (1998) 1173–1176
743. R. Manuca and R. Savit, Physica D **99** (1999) 134–161
744. J. Pijn and D. Velis and M. van der Heyden and J. DeGoede and C. van Veelen and F. Lopes da Silva, Brain Topogr. **9** (1997) 249–270
745. G. Widman and D. Bingmann and K. Lehnertz and C. Elger, Brain Res. **836** (1999) 156–163
746. L. Hively and P. Gailey and V. Protopopescu, Phys. Lett. A **258** (1999) 103–114
747. R. Manuca and M. Casdagli and R. Savit, Math. Biosci. **147** (1998) 1–22
748. P. So and E. Ott and T. Sauer and B. J. Gluckman and C. Grebogi and S. J. Schiff, Phys. Rev. Lett. **76** (1996) 4705–4708
749. J. Glanz, Science **277** (1997) 1758–1760
750. L. M. Hively and V. A. Protopopescu and P. C. Gailey, Chaos **10** (2001) 864–875
751. M. Le Van Quyen and J. Martinerie and M. Baulac and F. Varela, NeuroReport **10** (1999) 2149–2155
752. S. J. Schiff, Nature Med. **4** (1998) 1117–1118
753. H. H. Lange and J. P. Lieb and J. Engel, Jr. and P. H. Crandall, Electroenceph. Clin. Neurophysiol. **56** (1983) 543–555
754. Z. Rogowski and I. Gath and E. Bental, Biol. Cybern. **42** (1981) 9–15
755. A. Siegel and C. L. Grady and A. F. Minsky, Epilepsia **23** (1982) 47–60
756. T. Inouye and Y. Matsumoto and K. Shinosaki and A. Iyama and S. Toi, Neurosci. Lett. **173** (1994) 197–200
757. H. H. Jasper, Epilepsia **5** (1964) 1–20
758. G. Alarcon and C. D. Binnie and R. D. Elwes and C. E. Polkey, Electroenceph. Clin. Neurophysiol. **94** (1995) 326–337
759. M. E. Drake and H. Padamadan and S. A. Newell, Seizure **7** (1998) 39–41
760. J. Wang and H. G. Wieser, Epilepsia **35** (1994) 495–504
761. N. Wiener, *Cybernetics or control and communication in the animal and the machine* (The M.I.T. Press, Cambridge, MA, 1969)
762. B. W. Colder and R. C. Frysinger and C. L. Wilson and R. M. Harper and J. Engel, Jr., Epilepsia **37** (1996) 113–121
763. N. J. Mars and F. H. Lopes da Silva, Electroenceph. Clin. Neurophysiol. **56** (1983) 194–209
764. F. Lopes da Silva and J. P. Pijn and P. Boeijinga, Brain Topogr. **2** (1989) 9–18
765. B. Harris and I. Gath and G. Roudouin and C. Feuerstein, IEEE Trans. Biomed. Eng. **41** (1994) 820–829
766. K. Mann and P. Backer and J. Roschke, Int. J. Neurosci. **73** (1993) 161–169
767. J. Maier and G. Dagnelie and H. Spekreije and B. W. van Dijk, Vision Res. **27** (1987) 165–177
768. W. J. Freeman and B. W. van Dijk, Brain Res. **422** (1987) 267–276
769. J. M. Barrie and W. J. Freeman and M. D. Lenhart, J. Neurophysiol. **76** (1996) 520–539

770. J. M. Barrie and W. J. Freeman and M. D. Lenhart, J. Neurophysiol. **76** (1996) 520–539
771. M. Jobert and H. Escola and E. Poiseau and P. Gaillard, Biol. Cybern. **71** (1994) 197–207
772. J. E. Arruda and M. D. Weiler and D. Valentino and W. G. Willis and J. S. Rossi and R. A. Stern and S. M. Gold and L. Costa, Int. J. Psychophysiol. **23** (1996) 63–81
773. P. A. Goupillaud and P. Grossmann and J. Morlet, Geoexploration **23** (1984) 85–102
774. R. R. Coifman and Y. Meyer and S. Quake and M. V. Wickerhauser, Signal processing and compression with wave packets, in *Proc. Int. Conf. Wavelets, Marseilles, 1989*, ed. Y. Meyer, Masson, Paris, 1992
775. R. A. DeVore and B. Jaweth and B. J. Lucier, IEEE Trans. Inf. Theory **38** (1992) 719–746
776. F. Argoul and A. Anreodo and G. Grasseau and Y. Gagne and E. J. Hopfinger and U. Frisch, Nature **338** (1989) 51–52
777. S. Kadambe and G. F. Boudreaux-Bartels, IEEE Trans. Inf. Theory **38** (1992) 917–924
778. J. Gotman, Electroenceph. Clin. Neurophysiol. **54** (1982) 530–540
779. R. C. Eberhart and R. W. Dobbins and W. R. S. Webber, Computer Based Medical Systems (IEEE Symp.) (1989)
780. A. J. Gabor and R. R. Leach and F. U. Dowla, Electroenceph. Clin. Neurophysiol. **99** (1996) 257–266
781. A. J. Gabor and M. Seyal, Electroenceph. Clin. Neurophysiol. **83** (1992) 271–280
782. G. Jando and R. M. Siegel and Z. Horvath and G. Buzsaki, Electroenceph. Clin. Neurophysiol. **86** (1993) 100–109
783. O. Ozdamar and I. Yayali and P. Jayakar and C. N. Lopez, Multilevel neural network system for EEG spike detection, in *Computer-Based Detection Systems*, ed. I. N. Bankman and J. E. Tsitlik, IEEE Computer Society Press, Washington, DC, 1991, pp. 272–279
784. S. J. Schiff and A. Aldroubi and M. Unser and S. Sato, Electroenceph. Clin. Neurophysiol. **91** (1994) 442–455
785. S. J. Schiff and J. Milton and J. Heller and S. L. Weinstein, Opt. Eng. **33** (1994) 2162–2169
786. W. R. S. Webber and B. Litt and K. Wilson and R. P. Lesser, Electroenceph. Clin. Neurophysiol. **91** (1994) 104–204
787. I. Osorio and M. G. Frei and S. B. Wilkinson, Epilepsia **39** (1998) 615–627
788. I. Daubechies, *Ten Lectures on Wavelets* (Society for Industrial and Applied Mathematics, Philadelphia, 1992)
789. S. L. Bressler and R. Coppola and R. Nakamura, Nature **366** (1993) 153–156
790. J. Yordanova and V. Kolev and T. Demiralp, Neuroreport **8** (1997) 3999–4004
791. E. Rodriquez and N. George and J. P. Lachaux and J. Martinerie and B. Renault and F. J. Varela, Nature **397** (1999) 430–433
792. M. G. Rosenblum and A. S. Pikovsky and J. Kurths, Phys. Rev. Lett. **76** (1996) 1804–1807
793. I. Blekman, *Synchronization in Science and Technology* (ASME Press, New York, 1988)

794. P. R. Tass and M. G. Rosenblum and J. Weule and J. Kurths and A. Pikovksy and J. Volkmann and A. Schnitzler and H.-J. Freund, Phys. Rev. Lett. **81** (1998) 3291–3294
795. R. L. Stratonvich, *Topics in the Theory of Random Noise* (Gordon and Breach, New York, 1963)
796. P. Grassberger and I. Procaccia, Phys. Rev. Lett. **50** (1983) 346–349
797. A. Babloyantz and A. Destexhe, Proc. Natl. Acad. Sci. USA **83** (1986) 3513–3517
798. A. R. Osborne and A. Provenzale, Physica D **35** (1989) 357–381
799. J. P. Theiler, Phys. Lett. A **34** (1995) 335–341
800. K. Lehnertz and C. E. Elger, Electroenceph. Clin. Neurophysiol. **95** (1995) 108–117
801. J. Lachaux and L. Pezard and L. Garnero and C. Pelte and B. Renault and F. Varela and J. Martinerie, Human Brain Mapping **5** (1997) 24–40
802. D. Lerner, Physica D **97** (1996) 563–576
803. L. M. Pecora and T. L. Carroll and J. F. Heagy, Phys. Rev. E **52** (1995) 3420–3439
804. N. F. Rulkov and M. M. Sushchik and L. S. Tsimring and H. D. I. Abarbanel, Phys. Rev. E **512** (1995) 980–994
805. S. J. Schiff and T. Chang and R. E. Burke and T. Sauer, Phys. Rev. E **54** (1996) 6709–6724
806. M. Le Van Quyen and C. Adam and M. Baulac and J. Martinerie and F. J. Varela, Brain Res. **792** (1998) 24–40
807. K. K. Jerger and T. I. Netoff and J. T. Francis and T. Sauer and L. Pecora and S. L. Weinstein and S. J. Schiff, J. Clin. Neurophysiol. **18** (2001) 259–268
808. M. J. McKeown and S. Makeig and G. G. Brown and T. P. Jung and S. S. Kingermann and A. J, Bell and T. J. Sejnowski, Human Brain Mapping **6** (1998) 160–188
809. A. Bell and T. Sejnowski, Neural Comp. **7** (1995) 1129–1159
810. P. Comon, Signal Processing **36** (1994) 287–314
811. B. Scholkopf and A. Smola and K. Muller, Neural Comp. **10** (1998) 1299–1319
812. J. Engel, Jr., Can. J. Neurol. Sci. **23** (1986) 167–174
813. N. A. Buchwald and E. J. Wyers and C. W. Lauprecht and G. Heuser, Electroenceph. Clin. Neurophysiol. **13** (1961) 531–537
814. N. A. Buchwald and E. J. Wyers and T. Okuma and G. Heuser, Electroenceph. Clin. Neurophysiol. **13** (1961) 509–518
815. S. M. Butkhuzi, *Electrophysiological Study of Functions of the Nucleus Caudatus* (Metzniereba, Tbilisi, 1971)
816. J. G. Dusser de Barenne and W. S. McCulloch, J. Neurophysiol. **1** (1938) 364–377
817. R. Jung, Arch. Psychiat. Nervenkr. **183** (1949) 206–244
818. G. A. Khasabov, Epilepsy, Clinic, Pathogenesis, Treatment **1** (1972) 308–312
819. A. Kreindler, Epilepsia **3** (1962) 329–337
820. V. La Grutta and G. Amazo and M. T. Zacami, Electroenceph. Clin. Neurophysiol. **31** (1971) 57–69
821. J. K. Merlis, Electroenceph. Clin. Neurophysiol. **34** (1973) 695–533
822. R. Mutani and L. Bergamini and T. Doriguzzi, Epilepsia **10** (1969) 337–350
823. P. M. Saradzhishvili and V. M. Okudzhava and S. A. Chkhenkeli, Some preliminary results of the human brain electro-stimulation at epilepsy, in *Proc. of All-USSR Conf. on Epilepsy Problem*, Aiastan Publ., Erevan, 1976, pp. 73–77

824. M. Sramka and L. Ciganek and P. Nadvornik, Prof. Intern. Symposium on Stereotactic Treatment of Epilepsy (Bratislava) (1975) 78–79
825. W. Umbach, J. Neurosurg. **199** (1959) 553–572
826. J. M. van Buren, J. Neurosurg. **20** (1963) 148–157
827. J. M. van Buren and G. A. Ojemann, Electroenceph. Clin. Neurophysiol. **21** (1966) 114–130
828. G. L. Bekaya and G. G. Beradze and T. K. Janashia and L. U. Shelka, Dynamic of the hippocampal electrical activity at the electrical stimulation of the cerebellar cortex, in *Modern Problems of Physiology of Neural and Muscular Systems*, Tbilisi University Pub., Tbilisi, 1974, pp. 1770–1787
829. J. Bidzinski and T. Bacia and E. Ruzikowski, Abstracts of 9th World Congress of Neurol. Surg. (1989) 85
830. I. S. Cooper and I. Amin and R. S. Snider, *The Cerebellum, Epilepsy, and Behavior* (Plenum Press, New York, 1974)
831. R. Davis and S. Emmons, Abstract of XIIth Meeting of the World Society for Stereotactic and Functional Neurosurgery (Lyon) (1997) 137
832. R. Davis, Cerebellar stimulation for seizure control, in *Textbook of Stereotactic and Functional Neurosurgery*, ed. P. Gildenburg and R. R. Tasker, McGraw-Hill, New York, 1998, pp. 1945–1951
833. R. Davis and H. Engel and J. Kudzma and A. Dusnak, Appl. Neurophysiol. **45** (1982) 44–50
834. R. S. Dow and A. Fernandez-Guardiola and E. Manni, Electroenceph. Clin. Neurophysiol. **14** (1962) 383–398
835. R. G. Heath and C. W. Dempsey and C. J. Fontana and A. T. Fitzjarrell, Biol. Psychiatry **15** (1980) 541–556
836. L. F. Levy and W. C. Auchterlionie, Epilepsia **29** (1979) 235–245
837. J. K. Min and P. A. Valentine and G. C. Teskey, Epilepsia **39** (1998) 692–699
838. G. Moruzzi, Arch. Fisiol. **41** (1941) 157–182
839. N. S. Riabokon and B. P. Boznik and T. G. Iarkina, The results of the epilepsy treatment by the stimulation of cerebellar cortex, in *Surgical Treatment of Epilepsy: Proc. of Int. Symp. for Functional Neurosurgery*, ed. S. Chkhenkeli, Metzniereba, Tbilisi, 1980, pp. 83–85
840. R. S. Snider, Cerebellar modifications of abnormal discharges in cerebral sensory and motor areas, in *The Cerebellum, Epilepsy and Behavior*, ed. I. S. Cooper and M. Riklan and R. S. Snider, Raven Press, New York, 1974, pp. 3–18
841. R. S. Snider, Epilepsia **20** (1979) 115–125
842. R. S. Snider and A. Maiti, Exp. Neurol. **49** (1975) 529–539
843. G. M. Strain and W. G. van Meter and W. H. Brockman, Epilepsia **19** (1978) 493–504
844. G. M. Strain and T. L. Babb and H. V. Soper and K. M. Perryman and J. P. Lieb and P. H. Crandall, Epilepsia **20** (1979) 651–664
845. N. Tsuru and H. Kawasaki and S. Genda and K. Hara and H. Hashiguchi and Y. Ueda, Epilepsia **33** (1992) 213–221
846. M. Demetrescu and M. Demetrescu, Electroenceph. Clin. Neurophysiol. **14** (1962) 602–620
847. V. M. Okudzhava, *Basic Neurophysiological Mechanisms of Epileptic Activity* (Metzniereba Publ., Tbilisi, 1965)
848. O. J. Andy, Abstracts of IX Meeting of the World Society for Stereotactic and Functional Neurosurgery (Toronto) (1985) 158

849. L. P. Mestvirishvili and V. M. Okudzhava and T. I. Bagashvili, The nature of different desynchronizing influences on the cortical epileptic activity, in *Problems of Psychoneurology*, Azerb State Univ. Publ., Baku, 1977, pp. 359–361
850. F. Velasco and M. Velasco and C. Ogarrio and G. Fanghenel, Epilepsia **28** (1987) 421–430
851. D. M. Finch and R. E. Feld and T. L. Babb, Exp. Neurol. **61** (1978) 318–336
852. D. A. Kovacs and J. G. Zoll, Brain Res. **70** (1974) 165–169
853. K. M. Perryman and T. L. Babb and D. M. Finch and W. J. Brown and P. H. Crandall, Epilepsia **21** (1980) 479–487
854. M. Segal, Brain Res. **94** (1975) 115–131
855. M. V. Dumbadze and N. R. Mgaloblishvili, Bull. Georgian Acad. Sci. **108** (1962) 157–160
856. B. Feinstein and C. A. Gleason and B. Libet, Stereotactic and Functional Neurosurgery **52** (1989) 26–41
857. T. K. Ioseliani and K. G. Chokheli and N. R. Mgaloblishvili, Bull. Georgian Acad. Sci. **93** (1979) 169–172
858. E. A. Kafilludin and S. Trottier and P. Chauvel, Neurosci. Lett. Suppl. **1** (1978) 64
859. B. Libet and C. A. Gleason and E. W. Wright, Jr. and B. Feinstein, Epilepsia **18** (1977) 451–462
860. M. C. Narikashvili and D. B. Kajaya and S. I. Nanobashvili and I. G. Bilanishvili and S. P. Harikashvili, Bull. Georgian Academy Sci. **102** (1981) 685–688
861. R. S. Neumann, Epilepsia **27** (1986) 358–366
862. H. C. Sullivan and I. Osorio, Epilepsia **32** (1991) 591–596
863. H. C. Sullivan and I. Osorio and S. I. Harik, Epilepsia **30** (1989) 681
864. K. Morimoto and G. V. Goddard, Epilepsia **28** (1987) 1–10
865. R. S. Fisher and M. Mirski and G. L. Krauss, Brain stimulation, in *Epilepsy: A Comprehensive Textbook*, ed. J. Engel, Jr. and T. A. Pedley, Lippincott-Raven, Philadelphia, 1997, pp. 1867–1875
866. G. Heuser and N. A. Buchwald and E. J. Wyers, Electroenceph. Clin. Neurophysiol. **13** (1961) 519–524
867. G. Amato and G. Crescimanno and F. Sorbera and V. La Grutta, Exp. Neurol. **77** (1982) 422–504
868. V. N. Shelikhov and S. E. Petrenko, Zhurnal Nevropatologii I Psykhiatrii Imeni S-S-Korsakova **76** (1976) 356–361
869. F. Sorbera and G. Crescimanno and G. Amato and V. La Grutta, Bull. Soc. Ital. Biol. Sper. **57** (1981) 1901–1904
870. M. Benvenuti and S. Parigi, Lav. Neuropsichiat. **40** (1967) 73–131
871. R. G. Heath, Amer. J. Psychiat. **120** (1963) 571–577
872. A. Costin and J. Gutman and F. Bergmann, Electroenceph. Clin. Neurophysiol. **15** (1963) 997–1005
873. F. Ferillo and R. Amore and M. Palestini and G. Rosadini, Acta. Neurochir. **23 (Suppl.)** (1976) 9–14
874. R. Mutani and R. Fariello, Brain Res. **14** (1969) 749–753
875. G. Rosadini and B. Gavazza and F. Ferrillo and A. Siccard, Electroenceph. Clin. Neurophysiol. **36** (1974) 541
876. E. B. Arushanian, Zhurnal Nevropatologii i Psikhiatrii Imeni S-S-Korsakova **68** (1968) 279–285

877. P. M. Avakian and E. B. Arushanian, Fisiological Journal of the USSR **61** (1975) 480–487
878. S. A. Chkhenkeli, Principles and results of contemporary stereotactic treatment of complex forms of epileptic seizures, in *Surgical Treatment of Epilepsy (Proc. 1st Tbilisi Int. Symp. Epilepsy Surg. Funct. Neurosurg.)*, Metsniereba Publ., Tbilisi, 1980, pp. 60–65
879. S. A. Chkhenkeli, Proc. 8th World Congress Neurological Surgery (Toronto, Canada) (1985) 32
880. S. A. Chkhenkeli and I. S. Chkhenkeli, Epilepsia **38 (Suppl. 8)** (1997) 214
881. S. A. Chkhenkeli and I. S. Chkhenkeli, Epilepsia **39 (suppl. 6)** (1998) 200
882. M. R. Klee and H. D. Lux, Arch. Psychiat. Nervenkr. **203** (1962) 667–689
883. R. Spehlmann and O. Creutzfeld and R. Jung, Arch. Psychiat. Nervendr. **3** (1960) 332–354
884. T. L. Friguesi and D. P. Purpura, Exp. Neurol. **10** (1964) 305–324
885. A. I. Gorbachevskaja, *Caudate-thalamic and thalamocaudate connections in the cat* (Dissertation, University of Leningrad, 1972)
886. G. M. Krauthamer and D. Albe-Fessard, J. Neurophysiol. **28** (1965) 100–124
887. E. Shulman and J. Aver, Anat. Res. **127** (1957) 363–385
888. N. K. Totibadze, Arch. Anat. Histol. and Embriol. **59** (1969) 7–14
889. J. B. Johnson, J. Comp. Neurol. **30** (1961) 556–569
890. E. Smith, J. Anat. **53** (1919) 271–283
891. J. Omos, The amygdaloid field in the rat as studied with the cupri-silver method, in *The Neurobiology of the Amygdala*, ed. B. E. Eleftheriou, Plenum Press, New York, 1972, pp. 186–194
892. G. Raismann and W. M. Cowan and T. P. Powell, Brain **89** (1966) 83–108
893. M. O. Gurevich, Zhurnal Nevrpatologii I Psikhiari Imeni S-S-Korsakova **30** (1930) 105–116
894. M. Namba, Z. Hirnforsch. **3** (1957) 247–271
895. E. B. Arushanian and J. A. Belozertzev, Bull. Exp. Biol. Med. **75** (1963) 3–6
896. Li-Chon-Luh and R. A. Ratcherson, Exp. Neurol. **50** (1976) 134–145
897. S. A. Chkhenkeli, Modern Meth. Invest. Neurol. Psych. **2** (1977) 75–76
898. M. Sramka and S. A. Chkhenkeli, Bratislava. Slovak Academy of Sciences Pub. VEDA **2** (1993) 1–289
899. M. Sramka and P. Nadvornik, The influence of epileptic focus stimulation on the epilepsy, in *Proc. of the Int. Symp. on Epilepsy Surgery*, Metzniereba, Tbilisi, 1980, pp. 90–93
900. R. Fariello, Exp. Neurol. **51** (1976) 515–528
901. T. L. Babb and A. G. Mitchell, Jr. and P. H. Crandall, Electroenceph. Clin. Neurophysiol. **36** (1974) 141–154
902. P. M. Cooke and R. S. Snider, Electroenceph. Clin. Neurophysiol. **5** (1953) 563–569
903. P. M. Cooke and R. S. Snider, Epilepsia **4** (1955) 19–28
904. R. S. Dow, Epilepsia **6** (1965) 122–140
905. J. T. Hutton and J. D. Frost, Jr. and J. Foster, Epilepsia **13** (1972) 401–408
906. K. Iwata and R. S. Snider, Electroenceph. Clin. Neurophysiol. **11** (1964) 439–446
907. R. S. Snider and P. M. Cooke, Electroenceph. Clin. Neurophysiol. **53(Suppl. 3)** (1953) 78
908. G. Rossi, Arch. Fisiol. **10** (1912) 251–260

966. U. an der Heiden and A. Longtin and M. C. Mackey and J. G. Milton and R. Sholl, J. Diff. Eqns. **2** (1990) 423–449
967. S. A. Campbell and J. Bélair and T. Ohira and J. G. Milton, CHAOS **5** (1995) 640–645
968. S. A. Campbell and J. Bélair and T. Ohira and J. G. Milton, J. Diff. Eqns. **7** (1995) 213–236
969. I. Ikeda and K. Matsumoto, Physica D **29** (1987) 223–235
970. J. Losson and M. C. Mackey and A. Longtin, CHAOS **3** (1993) 167–176
971. M. C. Mackey and U. an der Heiden, J. Math. Biol. **19** (1984) 211–225
972. J. G. Milton and U. an der Heiden and A. Longtin and M. C. Mackey, Biomed. Biochim. Acta **49** (1990) 697–707
973. J. G. Milton and A. Longtin and A. Beuter and M. C. Mackey and L. Glass, J. Theor. Biol. **138** (1989) 129–147
974. J. K. Hale, *Theory of Functional Differential Equations* (Springer-Verlag, New York, 1977)
975. V. B. Kolmanovskii and V. R. Nosov, *Stability of Functional Differential Equations* (Academic Press, New York, 1986)
976. N. Bernstein, *Co-ordination and Regulation of Movement* (Pergamon Press, New York, 1967)
977. F. I. M. Craik and R. S. Lockhart, J. Verbal Learn. Verbal Beh. **11** (1972) 671–684
978. D. C. van Essen and C. H. Anderson and D. J. Fellman, Science **255** (1992) 419–423
979. D. A. Coulter, Thalamocortical anatomy and physiology, in *Epilepsy: A Comprehensive Textbook*, ed. J. Engel, Jr. and T. A. Pedley, Lippincott-Raven, Philadelphia, 1997, pp. 341–351
980. P. A. Schwartzkroin and D. C. McIntyre, Limbic anatomy and physiology, in *Epilepsy: A Comprehensive Textbook*, ed. J. Engel, Jr. and T. A. Pedley, Lippincott-Raven, Philadelphia, 1997, pp. 323–340
981. A. M. Castelfranco and H. W. Stech, SIAM J. Appl. Math. **47** (1987) 573–588
982. J. Foss and A. Longtin and B. Mensour and J. Milton, Phys. Rev. Lett. **76** (1996) 708–711
983. J. Foss and F. Moss and J. Milton, Phys. Rev. E **55** (1997) 4536–4543
984. B. Mensour and A. Longtin, Phys. Lett. A **205** (1995) 18–24
985. R. E. Plant, SIAM J. Appl. Math. **40** (1981) 150–162
986. J. Milton and J. Foss, Oscillations and multistability in delayed feedback control, in *Case Studies in Mathematical Modeling: Ecology, Physiology, and Cell Biology*, ed. H. G. Othmer and F. R. Adler and M. A. Lewis and J. C. Dallon, Prentice-Hall, Englewood, NJ, 1997, pp. 179–198
987. J. Foss, *Mulistability in Delayed Recurrent Loops* (The University of Chicago (Ph.D. thesis), Chicago, 1999)
988. K. T. Judd and K. Aihara, Neural Networks **6** (1993) 203–215
989. W. Maass and C. M. Bishop, *Pulsed Neural Networks* (The MIT Press, Cambridge, MA., 1999)
990. L. Glass and M. R. Guevara and J. Bélair and A. Shrier, Phys. Rev. A **29** (1984) 1348–1357
991. A. Areili and A. Sterkin and A. Grinvald and A. Aertson, Science **273** (1996) 1868–1871
992. W. H. Calvin and C. F. Stevens, J. Neurophysiol. **31** (1968) 574–587

993. C. W. Eurich and J. G. Milton, Phys. Rev. E **54** (1996) 6681–6684
994. A. Borsellino and A. De Marco and A. Allazetta and S. Rinesi and B. Bartolini, Kybernetik **10** (1972) 139–144
995. T. Yoshida and H. Mori and H. Shigematsu, J. Stat. Phys. **31** (1983) 279–308
996. S. Ito and S. Tanaka and H. Nakada, Tokyo J. Math. **2** (1979) 221–258
997. N. Provatas and M. C. Mackey, J. Stat. Phys. **63** (1991) 661–688
998. A. Lasota and M. C. Mackey, Physica D **28** (1987) 143–154
999. J. Losson and M. C. Mackey, Phys. Rev. E **52** (1995) 115–128
1000. M. C. Mackey and I. G. Nechayeva, J. Dynam. Diff. Eqns. **6** (1994) 395–426
1001. M. C. Mackey and I. G. Nechayeva, Phys. Rev. E **52** (1995) 3366–3376
1002. J. Losson and J. G. Milton and M. C. Mackey, Physica D **81** (1995) 177–203
1003. M. C. Mackey, *Time's Arrow: The origins of thermodynamic behavior* (Springer-Verlag, New York, 1992)
1004. A. Lasota and M. C. Mackey, *Chaos, fractals and noise: Stochastic aspects of dynamics* (Springer-Verlag, New York, 1994)
1005. N. Provatas and M. C. Mackey, Physica D **53** (1991) 295–318
1006. J. Losson and M. C. Mackey, Phys. Rev. E **50** (1994) 843–856
1007. U. Feudel and C. Grebogi, Chaos **7** (1997) 597–604
1008. S. Gadaleta and G. Dangelmayr, Phys. Rev. E **63** (2001) 6217–6226
1009. K. J. Aström and G. C. Goodwin and P. R. Kumar, *Adaptive Control, Filtering and Signal Processing* (Springer-Verlag, New York, 1996)
1010. J-J. E. Slotine and W. Li, *Applied Nonlinear Control* (Prentice-Hall, New York, 1991)
1011. G. L. Baker and J. P. Gollub, *Chaotic Dynamics: An Introduction* (Cambridge University Press, Cambridge, 1990)
1012. P. Cvitanovic, Phys. Rev. Lett. **61** (1988) 2729–2732
1013. D. Pierson and F. Moss, Phys Rev. Lett. **75** (1995) 2124–2127
1014. W. L. Ditto and S. N. Rauseo and M. L. Spano, Phys. Rev. Lett. **257** (1990) 3211–3214
1015. E. Ott and C. Grebogi and J. A. Yorke, Phys. Rev. Lett. **64** (1990) 1196–1199
1016. E. R. Hunt, Phys. Rev. Lett. **67** (1991) 1953–1959
1017. V. Petrov and V. Gaspar and J. Masere and K. Showalter, Nature **361** (1993) 240–243
1018. R. Roy and T. W. Murphy and T. D. Maier and Z. Gills and E. T. Hunt, Phys. Rev. Lett. **68** (1992) 1259–1262
1019. J. P. Eckmann and D. Ruelle, Rev. Mod. Phys. **57** (1985) 617–656
1020. T. Sauer and J. A. Yorke and M. Casdagli, J. Stat. Phys. **65** (1991) 579–616
1021. F. Takens, Detecting strange attractors in turbulence, in *Dynamical Systems and Turbulence (Lecture Notes in Biomathematics 898)*, ed. D. Rand and L. S. Young, Springer-Verlag, Berlin, 1981, pp. 336–381
1022. R. Shaw, *The Dripping Faucet as a Model Chaotic System* (Aerial Press, Santa Cruz, 1984)
1023. T. Sauer, Phys. Rev. Lett. **72** (1994) 3811–3814
1024. H. Braun and M. Dewald and K. Schäfer and K. Voigt and X. Pei and K. Dolan and F. Moss, J. Comp. Neurosci. **7** (1999) 17–32
1025. P. So and E. Ott and T. Sauer and B. J. Gluckman and C. Grebogi and S. J. Schiff, Phys. Rev. E **55** (1997) 5398–5417
1026. K. Dolan and A. Witt and M. L. Spano and A. Neiman and F. Moss, Phys. Rev. E **59** (1999) 5235–5241

1027. J. Theiler and S. Eubank and A. Longtin and B. Galdrikian, B. and J. D. Farmer, Physica D **58** (1992) 77–94
1028. G. E. Uhlenbeck and L. S. Ornstein, Phys Rev. **36** (1930) 823–841
1029. A. Neiman and L. Schimansky-Geier, Phys. Rev. Lett. **72** (1994) 2988–2991
1030. L. Schimansky-Geier and Ch. Zülike, Z. Phys. B **79** (1990) 451–460
1031. F. X. Witkowski and K. M. Kavanagh and P. A. Penkoske and R. Plonsey and M. L. Spano and W. L. Ditto and D. T. Kaplan, Phys. Rev. Lett. **75** (1995) 1230–1233
1032. F. X. Witkowski and P. A. Penkoske, Ann. N. Y. Acad. Sci. **591** (1990) 219–231
1033. W. L. Ditto and L. M. Pecora, Scientific American **269** (1993) 78–84
1034. E. Ott and M. L. Spano, Physics Today **48** (1995) 34–40
1035. A. Garfinkel and M. L. Spano and W. L. Ditto and J. N. Weiss, Science **257** (1992) 1230–1235
1036. S. J. Schiff and K. Jerger and D. H. Duong and C. Taeun and M. L. Spano and W. L. Ditto, Nature **370** (1994) 615–620
1037. W. L. Ditto and M. L. Spano and V. In and J. Neff and B. Meadows and J. J. Langberg and A. Bolmann and T. McTeague, Intl. J. Bif. Chaos **10** (2000) 593–601
1038. W. L. Ditto WL and M. L. Spano and V. In and J. Neff and B. Meadows and J. J. Langberg and A. Bolmann and K. McTeague, Int. J. Bifurc. Chaos **10** (2000) 593–601
1039. V. In and S. E. Mahan and W. L. Ditto and M. L. Spano, Phys. Rev. Lett. **74** (1995) 4420–4423
1040. V. In and M. L. Spano and M. Ding, Phys. Rev. Lett. **80** (1998) 700–703
1041. W. Yang and M. Ding and A. Mandell and E. Ott, Phys. Rev. E **51** (1995) 102–110
1042. V. In and M. L. Spano and J. D. Neff and W. L. Ditto and S. Daw and K. D. Edwards and K. Nguyen, Chaos **7** (1997) 605–613
1043. H. Kayyali and D. Durand, Exp. Neurol. **113** (1991) 249–254
1044. M. Nakagama and D. Durand, Brain Res. **567** (1991) 241–247
1045. R. J. Warren and D. M. Durand, Brain Res. **806** (1998) 186–195
1046. A. Rafiq and Y. F. Zhang and R. J. DeLorenzo and D. A. Coulter, J. Neurophysiol. **74** (1995) 2028–2042
1047. J. Gaito, Can. J. Neurol. Sci. **8** (1981) 249–253
1048. S. R. Weiss and X. L. Xi and J. B. Rosen and H. Li and T. Heynen and R. M. Post, Neuroreport **6** (1995) 2171–2176
1049. M. A. Mirski and R. S. Fisher, Epilepsia **35** (1994) 1309–1316
1050. R. S. Fisher and G. L. Krauss and E. Ramsey and K. Laxer and J. Gates, Neurology **49** (1997) 293–297
1051. M. Monnier and M. Kalberer and P. Krupp, Exp. Neurol. **2** (1960) 271–289
1052. N. M. Sussman and H. W. Goldman and R. A. Jackal and L. Kaplan and M. Callanan and J. Bergen and R. N. Harner, Epilepsia **29** (1988) 677
1053. R. P. Lesser and S. H. Kim and L. Beyderman and D. L. Miglioretti and W. R. Webber and M. Bare and B. Cysyk and G. Krauss and B. Gordon, Neurology **53** (1999) 2073–2081
1054. A. L. Velasco and M. Velasco and F. Velasco and D. Menes and F. Gordon and L. Rocha and M. Briones and I. Marquez, Arch. Med. Res. **31** (2000) 316–328
1055. S. M. Narayan and E. M. Santori and A. W. Toga, Cerebral Cortex **4** (1994) 195–204

1056. L. A. Geddes, Ann. Biomed. Eng. **25** (1997) 1–14
1057. P. Fromherz and A. Offenhausser and T. Vetter and J. Weis, Science **252** (1991) 1290–1293
1058. A. Offenhausser and C. Sprossler and M. Matsuzawa and W. Knoll, Biosensors & Bioelectronics **12** (1997) 819–826
1059. P. J. Rousche and R. A. Normann, J. Neurosci. Meth. **82** (1998) 1–15
1060. D. Pellinen and P. Rousche and K. Yoshida and D. Pivin and D. Kipke, Med. Phys. **27** (2000) 1377
1061. M. N. Friedman and S. W. Robinson and G. A. Gerhadt, Anal. Chem. **68** (1996) 2621–2628
1062. I. Ben-Orin and A. Lewis and B. Minke and L. Loew, Biophys. J. **71** (1996) 1616–1620
1063. R. Eckmiller, Ophthalmic Res. **29** (1997) 281–289
1064. A. B. Majji and M. S. Humayun and J. D. Weiland and S. Suzuki and S. A. D'Anna and E. de Juan, Jr., Invest. Ophthal. Vis. Sci. **40** (1999) 2073–2081
1065. L. Lapicque, J. Physiol. (Paris) **9** (1907) 622–635
1066. J. J. B. Jack and D. Noble and R. W. Tsien, *Electrical Current Flow in Excitable Cells* (Clarendon Press, Oxford, 1983)
1067. U. Egert and B. Schlosshauer and S. Fennich and W. Nisch and M. Fejtl and T. Knott and T. Muller and H. Hammerle, Brain Res. Protocols **2** (1998) 229–242
1068. M. Janders and U. Egert and M. Stelze and W. Nisch, Proc. 18th Ann. Conf. IEEE Eng. Med. Biol. Soc. (1997) 54
1069. W. Nisch and J. Boeck and U. Egert and H. Haemmerle and A. Mohr, Biosensors & Bioelectronics **9** (1994) 737–741
1070. R. LeAnder and J. Hetling, Med. Physics **27** (2000) 1377
1071. B. Schlosshauer and A. Hoff and E. Guenther and E. Zrenner and H. Hammerle, Biomed. Microdevices **2** (1999) 61–72
1072. H. Fischler and H. P. Schwan, Med. Biol. Eng. Comp. **19** (1981) 579–588
1073. C. Walton and S. Gergely and A. P. Economides, PACE **10** (1987) 87–99
1074. J. M. Ahn and D. W. Kang and H. C. Kim and B. G. Min, ASAIO J. **39** (1993) M208–M212

Index